T0220873

A Primer on Mapping Class Groups

Princeton Mathematical Series

EDITORS: PHILLIP A. GRIFFITHS, JOHN N. MATHER, AND ELIAS M. STEIN

A Primer on Mapping Class Groups

Benson Farb and Dan Margalit

PRINCETON UNIVERSITY PRESS

PRINCETON AND OXFORD

Published by Princeton University Press, 41 William Street, Princeton, New Jersey 08540

In the United Kingdom: Princeton University Press, 6 Oxford Street, Woodstock, Oxfordshire OX20 1TW

press.princeton.edu

ISBN 978-0-691-14794-9

Library of Congress Cataloging-in-Publication Data

Farb, Benson.
  A primer on mapping class groups / Benson Farb, Dan Margalit.
      p. cm. – (Princeton mathematical series)
  Includes bibliographical references and index.
  ISBN 978-0-691-14794-9 (hardback)
  1. Mappings (Mathematics) 2. Class groups (Mathematics) I. Margalit, Dan, 1976– II. Title. III. Series

  QA360.F37 2011
  512.7'4–dc22

                    2011008491

British Library Cataloging-in-Publication Data is available

This book has been composed in Times and Helvetica.

The publisher would like to acknowledge the authors of this volume for providing the camera-ready copy from which this book was printed.

10 9 8 7 6 5 4 3 2 1

To Amie and Kathleen

Contents

Preface

Our goal in this book is to explain as many important theorems, examples, and techniques as possible, as quickly and directly as possible, while at the same time giving (nearly) full details and keeping the text (nearly) self-contained. This book contains some simplifications of known approaches and proofs, the exposition of some results that are not readily available, and some new material as well. We have tried to incorporate many of the "greatest hits" of the subject, as well as its small quirks and gems.

There are a number of other references that cover various of the topics we cover here (and more). We would especially like to mention the books by Abikoff [1], Birman [24], Casson–Bleiler [44], Fathi–Laudenbach–Poénaru [61], and Hubbard [97], as well as the survey papers by Harer [84] and Ivanov [107]. The works of Bers [14, 15] on Teichmüller's theorems and on the Nielsen–Thurston classification theorem have had a particularly strong influence on this book.

The first author learned much of what he knows about these topics from his advisor Bill Thurston, his teacher Curt McMullen, and his collaborators Lee Mosher and Howard Masur. The second author's perspective on this subject was greatly influenced by his advisor Benson Farb, his mentors Mladen Bestvina and Joan Birman, and his collaborator Chris Leininger. This book in particular owes a debt to notes the first author took from a course given by McMullen at Berkeley in 1991.

Benson Farb and Dan Margalit
Chicago and Atlanta, January 2011

Acknowledgments

We would like to thank Mohammed Abouzaid, Yael Algom-Kfir, Javier Aramayona, Jessica Banks, Mark Bell, Jeff Brock, Xuanting Cai, Meredith Casey, Pallavi Dani, Florian Deloup, John Franks, Siddhartha Gadgil, Bill Goldman, Carlos Segovia González, Michael Handel, John Harer, Peter Horn, Nikolai Ivanov, Jamie Jorgensen, Krishna Kaipa, Keiko Kawamuro, Richard Kent, Sarah Kitchen, Mustafa Korkmaz, Maria Luisa, Feng Luo, Joseph Maher, Kathryn Mann, Vladimir Markovic, Howard Masur, Jon McCammond, Curt McMullen, Ben McReynolds, Guido Mislin, Mahan Mj, Jamil Mortada, Lee Mosher, Zbigniew Nitecki, Jeremy Pecharich, Rita Jimenez Rolland, Simon Rose, Keefe San Agustin, Travis Schedler, Paul Seidel, Ignat Soroko, Steven Spallone, Harold Sultan, Macky Suzuki, Genevieve Walsh, Richard Webb, Alexander Wickens, Bert Wiest, Jenny Wilson, Rebecca Winarski, Stefan Witzel, Scott Wolpert, Kevin Wortman, Alex Wright, and Bruno Zimmermann for comments and questions.

We would also like to thank Robert Bell, Tara Brendle, Jeff Carlson, Tom Church, Spencer Dowdall, Moon Duchin, David Dumas, Jeff Frazier, Daniel Groves, Asaf Hadari, Allen Hatcher, Thomas Koberda, Justin Malestein, Johanna Mangahas, Erika Meucci, Catherine Pfaff, Kasra Rafi, and Saul Schleimer for thorough comments on large portions of the book.

Stergios Antonakoudis read the penultimate version of this book and made numerous insightful comments and corrections on the entire manuscript. We are grateful to him for his efforts.

We would like to especially thank Mladen Bestvina, Joan Birman, Ken Bromberg, Chris Leininger, and Andy Putman for extensive discussions on the material in this book.

Both authors are grateful to Princeton University Press. We are particularly indebted to Ben Holmes, Vickie Kearn, Anna Pierrehumbert, and Stefani Wexler for their tireless efforts on behalf of this project.

The first author would like to thank the second author for everything he has done to bring this project to its current form. His hard work, thorough understanding of the material, and ability to simplify proofs and explain concepts clearly have been invaluable. The first author would also like to thank the second author for all he has taught him about mapping class groups.

The second author would like to thank the first author for introducing him to and teaching him about the subject of mapping class groups and for the opportunity to work together with him on this project. The second author could not thank the first author enough for his support, enthusiasm, and endless generosity with his time, energy, and knowledge over the years.

Finally, the authors would like to thank their families for their love, support, patience, and encouragement. This book would not have been possible without them.

A Primer on Mapping Class Groups

Overview

In this book we will consider two fundamental objects attached to a surface S: a group and a space. We will study these two objects and how they relate to each other.

The group. The group is the *mapping class group* of S, denoted by $\mathrm{Mod}(S)$. It is defined to be the group of isotopy classes of orientation-preserving diffeomorphisms of S (that restrict to the identity on ∂S if $\partial S \neq \emptyset$):

$$\mathrm{Mod}(S) = \mathrm{Diff}^+(S, \partial S) / \mathrm{Diff}_0(S, \partial S).$$

Here $\mathrm{Diff}_0(S, \partial S)$ is the subgroup of $\mathrm{Diff}^+(S, \partial S)$ consisting of elements that are isotopic to the identity. We will study the algebraic structure of the group $\mathrm{Mod}(S)$, the detailed structure of its individual elements, and the beautiful interplay between them.

The space. The space is the *Teichmüller space* of S. When $\chi(S) < 0$, this is the space of hyperbolic metrics on S up to isotopy:

$$\mathrm{Teich}(S) = \mathrm{HypMet}(S) / \mathrm{Diff}_0(S).$$

The space $\mathrm{Teich}(S)$ is a metric space homeomorphic to an open ball. The group $\mathrm{Diff}^+(S)$ acts on $\mathrm{HypMet}(S)$ by pullback. This action descends to an action of $\mathrm{Mod}(S)$ on $\mathrm{Teich}(S)$. A fundamental result in the theory is that this action is properly discontinuous. The quotient space

$$\mathcal{M}(S) = \mathrm{Teich}(S) / \mathrm{Mod}(S)$$

is the *moduli space of Riemann surfaces* homeomorphic to S. The space $\mathcal{M}(S)$ is one of the fundamental objects of mathematics. Since (as we will prove) $\mathcal{M}(S)$ is finitely covered by a closed aspherical manifold, the group $\mathrm{Mod}(S)$ encodes most of the topological features of $\mathcal{M}(S)$. Conversely, invariants such as the cohomology of $\mathrm{Mod}(S)$ are determined by the topology of $\mathcal{M}(S)$.

The appearance of $\mathrm{Mod}(S)$, $\mathrm{Teich}(S)$, and $\mathcal{M}(S)$ in mathematics is ubiquitous: from hyperbolic geometry to algebraic geometry to combinatorial group theory to symplectic geometry to 3-manifold theory to dynamics. In this book we will relate the algebraic structure of $\mathrm{Mod}(S)$, the geometry of $\mathrm{Teich}(S)$, and the topology of $\mathcal{M}(S)$. Underlying the connections between these structures is the combinatorial topology of the surface S. Indeed, one leitmotif of this book is the interplay of the "local" study of the geometry and topology of a single surface S and the "global" properties of the spaces $\mathrm{Teich}(S)$ and $\mathcal{M}(S)$. It is a beautiful thing to see how each informs the other.

The classification. The third player in our story is the Nielsen–Thurston classification theorem, which gives a particularly nice representative for each element of $\mathrm{Mod}(S)$. This is a nonlinear analogue of the Jordan canonical form for matrices; as such, it is a cornerstone of the theory. It is in Bers' proof of this theorem where the first two characters play off of each other: the key is to understand how elements of $\mathrm{Mod}(S)$ act on $\mathrm{Teich}(S)$ via isometries of the Teichmüller metric. Much of the usefulness of the Nielsen–Thurston classification comes from the fact that the typical element of $\mathrm{Mod}(S)$ has a pseudo-Anosov representative. Pseudo-Anosov homeomorphisms have very specific descriptions and exhibit many remarkable properties.

In light of the above discussion this book is divided into three parts. We now outline them, emphasizing what we consider to be some of the more important results and focusing for simplicity on the case of the closed surface S_g of genus g.

Part 1

Part 1 covers what might be called the core theory of mapping class groups. The central theme is the relationship between the algebraic structure of $\mathrm{Mod}(S)$ and the combinatorial topology of S.

Chapter 1. Just as one understands a linear transformation by its action on vectors, so one understands an element of $\mathrm{Mod}(S_g)$ by its action on simple closed curves in S. Chapter 1 explains the basics of working with simple closed curves. This is more difficult than it might sound, as the typical simple closed curve can be rather complicated (see Figure 1).

When $g \geq 2$, hyperbolic geometry enters as a useful tool since each homotopy class of simple closed curves has a unique geodesic representative. Following the linear algebra analogy, we introduce the geometric intersection number. This is the analogue of an inner product on a vector space and is a basic tool for working with simple closed curves in S_g. The chapter

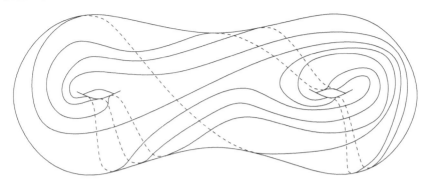

Figure 1 Thurston's typical curve.

ends with the change of coordinates principle. This principle plays the same role that change of basis plays for matrices, so it is not surprising that it is applied with great frequency.

Chapter 2. After defining the mapping class group $\mathrm{Mod}(S)$, we compute the examples that can be explicitly determined "by hand." We then introduce what we call the Alexander method, which gives an algorithm for determining whether or not two elements of $\mathrm{Mod}(S)$ are equal. In particular, this method is used for showing that an element of $\mathrm{Mod}(S)$ is nontrivial or for verifying relations in $\mathrm{Mod}(S)$. One of the computations we perform is the following classical fundamental theorem of Dehn.

Theorem 2.5 $\mathrm{Mod}(T^2) \approx \mathrm{SL}(2, \mathbb{Z})$.

Chapter 3. Dehn twists are the simplest infinite-order elements of $\mathrm{Mod}(S)$. They play the role of elementary matrices in linear algebra, so it is not surprising that they appear in much of what follows. We present an in-depth study of Dehn twists and their action on simple closed curves. As one application of this study, we prove that if two simple closed curves in S_g have geometric intersection number greater than 1, then the associated Dehn twists generate a free group of rank 2 in $\mathrm{Mod}(S)$. We also apply our knowledge of Dehn twists in order to prove the following basic theorem.

Theorem 3.10 *For* $g \geq 3$, *the center of* $\mathrm{Mod}(S_g)$ *is trivial.*

Chapter 4. At this point we have developed the nuts and bolts of the theory, and we start to expose some of the most basic algebraic structure of $\mathrm{Mod}(S)$. The following fundamental theorem of Dehn is analogous to the fact that $\mathrm{SL}(n, \mathbb{Z})$ is generated by elementary matrices.

Theorem 4.1 $\mathrm{Mod}(S_g)$ *is generated by finitely many Dehn twists.*

Theorem 4.1 is proved by induction on genus, and the Birman exact sequence is introduced as the key step for the induction. The key to the inductive step is to prove that the complex of curves $\mathcal{C}(S_g)$ is connected when $g \geq 2$. The simplicial complex $\mathcal{C}(S_g)$ is a useful combinatorial object that encodes intersection patterns of simple closed curves in S_g. More detailed structure of $\mathcal{C}(S_g)$ is then used to find various explicit generating sets for $\mathrm{Mod}(S_g)$, including those due to Lickorish and to Humphries.

A natural problem now arises: given a finite product of Dehn twists, is there an algorithm to determine whether the resulting element of $\mathrm{Mod}(S_g)$ is trivial or not? The next theorem says that the answer is yes.

Theorem 4.2 $\mathrm{Mod}(S_g)$ *has a solvable word problem.*

Chapter 5. After proving that a group G is finitely generated, the next invariant one wants to compute is the abelianization of G or, what is the same thing, its first homology $H_1(G; \mathbb{Z})$. Chapter 5 begins with a simple proof, due to Harer, of the following theorem of Mumford, Birman, and Powell.

Theorem 5.2 *If* $g \geq 3$, *then* $H_1(\mathrm{Mod}(S_g); \mathbb{Z}) = 1$.

The key ingredient in the proof of Theorem 5.2 is Theorem 4.1 together with the lantern relation, a beautiful relation between seven Dehn twists that was discovered by Dehn. We then apply a method from geometric group theory to prove the following theorem.

Theorem 5.7 $\mathrm{Mod}(S_g)$ *is finitely presentable.*

The geometric group theory technique converts the statement of Theorem 5.7 to a problem about the topology of a certain arc complex and an associated mapping class group action on it. The key in this case is a shockingly simple and beautiful proof by Hatcher that the arc complex is contractible. We also give explicit presentations of $\mathrm{Mod}(S_g)$, including those by Wajnryb and Gervais.

Hopf gave a formula for computing $H_2(G; \mathbb{Z})$ for any group G from a finite presentation for G. While this computation is usually too difficult to perform in practice, Pitsch discovered that one can use the Wajnryb presentation of $\mathrm{Mod}(S_g)$ to give an upper bound on the rank of $H_2(\mathrm{Mod}(S_g); \mathbb{Z})$. We use this method in proving the following deep theorem originally due to Harer.

Theorem 5.8 *If* $g \geq 4$, *then* $H_2(\mathrm{Mod}(S_g); \mathbb{Z}) \cong \mathbb{Z}$.

The lower bound in Theorem 5.8 is given by explicitly constructing nontrivial classes. We give a detailed construction of the the Euler class, the most basic invariant for surface bundles, as a 2-cocycle for the mapping class group of a punctured surface. At this point homological algebra, in the form of (a degenerate form of) the Hochschild–Serre spectral sequence, is used to deduce Theorem 5.8. The Meyer signature cocycle is also explained, as is the important connection of this circle of ideas with the theory of S_g-bundles. Indeed, understanding S_g-bundles and their invariants is a major motivation for computing $H_2(\mathrm{Mod}(S_g); \mathbb{Z})$. The strong connection between $\mathrm{Mod}(S_g)$ and S_g-bundles comes from the following bijection:

$$\left\{ \begin{array}{c} \text{Isomorphism classes} \\ \text{of oriented } S_g\text{-bundles} \\ \text{over } B \end{array} \right\} \longleftrightarrow \left\{ \begin{array}{c} \text{Conjugacy classes} \\ \text{of representations} \\ \rho : \pi_1(B) \to \mathrm{Mod}(S_g) \end{array} \right\}$$

for each fixed $g \geq 2$ and each fixed base B.

Chapter 6. Algebraic intersection number gives a $\mathrm{Mod}(S_g)$-invariant symplectic form on $H_1(S_g; \mathbb{Z})$, thus inducing a representation

$$\Psi : \mathrm{Mod}(S_g) \to \mathrm{Sp}(2g, \mathbb{Z})$$

with target the integral symplectic group. This symplectic representation of $\mathrm{Mod}(S_g)$ can be viewed as a kind of "linear approximation" to $\mathrm{Mod}(S_g)$. We present three different proofs of the surjectivity of Ψ, each illustrating a different theme. The usefulness of the symplectic representation is then illustrated by two applications to understanding the algebraic structure of $\mathrm{Mod}(S)$. First, we explain how Serre used this representation to prove the following.

Theorem 6.9 $\mathrm{Mod}(S_g)$ *has a torsion-free subgroup of finite index.*

The actual statement of Theorem 6.9 given below provides explicit torsion-free subgroups of $\mathrm{Mod}(S_g)$ that come from congruence subgroups of $\mathrm{Sp}(2g, \mathbb{Z})$. We then use the symplectic representation to prove, following Ivanov, the following theorem of Grossman.

Theorem 6.11 $\mathrm{Mod}(S_g)$ *is residually finite.*

The symplectic representation has a kernel, called the Torelli group, denoted $\mathcal{I}(S_g)$. This is an important but still poorly understood subgroup of $\mathrm{Mod}(S_g)$. The Torelli group supports a rich and beautiful theory with important connections to other parts of mathematics. We continue Chapter 6 by explaining some of the pioneering work of Dennis Johnson on $\mathcal{I}(S_g)$. In particular, we construct the Johnson homomorphism

$$\tau : \mathcal{I}(S_g^1) \to \wedge^3 H,$$

where S_g^1 is S_g minus an open disk and $H = H_1(S_g^1; \mathbb{Z})$. We then explain a few of the many applications of τ.

Chapter 7. What are the finite groups of topological symmetries of S? That is, what are the finite subgroups of $\mathrm{Mod}(S)$? A deep theorem of Kerckhoff states that each finite subgroup of $\mathrm{Mod}(S)$ comes from a group of orientation-preserving isometries for some hyperbolic metric on S. Such groups are highly constrained: using the Riemann–Hurwitz formula and basic facts about 2-dimensional orbifolds, we prove Hurwitz's $84(g-1)$ theorem, a nineteenth century classic.

Theorem 7.4 ($84(g-1)$ theorem) *If X is a hyperbolic surface homeomorphic to S_g, where $g \geq 2$, then*

$$|\mathrm{Isom}^+(X)| \leq 84(g-1).$$

We also prove a corresponding $4g+2$ theorem for cyclic subgroups of $\mathrm{Mod}(S_g)$. Later in the book we prove Kerckhoff's theorem for cyclic groups (i.e., "cyclic Nielsen realization") by using the action of $\mathrm{Mod}(S_g)$ on $\mathrm{Teich}(S_g)$.

The basic orbifold theory that we develop to prove Theorem 7.4 is then applied to prove that $\mathrm{Mod}(S)$ has only finitely many conjugacy classes of finite subgroups. On the other hand, we prove that there is enough torsion in $\mathrm{Mod}(S)$ to generate it with finitely many torsion elements, and indeed we can take these elements to have order 2.

Chapter 8. This chapter is an exposition of one of the most beautiful connections between topology and algebra in dimension 2: the Dehn–Nielsen–Baer theorem. Let $\mathrm{Out}(\pi_1(S))$ denote the group of outer automorphisms of $\pi_1(S)$ and let $\mathrm{Mod}^\pm(S)$ denote the extended mapping class group, which is the group of isotopy classes of all homeomorphisms of S (including the orientation-reversing ones).

Theorem 8.1 (Dehn–Nielsen–Baer theorem) *For $g \geq 1$, we have*

$$\mathrm{Mod}^\pm(S_g) \cong \mathrm{Out}(\pi_1(S_g)).$$

Theorem 8.1 equates a topologically defined group, $\mathrm{Mod}(S_g)$, with an algebraically defined group, $\mathrm{Out}(\pi_1(S_g))$. What is more, Dehn's original proof uses hyperbolic geometry! Both the theorem and the ideas in the proof foreshadowed the Mostow rigidity theorem nearly 50 years in advance.

Chapter 9. Part 1 ends with a brief introduction to braid groups B_n. The group B_n is isomorphic to the mapping class group of a disk with n marked points. Since disks are planar, the braid groups lend themselves to special pictorial representations. This gives the theory of braid groups its own special flavor within the theory of mapping class groups.

After presenting some classical facts about the algebraic structure of the braid group, we give a new proof of the Birman–Hilden theorem, which relates the braid groups to the mapping class groups of closed surfaces. Let $\mathrm{SMod}(S_g^1)$ denote the subgroup of $\mathrm{Mod}(S_g^1)$ consisting of elements with representative homeomorphisms that commute with some fixed hyperelliptic involution.

Theorem 9.2 (Birman–Hilden theorem) *Let $g \geq 1$. Then*

$$\mathrm{SMod}(S_g^1) \approx B_{2g+1}.$$

Part 2

Part 2 of the book is a concise introduction to Teichmüller theory and the moduli space of Riemann surfaces. We concentrate on those aspects of the theory that are most directly applicable to understanding $\mathrm{Mod}(S_g)$. Part 2 has a decidedly more analytic and geometric flavor than Part 1.

Chapter 10. We introduce Teichmüller space $\mathrm{Teich}(S_g)$ as the space of hyperbolic structures on S_g. After putting a natural topology on $\mathrm{Teich}(S_g)$ and giving two heuristic counts of its dimension, we prove the following classical result due to Fricke and Klein in 1897.

Theorem 10.6 *For $g \geq 2$ we have $\mathrm{Teich}(S_g) \cong \mathbb{R}^{6g-6}$.*

We prove Theorem 10.6 by giving explicit coordinates on $\mathrm{Teich}(S_g)$ coming from certain length and twist parameters for curves in a pants decomposition of S_g; these are the Fenchel–Nielsen coordinates on $\mathrm{Teich}(S_g)$. It is worth emphasizing how miraculous it is that the quotient $\mathrm{Teich}(S_g) = \mathrm{HypMet}(S_g)/\mathrm{Diff}_0(S_g)$ of an infinite-dimensional space by an infinite-dimensional group action gives a finite-dimensional manifold. The kind of "rigidity" behind this is in some sense contained in hyperbolic trigonometry, as can be seen in the proof of Theorem 10.6. The chapter ends with the following fundamental theorem about hyperbolic metrics on surfaces.

Theorem 10.7 *Let $g \geq 2$. There are $9g - 9$ specific homotopy classes of simple closed curves on S_g with the property that any hyperbolic metric* *on S_g is determined up to isotopy by the lengths of the geodesics in these homotopy classes.*

The key to the proof of Theorem 10.7 is a convexity result for the function "length of a" (where a is an isotopy class of simple closed curves) considered as a function on $\mathrm{Teich}(S_g)$.

Chapter 11. After determining the topology of $\mathrm{Teich}(S_g)$, we turn to its metric geometry. In order to do this, we first explain how one can think of $\mathrm{Teich}(S_g)$ as the space of complex structures on S_g.

Given a pair of points $\mathfrak{X}, \mathfrak{Y} \in \mathrm{Teich}(S_g)$, one associates a pair of Riemann surfaces X, Y and a homeomorphism $f : X \to Y$ well defined up to homotopy. While f is in general not conformal, it can always be chosen to be quasiconformal. This means that f distorts angles by at most a fixed bounded amount $K(f)$.

A natural extremal mapping problem then arises:

> *Given a homeomorphism of Riemann surfaces $f : X \to Y$, is there a quasiconformal map $X \to Y$ that minimizes quasiconformal dilatation among all maps homotopic to f?*

Teichmüller answered this question by finding a concrete, explicit mapping now called the Teichmüller map. Away from a finite number of points, a Teichmüller mapping locally looks like the linear map $(x, y) \mapsto (Kx, \frac{1}{K}y)$ for some K. In 1939 Teichmüller proved[1] that his maps solve the above extremal problem. What is more, he proved that his maps give the unique solution.

Theorems 11.8 and 11.9 (Teichmüller's existence and uniqueness theorems) *Let $g \geq 2$ and let $\mathfrak{X}, \mathfrak{Y} \in \mathrm{Teich}(S_g)$. Let $f : X \to Y$ be the associated homeomorphism of Riemann surfaces. Then there exists a Teichmüller mapping $h : X \to Y$ that is homotopic to f. The map h uniquely minimizes the quasiconformal dilatation among all homeomorphisms homotopic to f.*

The proof of Theorem 11.8 illustrates how the global point of view informs the local. Namely, in the course of proving the existence statement for a single $\mathfrak{Y} \in \mathrm{Teich}(S_g)$, we actually are led to proving the existence statement for all possible targets $\mathfrak{Y} \in \mathrm{Teich}(S_g)$ at the same time. Specifically, this is accomplished by proving the surjectivity of a certain map $\mathrm{QD}(X) \to \mathrm{Teich}(S_g)$, where $\mathrm{QD}(X)$ is the space of holomorphic quadratic differentials on a Riemann surface X. To prove this surjectivity,

---

[1] Actually, Ahlfors is usually credited with the first complete, understandable proof of this fact.

we use the global topology of $\mathrm{Teich}(S_g)$ via an application of the invariance of domain theorem. This proof is an example of the so-called method of continuity.

The solution to the extremal problem can be used to define a metric on $\mathrm{Teich}(S_g)$ called the Teichmüller metric. Let $h : X \to Y$ be the Teichmüller map associated to $\mathcal{X}, \mathcal{Y} \in \mathrm{Teich}(S_g)$ and let $K(h)$ be its dilatation. We prove that

$$d_{\mathrm{Teich}(S_g)}(\mathcal{X}, \mathcal{Y}) = \frac{1}{2} \log(K(h))$$

defines a complete metric on $\mathrm{Teich}(S_g)$. This is called the Teichmüller metric. In order to describe the geodesics in this metric, we explain the fundamental connection between Teichmüller's theorems, holomorphic quadratic differentials, and measured foliations. This description is a crucial ingredient in the proof of the Nielsen–Thurston classification theorem that we give later in the book.

Chapter 12. Let $g \geq 2$. The moduli space $\mathcal{M}(S_g)$ of genus g Riemann surfaces is defined to be

$$\mathcal{M}(S_g) = \mathrm{Teich}(S_g)/\mathrm{Mod}(S_g).$$

The space $\mathcal{M}(S_g)$ parameterizes many different kinds of structures on S_g. It can be viewed as any one of the following sets:

1. Isometry classes of constant curvature metrics on S_g

2. Conformal classes of Riemannian metrics on S_g

3. Biholomorphism classes of complex structures on S_g

4. Isomorphism classes of smooth, complex algebraic structures on S_g.

The natural bijective correspondences between these sets are derived from deep theorems, namely, the uniformization theorem and the Kodaira embedding theorem. As such, the bijections between the sets above are very difficult to access explicitly. The interplay between these different incarnations is one reason the study of $\mathcal{M}(S_g)$ is rich and often difficult.

The group $\mathrm{Mod}(S_g)$ and the space $\mathcal{M}(S_g)$ are tied together closely because of the following theorem due to Fricke.

Theorem 12.2 $\mathrm{Mod}(S_g)$ *acts properly discontinuously on* $\mathrm{Teich}(S_g)$.

In order to prove Theorem 12.2, we consider the raw length spectrum $\mathrm{rls}(X)$ of a hyperbolic surface $X \approx S_g$. The set $\mathrm{rls}(X)$ is defined to be the set of lengths of all closed geodesics in X. The crucial property is that $\mathrm{rls}(X)$ is a closed, discrete subset of $[0, \infty)$. The Wolpert lemma then tells us that nearby points in $\mathrm{Teich}(S_g)$ have nearly equal length spectra. From these two facts Theorem 12.2 follows easily.

Since $\mathrm{Mod}(S_g)$ acts properly discontinuously on $\mathrm{Teich}(S_g)$, the quotient space $\mathcal{M}(S_g)$ is an orbifold. By Theorem 6.9, $\mathcal{M}(S_g)$ is finitely covered by a manifold. Since $\mathrm{Teich}(S_g)$ is contractible (Theorem 10.6), we have the following.

Theorem 12.3 *For $g \geq 1$, the space $\mathcal{M}(S_g)$ is an aspherical orbifold and is finitely covered by an aspherical manifold.*

It is not hard to see that $\mathcal{M}(S_g)$ is not compact. Understanding this non-compactness is a central issue. The most basic theorem in this direction is the Mumford compactness criterion, which we think of as a generalization of the Mahler compactness criterion for lattices in $\mathbb{R}^n$. For a hyperbolic surface X we denote by $\ell(X)$ the length of the shortest essential closed curve in X.

Theorem 12.6 (Mumford's compactness criterion) *Let $g \geq 1$. For each $\epsilon > 0$, the space*

$$\mathcal{M}_\epsilon(S_g) = \{X \in \mathcal{M}(S_g) : \ell(X) \geq \epsilon\}$$

is compact.

Since the sets $\mathcal{M}_\epsilon(S_g)$ exhaust $\mathcal{M}(S_g)$, Theorem 12.6 tells us that the only way to leave every compact set in $\mathcal{M}(S_g)$ is to decrease the length of some closed geodesic. Mumford's compactness criterion leads us to study the topology of $\mathcal{M}(S_g)$ at infinity. Combining a number of ingredients, including connectedness of $\mathcal{C}(S_g)$ for $g \geq 2$, we prove the following.

Corollaries 12.11 and 12.12 *Let $g \geq 2$. Then $\mathcal{M}(S_g)$ has one end, and every loop in $\mathcal{M}(S_g)$ can be homotoped outside every compact set in $\mathcal{M}(S_g)$.*

We end the chapter by explaining one more of the (many) reasons for the importance of $\mathcal{M}(S_g)$ in mathematics: $\mathcal{M}(S_g)$ is very close to being a classifying space for S_g-bundles. By "very close" we mean that an analogous statement holds for any finite manifold cover of $\mathcal{M}(S_g)$. In particular, we prove that the rational cohomology of the space $\mathcal{M}(S_g)$ is isomorphic to the rational cohomology of the group $\mathrm{Mod}(S_g)$.

Part 3

Chapter 13. The main goal of Part 3 is to understand what individual elements of $\mathrm{Mod}(S_g)$ look like, in the same way that the Jordan canonical form of a matrix gives us a geometric picture of what a linear transformation looks like. The precise statement is the following.

Theorem 13.2 (Nielsen–Thurston classification) *Let $g \geq 2$. Each $f \in \mathrm{Mod}(S_g)$ has a representative $\phi \in \mathrm{Homeo}^+(S_g)$ of one of the following types.*

   1. Periodic: $\phi^m = Id$ for some $m > 0$.

   2. Reducible: ϕ leaves invariant a finite collection of pairwise disjoint simple closed curves in S_g.

   3. Pseudo-Anosov: there are transverse measured foliations $(\mathcal{F}^s, \mu_s)$ and $(\mathcal{F}^u, \mu_u)$ on S_g, and a real number $\lambda > 1$ so that

$$\phi \cdot (\mathcal{F}^u, \mu_u) = (\mathcal{F}^u, \lambda\mu_u) \quad and \quad \phi \cdot (\mathcal{F}^s, \mu_s) = (\mathcal{F}^s, \lambda^{-1}\mu_s).$$

Case 3 is exclusive from cases 1 and 2. The number λ associated to a pseudo-Anosov homeomorphism ϕ is called the stretch factor of ϕ. Away from a finite number of points, a pseudo-Anosov homeomorphism locally looks like the linear map $(x, y) \mapsto (\lambda x, \frac{1}{\lambda}y)$, just like a Teichmüller mapping.

Type 1 mapping classes are relatively easy to understand. For type 2 we can cut along the invariant collection of curves and reapply the theorem to each component of the cut surface. By doing this we obtain a "canonical form" for mapping classes: any mapping class can be reduced into finite order and pseudo-Anosov pieces. Thus the more we know about pseudo-Anosov homeomorphisms, the more we know about arbitrary homeomorphisms. Chapter 14 is completely devoted to studying properties of pseudo-Anosov homeomorphisms.

We present Bers' proof of Theorem 13.2. The proof uses many of the ideas and results proved earlier in the book, such as the proper discontinuity of the action of $\mathrm{Mod}(S_g)$ on $\mathrm{Teich}(S_g)$, the Mumford compactness criterion, and the structure of Teichmüller geodesics. The main idea is to prove that if a mapping class f is not of type 1 or type 2, then there is an f-invariant Teichmüller geodesic which one then interprets, using Teichmüller's theorems, to show that f is pseudo-Anosov.

Chapter 14. In this chapter we begin the study of pseudo-Anosov homeomorphisms in earnest. Although in some sense the typical mapping class is pseudo-Anosov, it is actually rather nontrivial to construct explicit examples. We begin by presenting five constructions of pseudo-Anosov homeomorphisms.

The simplest invariant of a pseudo-Anosov mapping class is its stretch factor λ, which is analogous to the largest eigenvalue of a linear map. The next theorem tells us that the set of pseudo-Anosov stretch factors is quite constrained.

Theorem 14.8 *Let $g \geq 2$. Let λ be the stretch factor associated to a pseudo-Anosov element of $\mathrm{Mod}(S_g)$. Then λ is an algebraic integer with degree bounded above by $6g - 6$.*

Each pseudo-Anosov mapping class has an invariant axis in $\mathrm{Teich}(S)$ and thus gives a geodesic loop in $\mathcal{M}(S)$. The length of this loop is the logarithm of the corresponding stretch factor. Thus the set of logarithms of stretch factors of pseudo-Anosov elements of $\mathrm{Mod}(S)$ can be thought of as the length spectrum of $\mathcal{M}(S)$. The following theorem of Arnoux–Yoccoz and Ivanov can thus be interpreted as implying that the length spectrum of $\mathcal{M}(S)$ is discrete.

Theorem 14.9 *Let $g \geq 2$. For any $C \geq 1$, there are only finitely many conjugacy classes in $\mathrm{Mod}(S_g)$ of pseudo-Anosov mapping classes with stretch factor at most C.*

Pseudo-Anosov homeomorphisms have a number of remarkable dynamical properties. Among them, we prove:

- Every pseudo-Anosov homeomorphism has a dense orbit.

- The periodic points of a pseudo-Anosov homeomorphism are dense.

- A pseudo-Anosov homeomorphism has the minimum number of periodic points, for each period, in its homotopy class.

In analogy with the behavior of the lengths of vectors under iteration of a linear transformation with a dominant eigenvalue, we also prove the following.

Theorem 14.23 *Let $g \geq 2$. Let $f \in \mathrm{Mod}(S_g)$ be pseudo-Anosov with stretch factor λ. If ρ is any Riemannian metric on S_g, and if a is any isotopy class of simple closed curves in S_g, then*

$$\lim_{n \to \infty} \sqrt[n]{\ell_\rho(f^n(a))} = \lambda.$$

Chapter 15. The final chapter begins with a description of Thurston's original path of discovery to the Nielsen–Thurston classification theorem. As Thurston wrote in his famous paper [207]:

> The nicest aspects of this theory I have been trying to sketch are not formal, but intuitive. If you draw pictures of a pseudo-Anosov diffeomorphism, you can understand geometrically what it *does*, something which has puzzled me for several years. ...it is pleasant to see something of this abstract origin made very concrete.

We begin by illustrating Thurston's approach via a beautiful and fundamental example. Thurston's first idea is that one can understand $f \in \text{Mod}(S_g)$ by iterating f on an isotopy class of essential simple closed curves c. In general, the sequence $f^n(c)$ gets very complicated very quickly. This is where the next idea comes in: one can encode a very complicated simple closed curve in a surface with a small amount of data called a train track. A train track in S_g is an embedded graph with some extra data attached, for example, each edge is labeled by a nonnegative integer. Under certain conditions, f preserves a train track (up to a certain equivalence) and acts linearly on its labels. When f is pseudo-Anosov, the corresponding matrix is a Perron–Frobenius matrix, and all of the information attached to f (stretch factor, stable foliation, etc.) can be easily determined by linear algebra.

Thus in this example the combinatorial device of train tracks converts the nonlinear problem of understanding a homeomorphism of a surface to a simple linear algebra problem. Thurston's remarkable discovery is that this linearization process works for all pseudo-Anosov homeomorphisms, and in fact it can be used to prove the Nielsen–Thurston classification.

We give a sketch of how all of this works in general and how Thurston proves the Nielsen–Thurston classification in this way. The idea is that the space $\mathcal{PMF}(S_g)$ of all projective classes of measured foliations on S_g can be used to give a compactification of $\text{Teich}(S_g)$ that is homeomorphic to a closed ball. Each element of $\text{Mod}(S_g)$ induces a homeomorphism on this ball, and so the Brouwer fixed point theorem can be applied. Analyzing the various possibilities for fixed points leads to the various cases of the classification theorem. As Thurston says:

> And there is a great deal of natural geometric structure on $\mathcal{PMF}$, relating to the structure on S, beautiful to contemplate.

PART 1
Mapping Class Groups

Chapter One

Curves, Surfaces, and Hyperbolic Geometry

A linear transformation of a vector space is determined by, and is best understood by, its action on vectors. In analogy with this, we shall see that an element of the mapping class group of a surface S is determined by, and is best understood by, its action on homotopy classes of simple closed curves in S. We therefore begin our study of the mapping class group by obtaining a good understanding of simple closed curves on surfaces.

Simple closed curves can most easily be studied via their geodesic representatives, and so we begin with the fact that every surface may be endowed with a constant-curvature Riemannian metric, and we study the relation between curves, the fundamental group, and geodesics. We then introduce the geometric intersection number, which we think of as an "inner product" for simple closed curves. A second fundamental tool is the change of coordinates principle, which is analogous to understanding change of basis in a vector space. After explaining these tools, we conclude this chapter with a discussion of some foundational technical issues in the theory of surface topology, such as homeomorphism versus diffeomorphism, and homotopy versus isotopy.

1.1 SURFACES AND HYPERBOLIC GEOMETRY

We begin by recalling some basic results about surfaces and hyperbolic geometry that we will use throughout the book. This is meant to be a brief review; see [208] or [119] for a more thorough discussion.

1.1.1 SURFACES

A *surface* is a 2-dimensional manifold. The following fundamental result about surfaces, often attributed to Möbius, was known in the mid-nineteenth century in the case of surfaces that admit a triangulation. Radò later proved, however, that every compact surface admits a triangulation. For proofs of both theorems, see, e.g., [204].

THEOREM 1.1 (Classification of surfaces) *Any closed, connected, orientable surface is homeomorphic to the connect sum of a 2-dimensional* *sphere with $g \geq 0$ tori. Any compact, connected, orientable surface is obtained from a closed surface by removing $b \geq 0$ open disks with disjoint closures. The set of homeomorphism types of compact surfaces is in bijective correspondence with the set $\{(g, b) : g, b \geq 0\}$.*

The g in Theorem 1.1 is the *genus* of the surface; the b is the number of *boundary components*. One way to obtain a noncompact surface from a compact surface S is to remove n points from the interior of S; in this case, we say that the resulting surface has n *punctures*.

Unless otherwise specified, when we say "surface" in this book, we will mean a compact, connected, oriented surface that is possibly punctured (of course, after we puncture a compact surface, it ceases to be compact). We can therefore specify our surfaces by the triple (g, b, n). We will denote by $S_{g,n}$ a surface of genus g with n punctures and empty boundary; such a surface is homeomorphic to the interior of a compact surface with n boundary components. Also, for a closed surface of genus g, we will abbreviate $S_{g,0}$ as S_g. We will denote by ∂S the (possibly disconnected) boundary of S.

Recall that the *Euler characteristic* of a surface S is

$$\chi(S) = 2 - 2g - (b + n).$$

It is a fact that $\chi(S)$ is also equal to the alternating sum of the Betti numbers of S. Since $\chi(S)$ is an invariant of the homeomorphism class of S, it follows that a surface S is determined up to homeomorphism by any three of the four numbers g, b, n, and $\chi(S)$.

Occasionally, it will be convenient for us to think of punctures as *marked points*. That is, instead of deleting the points, we can make them distinguished. Marked points and punctures carry the same topological information, so we can go back and forth between punctures and marked points as is convenient. On the other hand, all surfaces will be assumed to be without marked points unless explicitly stated otherwise.

If $\chi(S) \leq 0$ and $\partial S = \emptyset$, then the universal cover $\widetilde{S}$ is homeomorphic to $\mathbb{R}^2$ (see, e.g., [199, Section 1.4]). We will see that, when $\chi(S) < 0$, we can take advantage of a hyperbolic structure on $\widetilde{S}$.

1.1.2 THE HYPERBOLIC PLANE

Let $\mathbb{H}^2$ denote the hyperbolic plane. One model for $\mathbb{H}^2$ is the *upper half-plane model*, namely, the subset of $\mathbb{C}$ with positive imaginary part ($y > 0$), endowed with the Riemannian metric

$$ds^2 = \frac{dx^2 + dy^2}{y^2},$$

where $dx^2 + dy^2$ denotes the Euclidean metric on $\mathbb{C}$. In this model the geodesics are semicircles and half-lines perpendicular to the real axis.

It is a fact from Riemannian geometry that any complete, simply connected Riemannian 2-manifold with constant sectional curvature -1 is isometric to $\mathbb{H}^2$.

For the *Poincaré disk model* of $\mathbb{H}^2$, we take the open unit disk in $\mathbb{C}$ with the Riemannian metric

$$ds^2 = 4\frac{dx^2 + dy^2}{(1 - r^2)^2}.$$

In this model the geodesics are circles and lines perpendicular to the unit circle in $\mathbb{C}$ (intersected with the open unit disk).

Any Möbius transformation from the upper half-plane to the unit disk is an isometry between the upper half-plane model for $\mathbb{H}^2$ and the Poincaré disk model of $\mathbb{H}^2$. The group of orientation-preserving isometries of $\mathbb{H}^2$ is (in either model) the group of Möbius transformations taking $\mathbb{H}^2$ to itself. This group, denoted $\mathrm{Isom}^+(\mathbb{H}^2)$, is isomorphic to $\mathrm{PSL}(2, \mathbb{R})$. In the upper half-plane model, this isomorphism is given by the following map:

$$\pm \begin{pmatrix} a & b \\ c & d \end{pmatrix} \mapsto \left(z \mapsto \frac{az + b}{cz + d} \right).$$

The boundary of the hyperbolic plane. One of the central objects in the study of hyperbolic geometry is the *boundary at infinity* of $\mathbb{H}^2$, denoted by $\partial \mathbb{H}^2$. A point of $\partial \mathbb{H}^2$ is an equivalence class $[\gamma]$ of unit-speed geodesic rays where two rays $\gamma_1, \gamma_2 : [0, \infty) \to \mathbb{H}^2$ are equivalent if they stay a bounded distance from each other; that is, there exists $D > 0$ so that

$$d_{\mathbb{H}^2}(\gamma_1(t), \gamma_2(t)) \leq D \quad \text{for all } t \geq 0.$$

Actually, if γ_1 and γ_2 are equivalent, then they can be given unit-speed parameterizations so that

$$\lim_{t \to \infty} d_{\mathbb{H}^2}(\gamma_1(t), \gamma_2(t)) = 0.$$

We denote the union $\mathbb{H}^2 \cup \partial \mathbb{H}^2$ by $\overline{\mathbb{H}^2}$. The set $\overline{\mathbb{H}^2}$ is topologized via the following basis. We take the usual open sets of $\mathbb{H}^2$ plus one open set U_P for each open half-plane P in $\mathbb{H}^2$. A point of $\mathbb{H}^2$ lies in U_P if it lies in P, and a point of $\partial \mathbb{H}^2$ lies in U_P if every representative ray $\gamma(t)$ eventually lies in P, i.e., if there exists $T \geq 0$ so that $\gamma(t) \in P$ for all $t \geq T$.

In this topology $\partial \mathbb{H}^2$ is homeomorphic to S^1, and the union $\overline{\mathbb{H}^2}$ is homeomorphic to the closed unit disk. The space $\overline{\mathbb{H}^2}$ is a compactification of $\mathbb{H}^2$

and is called *the* compactification of $\mathbb{H}^2$. In the Poincaré disk model of $\mathbb{H}^2$, the boundary $\partial\mathbb{H}^2$ corresponds to the unit circle in $\mathbb{C}$, and $\overline{\mathbb{H}^2}$ is identified with the closed unit disk in $\mathbb{C}$.

Any isometry $f \in \mathrm{Isom}(\mathbb{H}^2)$ takes geodesic rays to geodesic rays, clearly preserving equivalence classes. Also, f takes half-planes to half-planes. It follows that f extends uniquely to a map $\overline{f} : \overline{\mathbb{H}^2} \to \overline{\mathbb{H}^2}$. As any pair of distinct points in $\partial\mathbb{H}^2$ are the endpoints of a unique geodesic in $\mathbb{H}^2$, it follows that $\overline{f}$ maps distinct points to distinct points. It is easy to check that in fact $\overline{f}$ is a homeomorphism.

Classification of isometries of $\mathbb{H}^2$. We can use the above setup to classify nontrivial elements of $\mathrm{Isom}^+(\mathbb{H}^2)$. Suppose we are given an arbitrary nontrivial element $f \in \mathrm{Isom}^+(\mathbb{H}^2)$. Since $\overline{f}$ is a self-homeomorphism of a closed disk, the Brouwer fixed point theorem gives that $\overline{f}$ has a fixed point in $\overline{\mathbb{H}^2}$. By considering the number of fixed points of $\overline{f}$ in $\overline{\mathbb{H}^2}$, we obtain a classification of isometries of $\mathbb{H}^2$ as follows.

Elliptic. If $\overline{f}$ fixes a point $p \in \mathbb{H}^2$, then f is called *elliptic*, and it is a rotation about p. Elliptic isometries have no fixed points on $\partial\mathbb{H}^2$. They correspond to elements of $\mathrm{PSL}(2, \mathbb{R})$ whose trace has absolute value less than 2.

Parabolic. If $\overline{f}$ has exactly one fixed point in $\partial\mathbb{H}^2$, then f is called *parabolic*. In the upper half-plane model, f is conjugate in $\mathrm{Isom}^+(\mathbb{H}^2)$ to $z \mapsto z \pm 1$. Parabolic isometries correspond to those nonidentity elements of $\mathrm{PSL}(2, \mathbb{R})$ with trace ± 2.

Hyperbolic. If $\overline{f}$ has two fixed points in $\partial\mathbb{H}^2$, then f is called *hyperbolic* or *loxodromic*. In this case, there is an f-invariant geodesic *axis* γ; that is, an f-invariant geodesic in $\mathbb{H}^2$ on which f acts by translation. On $\partial\mathbb{H}^2$ the fixed points act like a source and a sink, respectively. Hyperbolic isometries correspond to elements of $\mathrm{PSL}(2, \mathbb{R})$ whose trace has absolute value greater than 2.

It follows from the above classification that if $\overline{f}$ has at least three fixed points in $\overline{\mathbb{H}^2}$, then f is the identity.

Also, since commuting elements of $\mathrm{Isom}^+(\mathbb{H}^2)$ must preserve each other's fixed sets in $\overline{\mathbb{H}^2}$, we see that two nontrivial elements of $\mathrm{Isom}^+(\mathbb{H}^2)$ commute if and only if they have the same fixed points in $\overline{\mathbb{H}^2}$.

1.1.3 HYPERBOLIC SURFACES

The following theorem gives a link between the topology of surfaces and their geometry. It will be used throughout the book to convert topological problems to geometric ones, which have more structure and so are often easier to solve.

We say that a surface S *admits a hyperbolic metric* if there exists a complete, finite-area Riemannian metric on S of constant curvature -1 where the boundary of S (if nonempty) is totally geodesic (this means that the geodesics in ∂S are geodesics in S). Similarly, we say that S *admits a Euclidean metric*, or *flat metric* if there is a complete, finite-area Riemannian metric on S with constant curvature 0 and totally geodesic boundary.

If S has empty boundary and has a hyperbolic metric, then its universal cover $\widetilde{S}$ is a simply connected Riemannian 2-manifold of constant curvature -1. It follows that $\widetilde{S}$ is isometric to $\mathbb{H}^2$, and so S is isometric to the quotient of $\mathbb{H}^2$ by a free, properly discontinuous isometric action of $\pi_1(S)$. If S has nonempty boundary and has a hyperbolic metric, then $\widetilde{S}$ is isometric to a totally geodesic subspace of $\mathbb{H}^2$. Similarly, if S has a Euclidean metric, then $\widetilde{S}$ is isometric to a totally geodesic subspace of the Euclidean plane $\mathbb{E}^2$.

THEOREM 1.2 *Let S be any surface (perhaps with punctures or boundary). If $\chi(S) < 0$, then S admits a hyperbolic metric. If $\chi(S) = 0$, then S admits a Euclidean metric.*

A surface endowed with a fixed hyperbolic metric will be called a *hyperbolic surface*. A surface with a Euclidean metric will be called a *Euclidean surface* or *flat surface*.

Note that Theorem 1.2 is consistent with the Gauss–Bonnet theorem which, in the case of a compact surface S with totally geodesic boundary, states that the integral of the curvature over S is equal to $2\pi\chi(S)$.

One way to get a hyperbolic metric on a closed surface S_g is to construct a free, properly discontinuous isometric action of $\pi_1(S_g)$ on $\mathbb{H}^2$ (as above, this requires $g \geq 2$). By covering space theory and the classification of surfaces, the quotient will be homeomorphic to S_g. Since the action was by isometries, this quotient comes equipped with a hyperbolic metric. Another way to get a hyperbolic metric on S_g, for $g \geq 2$, is to take a geodesic $4g$-gon in $\mathbb{H}^2$ with interior angle sum 2π and identify opposite sides (such a $4g$-gon always exists; see Section 10.4 below). The result is a surface of genus g with a hyperbolic metric and, according to Theorem 1.2, its universal cover is $\mathbb{H}^2$.

We remark that while the torus T^2 admits a Euclidean metric, the once-punctured torus $S_{1,1}$ admits a hyperbolic metric.

Loops in hyperbolic surfaces. Let S be a hyperbolic surface. A *neighborhood of a puncture* is a closed subset of S homeomorphic to a once-punctured disk. Also, by a *free homotopy* of loops in S we simply mean an unbased homotopy. If a nontrivial element of $\pi_1(S)$ is represented by a loop that can be freely homotoped into the neighborhood of a puncture, then it follows that the loop can be made arbitrarily short; otherwise, we would find an embedded annulus whose length is infinite (by completeness) and where the length of each circular cross section is bounded from below, giving infinite area. The deck transformation corresponding to such an element of $\pi_1(S)$ is a parabolic isometry of the universal cover $\mathbb{H}^2$. This makes sense because for any parabolic isometry of $\mathbb{H}^2$, there is no positive lower bound to the distance between a point in $\mathbb{H}^2$ and its image. All other nontrivial elements of $\pi_1(S)$ correspond to hyperbolic isometries of $\mathbb{H}^2$ and hence have associated axes in $\mathbb{H}^2$.

We have the following fact, which will be used several times throughout this book:

> If S admits a hyperbolic metric, then the centralizer of any nontrivial element of $\pi_1(S)$ is cyclic. In particular, $\pi_1(S)$ has a trivial center.

To prove this we identify $\pi_1(S)$ with the deck transformation group of S for some covering map $\mathbb{H}^2 \to S$. Whenever two nontrivial isometries of $\mathbb{H}^2$ commute, it follows from the classification of isometries of $\mathbb{H}^2$ that they have the same fixed points in $\partial \mathbb{H}^2$. So if $\alpha \in \pi_1(S)$ is centralized by β, it follows that α and β have the same fixed points in $\partial \mathbb{H}^2$. By the discreteness of the action of $\pi_1(S)$, we would then have that the centralizer of α in $\pi_1(S)$ is infinite cyclic. If $\pi_1(S)$ had nontrivial center, it would then follow that $\pi_1(S) \approx \mathbb{Z}$. But then S would necessarily have infinite volume, a contradiction.

1.2 SIMPLE CLOSED CURVES

Our study of simple closed curves in a surface S begins with the study of all closed curves in S and the usefulness of geometry in understanding them.

1.2.1 CLOSED CURVES AND GEODESICS

By a *closed curve* in a surface S we will mean a continuous map $S^1 \to S$. We will usually identify a closed curve with its image in S. A closed curve is called *essential* if it is not homotopic to a point, a puncture, or a boundary component.

Closed curves and fundamental groups. Given an oriented closed curve $\alpha \in S$, we can identify α with an element of $\pi_1(S)$ by choosing a path from the basepoint for $\pi_1(S)$ to some point on α. The resulting element of $\pi_1(S)$ is well defined only up to conjugacy. By a slight abuse of notation we will denote this element of $\pi_1(S)$ by α as well.

There is a bijective correspondence:

$$\left\{ \begin{array}{c} \text{Nontrivial} \\ \text{conjugacy classes} \\ \text{in } \pi_1(S) \end{array} \right\} \longleftrightarrow \left\{ \begin{array}{c} \text{Nontrivial free} \\ \text{homotopy classes of oriented} \\ \text{closed curves in } S \end{array} \right\}$$

An element g of a group G is *primitive* if there does not exist any $h \in G$ so that $g = h^k$, where $|k| > 1$. The property of being primitive is a conjugacy class invariant. In particular, it makes sense to say that a closed curve in a surface is primitive.

A closed curve in S is a *multiple* if it is a map $S^1 \rightarrow S$ that factors through the map $S^1 \xrightarrow{\times n} S^1$ for $n > 1$. In other words, a curve is a multiple if it "runs around" another curve multiple times. If a closed curve in S is a multiple, then no element of the corresponding conjugacy class in $\pi_1(S)$ is primitive.

Let $p : \widetilde{S} \rightarrow S$ be any covering space. By a *lift* of a closed curve α to $\widetilde{S}$ we will always mean the image of a lift $\mathbb{R} \rightarrow \widetilde{S}$ of the map $\alpha \circ \pi$, where $\pi : \mathbb{R} \rightarrow S^1$ is the usual covering map. For example, if S is a surface with $\chi(S) \leq 0$, then a lift of an essential simple closed curve in S to the universal cover is a copy of $\mathbb{R}$. Note that a lift is different from a path lift, which is typically a proper subset of a lift.

Now suppose that $\widetilde{S}$ is the universal cover and α is a simple closed curve in S that is not a nontrivial multiple of another closed curve. In this case, the lifts of α to $\widetilde{S}$ are in natural bijection with the cosets in $\pi_1(S)$ of the infinite cyclic subgroup $\langle \alpha \rangle$. (Any nontrivial multiple of α has the same set of lifts as α but more cosets.) The group $\pi_1(S)$ acts on the set of lifts of α by deck transformations, and this action agrees with the usual left action of $\pi_1(S)$ on the cosets of $\langle \alpha \rangle$. The stabilizer of the lift corresponding to the coset $\gamma \langle \alpha \rangle$ is the cyclic group $\langle \gamma \alpha \gamma^{-1} \rangle$.

When S admits a hyperbolic metric and α is a primitive element of $\pi_1(S)$, we have a bijective correspondence:

$$\left\{ \begin{array}{c} \text{Elements of the conjugacy} \\ \text{class of } \alpha \text{ in } \pi_1(S) \end{array} \right\} \longleftrightarrow \left\{ \begin{array}{c} \text{Lifts to } \widetilde{S} \text{ of the} \\ \text{closed curve } \alpha \end{array} \right\}$$

More precisely, the lift of the curve α given by the coset $\gamma \langle \alpha \rangle$ corresponds to the element $\gamma \alpha \gamma^{-1}$ of the conjugacy class $[\alpha]$. That this is a bijective correspondence is a consequence of the fact that, for a hyperbolic surface S, the centralizer of any element of $\pi_1(S)$ is cyclic.

If α is any multiple, then we still have a bijective correspondence between elements of the conjugacy class of α and the lifts of α. However, if α is not primitive and not a multiple, then there are more lifts of α than there are conjugates. Indeed, if $\alpha = \beta^k$, where $k > 1$, then $\beta\langle\alpha\rangle \neq \langle\alpha\rangle$ while $\beta\alpha\beta^{-1} = \alpha$.

Note that the above correspondence does not hold for the torus T^2. This is so because each closed curve has infinitely many lifts, while each element of $\pi_1(T^2) \approx \mathbb{Z}^2$ is its own conjugacy class. Of course, $\pi_1(T^2)$ is its own center, and so the centralizer of each element is the whole group.

Geodesic representatives. A priori the combinatorial topology of closed curves on surfaces has nothing to do with geometry. It was already realized in the nineteenth century, however, that the mere existence of constant-curvature Riemannian metrics on surfaces has strong implications for the topology of the surface and of simple closed curves in it. For example, it is easy to prove that any closed curve α on a flat torus is homotopic to a geodesic: one simply lifts α to $\mathbb{R}^2$ and performs a straight-line homotopy. Note that the corresponding geodesic is not unique.

For compact hyperbolic surfaces we have a similar picture, and in fact the free homotopy class of any closed curve contains a unique geodesic. The existence is indeed true for any compact Riemannian manifold. Here we give a more hands-on proof of existence and uniqueness for any hyperbolic surface.

Proposition 1.3 *Let S be a hyperbolic surface. If α is a closed curve in S that is not homotopic into a neighborhood of a puncture, then α is homotopic to a unique geodesic closed curve γ.*

Proof. Choose a lift $\widetilde{\alpha}$ of α to $\mathbb{H}^2$. As above, $\widetilde{\alpha}$ is stabilized by some element of the conjugacy class of $\pi_1(S)$ corresponding to α; let ϕ be the corresponding isometry of $\mathbb{H}^2$. By the assumption on α, we have that ϕ is a hyperbolic isometry and so has an axis of translation A; see Figure 1.1.

Consider the projection of A to S and let γ_0 be a geodesic closed curve that travels around this projection once. Any equivariant homotopy from $\widetilde{\alpha}$ to A projects to a homotopy between α and a multiple of γ_0, which is the desired γ. One way to get such a homotopy is to simply take the homotopy that moves each point of $\widetilde{\alpha}$ along a geodesic segment to its closest-point projection in A. This completes the proof of the existence of γ. Note that we do not need to worry that the resulting parameterization of γ is geodesic since any two parameterizations of the same closed curve are homotopic as parameterized maps.

To prove uniqueness, suppose we are given a homotopy $S^1 \times I \to S$

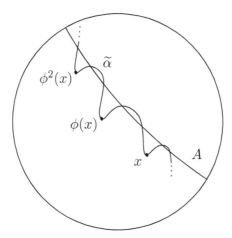

Figure 1.1 A lift $\widetilde{\alpha}$ of a closed curve α and the axis A for the corresponding isometry ϕ.

from α to a multiple γ' of some simple closed geodesic γ_0'. By compactness of $S^1 \times I$, there exists a constant $C \geq 0$ such that no point of α is moved a distance greater than C by the homotopy. In the universal cover $\mathbb{H}^2$, the homotopy lifts to a homotopy from the lift $\widetilde{\alpha}$ of α to a geodesic lift $\widetilde{\gamma}_0'$ of γ_0', and points of $\widetilde{\alpha}$ are moved a distance at most C. It follows that the endpoints of $\widetilde{\alpha}$ in $\partial \mathbb{H}^2$ are the same as those of $\widetilde{\gamma}_0'$. Since a geodesic in $\mathbb{H}^2$ is uniquely determined by its endpoints in $\partial \mathbb{H}^2$, this proves that the geodesic closed curve γ_0' is the same as γ_0 up to sign. The closed curve γ' is then specified by which multiple of γ_0 it is. But different multiples of γ_0 correspond to conjugacy classes in $\mathrm{Isom}^+(\mathbb{H}^2)$ that have different translation lengths and/or translation directions. Conjugacy classes with differing translation lengths are distinct, and so distinct multiples of γ_0 do not lie in the same free homotopy class. $\qquad \square$

It follows from Proposition 1.3 that for a compact hyperbolic surface we have a bijective correspondence:

$$\left\{ \begin{array}{c} \text{Conjugacy classes} \\ \text{in } \pi_1(S) \end{array} \right\} \longleftrightarrow \left\{ \begin{array}{c} \text{Oriented geodesic} \\ \text{closed curves in } S \end{array} \right\}$$

1.2.2 SIMPLE CLOSED CURVES

A closed curve in S is *simple* if it is embedded, that is, if the map $S^1 \to S$ is injective. Among the reasons for the particular importance of simple closed curves is that we can easily classify them up to homeomorphism of S (see Section 1.3), we can cut along them (see Section 1.3), and we can twist along them (see Section 3.1). As mentioned above, we will study homeomorphisms of surfaces via their actions on simple closed curves.

Any closed curve α can be approximated by a smooth closed curve, and a close enough approximation α' of α is homotopic to α. What is more, if α is simple, then α' can be chosen to be simple. Smooth curves are advantageous for many reasons. For instance, smoothness allows us to employ the notion of transversality (general position). When convenient, we will assume that our curves are smooth, sometimes without mention.

Simple closed curves are also natural to study because they represent primitive elements of $\pi_1(S)$.

Proposition 1.4 *Let α be a simple closed curve in a surface S. If α is not null homotopic, then each element of the corresponding conjugacy class in $\pi_1(S)$ is primitive.*

Proof. We give the proof for the case when S is hyperbolic. Fix a covering map $\mathbb{H}^2 \to S$ and let $\phi \in \mathrm{Isom}^+(\mathbb{H}^2)$ be the hyperbolic isometry corresponding to some element of the conjugacy class of α. The primitivity of the elements of the conjugacy class of α is equivalent to the primitivity of ϕ in the deck transformation group.

Assume that $\phi = \psi^n$, where ψ is another element of the deck transformation group and $n \in \mathbb{Z}$. In any group, powers of the same element commute, and so ϕ commutes with ψ. Thus ϕ and ψ have the same set of fixed points in $\partial \mathbb{H}^2$.

Let $\widetilde{\alpha}$ be the lift of the closed curve α that has the same endpoints in $\partial \mathbb{H}^2$ as the axis for ϕ. We claim that $\psi(\widetilde{\alpha}) = \widetilde{\alpha}$. We know that $\psi(\widetilde{\alpha})$ is some lift of α. Since α is simple, all of its lifts are disjoint and no two lifts of α have the same endpoints in $\partial \mathbb{H}^2$. Thus $\psi(\widetilde{\alpha})$ and $\widetilde{\alpha}$ are disjoint and have distinct endpoints. Now, we know that $\psi^{n-1}(\psi(\widetilde{\alpha})) = \phi(\widetilde{\alpha}) = \widetilde{\alpha}$. Since the fixed points in $\partial \mathbb{H}^2$ of ψ^{n-1} are the same as the endpoints of $\widetilde{\alpha}$, the only way $\psi^{n-1}(\psi(\widetilde{\alpha}))$ can have the same endpoints at infinity as $\widetilde{\alpha}$ is if $\psi(\widetilde{\alpha})$ does. This is to say that $\psi(\widetilde{\alpha}) = \widetilde{\alpha}$, and the claim is proven.

Thus the restriction of ψ to $\widetilde{\alpha}$ is a translation. As $\phi = \psi^n$, the closed curve α travels n times around the closed curve in S given by $\widetilde{\alpha}/\langle\psi\rangle$. Since α is simple, we have $n = \pm 1$, which is what we wanted to show. $\qquad\square$

Simple closed curves in the torus. We can classify the set of homotopy classes of simple closed curves in the torus T^2 as follows. Let $\mathbb{R}^2 \to T^2$ be the usual covering map, where the deck transformation group is generated by the translations by $(1,0)$ and $(0,1)$. We know that $\pi_1(T^2) \approx \mathbb{Z}^2$, and if we base $\pi_1(T^2)$ at the image of the origin, one way to get a representative for (p, q) as a loop in T^2 is to take the straight line from $(0, 0)$ to (p, q) in $\mathbb{R}^2$ and project it to T^2.

Let γ be any oriented simple closed curve in T^2. Up to homotopy, we can assume that γ passes through the image in T^2 of $(0, 0)$ in $\mathbb{R}^2$. Any path lifting of γ to $\mathbb{R}^2$ based at the origin terminates at some integral point (p, q). There is then a homotopy from γ to the standard straight-line representative of $(p, q) \in \pi_1(T^2)$; indeed, the straight-line homotopy from the lift of γ to the straight line through $(0, 0)$ and (p, q) is equivariant with respect to the group of deck transformations and thus descends to the desired homotopy.

Now, if a closed curve in T^2 is simple, then its straight-line representative is simple. Thus we have the following fact.

Proposition 1.5 *The nontrivial homotopy classes of oriented simple closed curves in T^2 are in bijective correspondence with the set of primitive elements of $\pi_1(T^2) \approx \mathbb{Z}^2$.*

An element (p, q) of $\mathbb{Z}^2$ is primitive if and only if $(p, q) = (0, \pm 1)$, $(p, q) = (\pm 1, 0)$, or $\gcd(p, q) = 1$.

We can classify homotopy classes of essential simple closed curves in other surfaces. For example, in S^2, $S_{0,1}$, $S_{0,2}$, and $S_{0,3}$, there are no essential simple closed curves. The homotopy classes of simple closed curves in $S_{1,1}$ are in bijective correspondence with those in T^2. In Section 2.2 below, we will show that there is a natural bijection between the homotopy classes of essential simple closed curves in $S_{0,4}$ and the homotopy classes in T^2.

Closed geodesics. For hyperbolic surfaces geodesics are the natural representatives of each free homotopy class in the following sense.

Proposition 1.6 *Let S be a hyperbolic surface. Let α be a closed curve in S not homotopic into a neighborhood of a puncture. Let γ be the unique geodesic in the free homotopy class of α guaranteed by Proposition 1.3. If α is simple, then γ is simple.*

Proof. We begin by applying the following fact.

> *A closed curve β in a hyperbolic surface S is simple if and only if the following properties hold:*
>
> *1. Each lift of β to $\mathbb{H}^2$ is simple.*
>
> *2. No two lifts of β intersect.*

> *3. β is not a nontrivial multiple of another closed curve.*

Thus if α is simple, then no two of its lifts to $\mathbb{H}^2$ intersect. It follows that for any two such lifts, their endpoints are not linked in $\partial\mathbb{H}^2$. But each lift of γ shares both endpoints with some lift of α. Thus no two lifts of γ have endpoints that are linked in $\partial\mathbb{H}^2$. Since these lifts are geodesics, it follows that they do not intersect. Further, by Proposition 1.4, any element of $\pi_1(S)$ corresponding to α is primitive. The same is then true for γ, and so γ cannot be a multiple. Since geodesics in $\mathbb{H}^2$ are always simple, we conclude that γ is simple. $\square$

1.2.3 INTERSECTION NUMBERS

There are two natural ways to count the number of intersection points between two simple closed curves in a surface: signed and unsigned. These correspond to the algebraic intersection number and geometric intersection number, respectively.

Let α and β be a pair of transverse, oriented, simple closed curves in S. Recall that the *algebraic intersection number* $\hat{i}(\alpha, \beta)$ is defined as the sum of the indices of the intersection points of α and β, where an intersection point is of index $+1$ when the orientation of the intersection agrees with the orientation of S and is -1 otherwise. Recall that $\hat{i}(\alpha, \beta)$ depends only on the homology classes of α and β. In particular, it makes sense to write $\hat{i}(a, b)$ for a and b, the free homotopy classes (or homology classes) of closed curves α and β.

The most naive way to count intersections between homotopy classes of closed curves is to simply count the minimal number of unsigned intersections. This idea is encoded in the concept of geometric intersection number. The *geometric intersection number* between free homotopy classes a and b of simple closed curves in a surface S is defined to be the minimal number of intersection points between a representative curve in the class a and a representative curve in the class b:

$$i(a, b) = \min\{|\alpha \cap \beta| : \alpha \in a, \ \beta \in b\}.$$

We sometimes employ a slight abuse of notation by writing $i(\alpha, \beta)$ for the intersection number between the homotopy classes of simple closed curves α and β.

We note that geometric intersection number is symmetric, while algebraic intersection number is skew-symmetric: $i(a, b) = i(b, a)$, while $\hat{i}(a, b) = -\hat{i}(b, a)$. While algebraic intersection number is well defined on homology classes, geometric intersection number is well defined only on free homotopy classes. Geometric intersection number is a useful invariant but, as we will see, it is more difficult to compute than algebraic intersection number.

Observe that $i(a, a) = 0$ for any homotopy class of simple closed curves a. If α separates S into two components, then for any β we have $\hat{i}(\alpha, \beta) = 0$ and $i(\alpha, \beta)$ is even. In general, i and $\hat{i}$ have the same parity.

Intersection numbers on the torus. As noted above, the nontrivial free homotopy classes of oriented simple closed curves in T^2 are in bijective correspondence with primitive elements of $\mathbb{Z}^2$. For two such homotopy classes (p, q) and (p', q'), we have

$$\hat{i}((p, q), (p', q')) = pq' - p'q$$

and

$$i((p, q), (p', q')) = |pq' - p'q|.$$

To verify these formulas, one should first check the case where $(p, q) = (1, 0)$ (exercise). For the general case, we note that if (p, q) represents an essential oriented simple closed curve, that is, if it is primitive, then there is a matrix $A \in \mathrm{SL}(2, \mathbb{Z})$ with $A((p, q)) = (1, 0)$. Since A is a linear, orientation-preserving homeomorphism of $\mathbb{R}^2$ preserving $\mathbb{Z}^2$, it induces an orientation-preserving homeomorphism of the torus $T^2 = \mathbb{R}^2/\mathbb{Z}^2$ whose action on $\pi_1(T^2) \approx \mathbb{Z}^2$ is given by A. Since orientation-preserving homeomorphisms preserve both algebraic and geometric intersection numbers, the general case of each formula follows.

Minimal position. In practice, one computes the geometric intersection number between two homotopy classes a and b by finding representatives α and β that realize the minimal intersection in their homotopy classes, so that $i(a, b) = |\alpha \cap \beta|$. When this is the case, we say that α and β are in *minimal position*.

Two basic questions now arise.

1. Given two simple closed curves α and β, how can we tell if they are in minimal position?

2. Given two simple closed curves α and β, how do we find homotopic simple closed curves that are in minimal position?

While the first question is a priori a minimization problem over an infinite-dimensional space, we will see that the question can be reduced to a finite check—the bigon criterion given below. For the second question, we will see that geodesic representatives of simple closed curves are always in minimal position.

1.2.4 THE BIGON CRITERION

We say that two transverse simple closed curves α and β in a surface S form a *bigon* if there is an embedded disk in S (the bigon) whose boundary is the union of an arc of α and an arc of β intersecting in exactly two points; see Figure 1.2.

Figure 1.2 A bigon.

The following proposition gives a simple, combinatorial condition for deciding whether or not two simple closed curves are in minimal position. It therefore gives a method for determining the geometric intersection number of two simple closed curves.

Proposition 1.7 (The bigon criterion) *Two transverse simple closed curves in a surface S are in minimal position if and only if they do not form a bigon.*

One immediate and useful consequence of the bigon criterion is the following:

> *Any two transverse simple closed curves that intersect exactly once are in minimal position.*

Before proving Proposition 1.7, we need a lemma.

Lemma 1.8 *If transverse simple closed curves α and β in a surface S do not form any bigons, then in the universal cover of S, any pair of lifts $\widetilde{\alpha}$ and $\widetilde{\beta}$ of α and β intersect in at most one point.*

Proof. Assume $\chi(S) \leq 0$, so the universal cover $\widetilde{S}$ is homeomorphic to $\mathbb{R}^2$ (the case of $\chi(S) > 0$ is an exercise). Let $p \colon \widetilde{S} \to S$ be the covering map.

Suppose the lifts $\widetilde{\alpha}$ and $\widetilde{\beta}$ of α and β intersect in at least two points. It follows that there is an embedded disk D_0 in $\widetilde{S}$ bounded by one subarc of $\widetilde{\alpha}$ and one subarc of $\widetilde{\beta}$.

By compactness and transversality, the intersection $(p^{-1}(\alpha) \cup p^{-1}(\beta)) \cap D_0$ is a finite graph if we think of the intersection points as vertices. Thus there is an *innermost disk*, that is, an embedded disk D in $\widetilde{S}$ bounded by one arc of $p^{-1}(\alpha)$ and one arc of $p^{-1}(\beta)$ and with no arcs of $p^{-1}(\alpha)$ or $p^{-1}(\beta)$ passing through the interior of the D (see Figure 1.3). Denote the two vertices of D by v_1 and v_2, and the two edges of D by $\widetilde{\alpha}_1$ and $\widetilde{\beta}_1$.

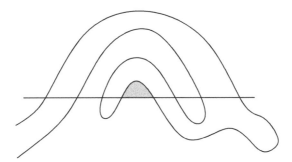

Figure 1.3 An innermost disk between two lifts.

We first claim that the restriction of p to ∂D is an embedding. The points v_1 and v_2 certainly map to distinct points in S since $\widetilde{\alpha}$ and $\widetilde{\beta}$ intersect with opposite orientations at these points. If a point of $\widetilde{\alpha}_1$ and a point of $\widetilde{\beta}_1$ have the same image in S, then both points would be an intersection of $p^{-1}(\alpha)$ with $p^{-1}(\beta)$, violating the assumption that D is innermost. If two points of $\widetilde{\alpha}_1$ (or two points of $\widetilde{\beta}_1$) map to the same point in S, then there is a lift of $p(v_1)$ between these two points, also contradicting the assumption that D is innermost.

We can now argue that D projects to an embedded disk in S. Indeed, if x and y in D project to the same point in S, then $x = \phi(y)$ for some deck transformation ϕ. Since ∂D embeds under the covering map, $\phi(\partial D) \cap \partial D$ is either empty or all of ∂D (in the case that ϕ is the identity). By the Jordan curve theorem, we then see that either $\phi(D)$ or $\phi^{-1}(D)$ must be contained in D. Now, by the Brouwer fixed point theorem, ϕ has a fixed point, which is a contradiction unless ϕ is the identity.   □

We give two proofs of the bigon criterion. One proof uses hyperbolic geometry, and one proof uses only topology. We give both proofs since each of the techniques will be important later in this book.

First proof of Proposition 1.7. First suppose that two curves α and β form a bigon. It should be intuitive that there is a homotopy of α that reduces its intersection with β by 2, but here we provide a formal proof. We can choose a small closed neighborhood of this bigon that is homeomorphic to a disk, and so the intersection of $\alpha \cup \beta$ with this disk looks like Figure 1.2. More precisely, the intersection of $\alpha \cup \beta$ with this closed disk consists of one subarc α' of α and one subarc β' of β intersecting in precisely two points. Since the disk is simply connected and since the endpoints of α' lie on the same side of β', we may modify α by a homotopy in the closed disk so that, inside this disk, α and β are disjoint. This implies that the original curves were not in minimal position.

For the other direction, we treat only the case $\chi(S) < 0$. The case $\chi(S) = 0$ is similar, and the case $\chi(S) > 0$ is easy. Assume that simple closed curves α and β form no bigons. Let $\tilde\alpha$ and $\tilde\beta$ be nondisjoint lifts of α and β. By Lemma 1.8, $\tilde\alpha$ intersects $\tilde\beta$ in exactly one point x.

It cannot be that the axes of the hyperbolic isometries corresponding to $\tilde\alpha$ and $\tilde\beta$ share exactly one endpoint at $\partial \mathbb{H}^2$ because this would violate the discreteness of the action of $\pi_1(S)$ on $\mathbb{H}^2$; indeed, in this case the commutator of these isometries is parabolic and the conjugates of this parabolic isometry by either of the original hyperbolic isometries have arbitrarily small translation length. Further, these axes cannot share two endpoints on $\partial \mathbb{H}^2$, for then the corresponding hyperbolic isometries would have the same axis, and so they would have to have a common power ϕ (otherwise the action of $\pi_1(S)$ on this axis would be nondiscrete). But then $\phi^n(x)$ would be an intersection point between $\tilde\alpha$ and $\tilde\beta$ for each n.

We conclude that any lift of α intersects any lift of β at most once and that any such lifts have distinct endpoints on $\partial \mathbb{H}^2$. But we can now see that there is no homotopy that reduces intersection. Indeed, if $\tilde\alpha$ is a particular lift of α, then each fundamental domain of $\tilde\alpha$ intersects the set of lifts of β in $|\alpha \cap \beta|$ points. Now, any homotopy of β changes this π_1-equivariant picture in an equivariant way, so since the lifts of α and β are already intersecting minimally in $\mathbb{H}^2$, there is no homotopy that reduces intersection. $\square$

Second proof of Proposition 1.7. We give a different proof that two curves not in minimal position must form a bigon. Let α and β be two simple closed curves in S that are not in minimal position and let $H : S^1 \times [0,1] \to S$ be a homotopy of α that reduces intersection with β (this is possible by the definition of minimal position). We may assume without loss of generality that α and β are transverse and that H is transverse to β (in particular, all maps are assumed to be smooth). Thus the preimage $H^{-1}(\beta)$ in the annulus $S^1 \times [0,1]$ is a 1-submanifold.

There are various possibilities for a connected component of $H^{-1}(\beta)$: it could be a closed curve, an arc connecting distinct boundary components, or an arc connecting one boundary component to itself. Since H reduces the intersection of α with β, there must be at least one component δ connecting $S^1 \times \{0\}$ to itself. Together with an arc δ' in $S^1 \times \{0\}$, the arc δ bounds a disk Δ in $S^1 \times [0, 1]$. Now, $H(\delta \cup \delta')$ is a closed curve in S that lies in $\alpha \cup \beta$. This closed curve is null homotopic—indeed, $H(\Delta)$ is the null homotopy. It follows that $H(\delta \cup \delta')$ lifts to a closed curve in the universal cover $\widetilde{S}$; what is more, this lift has one arc in a lift of α and one arc in a lift of β. Thus these lifts intersect twice, and so Lemma 1.8 implies that α and β form a bigon. $\square$

Geodesics are in minimal position. Note that if two geodesic segments on a hyperbolic surface S together bounded a bigon, then, since the bigon is simply connected, one could lift this bigon to the universal cover $\mathbb{H}^2$ of S. But this would contradict the fact that the geodesic between any two points of $\mathbb{H}^2$ is unique. Hence by Proposition 1.7 we have the following.

Corollary 1.9 *Distinct simple closed geodesics in a hyperbolic surface are in minimal position.*

The bigon criterion gives an algorithmic answer to the question of how to find representatives in minimal position: given any pair of transverse simple closed curves, we can remove bigons one by one until none remain and the resulting curves are in minimal position. Corollary 1.9, together with Proposition 1.3, gives a qualitative answer to the question.

Multicurves. A *multicurve* in S is the union of a finite collection of disjoint simple closed curves in S. The notion of intersection number extends directly to multicurves. A slight variation of the proof of the bigon criterion (Proposition 1.7) gives a version of the bigon criterion for multicurves: two multicurves are in minimal position if and only if no two component curves form a bigon.

Proposition 1.3 and Corollary 1.9 together have the consequence that, given any number of distinct homotopy classes of essential simple closed curves in S, we can choose a single representative from each class (e.g. the geodesic) so that each pair of curves is in minimal position.

1.2.5 HOMOTOPY VERSUS ISOTOPY FOR SIMPLE CLOSED CURVES

Two simple closed curves α and β are *isotopic* if there is a homotopy

$$H : S^1 \times [0, 1] \to S$$

from α to β with the property that the closed curve $H(S^1 \times \{t\})$ is simple for each $t \in [0, 1]$.

In our study of mapping class groups, it will often be convenient to think about isotopy classes of simple closed curves instead of homotopy classes. One way to explain this is as follows. If $H : S^1 \times I \to S$ is an isotopy of simple closed curves, then the pair $(S, H(S^1 \times \{t\}))$ "looks the same" for all t (cf. Section 1.3).

When we appeal to algebraic topology for the existence of a homotopy, the result is in general not an isotopy. We therefore want a method for converting homotopies to isotopies whenever possible.

We already know $i(a, b)$ is realized by geodesic representatives of a and b. Thus, in order to apply the above results on geometric intersection numbers to isotopy classes of curves, it suffices to prove the following fact originally due to Baer.

Proposition 1.10 *Let α and β be two essential simple closed curves in a surface S. Then α is isotopic to β if and only if α is homotopic to β.*

Proof. One direction is vacuous since an isotopy is a homotopy. So suppose that α is homotopic to β. We immediately have that $i(\alpha, \beta) = 0$. By performing an isotopy of α, we may assume that α is transverse to β. If α and β are not disjoint, then by the bigon criterion they form a bigon. A bigon prescribes an isotopy that reduces intersection. Thus we may remove bigons one by one by isotopy until α and β are disjoint.

In the remainder of the proof, we assume $\chi(S) < 0$; the case $\chi(S) = 0$ is similar, and the case $\chi(S) > 0$ is easy. Choose lifts $\tilde{\alpha}$ and $\tilde{\beta}$ of α and β that have the same endpoints in $\partial \mathbb{H}^2$. There is a hyperbolic isometry ϕ that leaves $\tilde{\alpha}$ and $\tilde{\beta}$ invariant and acts by translation on these lifts. As $\tilde{\alpha}$ and $\tilde{\beta}$ are disjoint, we may consider the region R between them. The quotient $R' = R/\langle\phi\rangle$ is an annulus; indeed, it is a surface with two boundary components with an infinite cyclic fundamental group. A priori, the image R'' of R in S is a further quotient of R'. However, since the covering map $R' \to R''$ is single-sheeted on the boundary, it follows that $R' \approx R''$. The annulus R'' between α and β gives the desired isotopy. $\square$

1.2.6 EXTENSION OF ISOTOPIES

An isotopy of a surface S is a homotopy $H : S \times I \to S$ so that, for each $t \in [0, 1]$, the map $H(S, t) : S \times \{t\} \to S$ is a homeomorphism. Given an isotopy between two simple closed curves in S, it will often be useful to promote this to an isotopy of S, which we call an *ambient isotopy* of S.

Proposition 1.11 *Let S be any surface. If $F : S^1 \times I \to S$ is a smooth isotopy of simple closed curves, then there is an isotopy $H : S \times I \to S$ so that $H|_{S \times 0}$ is the identity and $H|_{F(S^1 \times 0) \times I} = F$.*

Proposition 1.11 is a standard fact from differential topology. Suppose that the two curves are disjoint. To construct the isotopy, one starts by finding a smooth vector field that is supported on a neighborhood of the closed annulus between the two curves and that carries one curve to the other. One then obtains the isotopy of the surface S by extending this vector field to S and then integrating it. For details of this argument see, e.g., [95, Chapter 8, Theorem 1.3].

1.2.7 ARCS

In studying surfaces via their simple closed curves, we will often be forced to think about arcs. For instance, many of our inductive arguments involve cutting a surface along some simple closed curve in order to obtain a "simpler" surface. Simple closed curves in the original surface either become simple closed curves or collections of arcs in the cut surface. Much of the discussion about curves carries over to arcs, so here we take a moment to highlight the necessary modifications.

We first pin down the definition of an arc. This is one place where marked points are more convenient than punctures. So assume S is a compact surface, possibly with boundary and possibly with finitely many marked points in the interior. Denote the set of marked points by $\mathcal{P}$.

A *proper arc* in S is a map $\alpha : [0, 1] \rightarrow S$ such that $\alpha^{-1}(\mathcal{P} \cup \partial S) = \{0, 1\}$. As with curves, we usually identify an arc with its image; in particular, this makes an arc an unoriented object. The arc α is *simple* if it is an embedding on its interior. The homotopy class of a proper arc is taken to be the homotopy class within the class of proper arcs. Thus points on ∂S cannot move off the boundary during the homotopy; all arcs would be homotopic to a point otherwise. But there is still a choice to be made: a homotopy (or isotopy) of an arc is said to be *relative to the boundary* if its endpoints stay fixed throughout the homotopy. An arc in a surface S is *essential* if it is neither homotopic into a boundary component of S nor a marked point of S.

The bigon criterion (Proposition 1.7) holds for arcs, except with one extra subtlety illustrated in Figure 1.4. If we are considering isotopies relative to the boundary, then the arcs in the figure are in minimal position, but if we are considering general isotopies, then the half-bigon shows that they are not in minimal position.

Corollary 1.9 (geodesics are in minimal position) and Proposition 1.3 (existence and uniqueness of geodesic representatives) work for arcs in surfaces with punctures and/or boundary. Here we switch back from marked points to punctures to take advantage of hyperbolic geometry. Proposition 1.10 (homotopy versus isotopy for curves) and Theorem 1.13 (extension of iso-

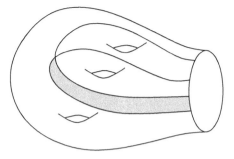

Figure 1.4 The shaded region is a half-bigon.

topies) also work for arcs.

1.3 THE CHANGE OF COORDINATES PRINCIPLE

We now describe a basic technique that is used quite frequently in the theory of mapping class groups, often without mention. We call this technique the *change of coordinates principle*. One example of this principle is that, in order to prove a topological statement about an arbitrary nonseparating simple closed curve, we can prove it for any specific simple closed curve. We will see below that this idea applies to any configuration of simple closed curves that is given by topological data.

1.3.1 CLASSIFICATION OF SIMPLE CLOSED CURVES

As a prelude to our explanation of the change of coordinates principle, we present a classification of simple closed curves in a surface.

We first need to introduce an essential concept. Given a simple closed curve α in a surface S, the surface obtained by *cutting* S along α is a compact surface S_α equipped with a homeomorphism h between two of its boundary components so that

1. the quotient $S_\alpha/(x \sim h(x))$ is homeomorphic to S, and

2. the image of these distinguished boundary components under this quotient map is α.

It also makes sense to cut a surface with boundary or marked points along a simple proper arc; the definition is analogous. Similarly, one can cut along a finite collection of curves and arcs. There are several distinct situations for cutting along a single arc, depending on whether the endpoints of the arc lie on a boundary component or a puncture, for instance, and the cut surface is allowed to have marked points on its boundary.

We remark that the cutting procedure is one place where it is convenient to assume that all curves under consideration are smooth. Indeed, if γ is a smooth simple closed curve in a surface S, then the pair (S, γ) is locally diffeomorphic to $(\mathbb{R}^2, \mathbb{R})$, and one can immediately conclude that the surface obtained from S by cutting along γ is again a surface, now with two additional boundary components. Hence the classification of surfaces can be applied to the cut surface.

We say that a simple closed curve α in the surface S is *nonseparating* if the cut surface S_α is connected. We claim the following.

> *If α and β are any two nonseparating simple closed curves in a surface S, then there is a homeomorphism $\phi : S \to S$ with $\phi(\alpha) = \beta$.*

In other words, up to homeomorphism, there is only one nonseparating simple closed curve in S. This statement follows from the classification of surfaces, as follows. The cut surfaces S_α and S_β each have two boundary components corresponding to α and β, respectively. Since S_α and S_β have the same Euler characteristic, number of boundary components, and number of punctures, it follows that S_α is homeomorphic to S_β. We can choose a homeomorphism $S_\alpha \to S_\beta$ that respects the equivalence relations on the distinguished boundary components. Such a homeomorphism gives the desired homeomorphism of S taking α to β. If we want an orientation-preserving homeomorphism, we can ensure this by postcomposing by an orientation-reversing homeomorphism fixing β if necessary.

A simple closed curve β is *separating* in S if the cut surface S_β is not connected. Note that when S is closed, β is separating if and only if it is the boundary of some subsurface of S. This is equivalent to the vanishing of the homology class of β in $H_1(S, \mathbb{Z})$. By the "classification of disconnected surfaces," we see that there are finitely many separating simple closed curves in S up to homeomorphism.

The above arguments give the following general classification of simple closed curves on a surface:

> *There is an orientation-preserving homeomorphism of a surface taking one simple closed curve to another if and only if the corresponding cut surfaces (which may be disconnected) are homeomorphic.*

The existence of such a homeomorphism is clearly an equivalence relation. The equivalence class of a simple closed curve or a collection of simple closed curves is called its *topological type*. For example, a separating simple closed curve in the closed surface S_g divides S_g into two disjoint subsurfaces of, say, genus k and $g - k$. The minimum of $\{k, g - k\}$ is called the *genus* of the separating simple closed curve. By the above, the genus of a curve determines and is determined by its topological type. Note that there are $\lfloor \frac{g}{2} \rfloor$ topological types of essential separating simple closed curves in a closed surface.

The uninitiated may have trouble visualizing separating simple closed curves that are not the obvious ones. We present a few in Figure 1.5, and we encourage the reader to draw even more complicated separating simple closed curves.

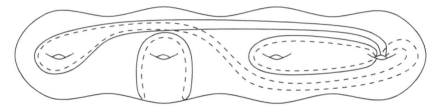

Figure 1.5 Some nonobvious separating simple closed curves.

1.3.2 THE CHANGE OF COORDINATES PRINCIPLE

The change of coordinates principle is a kind of change of basis for curves in a surface S. It roughly states that any two collections of simple closed curves in S with the same intersection pattern can be taken to each other via an orientation-preserving homeomorphism of S. In this way an arbitrary configuration can be transformed into a standard configuration. The classification of simple closed curves in surfaces given above is the simplest example.

We illustrate the principle with two sample questions. Suppose α is *any* nonseparating simple closed curve α on a surface S.

1. Is there a simple closed curve γ in S so that α and γ fill S, that is, α and γ are in minimal position and the complement of $\alpha \cup \gamma$ is a union of topological disks?

2. Is there a simple closed curve δ in S with $i(\alpha, \delta) = 0$? $i(\alpha, \delta) = 1$? $i(\alpha, \delta) = k$?

Even for the genus 2 surface S_2, it is not immediately obvious how to answer either question for the nonseparating simple closed curve α shown

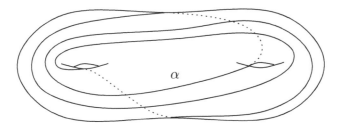

Figure 1.6 A simple closed curve on a genus 2 surface.

in Figure 1.6. However, we claim that Figure 1.7 gives proof that the answer to the first question is yes in this case, as we now show. The curves β and γ in Figure 1.7 fill the surface (check this!). By the classification of simple closed curves in a surface, there is a homeomorphism $\phi : S_2 \to S_2$ with $\phi(\beta) = \alpha$. Since filling is a topological property, it follows that $\phi(\gamma)$ is the curve we are looking for since it together with $\alpha = \phi(\beta)$ fills S_2.

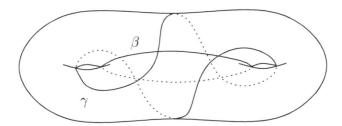

Figure 1.7 Two simple closed curves that fill a genus 2 surface.

We think of ϕ as changing coordinates so that the complicated curve α becomes the easy-to-see curve β. The second question can be answered similarly.

1.3.3 Examples of the Change of Coordinates Principle

The change of coordinates principle applies to more general situations. We give several examples here. Most of the proofs are minor variations of the above arguments and so are left to the reader.

1. Pairs of simple closed curves that intersect once. Suppose that α_1 and β_1 form such a pair in a surface S. Let S_{α_1} be the surface obtained by cutting S along α_1. There are two boundary components of S_{α_1} corresponding to the two sides of α_1. The image of β_1 in S_{α_1} is a simple arc connecting these boundary components to each other. We can cut S_{α_1} along this arc to obtain a surface $(S_{\alpha_1})_{\beta_1}$. The latter is a surface with one boundary component that is naturally subdivided into four arcs—two coming from α_1 and two coming from β_1. The equivalence relation coming from the definition of a cut surface identifies these arcs in order to recover the surface S with its curves α_1 and β_1.

If α_2 and β_2 are another such pair, there is an analogous cut surface $(S_{\alpha_2})_{\beta_2}$. By the classification of surfaces, $(S_{\alpha_2})_{\beta_2}$ is homeomorphic to $(S_{\alpha_1})_{\beta_1}$, and moreover there is a homeomorphism that preserves equivalence classes on the boundary. Any such homeomorphism descends to a homeomorphism of S taking the pair $\{\alpha_1, \beta_1\}$ to the pair $\{\alpha_2, \beta_2\}$.

2. Bounding pairs of a given genus. A *bounding pair* is a pair of disjoint, homologous, nonseparating simple closed curves in a closed surface. Figure 1.8 shows one example, but we again encourage the reader to find more complicated examples. The genus of a bounding pair in a closed surface is defined similarly to the genus of a separating simple closed curve.

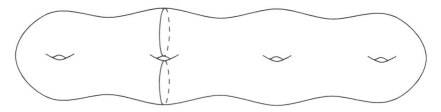

Figure 1.8 A genus 1 bounding pair.

3. Pairs (or k-tuples) of disjoint simple closed curves whose union does not separate.

4. Pairs of simple closed curves $\{\alpha, \beta\}$ *with* $i(\alpha, \beta) = |\alpha \cap \beta| = 2$ *and* $\hat{i}(\alpha, \beta) = 0$ *and whose union does not separate.*

5. Nonseparating simple proper arcs in a surface S that meet the same number of components of ∂S.

6. Chains of simple closed curves. A *chain* of simple closed curves in a surface S is a sequence $\alpha_1, \ldots, \alpha_k$ with the properties that $i(\alpha_i, \alpha_{i+1}) = 1$ for each i and $i(\alpha_i, \alpha_j) = 0$ whenever $|i - j| > 1$. A chain is *nonseparating* if the union of the curves does not separate the surface.

Any two nonseparating chains of simple closed curves with the same number of curves are topologically equivalent. This can be proved by induction. The starting point is the case of nonseparating simple closed curves, and the inductive step is example 5: cutting along the first few arcs, the next arc becomes a nonseparating arc on the cut surface. Note that example 1 is the case $k = 2$. One can also prove by induction that every chain in S_g of even length is nonseparating, and so such chains must be topologically equivalent.

We remark that the homeomorphism representing the change of coordinates in each of the six examples above can be taken to be orientation-preserving.

1.4 THREE FACTS ABOUT HOMEOMORPHISMS

In this subsection we collect three useful facts from surface topology. Each allows us to replace one kind of map with a better one: a homotopy of homeomorphisms can be improved to an isotopy; a homeomorphism of a surface can be promoted to a diffeomorphism; and $\mathrm{Homeo}_0(S)$ is contractible, so in particular any isotopy from the identity homeomorphism to itself is homotopic to the constant isotopy.

1.4.1 HOMOTOPY VERSUS ISOTOPY FOR HOMEOMORPHISMS

When are two homotopic homeomorphisms isotopic? Let us look at two of the simplest examples: the closed disk D^2 and the closed annulus A. On D, any orientation-reversing homeomorphism f induces a degree -1 map on $S^1 = \partial D^2$, and from this follows that f is not isotopic to the identity. However, the straight-line homotopy gives a homotopy between f and the identity. On $A = S^1 \times I$, the orientation-reversing map that fixes the S^1 factor and reflects the I factor is homotopic but not isotopic to the identity.

It turns out that these two examples are the only examples of homotopic homeomorphisms that are not isotopic. This was proved in the 1920s by Baer using Proposition 1.10 (see [8, 9] and also [56]).

THEOREM 1.12 *Let S be any compact surface and let f and g be homotopic homeomorphisms of S. Then f and g are isotopic unless they are one of the two examples described above (on $S = D^2$ and $S = A$). In particular, if f and g are orientation-preserving, then they are isotopic.*

In fact, a stronger, relative result holds: if two homeomorphisms are homotopic relative to ∂S, then they are isotopic relative to ∂S. Theorem 1.12 can be proven using ideas from the proof of Proposition 2.8.

Theorem 1.12 also holds when S has finitely many marked points. In that case, we need to expand our list of counterexamples to include a sphere with one or two marked points.

1.4.2 HOMEOMORPHISMS VERSUS DIFFEOMORPHISMS

It is sometimes convenient to work with homeomorphisms and sometimes convenient to work with diffeomorphisms. For example, it is easier to construct the former, but we can apply differential topology to the latter. The following theorem will allow us to pass back and forth between homeomorphisms and diffeomorphisms of surfaces.

THEOREM 1.13 *Let S be a compact surface. Then every homeomorphism of S is isotopic to a diffeomorphism of S.*

It is a general fact that any homeomorphism of a smooth manifold can be approximated arbitrarily well by a smooth map. By taking a close enough approximation, the resulting smooth map is homotopic to the original homeomorphism. However, this general fact, which is easy to prove, is much weaker than Theorem 1.13 because the resulting smooth map might not be smoothly invertible; indeed, it might not be invertible at all.

Theorem 1.13 was proven in the 1950s by Munkres [167, Theorem 6.3], Smale, and Whitehead [213, Corollary 1.18]. In part, this work was prompted by Milnor's discovery of the "exotic" (nondiffeomorphic) smooth structures on S^7.

Theorem 1.13 gives us a way to replace homeomorphisms with diffeomorphisms. We can also replace isotopies with smooth isotopies. In other words, if two diffeomorphisms are isotopic, then they are smoothly isotopic; see, for example, [30].

In this book, we will switch between the topological setting and the smooth setting as is convenient. For example, when defining a map of a surface to itself (either by equations or by pictures), it is often easier to write down a homeomorphism than a smooth map. On the other hand, when we need to appeal to transversality, extension of isotopy, and so on, we will need to assume we have a diffeomorphism.

One point to make is that we will actually be forced to consider self-maps of a surface that are not smooth; pseudo-Anosov homeomorphisms, which are central to the theory, are special maps of a surface that are never smooth (cf. Chapter 13).

1.4.3 CONTRACTIBILITY OF COMPONENTS OF Homeo(S)

The following theorem was proven by Hamstrom in a series of papers [77, 78, 79] in the 1960s. In the statement, $\mathrm{Homeo}_0(S)$ is the connected component of the identity in the space of homeomorphisms of a surface S.

THEOREM 1.14 *Let S be a compact surface, possibly minus a finite number of points from the interior. Assume that S is not homeomorphic to S^2, $\mathbb{R}^2$, D^2, T^2, the closed annulus, the once-punctured disk, or the once-punctured plane. Then the space $\mathrm{Homeo}_0(S)$ is contractible.*

The fact that $\mathrm{Homeo}_0(S)$ is simply connected is of course an immediate consequence of Theorem 1.14. This fact will be used, among other places, in Section 4.2 in the proof of the Birman exact sequence. There is a smooth version of Theorem 1.14; see [53] or [73].

Chapter Two

Mapping Class Group Basics

In this chapter we begin our study of the mapping class group of a surface. After giving the definition, we compute the mapping class group in essentially all of the cases where it can be computed directly. This includes the case of the disk, the annulus, the torus, and the pair of pants. An important method, which we call the Alexander method, emerges as a tool for such computations. It answers the fundamental question: how can one prove that a homeomorphism is or is not homotopically trivial? Equivalently, how can one decide when two homeomorphisms are homotopic or not?

2.1 DEFINITION AND FIRST EXAMPLES

Let S be a surface. As in Chapter 1, we assume that S is the connect sum of $g \geq 0$ tori with $b \geq 0$ disjoint open disks removed and $n \geq 0$ points removed from the interior. Let $\mathrm{Homeo}^+(S, \partial S)$ denote the group of orientation-preserving homeomorphisms of S that restrict to the identity on ∂S. We endow this group with the compact-open topology.

The *mapping class group* of S, denoted $\mathrm{Mod}(S)$, is the group

$$\mathrm{Mod}(S) = \pi_0(\mathrm{Homeo}^+(S, \partial S)).$$

In other words, $\mathrm{Mod}(S)$ is the group of isotopy classes of elements of $\mathrm{Homeo}^+(S, \partial S)$, where isotopies are required to fix the boundary pointwise. If $\mathrm{Homeo}_0(S, \partial S)$ denotes the connected component of the identity in $\mathrm{Homeo}^+(S, \partial S)$, then we can equivalently write

$$\mathrm{Mod}(S) = \mathrm{Homeo}^+(S, \partial S) / \mathrm{Homeo}_0(S, \partial S).$$

The mapping class group was first studied by Dehn. He gave a lecture on this topic to the Breslau Mathematics Colloquium on February 22, 1922; see [49]. The notes from this lecture have been translated to English by Stillwell [51, Chapter 7].

There are several possible variations in the definition of $\mathrm{Mod}(S)$. For example, we could consider diffeomorphisms instead of homeomorphisms, or homotopy classes instead of isotopy classes. By the theorems in Section 1.4, these definitions would result in isomorphic groups. To summarize, we have

$$
\begin{aligned}
\mathrm{Mod}(S) \quad &= \pi_0(\mathrm{Homeo}^+(S, \partial S)) \\
&\approx \mathrm{Homeo}^+(S, \partial S) \,/\, \mathrm{homotopy} \\
&\approx \pi_0(\mathrm{Diff}^+(S, \partial S)) \\
&\approx \mathrm{Diff}^+(S, \partial S) \,/\sim,
\end{aligned}
$$

where $\mathrm{Diff}^+(S, \partial S)$ is the group of orientation-preserving diffeomorphisms of S that are the identity on the boundary and $\sim$ can be taken to be either smooth homotopy relative to the boundary or smooth isotopy relative to the boundary.

The terminology $\mathrm{Mod}(S)$ is meant to stand for "modular group." Fricke called the mapping class group the "automorphic modular group" since, as we will later see, it can be viewed as a generalization of the classical modular group $\mathrm{SL}(2, \mathbb{Z})$ of 2×2 integral matrices with determinant 1.

Elements of $\mathrm{Mod}(S)$ are called *mapping classes*. We use the convention of functional notation, namely,

Elements of the mapping class group are applied right to left.

Other definitions and notations. In the literature, there are various other notations for the mapping class group, for instance: $\mathrm{MCG}(S)$, $\mathrm{Map}(S)$, $\mathcal{M}(S)$, and $\Gamma_{g,n}$. As a general rule, the term "mapping class group" refers to the group of homotopy classes of homeomorphisms of a surface, but there are plenty of variations: one can consider homeomorphisms that do not necessarily preserve the orientation of the surface or that do not act as the identity on the boundary or that fix each puncture individually, and so on.

Punctures versus marked points. If S is a surface with punctures, then it is sometimes more convenient to think of (some of) the punctures as marked points on S. Then, $\mathrm{Mod}(S)$ is the group of homeomorphisms of S that leave the set of marked points invariant, modulo isotopies that leave the set of marked points invariant. Here, one has to be careful when using homotopies instead of isotopies: a homotopy of surfaces with marked points must not only send marked points to marked points at all times but must also send unmarked points to unmarked points at all times.

Punctures versus boundary. One difference between a surface with punctures and a surface with boundary is that, as an artifact of our definitions, a mapping class is allowed to permute punctures on a surface, but it must preserve the individual boundary components pointwise. Also, isotopies must fix each boundary component pointwise, while on the other hand, isotopies

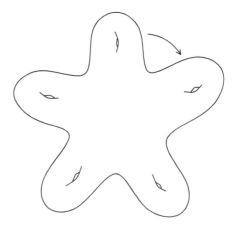

Figure 2.1 An order 5 element of $\mathrm{Mod}(S_5)$.

can rotate a neighborhood of a puncture.

Exceptional surfaces. Recall from Section 1.4 that there are four surfaces for which homotopy is not the same as isotopy: the disk D^2, the annulus A, the once-punctured sphere $S_{0,1}$, and the twice-punctured sphere $S_{0,2}$. Also recall that in these cases, homotopy is the same as isotopy for orientation-preserving homeomorphisms. Thus, even in these cases, the various definitions of $\mathrm{Mod}(S)$ are still equivalent.

2.1.1 FIRST EXAMPLES OF MAPPING CLASSES

As a first example of a nontrivial element of $\mathrm{Mod}(S_g)$, one can take the order g homeomorphism ϕ of S_g indicated in Figure 2.1 for $g = 5$. The mapping class represented by ϕ also has order g. To see this, look for a simple closed curve α in S_g so that $\alpha, \phi(\alpha), \phi^2(\alpha), \ldots, \phi^{g-1}(\alpha)$ are pairwise nonisotopic.

If we represent S_g as a $(4g + 2)$-gon with opposite sides identified (Figure 2.2 shows the case $g = 2$), we can get elements of $\mathrm{Mod}(S_g)$ by rotating the $(4g + 2)$-gon by any number of "clicks." For example, if we rotate by an angle π (i.e., $2g + 1$ clicks) we get an important example of a mapping class called a hyperelliptic involution (see Sections 7.4 and 9.4 for further discussion of hyperelliptic involutions).

It is possible to realize a hyperelliptic involution as a rigid rotation of S_g in $\mathbb{R}^3$, namely, the rotation by π about the axis indicated in Figure 2.3 (it is not obvious that this is indeed a hyperelliptic involution). Other elements of $\mathrm{Mod}(S_g)$ obtained by rotating a $(4g + 2)$-gon are less easy to visualize; for

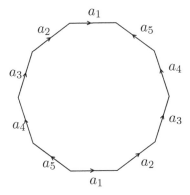

Figure 2.2 Rotation by $2\pi/10$ gives an order 10 element of $\mathrm{Mod}(S_2)$.

Figure 2.3 The rotation by π about the indicated axis is a hyperelliptic involution.

example, what does an order 5 symmetry of S_2 look like with respect to the standard picture of S_2 embedded in $\mathbb{R}^3$?

Unlike the preceding examples, most elements of the mapping class group have infinite order. The simplest such elements are Dehn twists, which are defined and studied in detail in Chapter 3.

2.2 COMPUTATIONS OF THE SIMPLEST MAPPING CLASS GROUPS

In this section we give complete descriptions of the mapping class groups of the simplest surfaces, working directly from the definitions.

2.2.1 THE ALEXANDER LEMMA

Our first computation is the mapping class group $\mathrm{Mod}(D^2)$ of the closed disk D^2. This simple result underlies most computations of mapping class groups.

Lemma 2.1 (Alexander lemma) *The group* $\mathrm{Mod}(D^2)$ *is trivial.*

In other words, Lemma 2.1 states that given any homeomorphism ϕ of D^2 that is the identity on the boundary ∂D^2, there is an isotopy of ϕ to the identity through homeomorphisms that are the identity on ∂D^2.

Proof. Identify D^2 with the closed unit disk in $\mathbb{R}^2$. Let $\phi : D^2 \to D^2$ be a homeomorphism with $\phi|_{\partial D^2}$ equal to the identity. We define

$$F(x,t) = \begin{cases} (1-t)\phi\left(\frac{x}{1-t}\right) & 0 \le |x| < 1-t \\ x & 1-t \le |x| \le 1 \end{cases}$$

for $0 \le t < 1$, and we define $F(x,1)$ to be the identity map of D^2. The result is an isotopy F from ϕ to the identity. □

We can think of combining the $\{F(x,t)\}$ from the proof into a level-preserving homeomorphism of a cylinder whose support is a cone; see Figure 2.4. The individual $F(\star, t)$ homeomorphisms appear at horizontal slices.

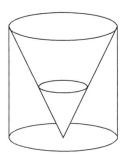

Figure 2.4 The Alexander trick.

The isotopy given by the proof can be thought of as follows: at time t, do the original map ϕ on the disk of radius $1 - t$ and apply the identity map outside this disk. This clever proof is called the Alexander trick.

The reader will notice that the Alexander trick works in all dimensions. However, this is one place where it is convenient to think about homeomorphisms instead of diffeomorphisms. The smooth version of the Alexander lemma in dimension 2 is not nearly as simple, although in this case Smale proved the stronger statement that $\text{Diff}(D^2, \partial D^2)$ is contractible [197]. In higher dimensions, the situation is worse: it is not known if $\text{Diff}(D^4, \partial D^4)$ is connected, and for infinitely many n we have that $\text{Diff}(D^n, \partial D^n)$ is not connected.

The proof of Lemma 2.1 also holds with D^2 replaced by a once-punctured disk (take the puncture/marked point to lie at the origin), and hence we also have the following:

The mapping class group of a once-punctured disk is trivial.

The sphere and the once-punctured sphere. There are two other mapping class groups $\mathrm{Mod}(S_{g,n})$ that are trivial, namely, $\mathrm{Mod}(S_{0,1})$ and $\mathrm{Mod}(S^2)$. For the former, we can identify $S_{0,1}$ with $\mathbb{R}^2$ and use that fact that every orientation-preserving homeomorphism of $\mathbb{R}^2$ is homotopic to the identity via the straight-line homotopy. For S^2, any homeomorphism can be modified by isotopy so that it fixes a point, and so we can apply the previous example.

2.2.2 THE MAPPING CLASS GROUP OF THE THRICE-PUNCTURED SPHERE

Our next example, the mapping class group of $S_{0,3}$, illustrates an important idea in the theory of mapping class groups. The way we will compute $\mathrm{Mod}(S_{0,3})$ is to understand its action on some fixed arc in $S_{0,3}$. The surface obtained by cutting $S_{0,3}$ along this arc is a punctured disk, and so we will be able to apply the Alexander lemma. This is in general how we use the cutting procedure for surfaces in order to perform inductive arguments.

In this section it will be convenient to think of $S_{0,3}$ as a sphere with three marked points (instead of three punctures). In order to determine $\mathrm{Mod}(S_{0,3})$ we first need to understand simple proper arcs in $S_{0,3}$.

Proposition 2.2 *Any two essential simple proper arcs in $S_{0,3}$ with the same endpoints are isotopic. Any two essential arcs that both start and end at the same marked point of $S_{0,3}$ are isotopic.*

Proof. Let α and β be two simple proper arcs in $S_{0,3}$ connecting the same two distinct marked points. We can modify α by isotopy so that it has general position intersections with β. By thinking of the third marked point as being the point at infinity, we can think of α and β as arcs in the plane. As in the proof of Lemma 1.8, if α and β are not disjoint, then we can find an innermost disk bounded by an arc of α and an arc of β. Pushing α by isotopy across such disks, we may reduce intersection until α and β have disjoint interiors. At this point, we can cut $S_{0,3}$ along $\alpha \cup \beta$. By the classification of surfaces, the resulting surface is the disjoint union of a disk (with two marked points on the boundary) and a once-marked disk (with two additional marked points on the boundary). Thus α and β bound an embedded disk in $S_{0,3}$, and so they are isotopic.

The case where α and β are essential simple proper arcs where all four endpoints lie on the same marked point of $S_{0,3}$ is similar. $\qquad\square$

We are now ready to compute $\mathrm{Mod}(S_{0,3})$. Let Σ_3 denote the group of permutations of three elements.

Proposition 2.3 *The natural map*

$$\mathrm{Mod}(S_{0,3}) \to \Sigma_3$$

given by the action of $\mathrm{Mod}(S_{0,3})$ *on the set of marked points of* $S_{0,3}$ *is an isomorphism.*

Proof. The map in the statement is obviously a surjective homomorphism. Thus it suffices to show that if a homeomorphism ϕ of $S_{0,3}$ fixes the three marked points—call them p, q, and r—then ϕ is homotopic to the identity. Choose an arc α in $S_{0,3}$ with distinct endpoints, say p and q. Since ϕ fixes the marked points p, q, and r, the endpoints of $\phi(\alpha)$ are again p and q. By Proposition 2.2, we have that $\phi(\alpha)$ is isotopic to α. It follows that ϕ is isotopic to a map (which we also call ϕ) that fixes α pointwise (Proposition 1.11).

We can cut $S_{0,3}$ along α so as to obtain a disk with one marked point (the boundary comes from α, and the marked point comes from r). Since ϕ preserves the orientations of $S_{0,3}$ and of α, it follows that ϕ induces a homeomorphism $\overline{\phi}$ of this disk which is the identity on the boundary (the map $\overline{\phi}$ is the unique set map on the cut-open surface inducing ϕ). By Lemma 2.1, the mapping class group of a once-marked disk is trivial, and so $\overline{\phi}$ is homotopic to the identity. The homotopy induces a homotopy from ϕ to the identity. $\square$

Pairs of pants. The surface $S_{0,3}$ is homeomorphic to the interior of a *pair of pants*[1] P, which is the compact surface obtained from S^2 by removing three open disks with embedded, disjoint closures. Pairs of pants are important because all compact hyperbolic surfaces can be built from pairs of pants (cf. Section 10.5). In Section 3.6, we will apply Proposition 2.3 to show that $\mathrm{Mod}(P) \approx \mathbb{Z}^3$.

The twice-punctured sphere. There is a homomorphism $\mathrm{Mod}(S_{0,2}) \to \mathbb{Z}/2\mathbb{Z}$ given by the action on the two marked points. An analogous proof to that of Proposition 2.3 gives that $\mathrm{Mod}(S_{0,2}) \approx \mathbb{Z}/2\mathbb{Z}$.

2.2.3 THE MAPPING CLASS GROUP OF THE ANNULUS

We now come to the simplest infinite-order mapping class group, that of the annulus A. The basic procedure we use to compute $\mathrm{Mod}(A)$ is similar to the one we used for $S_{0,3}$. That is, we find an arc in A so that when we cut

---

[1] Möbius used the term "trinion" for a pair of pants (he called an annulus a "binion" and a disk a "union").

A along that arc, we obtain a closed disk. If we can understand the action of a homeomorphism on the arc, then we can completely understand the homeomorphism up to homotopy.

Proposition 2.4 $\mathrm{Mod}(A) \approx \mathbb{Z}$.

Proof. First we construct a map $\rho : \mathrm{Mod}(A) \to \mathbb{Z}$. Let $f \in \mathrm{Mod}(A)$ and let $\phi : A \to A$ be any homeomorphism representing f. The universal cover of A is the infinite strip $\widetilde{A} \approx \mathbb{R} \times [0,1]$, and ϕ has a preferred lift $\widetilde{\phi} : \widetilde{A} \to \widetilde{A}$ fixing the origin. Let $\widetilde{\phi}_1 : \mathbb{R} \to \mathbb{R}$ denote the restriction of $\widetilde{\phi}$ to $\mathbb{R} \times \{1\}$, which is canonically identified with $\mathbb{R}$. Since $\widetilde{\phi}_1$ is a lift to $\mathbb{R}$ of the identity map on one of the boundary components of A, it is an integer translation. We define $\rho(f)$ to be $\widetilde{\phi}_1(0)$. If we identify $\mathbb{Z}$ with the group of integer translations of $\mathbb{R}$, then the map $\widetilde{\phi}_1$ itself is an element of $\mathbb{Z}$, and we can write $\rho(f) = \widetilde{\phi}_1 \in \mathbb{Z}$. From this point of view, it is clear that ρ is a homomorphism since compositions of maps of A are sent to compositions of translations of $\mathbb{R}$.

We can give an equivalent definition of ρ as follows. Let δ be an oriented simple proper arc that connects the two boundary components of A. Given f and ϕ as above, the concatenation $\phi(\delta) * \delta^{-1}$ is a loop based at $\delta(0)$, and $\rho(f)$ equals $[\phi(\delta) * \delta^{-1}] \in \pi_1(A, \delta(0)) \approx \mathbb{Z}$. Yet another equivalent way to define ρ is to let $\widetilde{\delta}$ be the unique lift of δ to $\widetilde{A}$ based at the origin and to set $\rho(f)$ to be the endpoint of $\widetilde{\phi}(\widetilde{\delta})$ in $\mathbb{R} \times \{1\} \approx \mathbb{R}$.

We now show that ρ is surjective. The linear transformation of $\mathbb{R}^2$ given by the matrix

$$M = \begin{pmatrix} 1 & n \\ 0 & 1 \end{pmatrix}$$

preserves $\mathbb{R} \times [0,1]$ and is equivariant with respect to the group of deck transformations. Thus the restriction of the linear map M to $\mathbb{R} \times [0,1]$ descends to a homeomorphism ϕ of A. The action of this homeomorphism on δ is depicted in Figure 2.5 for the case $n = -1$. It follows from the definition of ρ that $\rho([\phi]) = n$.

It remains to show that ρ is injective. Let $f \in \mathrm{Mod}(A)$ be an element of the kernel of ρ and say that f is represented by a homeomorphism ϕ. Let $\widetilde{\phi}$ be the preferred lift of ϕ. Since $\rho(f) = 0$, we have that $\widetilde{\phi}$ acts as the identity on $\partial \widetilde{A}$. We claim that the straight-line homotopy from $\widetilde{\phi}$ to the identity map of $\widetilde{A}$ is equivariant. For this, it suffices to show that

$$\widetilde{\phi}(\tau \cdot x) = \tau \cdot \widetilde{\phi}(x)$$

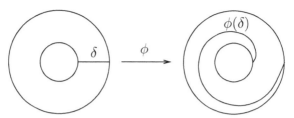

Figure 2.5 A generator for $\mathrm{Mod}(A)$.

for any deck transformation τ and for any $x \in \widetilde{A}$. It follows from general covering space theory that

$$\widetilde{\phi}(\tau \cdot x) = \phi_*(\tau) \cdot \widetilde{\phi}(x).$$

But because ϕ fixes ∂A pointwise, it follows that ϕ_* is the identity automorphism of $\pi_1(A) \approx \mathbb{Z}$, and so $\phi_*(\tau) = \tau$, and the claim is proven.

We have that the straight-line homotopy from $\widetilde{\phi}$ to the identity is equivariant and it fixes the boundary of $\widetilde{A}$, so it descends to a homotopy between ϕ and the identity map of A that fixes the boundary of A pointwise. Thus f is the identity, and so ρ is injective. $\square$

We remark that in the proof of Proposition 2.4 we took advantage of the fact that we can conflate homotopy with isotopy.

The homeomorphism of A induced by the matrix

$$\begin{pmatrix} 1 & -1 \\ 0 & 1 \end{pmatrix}$$

is called a Dehn twist. Since any surface contains an annulus, we can perform a Dehn twist in any surface. Dehn twists are important elements of the mapping class group. In fact, the next chapter is entirely devoted to their study.

2.2.4 THE MAPPING CLASS GROUP OF THE TORUS

The torus T^2 acts as a guidepost in the study of mapping class groups. While it has an explicit description as a group of integral matrices, and while it is much easier to understand than mapping class groups of higher-genus surfaces, it still exhibits enough richness to give us a hint of what to expect in the higher-genus case. This is a recurring theme in this book.

THEOREM 2.5 *The homomorphism*

$$\sigma: \mathrm{Mod}(T^2) \longrightarrow \mathrm{SL}(2, \mathbb{Z})$$

given by the action on $H_1(T; \mathbb{Z}) \approx \mathbb{Z}^2$ is an isomorphism.

Proof. Any homeomorphism ϕ of T^2 induces a map $\phi_* : H_1(T^2; \mathbb{Z}) \to H_1(T^2; \mathbb{Z})$. Since ϕ is invertible, ϕ_* is an automorphism of $H_1(T^2; \mathbb{Z}) \approx \mathbb{Z}^2$. Homotopic maps induce the same map on homology, and so the map $\phi \mapsto \phi_*$ induces a map $\sigma : \mathrm{Mod}(T^2) \to \mathrm{Aut}(\mathbb{Z}^2) \approx \mathrm{GL}(2, \mathbb{Z})$ (the exact identification of $\sigma(f)$ with a 2×2 matrix depends on the particular identification of $H_1(T^2; \mathbb{Z})$ with $\mathbb{Z}^2$). The fact that $\sigma(f)$ is an element of $\mathrm{SL}(2, \mathbb{Z})$ can be seen directly from the fact that the algebraic intersection numbers in T^2 correspond to determinants (see Section 1.2) and the fact that orientation-preserving homeomorphisms preserve algebraic intersection number.

We next prove that σ is surjective. Any element M of $\mathrm{SL}(2, \mathbb{Z})$ induces an orientation-preserving linear homeomorphism of $\mathbb{R}^2$ that is equivariant with respect to the deck transformation group $\mathbb{Z}^2$ and thus descends to a linear homeomorphism ϕ_M of the torus $T^2 = \mathbb{R}^2/\mathbb{Z}^2$. Because of our identification of primitive vectors in $\mathbb{Z}^2$ with homotopy classes of oriented simple closed curves in T^2, it follows that $\sigma([\phi_M]) = M$, and so σ is surjective.

Finally, we prove that σ is injective. Since T^2 is a $K(G, 1)$-space, there is a correspondence:

$$\left\{ \begin{array}{c} \text{Homotopy classes of} \\ \text{based maps } T^2 \to T^2 \end{array} \right\} \longleftrightarrow \left\{ \begin{array}{c} \text{Homomorphisms} \\ \mathbb{Z}^2 \to \mathbb{Z}^2 \end{array} \right\}$$

(see [91, Proposition 1B.9]). What is more, any element f of $\mathrm{Mod}(T^2)$ has a representative ϕ that fixes a basepoint for T^2. Thus, if $f \in \ker(\sigma)$, then ϕ is homotopic (as a based map) to the identity, so σ is injective. Actually, we can construct the homotopy of ϕ to the identity explicitly. As in the case of the annulus, the straight-line homotopy between the identity map of $\mathbb{R}^2$ and any lift of ϕ is equivariant and hence descends to a homotopy between ϕ and the identity. $\qquad\square$

The annulus versus the torus. The reader will notice that our proof of the injectivity of $\sigma : \mathrm{Mod}(T^2) \to \mathrm{SL}(2, \mathbb{Z})$ was actually easier than our proof of the injectivity of $\rho : \mathrm{Mod}(A) \to \mathbb{Z}$. The reason for this is that if we apply $K(G, 1)$-theory to two homeomorphisms of A that induce the same map on $\pi_1(A)$, then the theory gives that the two homeomorphisms are homotopic but not necessarily via a homotopy that fixes the boundary. That is why we needed to construct the homotopy by hand in the case of the annulus.

Hands-on proof of Theorem 2.5. We can give another, more hands-on proof of the injectivity of $\sigma : \mathrm{Mod}(T^2) \to \mathrm{SL}(2, \mathbb{Z})$. Suppose that $\sigma(f)$ is the identity matrix in $\mathrm{SL}(2, \mathbb{Z})$ and let ϕ be a representative of f. If α and β are simple closed curves corresponding to the elements $(1, 0)$ and $(0, 1)$ of $\pi_1(T^2)$, then it follows that $\phi(\alpha)$ is homotopic to α and $\phi(\beta)$ is homotopic to β. We proceed in two steps to show that ϕ is isotopic to the identity.

1. By Proposition 1.10, we know that $\phi(\alpha)$ is isotopic to α (as a map), and by Proposition 1.11 any such isotopy can be extended to an isotopy of T^2. Thus, up to isotopy, we may assume that ϕ fixes α pointwise. As ϕ is orientation-preserving, we also know that ϕ preserves the two sides of α.

2. Let A be the annulus obtained from T^2 by cutting along α. Given that ϕ fixes α pointwise and that ϕ preserves the two sides of α, we have that ϕ induces a homeomorphism $\overline{\phi}$ of A which represents an element $\overline{f}$ of $\mathrm{Mod}(A)$. We can think of β and $\overline{\phi}(\beta)$ as arcs in A. Since $\phi(\beta)$ is isotopic to β in T^2, we see that $\rho(\overline{f}) = 0$, where $\rho : \mathrm{Mod}(A) \to \mathbb{Z}$ is the map from Proposition 2.4.

3. At this point, we can simply quote Proposition 2.4, which gives that $\overline{f} = 1$. This means that $\overline{\phi}$ is isotopic to the identity map of A via an isotopy fixing ∂A pointwise. But then ϕ is also isotopic to the identity.

In the last step, instead of quoting Proposition 2.4 one can continue the line of thought to give a hands-on proof of that proposition. As we shall see in Section 2.3, these hands-on proofs lead to a method for understanding mapping classes of arbitrary surfaces.

The once-punctured torus. For the once-punctured torus $S_{1,1}$, we have $H_1(S_{1,1}; \mathbb{Z}) \approx H_1(T^2; \mathbb{Z}) \approx \mathbb{Z}^2$. Therefore, as in the case of T^2, there is a homomorphism $\sigma : \mathrm{Mod}(S_{1,1}) \to \mathrm{SL}(2, \mathbb{Z})$. The map σ is surjective since any element of $\mathrm{SL}(2, \mathbb{Z})$ can be realized as a map of $\mathbb{R}^2$ that is equivariant with respect to $\mathbb{Z}^2$ and that fixes the origin; such a map descends to a homeomorphism of $S_{1,1}$ with the desired action on homology.

To prove that σ is injective, we can apply a version of the hands-on proof we used in the case of the torus, as follows. Let α and β be simple closed curves in $S_{1,1}$ that intersect in one point. If $f \in \ker(\sigma)$ is represented by ϕ, then $\phi(\alpha)$ and $\phi(\beta)$ are isotopic to α and β. We can then modify ϕ by isotopy so that it fixes α and β pointwise. If we cut $S_{1,1}$ along $\alpha \cup \beta$, we obtain a once-punctured disk, and ϕ induces a homeomorphism of this disk fixing the boundary. By the Alexander trick, this homeomorphism of the punctured disk is homotopic to the identity by a homotopy that fixes the boundary. It follows that ϕ is homotopic to the identity, as desired.

2.2.5 THE MAPPING CLASS GROUP OF THE FOUR-TIMES-PUNCTURED SPHERE

In the theory of mapping class groups, there is a strong relationship between the torus and the sphere with four punctures. Recall that if we think of the torus as a square (or hexagon) with opposite sides identified, then the hyperelliptic involution ι is the map that rotates about the center of the square (or hexagon) by an angle of π. The map ι has four fixed points, and so the quotient, which is topologically a sphere, has four distinguished points. We identify this quotient with $S_{0,4}$. Since every linear map of T^2 (fixing the image of the origin in $\mathbb{R}^2$) commutes with ι, each element of $\mathrm{Mod}(T^2)$ induces an element of $\mathrm{Mod}(S_{0,4})$. We will now exploit this relationship in order to compute $\mathrm{Mod}(S_{0,4})$.

We begin by classifying simple closed curves in $S_{0,4}$ up to homotopy.

Proposition 2.6 *The hyperelliptic involution induces a bijection between the set of homotopy classes of essential simple closed curves in T^2 and the set of homotopy classes of essential simple closed curves in $S_{0,4}$.*

Proof. Proposition 1.5 gives a bijection between the set of homotopy classes of essential simple closed curves in T^2 and the set of primitive elements of $\mathbb{Z}^2$. Given a primitive element of $\mathbb{Z}^2$, we obtained a (p, q)-curve by projecting a line of slope q/p to T^2.

We will give a different construction of (p, q)-curves in T^2, and we will give a construction of (p, q)-curves in $S_{0,4}$, and then we will observe that the lift of a (p, q)-curve in $S_{0,4}$ to T^2 is a (p, q)-curve in T^2.

Let α and β be two simple closed curves in T^2 that intersect each other in one point. We identify α with $(1, 0) \in \mathbb{Z}^2$ and β with $(0, 1) \in \mathbb{Z}^2$. Let (p, q) be a primitive element of $\mathbb{Z}^2$. A simple closed curve γ in T^2 is a (p, q)-curve if we have $(\hat{i}(\gamma, \beta), \hat{i}(\gamma, \alpha)) = \pm(p, q)$. To construct the (p, q)-curve, we start by taking p parallel copies of α, and we modify this collection by a $2\pi/q$ twist along β.

Up to homotopy in T^2, we may assume that α and β project via ι to simple closed curves $\bar{\alpha}$ and $\bar{\beta}$ in $S_{0,4}$ that intersect in two points, as in Figure 2.6. We can then perform an analogous construction of a (p, q)-curve in $S_{0,4}$. We take p parallel copies of $\bar{\alpha}$ and twist along $\bar{\beta}$ by π/q.

We need to check that every homotopy class of essential simple closed curves in $S_{0,4}$ comes from our construction. Let γ be an arbitrary essential simple closed curve in $S_{0,4}$. Up to homotopy, we may assume that γ is in minimal position with respect to α. If we cut $S_{0,4}$ along β, we obtain two twice-punctured disks, and γ and α both give collections of disjoint arcs on each. By the assumptions on minimal position, these arcs are all essential. By Proposition 2.2, the arcs coming from α and the arcs coming from γ are freely homotopic. It follows that the homotopy class of γ comes from our construction.

The preimage of a (p, q)-curve in $S_{0,4}$ in T^2 is a $(2p, 2q)$–curve, that is, two parallel copies of a (p, q)-curve in T^2. That is to say, the identification of (p, q)-curves in the two surfaces is induced by ι. □

Proposition 2.7 $\mathrm{Mod}(S_{0,4}) \approx \mathrm{PSL}(2, \mathbb{Z}) \ltimes (\mathbb{Z}/2\mathbb{Z} \times \mathbb{Z}/2\mathbb{Z})$.

Proof. We first construct a homomorphism $\overline{\sigma} : \mathrm{Mod}(S_{0,4}) \to \mathrm{PSL}(2, \mathbb{Z})$ together with a right inverse. Then we will show that the kernel is isomorphic to $\mathbb{Z}/2\mathbb{Z} \times \mathbb{Z}/2\mathbb{Z}$.

Let ϕ be a homeomorphism representing a given $f \in \mathrm{Mod}(S_{0,4})$. There are two lifts of ϕ to $\mathrm{Homeo}^+(T^2)$, say $\widetilde{\phi}$ and $\iota\widetilde{\phi}$. We define $\overline{\sigma}(f)$ to be the element of $\mathrm{PSL}(2, \mathbb{Z})$ represented by the matrix $\sigma([\widetilde{\phi}])$, where $\sigma : \mathrm{Mod}(T^2) \to \mathrm{SL}(2, \mathbb{Z})$ is the homomorphism from Theorem 2.5. This is well defined since the two lifts of ϕ differ by ι, and $\sigma(\iota) = -I$.

Next we construct the right inverse of $\overline{\sigma}$. An element of $\mathrm{PSL}(2, \mathbb{Z})$ induces an orientation-preserving, linear homeomorphism of T^2 that is well defined up to multiplication by ι. Any such map of T^2 commutes with ι and hence induces an orientation-preserving homeomorphism of $S_{0,4}$. In this way we have defined a map $\mathrm{PSL}(2, \mathbb{Z}) \to \mathrm{Mod}(S_{0,4})$; it is a right inverse of $\overline{\sigma}$ by construction.

The order 2 homeomorphisms of $S_{0,4}$ indicated in Figure 2.6 are called *hyperelliptic involutions* of $S_{0,4}$. The corresponding mapping classes ι_1 and ι_2 generate a subgroup of $\mathrm{Mod}(S_{0,4})$ isomorphic to $\mathbb{Z}_2 \times \mathbb{Z}_2$. The hyperelliptic involutions each lift to a homeomorphism of $T^2 \approx S^1 \times S^1$ that rotates one of the factors by π. Hence $\langle \iota_1, \iota_2 \rangle$ is contained in the kernel of $\overline{\sigma}$.

We will show that $\langle \iota_1, \iota_2 \rangle$ is the entire kernel of $\overline{\sigma}$. Let $f \in \ker(\overline{\sigma})$. By definition of $\overline{\sigma}$, any lift of a representative of f to $\mathrm{Homeo}^+(T^2)$ acts by $\pm I$ on $H_1(T^2; \mathbb{Z})$ and hence acts trivially on the set of homotopy classes of simple closed curves in T^2. By the natural bijection given by Proposition 2.6, it follows that f acts trivially on the set of homotopy classes of simple closed curves in $S_{0,4}$. In particular, f fixes the homotopy classes of $\overline{\alpha}$ and $\overline{\beta}$. It follows that we can precompose f with an element of $k \in \langle \iota_1, \iota_2 \rangle$ so that fk fixes the four marked points of $S_{0,4}$.

Our goal now is to show that fk is the identity. Say that fk is represented by a homeomorphism ϕ. As in the proof of Theorem 2.5, we can modify ϕ

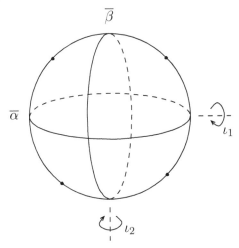

Figure 2.6 The hyperelliptic involutions of $S_{0,4}$.

so that it fixes $\overline{\alpha}$ and $\overline{\beta}$. Since ϕ fixes the four marked points, we have that ϕ induces relative homeomorphisms of the four once-marked disks obtained when we cut $S_{0,4}$ along $\overline{\alpha}$ and $\overline{\beta}$. At this point, we can once again apply the Alexander lemma to show that fk is the identity. $\qquad\square$

Two splittings of $\mathrm{Mod}^{\pm}(S_{0,4})$. Let $\mathrm{Mod}^{\pm}(S_{0,4})$ denote a group of homotopy classes of all homeomorphisms of $S_{0,4}$, including the orientation-reversing ones (see Chapter 8 for more about this group). It follows from Theorem 2.5 and the argument of Proposition 2.7 that

$$\mathrm{Mod}^{\pm}(S_{0,4}) \approx \mathrm{PGL}(2,\mathbb{Z}) \ltimes (\mathbb{Z}/2\mathbb{Z} \times \mathbb{Z}/2\mathbb{Z}).$$

We can give another description of $\mathrm{Mod}^{\pm}(S_{0,4})$ as a semidirect product. There is a short exact sequence

$$1 \to \mathrm{PMod}^{\pm}(S_{0,4}) \to \mathrm{Mod}^{\pm}(S_{0,4}) \to \Sigma_4 \to 1,$$

where Σ_4 is the symmetric group on the four punctures, the map $\mathrm{Mod}^{\pm}(S_{0,4}) \to \Sigma_4$ is given by the action on the punctures, and $\mathrm{PMod}^{\pm}(S_{0,4})$ is the subgroup of $\mathrm{Mod}^{\pm}(S_{0,4})$ consisting of those elements fixing each of the punctures (one is tempted to write a sequence with $\mathrm{Mod}(S_{0,4})$ surjecting onto the alternating group A_4, but the image of $\mathrm{Mod}(S_{0,4})$ is all of Σ_4). Thinking of $S_{0,4}$ as the 2-skeleton of a tetrahedron minus its vertices, we see that there is a section $\Sigma_4 \to \mathrm{Mod}^{\pm}(S_{0,4})$, and so the group $\mathrm{Mod}^{\pm}(S_{0,4})$ is isomorphic to the semidirect product

$\mathrm{PMod}^{\pm}(S_{0,4}) \rtimes \Sigma_4$. It follows from the results in Section 4.2 below that $\mathrm{PMod}^{\pm}(S_{0,4}) \approx F_2 \rtimes \mathbb{Z}/2\mathbb{Z}$, and so

$$\mathrm{Mod}^{\pm}(S_{0,4}) \approx (F_2 \rtimes \mathbb{Z}/2\mathbb{Z}) \rtimes \Sigma_4.$$

2.3 THE ALEXANDER METHOD

Our computations of the mapping class groups of $S_{0,3}$, $S_{0,2}$, A, T^2, $S_{1,1}$, and $S_{0,4}$ all follow the same general scheme: find a collection of curves and/or arcs that cut the surface into disks and apply the Alexander lemma in order to say that the action of the mapping class group on the surface is completely determined by the action on the isotopy classes of these curves and arcs.

It turns out that this basic setup works for a general surface. The Alexander method (given below) states that, for any S, an element of $\mathrm{Mod}(S)$ is often determined by its action on a well-chosen collection of curves and arcs in S. Thus, there is a concrete way to determine when two homeomorphisms $f, g \in \mathrm{Homeo}^+(S)$ represent the same element of $\mathrm{Mod}(S)$.

Before we give the precise statement, we point out that the situation is more subtle than one might think at first. It is simply not true in general that if a homeomorphism of a surface S fixes a collection of curves and arcs that cut S into disks, then it represents the trivial mapping class. For instance, the hyperelliptic involution of S_g fixes the $2g + 1$ simple closed curves shown in Figure 2.7; on the other hand, we know that the hyperelliptic involution represents a nontrivial mapping class since it acts nontrivially on $H_1(S_g; \mathbb{Z})$. Even worse, the hyperelliptic involutions in $\mathrm{Mod}(T^2)$ and $\mathrm{Mod}(S_2)$ fix *every* isotopy class of simple closed curves (cf. Section 3.4). What is happening in the case of the hyperelliptic involution, and what can happen in general, is that a homeomorphism of a surface can fix a collection of curves while still permuting or rotating the complementary disks.

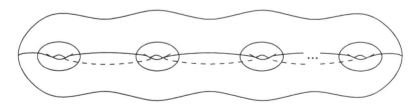

Figure 2.7 A collection of simple closed curves that is fixed by the hyperelliptic involution.

In view of the example of the hyperelliptic involution, one is tempted to simply add the hypothesis that the curves and arcs are fixed with their orientations. But this is still not right: the hyperelliptic involution in $\mathrm{Mod}(S_2)$ fixes the orientation of every isotopy class of separating simple closed curves in S_2, and certainly there are enough of these curves to cut S_2 into disks; see Figure 2.8 for such a configuration.

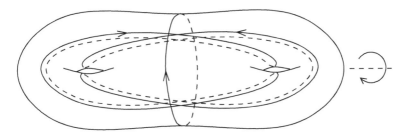

Figure 2.8 The hyperelliptic involution fixes the isotopy class of every simple closed curve in S_2 and even fixes the orientation of each separating isotopy class. However, it is a nontrivial mapping class.

We finally arrive at the following statement, which we call the *Alexander method*. To simplify the discussion we consider only compact surfaces, possibly with finitely many marked points in the interior. Again, for all intents and purposes, marked points play the same role as punctures in the theory of mapping class groups. For a surface S with marked points, we say that a collection $\{\gamma_i\}$ of curves and arcs *fills* S if the surface obtained from S by cutting along all γ_i is a disjoint union of disks and once-marked disks.

Proposition 2.8 (Alexander method) *Let S be a compact surface, possibly with marked points, and let $\phi \in \mathrm{Homeo}^+(S, \partial S)$. Let $\gamma_1, \ldots, \gamma_n$ be a collection of essential simple closed curves and simple proper arcs in S with the following properties.*

1. *The γ_i are pairwise in minimal position.*

2. *The γ_i are pairwise nonisotopic.*

3. *For distinct i, j, k, at least one of $\gamma_i \cap \gamma_j$, $\gamma_i \cap \gamma_k$, or $\gamma_j \cap \gamma_k$ is empty.*

(1) If there is a permutation σ of $\{1, \ldots, n\}$ so that $\phi(\gamma_i)$ is isotopic to $\gamma_{\sigma(i)}$ relative to ∂S for each i, then $\phi(\cup\gamma_i)$ is isotopic to $\cup\gamma_i$ relative to ∂S.

If we regard $\cup\gamma_i$ as a (possibly disconnected) graph Γ in S, with vertices at the intersection points and at the endpoints of arcs, then the composition of ϕ with this isotopy gives an automorphism ϕ_* of Γ.

(2) Suppose now that $\{\gamma_i\}$ fills S. If ϕ_* fixes each vertex and each edge of Γ with orientations, then ϕ is isotopic to the identity. Otherwise, ϕ has a nontrivial power that is isotopic to the identity.

The power of the Alexander method is that it converts the computation of a mapping class into a finite combinatorial problem. We will use this frequently, for example:

1. to compute the center of the mapping class group (see Section 3.3)

2. to prove the Dehn–Nielsen–Baer theorem (see Chapter 8)

3. to show that $\mathrm{Mod}(S)$ has a solvable word problem (see Chapter 4)

4. to verify that certain relations hold in $\mathrm{Mod}(S)$ (see, e.g., Proposition 5.1).

We leave it as an exercise to check that every compact surface S has a collection $\{\gamma_i\}$ as in the statement of Proposition 2.8.

A priori the Alexander method allows us to determine a mapping class only up to a finite power. However, on almost every surface, it is possible to choose the $\{\gamma_i\}$ so that mapping classes are determined uniquely by their action on the $\{\gamma_i\}$; that is, on almost every surface one can choose the γ_i so that whenever a homeomorphism ϕ fixes each γ_i up to homotopy, then the induced map ϕ_* of the graph Γ is necessarily the identity. One example of such a collection is used in the proof of Theorem 3.10.

One would like to strengthen statement 2 of the Alexander method to say that ϕ is isotopic to a nontrivial finite-order homeomorphism. Indeed, it is a general fact that if a homeomorphism of a surface has a power that is isotopic to the identity, then the homeomorphism itself is isotopic to a finite-order homeomorphism. This fact is stated precisely in Chapter 7 and is proven in Section 13.2.

The condition on triples in the statement of the Alexander method is crucial. This is because there is not, in general, a canonical minimal position configuration for a triple of curves that intersect pairwise. Therefore, there is no canonical way to construct the graph Γ. Consider, for instance, the configuration shown in Figure 2.9; the three arcs are individually isotopic, but there is no isotopy from the first union of arcs to the second.

We point out the following slight (but useful) improvement of the Alexander method. Consider the graph[2]

$$\Gamma' = (\cup \gamma_i) \cup \partial S \cup \{\text{marked points}\}.$$

Since Γ' is in general larger than Γ, it gives more information. For instance, say Γ is a chain of three simple closed curves γ_1, γ_2, and γ_3 in $S_{1,2}$. By the Alexander method, if $f \in \mathrm{Mod}(S)$ fixes the isotopy classes of each γ_i, then

---

[2]Technically, if some component of ∂S does not meet $\cup \gamma_i$, then we need to add a marked point on that component in order to obtain a graph.

one can deduce that f is either the identity or the hyperelliptic involution (see Figure 3.8 below). If we know that f also fixes the two marked points of $S_{1,2}$, then it is immediate from the action of f on Γ' that $f = 1$.

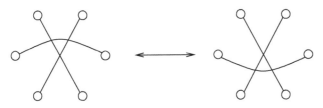

Figure 2.9 There is no canonical way to arrange these three arcs without creating a triple point.

Statement 1 of the Alexander method is an immediate consequence of the following lemma which, in addition to being slightly more general, is also notationally simpler.

Lemma 2.9 *Let S be a compact surface, possibly with marked points, and let $\gamma_1, \ldots, \gamma_n$ be a collection of essential simple closed curves and simple proper arcs in S that satisfy the three properties from Proposition 2.8. If $\gamma_1', \ldots, \gamma_n'$ is another such collection so that γ_i' is isotopic to γ_i relative to ∂S for each i, then there is an isotopy of S relative to ∂S that takes γ_i' to γ_i for all i simultaneously and hence takes $\cup \gamma_i$ to $\cup \gamma_i'$.*

Our proof of this lemma was greatly simplified by Allen Hatcher.

Proof. We will work by induction on n; that is, we assume that we can construct an isotopy of S that takes γ_i' to γ_i for $i = 1, \ldots, k-1$, and we will construct a relative homotopy of S that fixes the set $\Delta_{k-1} = \gamma_1 \cup \cdots \cup \gamma_{k-1}$ throughout the isotopy and takes γ_k' to γ_k. We can take the base case to be $k = 0$, which is vacuous.

First we perform a relative isotopy of S that fixes Δ_{k-1} and perturbs γ_k' to have general position intersections with γ_k as follows. By the hypothesis on triples $\{\gamma_i', \gamma_j', \gamma_k'\}$ and the fact that Δ_{k-1} is equal to $\gamma_1' \cup \cdots \cup \gamma_{k-1}'$, we have that γ_k' is disjoint from the vertices of the graph Δ_{k-1}. Thus there is a relative isotopy of S that fixes Δ_{k-1} and makes γ_k' disjoint from γ_k along the edges of Δ_{k-1}. Finally, we perform a relative isotopy of S that is the identity in a neighborhood of Δ_{k-1} and perturbs γ_k' to intersect γ_k transversely in the complement of Δ_{k-1}.

Next we perform a relative isotopy of S that fixes Δ_{k-1} and takes γ_k' to be disjoint from γ_k. If γ_k and γ_k' are not already disjoint, then by the bigon criterion they form a bigon (since γ_k and γ_k' are isotopic relative to ∂S, they

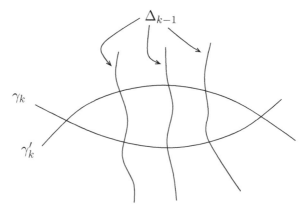

Figure 2.10 The intersection of Δ_{k-1} with a bigon formed by γ_k and γ'_k.

have the same endpoints and hence cannot form any half-bigons). By the hypothesis on triples, the intersection of Δ_{k-1} with this bigon is a collection of disjoint arcs. By the assumption on minimal position, each such arc connects one boundary arc of the bigon to the other; see Figure 2.10. It follows that there is an isotopy of S that fixes Δ_{k-1} as a set and pushes γ'_k across this bigon, thus reducing its intersection with γ_k. Repeating this process a finite number of times, we obtain the desired isotopy.

Finally, we are in the situation that γ'_k is disjoint from γ_k. As in the proof of Proposition 1.10, the region between γ_k and γ'_k is either an annulus or a disk, depending on whether γ_k and γ'_k are simple closed curves or simple proper arcs. The intersection of Δ_{k-1} with this region, if nonempty, is again a collection of disjoint arcs, each connecting γ_k to γ'_k. Thus, as above, there is a relative isotopy of S that fixes Δ_{k-1} and takes γ'_k to γ_k. $\square$

We can now complete the proof of the Alexander method.

Proof of Proposition 2.8. Let $\{\gamma_1, \dots, \gamma_n\}$ be as in statement 1, and for each i let γ'_i be the simple closed curve $\phi(\gamma_{\sigma^{-1}(i)})$. Applying Lemma 2.9 to the collections $\{\gamma_i\}$ and $\{\gamma'_i\}$, we can construct an isotopy of S that takes γ'_i to γ_i for each i and hence takes $\cup\gamma'_i$ to $\cup\gamma_i$. This proves statement 1.

It now follows, as in the statement of the proposition, that ϕ induces an automorphism ϕ_* of $\Gamma = \cup\gamma_i$. Since the automorphism group of a finite graph is necessarily finite, we may choose a power r so that ϕ_*^r is the identity automorphism, that is, it fixes each vertex, and fixes each edge with orientation. Since ϕ is orientation-preserving, it follows that ϕ also preserves the sides in S of each edge of Γ. It follows that ϕ^r, after possibly modifying it by an isotopy, fixes Γ pointwise and sends each complementary region into itself;

indeed, a complementary region is completely determined by the oriented edges of Γ that make up its boundary.

Now assume that the γ_i fill S, as in statement 2. In other words, the surface obtained by cutting S along Γ is a collection of closed disks, each possibly with one marked point. By applying the Alexander lemma (Lemma 2.1) to each of these disks, we see that ϕ^r is isotopic to the identity homeomorphism of S. Obviously, in the case $r = 1$ we have that ϕ is isotopic to the identity. In the case $r > 1$, we have only obtained that ϕ^r is isotopic to the identity. This proves statement 2. $\qquad\square$

Chapter Three

Dehn Twists

In this chapter we study a particular type of mapping class called a Dehn twist. Dehn twists are the simplest infinite-order mapping classes in the sense that they have representatives with the smallest possible supports. Dehn twists play the role for mapping class groups that elementary matrices play for linear groups. We begin by defining Dehn twists in S and proving that they have infinite order in $\mathrm{Mod}(S)$. We determine many of the basic properties of Dehn twists by studying their action on simple closed curves. As one consequence, we compute the center of $\mathrm{Mod}(S)$. At the end of the chapter, we determine all relations that can occur between two Dehn twists.

3.1 DEFINITION AND NONTRIVIALITY

In this section we define Dehn twists and prove they are nontrivial elements of the mapping class group.

3.1.1 DEHN TWISTS AND THEIR ACTION ON CURVES

Consider the annulus $A = S^1 \times [0, 1]$. To orient A we embed it in the (θ, r)-plane via the map $(\theta, t) \mapsto (\theta, t+1)$ and take the orientation induced by the standard orientation of the plane.

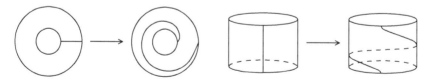

Figure 3.1 Two views of a Dehn twist.

Let $T : A \to A$ be the twist map of A given by the formula

$$T(\theta, t) = (\theta + 2\pi t, t).$$

The map T is an orientation-preserving homeomorphism that fixes ∂A pointwise. Note that instead of using $\theta + 2\pi t$ we could have used $\theta - 2\pi t$. Our choice is a left twist, while the other is a right twist.

Figure 3.1 gives two descriptions of the twist map T. We have seen the picture on the left-hand side before in our proof of Proposition 2.4. Indeed, the twist map T here is the same as the map used to show that $\mathrm{Mod}(A)$ surjects onto $\mathbb{Z}$.

Now let S be an arbitrary (oriented) surface and let α be a simple closed curve in S. Let N be a regular neighborhood of α and choose an orientation-preserving homeomorphism $\phi : A \to N$. We obtain a homeomorphism $T_\alpha : S \to S$, called a *Dehn twist about* α, as follows:

$$T_\alpha(x) = \begin{cases} \phi \circ T \circ \phi^{-1}(x) & \text{if } x \in N \\ x & \text{if } x \in S \setminus N. \end{cases}$$

In other words, the instructions for T_α are "perform the twist map T on the annulus N and fix every point outside of N."

The Dehn twist T_α depends on the choice of N and the homeomorphism ϕ. However, by the uniqueness of regular neighborhoods, the isotopy class of T_α does not depend on either of these choices. What is more, T_α does not depend on the choice of the simple closed curve α within its isotopy class. Thus, if a denotes the isotopy class of α, then T_a is well defined as an element of $\mathrm{Mod}(S)$, called the *Dehn twist about* a. We will sometimes abuse notation slightly and write T_α for the mapping class T_a.

The Dehn twist was introduced by Max Dehn. He originally used the term Schraubungen, which can be translated as "screw map" [50, Section 2b].

Dehn twists on the torus. Via the isomorphism of Theorem 2.5, the Dehn twists about the $(1, 0)$-curve and the $(0, 1)$-curve in T^2 map to the matrices

$$\begin{pmatrix} 1 & -1 \\ 0 & 1 \end{pmatrix} \quad \text{and} \quad \begin{pmatrix} 1 & 0 \\ 1 & 1 \end{pmatrix}.$$

Thus these two Dehn twists generate $\mathrm{Mod}(T^2) \approx \mathrm{SL}(2, \mathbb{Z})$. We will see in Chapter 4 that in fact for every $g \geq 0$ the group $\mathrm{Mod}(S_g)$ is generated by a finite number of Dehn twists.

Dehn twists via cutting and gluing. Here is another way to think about the Dehn twist T_α. We can cut S along α, twist a neighborhood of one boundary component through an angle of 2π, and then reglue; see Figure 3.2. This procedure gives a well-defined homeomorphism of S which is equivalent to T_α. If α is a separating simple closed curve, these instructions do *not* say

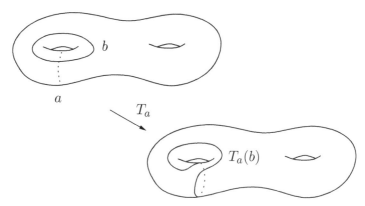

Figure 3.2 A Dehn twist via cutting and gluing.

to cut along α, twist one of the two pieces of the cut surface by 2π, and then reglue; this would give the identity homeomorphism of S. The key is to twist just the neighborhood of one boundary component.

Dehn twists via the inclusion homomorphism. In general, if S is a closed subsurface of a surface S', there is an induced homomorphism $\mathrm{Mod}(S) \rightarrow \mathrm{Mod}(S')$; see Theorem 3.18 below. Given any inclusion of the annulus A into a surface S, we obtain a homomorphism $\mathrm{Mod}(A) \rightarrow \mathrm{Mod}(S)$. The image of a generator of $\mathrm{Mod}(A)$ is a Dehn twist in $\mathrm{Mod}(S)$.

Action on simple closed curves. We can understand T_a by examining its action on the isotopy classes of simple closed curves on S. If b is an isotopy class with $i(a, b) = 0$, then $T_a(b) = b$. In the case that $i(a, b) \neq 0$ the isotopy class $T_a(b)$ is determined by the following rule: given particular representatives β and α of b and a, respectively, each segment of β crossing α is replaced with a segment that turns left, follows α all the way around, and then turns right. This is true no matter which way we orient β; the reason that we can distinguish left from right is that the map ϕ used in the definition of T_a is taken to be orientation-preserving.

Left versus right. We emphasize that, once an orientation of S is fixed, the direction of a twist T_a does not depend on any sort of orientation on a. This is because turning left is well defined on an oriented surface. (Similarly, a left-handed screw is still a left-handed screw when it is turned upside-down.) The inverse map T_a^{-1} is simply the twist about a in the other direction; it is defined similarly to T_a, with the twist map T replaced by its inverse T^{-1}.

The action on curves via surgery. If $i(a, b)$ is large (say, more than 2), it can be difficult to draw a picture of $T_a(b)$ using the turn left–turn right procedure given above. It is hard to plan ahead and leave enough room for all of the strands of $T_a(b)$ that run around a. A convenient way to draw $T_a(b)$ in practice is as follows. Start with one curve β in the class b and $i(a, b)$ parallel curves α_i, each in the class a, each in minimal position with β (one can also take the α_i to not have minimal position with β, but then one must take $|\alpha_i \cap \beta|$ parallel curves α_i). Of course, the result is not a simple closed curve. At each intersection point between β and some α_i, we do surgery as in Figure 3.3. The rule for the surgery is to resolve the intersection in the unique way so that if we follow an arc of β toward the intersection, the surgered arc turns left at the intersection. Again, this does not rely on any orientation of α_i or of β but rather on the orientation of the surface. After performing this surgery at each intersection, the result is a simple closed curve in the class $T_a(b)$.

Figure 3.3 Dehn twists via surgery.

3.1.2 NONTRIVIALITY OF DEHN TWISTS

If a is the isotopy class of a simple closed curve that is homotopic to a point or a puncture, then T_a is trivial in $\mathrm{Mod}(S)$—whatever twisting is done on the annulus can be undone by untwisting the disk or once-punctured disk inside. We can use the action of a Dehn twist on simple closed curves to prove that all other Dehn twists are nontrivial.

Proposition 3.1 *Let a be the isotopy class of a simple closed curve α in a surface S. If α is not homotopic to a point or a puncture of S, then the Dehn twist T_a is a nontrivial element of $\mathrm{Mod}(S)$.*

Proof. If α is a nonseparating simple closed curve, then by change of coordinates we can find a simple closed curve β with $i(\alpha, \beta) = 1$. Denote the isotopy class of β by b. As in Figure 3.2, one can draw a representative of $T_a(b)$ that intersects β once transversely. By the bigon criterion, $i(T_a(b), b)$

is actually equal to 1 (a bigon requires two intersections). Therefore, $T_a(b)$ is not the same as b, and so T_a is nontrivial in $\mathrm{Mod}(S)$.

Perhaps a simpler way to phrase the proof in the case that a is nonseparating is to check that T_a acts nontrivially on $H_1(S; \mathbb{Z})$; see Chapter 6 for more on this homology action. If α is a separating essential simple closed curve, then the action of T_a on $H_1(S; \mathbb{Z})$ is trivial, and so we are forced to use the more subtle machinery of the change of coordinates principle and the bigon criterion.

By the change of coordinates principle, an essential separating curve α is as depicted in Figure 3.4 (possibly with different genera and different numbers of punctures/boundary on the two sides of α). We can thus choose an isotopy class b with $i(a, b) = 2$, and we consider the isotopy class $T_a(b)$. We claim that $T_a(b) \neq b$, from which it follows that T_a is nontrivial.

We now prove the claim. On the right-hand side of Figure 3.4, we show representatives β and β' of b and $T_a(b)$; the given representatives intersect four times. We will use the bigon criterion to check that all intersections are essential and so $i(T_a(b), b) = 4$, from which it follows that $T_a(b) \neq b$. To do this, note that β cuts β' into four arcs, β'_1, β'_2, β'_3, and β'_4, and similarly β' cuts β into four arcs β_1, β_2, β_3, and β_4. For each β_i there is a unique β'_j that has the same pair of endpoints on $\beta \cap \beta'$. This gives four candidates for bigons. But each of these four candidate bigons $\beta_i \cup \beta'_j$ is a nonseparating simple closed curve, and so none is an actual bigon. This proves the claim, and so T_a is nontrivial.

The remaining case is that α is homotopic to a boundary component of S and that α is neither homotopic to a point or a puncture. It follows that S is some surface with boundary other than the disk or the once-punctured disk. Let $\overline{S}$ denote the *double* of S, obtained by taking two copies of S and identifying corresponding boundary components. In $\overline{S}$, the curve α becomes essential. By our definition of the mapping class group for a surface with boundary, if T_a were trivial in $\mathrm{Mod}(S)$, it would be trivial in $\mathrm{Mod}(\overline{S})$, contradicting the previous cases.   □

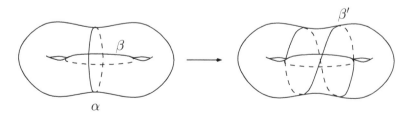

Figure 3.4 Checking that a Dehn twist about a separating simple closed curve is nontrivial.

3.2 DEHN TWISTS AND INTERSECTION NUMBERS

We have already seen the effectiveness of analyzing Dehn twists (and other mapping classes) via their actions on simple closed curves. We now give two explicit formulas for this action.

Proposition 3.2 *Let a and b be arbitrary isotopy classes of essential simple closed curves in a surface and let k be an arbitrary integer. We have*

$$i(T_a^k(b), b) = |k| i(a, b)^2.$$

We remark that, as an important consequence of Proposition 3.2, we have the following:

> *Dehn twists have infinite order.*

The only observation needed to prove this fact is that given an isotopy class a of essential simple closed curves, one can find an isotopy class b with $i(a, b) > 0$. As in the proof of Proposition 3.1, this is accomplished with the change of coordinates principle. Thus Proposition 3.2 is a generalization of Proposition 3.1. What is more, the proof of Proposition 3.2 is a generalization of the proof of Proposition 3.1.

Proof. We choose representative simple closed curves α and β in minimal position and form a simple closed curve β' in the class of $T_a(b)$ using the surgical recipe given above. More specifically, we take $k\,i(a, b)$ parallel copies of α lying to one side of α and one copy of β lying parallel to β, and then we surger as in Figure 3.3; see the left-hand side of Figure 3.5 for the case of $i(a, b) = 3$ and $k = 1$.

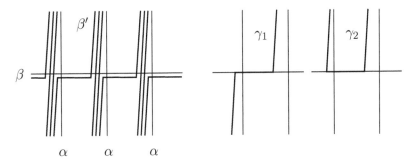

Figure 3.5 The simple closed curves in the proof of Proposition 3.2.

Simply by counting, we see that

$$|\beta \cap \beta'| = |k| i(a, b)^2.$$

Thus it suffices to show that β and β' are in minimal position. By the bigon criterion, we only need to check that they do not form any bigons.

We cut β and β' at the points of $\beta \cap \beta'$ and call the resulting closed arcs $\{\beta_i\}$ and $\{\beta'_i\}$. We see that there are two types of candidate bigons, that is, simple closed curves that can be formed from one arc β_i and one arc β'_j: either the orientations of the two intersection points are the same, as for the curve γ_1 on the right-hand side of Figure 3.5, or the orientations of the intersection points are different, as for γ_2 in the same figure. In a true bigon, the orientations at the two intersection points are different, and so the simple closed curve γ_1 in the first case cannot be a bigon. In the second case, if γ_2 were a bigon, then since the vertical arcs of β' are parallel to arcs of α, we see that α and β form a bigon, contrary to assumption.   □

Proposition 3.4 below is a useful generalization of Proposition 3.2. In order to prove it, we require the following lemma.

Lemma 3.3 *Let α and β be simple closed curves in a surface. Suppose that α and β are in minimal position. Given a third simple closed curve γ, there exists a simple closed curve γ' that is homotopic to γ and that is in minimal position with respect to both α and β.*

Proof. By perturbing γ by isotopy if necessary, we may assume that γ is transverse to both α and β. If γ is not in minimal position with α, say, then by the bigon criterion α and γ form a bigon. We can take this bigon to be innermost with respect to α and γ. By the assumption that α and β are in minimal position, any arc of intersection of β with this bigon either connects the α-side of the bigon to the γ-side, or the γ-side to itself. In the latter case, we have a bigon formed by β and γ that is contained inside the original bigon.

Continuing in this way, we can find either a bigon formed by α and γ or a bigon formed by β and γ that is innermost among all such bigons. Say the innermost bigon is formed by α and γ. As above, any intersection of β with this bigon is an arc connecting one side to the other. Thus we can push γ by homotopy across the bigon, reducing the number of intersection points with α by 2 and preserving the number of intersection points with β. We can repeat this process until all bigons are eliminated, and the lemma is proved. □

Another approach to Lemma 3.3 is the following: one can show that there exists a hyperbolic metric on the surface so that the curves α and β are geodesics [61, Exposé 3, Proposition 10]. Then the curve γ' can be taken to be the geodesic in the free homotopy class of γ.

Proposition 3.4 *Let $a_1, \ldots, a_n$ be a collection of pairwise disjoint isotopy classes of simple closed curves in a surface S and let $M = \prod_{i=1}^{n} T_{a_i}^{e_i}$. Suppose that $e_i > 0$ for all i or $e_i < 0$ for all i. If b and c are arbitrary isotopy classes of simple closed curves in S, then*

$$\left| i(M(b), c) - \sum_{i=1}^{n} |e_i| i(a_i, b) i(a_i, c) \right| \leq i(b, c).$$

Setting $n = 1$, $e_1 = k$, and $c = b$ gives Proposition 3.2 as a special case. There is a version of Proposition 3.4 where the e_i are allowed to have arbitrary signs, but the proof is not as straightforward; we refer the reader to [106, Lemma 4.2].

Proof. We start by forming a representative β' of $M(b)$ as in the proof of Proposition 3.2. As in that proof, it follows from the bigon criterion that β and β' are in minimal position. This uses the fact that all of the twists are in the same direction, that is, the e_i all have the same sign. By Lemma 3.3, there is a representative γ of c that is in minimal position with both β and β'. By perturbing γ if necessary, we can assume that it does not pass through $\beta \cap \beta'$.

There is a continuous map of the disjoint union of $\sum |e_i| i(a_i, b)$ copies of S^1 into S with image $\beta \cup \beta'$ and where the images of $|e_i|$ copies of S^1 lie in the class a_i. Each copy of a_i intersects γ in at least $i(a_i, c)$ points, by the definition of geometric intersection number. Since γ is in minimal position with β and β', we obtain

$$\sum |e_i| i(a_i, b) i(a_i, c) \leq |(\beta \cup \beta') \cap \gamma| = i(M(b), c) + i(b, c).$$

It remains to prove that

$$i(M(b), c) \leq \sum |e_i| i(a_i, b) i(a_i, c) + i(b, c).$$

For this it suffices to find representatives of $M(b)$ and c whose intersection consists of $\sum |e_i| i(a_i, b) i(a_i, c) + i(b, c)$ points. The most natural representatives satisfy this property. Precisely, for $M(b)$ we can choose a curve that lies in the union of the curve β and small regular neighborhoods of disjoint representatives α_i of the a_i. Then, for c, we take a curve that cuts across each α_i-annulus in $i(a_i, c)$ arcs and intersects β in $i(b, c)$ points not contained in the α_i-annuli. □

Pairs of filling curves. We now give one useful consequence of Proposition 3.4. Say that a pair of isotopy classes $\{a, b\}$ of simple closed curves in a surface S *fill* if any pair of minimal position representatives fill (i.e., the complement of the representatives in the surface is a collection of disks and once-punctured disks). This is the same as saying that for every isotopy class c of essential simple closed curves in the surface, either $i(a, c) > 0$ or $i(b, c) > 0$.

Proposition 3.5 *Let $g, n \geq 0$ and assume that $\chi(S_{g,n}) < 0$. There exists a pair of simple closed curves in $S_{g,n}$ that fill $S_{g,n}$.*

Proof. Choose a maximal collection $\{\alpha_1, \dots, \alpha_k\}$ of pairwise disjoint, non-homotopic, essential simple closed curves in $S_{g,n}$. When we cut $S_{g,n}$ along the α_i, we obtain a collection of surfaces. Each of these surfaces is a sphere with b boundary components and p punctures with $b + p = 3$ (cf. Section 8.3).

We claim that there is a simple closed curve β in $S_{g,n}$ so that $i(\beta, \alpha_i) > 0$ for each i. We can construct β as follows. First, we cut $S_{g,n}$ along the α_i. On each component of the cut surface, we then connect by an arc each pair of distinct boundary components coming from the α_i. We can take these arcs to be disjoint. In $S_{g,n}$, these arcs can be pasted together in an arbitrary fashion in order to obtain a collection $\beta_1, \dots, \beta_k$ of pairwise disjoint simple closed curves in $S_{g,n}$.

By the bigon criterion, each β_j is in minimal position with respect to each α_i and each α_i intersects either one or two of the β_j. Suppose that β_j and $\beta_{j'}$ intersect α_i and that β_j and $\beta_{j'}$ are distinct. Then we can perform a half-twist about α_i so that β_j and $\beta_{j'}$ become a single curve. Since this process does not create any bigons, the resulting collection $\{\beta_j\}$ is still in minimal position with each α_i. Continuing in this way, we obtain a single simple closed curve β that intersects each α_i and is in minimal position with respect to each α_i, as desired.

Let $M = T_{\alpha_1} \cdots T_{\alpha_k}$. We claim that β and $M(\beta)$ fill $S_{g,n}$. Indeed, let γ be an arbitrary isotopy class of simple closed curves in $S_{g,n}$. We wish to show that either $i(\beta, \gamma) > 0$ or $i(M(\beta), \gamma) > 0$. By Proposition 3.4, we have

$$\left| i(M(\beta), \gamma) - \sum_{i=1}^{k} i(\alpha_i, \beta) i(\alpha_i, \gamma) \right| \leq i(\beta, \gamma).$$

If $i(\beta, \gamma)$ and $i(M(\beta), \gamma)$ are both equal to zero, then this immediately implies that $i(\alpha_i, \gamma) = 0$ for each i. This means that γ is isotopic to some α_i. But then $i(\gamma, \beta) > 0$ by the construction of β, and so we have a contradiction. $\qquad\square$

3.3 BASIC FACTS ABOUT DEHN TWISTS

In this section we prove some fundamental facts about Dehn twists that will be used repeatedly throughout this book. Throughout this section a and b denote arbitrary (unoriented) isotopy classes of simple closed curves.

Fact 3.6 $T_a = T_b \iff a = b$.

We have already addressed the reverse implication of Fact 3.6, which says that Dehn twists are well-defined mapping classes. For the forward implication, we start by noting that the statement is not as obvious as it seems. Indeed, suppose we know that $T_a = T_b$. Then we know that, given any two representatives of T_a and T_b with annular supports (neighborhoods of simple closed curves in the classes a and b), there is an isotopy between the representative homeomorphisms. One would then like to say that there is an induced isotopy from one annular support to the other and hence an isotopy between curves. But partway through the isotopy of homeomorphisms, the support might become something other than an annulus—perhaps the whole surface, even—and we have lost any information we had about simple closed curves.

So assume now that $a \neq b$. We will show that $T_a \neq T_b$. We start by finding an isotopy class c of simple closed curves so that $i(a, c) = 0$ and $i(b, c) \neq 0$. There are two cases. First, if $i(a, b) \neq 0$, then we can take $c = a$. If $i(a, b) = 0$, then one can use change of coordinates to easily find c (there are several cases, depending on the separation properties of the curves). Given any such choice of c, we apply Proposition 3.2 and find

$$i(T_a(c), c) = i(a, c)^2 = 0 \neq i(b, c)^2 = i(T_b(c), c).$$

It follows that $T_a(c) \neq T_b(c)$, and so $T_a \neq T_b$.

We have the following formula for the conjugate of a Dehn twist.

Fact 3.7 *For any $f \in \mathrm{Mod}(S)$ and any isotopy class a of simple closed curves in S we have*

$$T_{f(a)} = fT_af^{-1}.$$

Fact 3.7 can be checked directly as follows. First, recall that we apply elements of the mapping class group from right to left. Let ϕ denote a representative of f, let α denote a representative of a, and let ψ_α denote a representative of T_a whose support is an annulus. Note that ϕ^{-1} takes a regular neighborhood of $\phi(\alpha)$ to a regular neighborhood of α (preserving the orientation), then ψ_α twists the neighborhood of α, and ϕ takes this twisted neighborhood of α back to a neighborhood of $\phi(\alpha)$ (again preserving the orientation). So the net result is a Dehn twist about $\phi(\alpha)$.

From the previous facts we obtain the following.

Fact 3.8 *For any $f \in \mathrm{Mod}(S)$ and any isotopy class a of simple closed curves in S, we have*

$$f \text{ commutes with } T_a \iff f(a) = a.$$

Indeed, by Facts 3.7 and 3.6, we have

$$fT_a = T_a f \iff fT_a f^{-1} = T_a$$
$$\iff T_{f(a)} = T_a$$
$$\iff f(a) = a.$$

By the classification of simple closed curves in S (see Section 1.3), given any two nonseparating simple closed curves a and b in S, there exists $h \in \mathrm{Mod}(S)$ with $h(a) = b$. Hence Fact 3.7 also gives the following.

> *If a and b are nonseparating simple closed curves in S, then T_a and T_b are conjugate in $\mathrm{Mod}(S)$.*

The last statement can be generalized, using change of coordinates, to twists about any two simple closed curves of the same topological type.

The next fact follows from Proposition 3.2 and Fact 3.8.

Fact 3.9 *For any two isotopy classes a and b of simple closed curves in a surface S, we have*

$$i(a,b) = 0 \iff T_a(b) = b \iff T_a T_b = T_b T_a.$$

The only nontrivial part of the proof of Fact 3.9 is that the second statement implies the first. But if $T_a(b) = b$, then $i(T_a(b), b) = i(b, b) = 0$. By Proposition 3.2, $i(T_a(b), b) = i(a, b)^2$, and it follows that $i(a, b) = 0$.

Powers of Dehn twists. There are analogues of each of the above facts for powers of Dehn twists. For $f \in \mathrm{Mod}(S)$, we have

$$fT_a^j f^{-1} = T_{f(a)}^j,$$

and so f commutes with T_a^j if and only if $f(a) = a$. Also, for nontrivial Dehn twists T_a, T_b and nonzero integers j, k, we have

$$T_a^j = T_b^k \iff a = b \text{ and } j = k$$
$$T_a^j T_b^k = T_b^k T_a^j \iff i(a, b) = 0.$$

In each case the proof is essentially the same as the cases when $j = k = 1$.

In the remainder of this section, we give three applications of the Alexander method and our basic facts about Dehn twists: we compute the center of the mapping class group, we derive some geometrically induced homomorphisms between mapping class groups, and we give computations of mapping class groups of certain surfaces with boundary.

3.4 THE CENTER OF THE MAPPING CLASS GROUP

Recall that the *center* $Z(G)$ of a group G is the subgroup of G consisting of those elements that commute with every element of G. We will apply Fact 3.8 and the Alexander method to compute the center of $\mathrm{Mod}(S)$.

THEOREM 3.10 *For $g \geq 3$, the group $Z(\mathrm{Mod}(S_g))$ is trivial.*

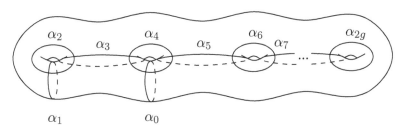

Figure 3.6 The simple closed curves used to determine the center of $\mathrm{Mod}(S)$.

Proof. By Fact 3.8, any central element f of $\mathrm{Mod}(S_g)$ must fix every isotopy class of simple closed curves in S_g. Consider the simple closed curves $\alpha_0, \ldots, \alpha_{2g}$ shown in Figure 3.6. By statement 1 of the Alexander method, f has a representative ϕ that fixes the graph $\cup \alpha_i$, and thus ϕ induces a map ϕ_* of this graph.

The graph $\cup \alpha_i$ is isomorphic to the abstract graph Γ shown in Figure 3.7 for the case $g = 4$. For $g \geq 3$, the only automorphisms of Γ come from flipping the three edges that form loops and swapping pairs of edges that

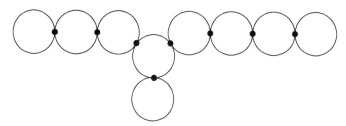

Figure 3.7 The collection of simple closed curves in Figure 3.6 form a graph in S_4 that is abstractly isomorphic to the graph Γ shown here for the case $g = 4$.

form a loop. In particular, any automorphism of Γ must fix the three edges coming from α_4. Thus we see that ϕ preserves the orientation of α_4, and so since ϕ is orientation-preserving, it must also preserve the two sides of α_4. It follows that ϕ_* does not flip the edge of Γ coming from α_0, and it does not interchange the two edges coming from α_3 or the two coming from α_5. Inductively, we see that ϕ_* fixes each edge of Γ with orientation. By statement 2 of the Alexander method, plus the fact that the $\{\alpha_i\}$ fill S_g, we have that ϕ is isotopic to the identity; that is, f is the identity.   □

The proof of Theorem 3.10 actually shows that the center of any finite index subgroup of $\mathrm{Mod}(S_g)$ is trivial when $g \geq 3$ since a finite-index subgroup contains some power of each Dehn twist and since Fact 3.8 applies to powers of Dehn twists.

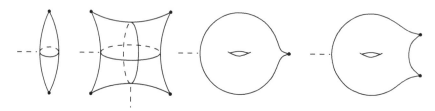

Figure 3.8 Rotations by π about the indicated axes give hyperelliptic involutions of the punctured surfaces $S_{0,2}$, $S_{0,4}$, $S_{1,1}$, and $S_{1,2}$.

By choosing appropriate configurations of simple closed curves on other surfaces, the method of proof of Theorem 3.10 shows that the only candidates for nontrivial central elements of (finite-index subgroups of) $\mathrm{Mod}(S_{g,n})$ are the hyperelliptic involutions of T^2 and S_2, as well as the hyperelliptic involutions shown in Figure 3.8. So the order of $Z(\mathrm{Mod}(S_{g,n}))$ is at most 2 when $S_{g,n}$ is one of the punctured surfaces $S_{0,2}$, $S_{1,0}$, $S_{1,1}$, $S_{1,2}$, or $S_{2,0}$, the order of $Z(\mathrm{Mod}(S_{0,4}))$ is at most 4, and $Z(\mathrm{Mod}(S_{g,n}))$ is trivial in all other cases. In the case of $\text{Mod}(S_{0,4})$, the center is trivial since the the subgroup generated by the hyperelliptic involutions acts faithfully on the four punctures, and the symmetric group on the four punctures is centerless.

On the other hand, to show that a mapping class z really is an element of $Z(\text{Mod}(S))$, it suffices to choose a generating set of Dehn twists and half-twists for $\text{Mod}(S)$ and show that z fixes each of the corresponding isotopy classes of simple closed curves and simple arcs (see Corollary 4.15). In this way, we find that $Z(\text{Mod}(S_{g,n})) \approx \mathbb{Z}/2\mathbb{Z}$ when $S_{g,n}$ is $S_{0,2}$, $S_{1,0}$, $S_{1,1}$, $S_{1,2}$, or $S_{2,0}$. By the same argument, for a surface with boundary, the Dehn twist about any boundary component is central.

We summarize the results for punctured surfaces in the following table.

Surface (with punctures)	$Z(\text{Mod}(S))$
$S_{0,2}$, $S_{1,0}$, $S_{1,1}$, $S_{1,2}$, $S_{2,0}$	$\mathbb{Z}_2$
All other $S_{g,n}$	1

As stated in the proof of Theorem 3.10, these nontrivial central elements have the property that they fix the isotopy class of every simple closed curve.

3.5 RELATIONS BETWEEN TWO DEHN TWISTS

The goal of this section is to answer the question: what algebraic relations can occur between two Dehn twists? In fact, we answer the more general question where powers of Dehn twists are allowed. We have already seen that Dehn twists about disjoint curves commute in the mapping class group. The next most basic relation between twists is the braid relation. Except in a few cases, we will see that there are no other relations between Dehn twists.

3.5.1 THE BRAID RELATION

The following proposition gives a basic relation between Dehn twists in $\text{Mod}(S)$ called the *braid relation*.

Proposition 3.11 (Braid relation) *If a and b are isotopy classes of simple closed curves with $i(a, b) = 1$, then*

$$T_a T_b T_a = T_b T_a T_b.$$

Proof. The relation

$$T_a T_b T_a = T_b T_a T_b$$

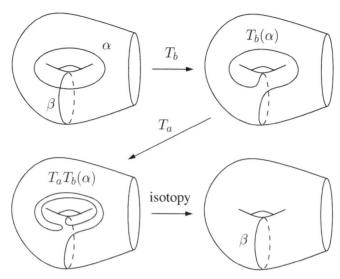

Figure 3.9 The proof of Proposition 3.12.

is equivalent to the relation

$$(T_a T_b) T_a (T_a T_b)^{-1} = T_b.$$

By Fact 3.7, this is equivalent to the relation

$$T_{T_a T_b(a)} = T_b.$$

Applying Fact 3.6, this is equivalent to the equality

$$T_a T_b(a) = b.$$

By the change of coordinates principle, it suffices to check the last statement for any two isotopy classes a and b with $i(a, b) = 1$. The computation is shown in Figure 3.9, where α is some representative of a and β is some representative of b. □

If a is the $(1, 0)$-curve and b is the $(0, 1)$-curve on the torus T^2, then via the isomorphism of Theorem 2.5 the braid relation corresponds to the familiar relation in $\mathrm{SL}(2, \mathbb{Z})$:

$$\begin{pmatrix} 1 & -1 \\ 0 & 1 \end{pmatrix} \begin{pmatrix} 1 & 0 \\ 1 & 1 \end{pmatrix} \begin{pmatrix} 1 & -1 \\ 0 & 1 \end{pmatrix} = \begin{pmatrix} 1 & 0 \\ 1 & 1 \end{pmatrix} \begin{pmatrix} 1 & -1 \\ 0 & 1 \end{pmatrix} \begin{pmatrix} 1 & 0 \\ 1 & 1 \end{pmatrix}$$

The next proposition records our rephrasing of the braid relation for use in the proof of Theorem 4.1 below.

Proposition 3.12 *If a and b are isotopy classes of simple closed curves that satisfy $i(a, b) = 1$, then $T_a T_b(a) = b$.*

The braid relation gets its name from the analogous relation in the braid group (see Section 9.4).

One can ask for a converse to the braid relation: if two Dehn twists satisfy the braid relation algebraically, then do the corresponding curves necessarily have intersection number one? McCarthy gave the following proof that the answer is yes [144]. Theorem 3.14 below is a much more general fact; we consider Proposition 3.13 as a warmup.

Proposition 3.13 *If a and b are distinct isotopy classes of simple closed curves and the Dehn twists T_a and T_b satisfy $T_a T_b T_a = T_b T_a T_b$, then $i(a, b) = 1$.*

Proof. As in the proof of Proposition 3.11, the relation $T_a T_b T_a = T_b T_a T_b$ is equivalent to the statement that $T_a T_b(a) = b$, which implies

$$i(a, T_a T_b(a)) = i(a, b).$$

Applying T_a^{-1} to both curves on the left-hand side of the equation, we see that

$$i(a, T_b(a)) = i(a, b).$$

Now, by Proposition 3.2, we have that

$$i(a, b)^2 = i(a, b).$$

And so $i(a, b)$ is either equal to 0 or 1. If $i(a, b)$ were 0, an application of Fact 3.9 reduces the relation to $T_a = T_b$, which, by Fact 3.6, contradicts the assumption $a \neq b$. Thus $i(a, b) = 1$. $\square$

We note that the same proof really shows the stronger result that if $a \neq b$ and $T_a^j T_b^k T_a^j = T_b^k T_a^j T_b^k$, then $i(a, b) = 1$ and $j = k = \pm 1$.

3.5.2 GROUPS GENERATED BY TWO DEHN TWISTS

Now that we know the braid relation it is natural to try to find other relations between two Dehn twists. In this subsection we will give a complete classification of such relations. We begin with the following.

THEOREM 3.14 *Let a and b be two isotopy classes of simple closed curves in a surface S. If $i(a, b) \geq 2$, then the group generated by T_a and T_b is isomorphic to the free group F_2 of rank 2.*

We can also say what happens in the other cases. If $a = b$, then $\langle T_a, T_b \rangle \approx$ $\mathbb{Z}$ since $T_a^j = T_b^k$ if and only if $a = b$ and $j = k$. If $a \neq b$ and $i(a, b) = 0$, then $\langle T_a, T_b \rangle$ is isomorphic to $\mathbb{Z}^2$ by Fact 3.9 plus the fact that $T_a^j = T_b^k$ if and only if $a = b$ and $j = k$. When $i(a, b) = 1$, we have that

$$\langle T_a, T_b \rangle \approx \mathrm{Mod}(S_1^1) \approx \langle x, y \mid xyx = yxy \rangle,$$

where S_1^1 is a torus with an open disk removed (see above).

We remark that the question of which groups can be generated by three Dehn twists is completely open. See Section 5.1 for one relation between three Dehn twists.

Below we give the proof of Theorem 3.14 published by Ishida and Hamidi-Tehrani [76, 103]. The theorem, though, was apparently known to Ivanov (and perhaps others) in the early 1980s. We first introduce the ping pong lemma, which is a basic and fundamental tool from geometric group theory. It is a method to prove that a group is free by understanding how it acts on a set. Poincaré used this method to prove that if two hyperbolic translations have different axes, then sufficiently high powers of these elements generate a free group of rank 2.

Lemma 3.15 (Ping pong lemma) *Let G be a group acting on a set X. Let $g_1, \ldots, g_n$ be elements of G. Suppose that there are nonempty, disjoint subsets $X_1, \ldots, X_n$ of X with the property that, for each i and each $j \neq i$, we have $g_i^k(X_j) \subset X_i$ for every nonzero integer k. Then the group generated by the g_i is a free group of rank n.*

Proof. We need to show that any nontrivial freely reduced word in the g_i represents a nontrivial element of G. First suppose that w is a freely reduced word that starts and ends with a nontrivial power of g_1. Then for any $x \in X_2$, we have $w(x) \in X_1$, and so $w(x) \neq x$ since $X_1 \cap X_2 = \emptyset$. Thus w represents a nontrivial element of g. Since any other freely reduced word in the g_i is conjugate to a word that starts and ends with g_1, every freely reduced word in the g_i represents an element of G that is conjugate to a nontrivial element and hence is itself nontrivial. $\square$

Proof of Theorem 3.14. Suppose that $i(a, b) \geq 2$. Let G be the group generated by $g_1 = T_a$ and $g_2 = T_b$ and let X be the set of isotopy classes of simple closed curves in S. The group G acts on X. With the ping pong lemma in mind, we define sets X_a and X_b as follows:

$$X_a = \{c \in X : i(c, b) > i(c, a)\},$$
$$X_b = \{c \in X : i(c, a) > i(c, b)\}.$$

These sets are obviously disjoint, and they are nonempty since $a \in X_a$ and $b \in X_b$.

By the ping pong lemma, the proof is reduced to checking that $T_a^k(X_b) \subset X_a$ and $T_b^k(X_a) \subset X_b$ for $k \neq 0$. By symmetry, we need to check only the former inclusion.

Setting $M = T_a^k$ in Proposition 3.4 yields

$$\left| i(T_a^k(c), b) - |k| i(a, b) i(a, c) \right| \leq i(b, c),$$

and so

$$-i(b, c) \leq i(T_a^k(c), b) - |k| i(a, b) i(a, c) \leq i(b, c).$$

If $c \in X_b$, then $i(a, c) > i(b, c)$. Since $k \neq 0$, the left-hand inequality implies

$$
\begin{aligned}
i(T_a^k(c), b) &\geq |k| i(a, b) i(a, c) - i(b, c) \\
&\geq 2|k| i(a, c) - i(b, c) \\
&> 2|k| i(a, c) - i(a, c) \\
&= (2|k| - 1) i(a, c) \\
&\geq i(a, c) \\
&= i(T_a^k(a), T_a^k(c)) \\
&= i(a, T_a^k(c)).
\end{aligned}
$$

Thus $i(T_a^k(c), b) > i(T_a^k(c), a)$, and so $T_a^k(c) \in X_a$, as desired. $\square$

A free group in $\mathbf{SL(2, \mathbb{Z})}$. The proof of Theorem 3.14 given above is inspired by a proof that the matrices

$$\begin{pmatrix} 1 & n \\ 0 & 1 \end{pmatrix} \quad \text{and} \quad \begin{pmatrix} 1 & 0 \\ n & 1 \end{pmatrix}$$

generate a free subgroup of $SL(2, \mathbb{Z})$ for $n \geq 2$ (this fact is originally due to Magnus [136]). In this case, the sets used for the ping pong lemma are $\{(x, y) \in \mathbb{Z}^2 : |x| > |y|\}$ and $\{(x, y) \in \mathbb{Z}^2 : |y| > |x|\}$.

The classification of groups generated by two Dehn twists. With a little more care, the method of proof of Theorem 3.14 can be applied to give the stronger statement that $\langle T_a^j, T_b^k \rangle \approx F_2$ except if $i(a, b) = 0$ or if $i(a, b) = 1$ and the set $\{j, k\}$ is equal to $\{1\}$, $\{1, 2\}$, or $\{1, 3\}$. When $j = k = 1$, we already know that we have the braid relation. And in the other exceptional cases, there exist nontrivial relations as well. For instance, if $i(a,b) = 1$, then T_a^2 and T_b satisfy the relation

$$T_a^2 T_b T_a^2 T_b = T_b T_a^2 T_b T_a^2$$

and T_a^3 and T_b satisfy

$$T_a^3 T_b T_a^3 T_b T_a^3 T_b = T_b T_a^3 T_b T_a^3 T_b T_a^3.$$

What is more, it turns out that these are the defining relations for the groups $\langle T_a^2, T_b \rangle$ and $\langle T_a^3, T_b \rangle$. The group $\langle T_a^2, T_b \rangle$ corresponds to a well-known index 3 subgroup of B_3 (the subgroup fixing the first strand). The group $\langle T_a^3, T_b \rangle$ does not seem to be a well-known subgroup of B_3. Luis Paris has explained to us that this is an index 8 subgroup of B_3, and he has used the Reidemeister–Schreier algorithm to give an elementary proof that the stated relation is the unique defining relation; see [174].

Combining the results from this section, we can completely list all possibilities for groups generated by powers of two Dehn twists. In the table we assume that a and b are essential, that $j \geq k > 0$, and that the underlying surface is not T^2 or $S_{1,1}$.

	Group generated by T_a^j, T_b^k
$i(a,b) = 0, a = b$	$\langle T_a^j, T_b^k \rangle \approx \langle x, y \mid x = y \rangle \approx \mathbb{Z}$
$i(a,b) = 0, a \neq b$	$\langle T_a^j, T_b^k \rangle \approx \langle x, y \mid xy = yx \rangle \approx \mathbb{Z}^2$
$i(a,b) = 1$	$\langle T_a, T_b \rangle \approx \langle x, y \mid xyx = yxy \rangle$
	$\langle T_a^2, T_b \rangle \approx \langle x, y \mid xyxy = yxyx \rangle$
	$\langle T_a^3, T_b \rangle \approx \langle x, y \mid xyxyxy = yxyxyx \rangle$
	$\langle T_a^j, T_b^k \rangle \approx \langle x, y \mid \rangle \approx F_2$ otherwise
$i(a,b) \geq 2$	$\langle T_a^j, T_b^k \rangle \approx \langle x, y \mid \rangle \approx F_2$

If the surface is T^2 or $S_{1,1}$ and $i(a,b) = 1$, we have the added relations $(T_a T_b)^6 = 1$, $(T_a^2 T_b)^4 = 1$, and $(T_a^3 T_b)^3 = 1$.

3.6 CUTTING, CAPPING, AND INCLUDING

In this section we apply our knowledge about Dehn twists to address a basic general question about mapping class groups: when does a geometric operation on a surface induce an algebraic operation on the corresponding mapping class group? We investigate three such operations: including a surface into another surface, capping a boundary component of a surface with a punctured disk, and deleting a simple closed curve from a surface. We will see that in each case there is indeed an induced homomorphism on the level of mapping class groups.

The results in this section are somewhat technical but very useful. The reader might consider skipping the proofs on a first reading.

3.6.1 THE INCLUSION HOMOMORPHISM

When S is a closed subsurface of a surface S', there is a natural homomorphism $\eta : \mathrm{Mod}(S) \to \mathrm{Mod}(S')$. For $f \in \mathrm{Mod}(S)$, we represent it by some $\phi \in \mathrm{Homeo}^+(S, \partial S)$. Then, if $\widehat{\phi}$ is the element of $\mathrm{Homeo}^+(S', \partial S')$ that agrees with ϕ on S and is the identity outside of S, we define $\eta(f)$ to be the class of $\widehat{\phi}$. The map η is well defined because any homotopy between two elements of $\phi \in \mathrm{Homeo}^+(S, \partial S)$ gives a homotopy between the corresponding elements of $\mathrm{Homeo}^+(S', \partial S')$.

Our goal in this subsection is to describe the kernel of η (Theorem 3.18 below). We begin with a simple lemma.

Lemma 3.16 *Let $\alpha_1, \ldots, \alpha_n$ be a collection of homotopically distinct simple closed curves in a surface S, each not homotopic to a point in S. Let β and β' be simple closed curves in S that are both disjoint from $\cup \alpha_i$ and are homotopically distinct from each α_i. If β and β' are isotopic in S, then they are isotopic in $S - \cup \alpha_i$.*

Proof. It suffices to find an isotopy from β to β' in S that avoids $\cup \alpha_i$. First, we may modify β so that it is transverse to β' and is still disjoint from $\cup \alpha_i$. If $\beta \cap \beta' = \emptyset$, then β and β' form the boundary of an annulus A in S. Since β (and β') is not homotopic to any α_i, it cannot be that any α_i are contained in A. The annulus A gives the desired isotopy from β to β'.

If $\beta \cap \beta' \neq \emptyset$, then by the bigon criterion they form a bigon. Since the α_i are not homotopic to a point and $(\cup \alpha_i) \cap (\beta \cup \beta') = \emptyset$, the intersection of $\cup \alpha_i$ with the bigon is empty. We can thus push β across the bigon, keeping β disjoint from $\cup \alpha_i$ throughout the isotopy. By induction, we reduce to the case where β and β' are disjoint. This completes the proof. $\square$

LEMMA 3.17 *Let $\{a_1, \ldots, a_m\}$ be a collection of distinct nontrivial isotopy classes of simple closed curves in a surface S and assume that $i(a_i, a_j) = 0$ for all i, j. Let $\{b_1, \ldots, b_n\}$ be another such collection. Let $p_i, q_i \in \mathbb{Z} - \{0\}$. If*

$$T_{a_1}^{p_1} T_{a_2}^{p_2} \cdots T_{a_m}^{p_m} = T_{b_1}^{q_1} T_{b_2}^{q_2} \cdots T_{b_n}^{q_n}$$

in $\mathrm{Mod}(S)$, *then* $m = n$ *and the sets* $\{T_{a_i}^{p_i}\}$ *and* $\{T_{b_i}^{q_i}\}$ *are equal. In particular,*

$$\langle T_{a_1}, T_{a_2}, \ldots, T_{a_m} \rangle \approx \mathbb{Z}^m.$$

A mapping class $\prod T_{a_i}^{p_i}$ as in Lemma 3.17 is called a *multitwist*. Lemma 3.17 is a generalization of Fact 3.6, and in fact the proof is also a straightforward generalization. Note that in the statement the a_i and b_i are allowed to be peripheral.

THEOREM 3.18 (The inclusion homomorphism) *Let S be a closed subsurface of a surface S'. Assume that S is not homeomorphic to a closed annulus and that no component of $S' - S$ is an open disk. Let $\eta : \mathrm{Mod}(S) \to \mathrm{Mod}(S')$ be the induced map. Let $\alpha_1, \ldots, \alpha_m$ denote the boundary components of S that bound once-punctured disks in $S' - S$ and let $\{\beta_1, \gamma_1\}, \ldots, \{\beta_n, \gamma_n\}$ denote the pairs of boundary components of S that bound annuli in $S' - S$. Then the kernel of η is the free abelian group*

$$\ker(\eta) = \langle T_{\alpha_1}, \ldots, T_{\alpha_m}, T_{\beta_1} T_{\gamma_1}^{-1}, \ldots, T_{\beta_n} T_{\gamma_n}^{-1} \rangle.$$

In particular, if no connected component of $S' - S$ is an open annulus, an open disk, or an open once-marked disk, then η is injective.

The annulus is a special case for Theorem 3.18 for the simple fact that it has two boundary components that are isotopic. If S is an annulus, then η is injective unless S' is obtained from S by capping one or both boundary components with disks or once-punctured disks.

Proof. Let $f \in \ker(\eta)$ and let $\phi \in \mathrm{Homeo}^+(S, \partial S)$ be a representative. As above, we may extend ϕ by the identity in order to obtain $\widehat{\phi} \in \mathrm{Homeo}^+(S', \partial S')$. By definition, $\widehat{\phi}$ represents $\eta(f)$. Therefore, $\widehat{\phi}$ lies in the connected component of the identity in $\mathrm{Homeo}^+(S', \partial S')$.

Let δ be an arbitrary oriented simple closed curve in S. Since $\widehat{\phi}$ is isotopic to the identity, we have that $\widehat{\phi}(\delta)$ is isotopic to δ in S'. Since $\widehat{\phi}$ agrees with ϕ on S, we have that $\phi(\delta)$ is isotopic to δ in S'. By Lemma 3.16 and the assumption on $S' - S$, we have that $\phi(\delta)$ is isotopic to δ in S.

We can choose a collection of simple closed curves $\delta_1, \ldots, \delta_k$ in S that satisfy the three properties in the statement of the Alexander method (pairwise minimal position, pairwise nonisotopic, no triple intersections) and so that the surface obtained from S by cutting along $\cup \delta_i$ is a collection of disks, once-punctured disks, and closed annular neighborhoods N_i of the boundary components. Moreover, we can choose $\{\delta_i\}$ so that any homeomorphism that fixes $\cup \delta_i \cup \partial S$ necessarily preserves the complementary regions.

By the first statement of the Alexander method, ϕ is isotopic (in S) to a homeomorphism of S that fixes $\cup \delta_i \cup \partial S$. Since $\text{Mod}(D^2) = 1$ and $\text{Mod}(D^2 - \text{point}) = 1$ (Lemma 2.1), it follows that f has a representative that is supported in the N_i. Since $\text{Mod}(A) \approx \mathbb{Z}$ (Proposition 2.4), it follows that f is a product of Dehn twists about boundary components. By Lemma 3.17, f must become the trivial multitwist in S'. The theorem follows. □

The proof of Theorem 3.18 extends to the case where S is disconnected and $\text{Mod}(S)$ is taken to be the direct product of the mapping class groups of its connected components.

3.6.2 THE CAPPING HOMOMORPHISM

One particularly useful special case of Theorem 3.18 is the case where $S' - S$ is a once-punctured disk. We say that S' is the surface obtained from S by *capping* one boundary component. In this case we have the following statement.

Proposition 3.19 (The capping homomorphism) *Let S' be the surface obtained from a surface S by capping the boundary component β with a once-marked disk; call the marked point in this disk p_0. Denote by $\text{Mod}(S, \{p_1, \ldots, p_k\})$ the subgroup of $\text{Mod}(S)$ consisting of elements that fix the punctures $p_1, \ldots, p_k$, where $k \geq 0$. Let $\text{Mod}(S', \{p_0, \ldots, p_k\})$ denote the subgroup of $\text{Mod}(S')$ consisting of elements that fix the marked points $p_0, \ldots, p_k$ and then let $Cap : \text{Mod}(S, \{p_1, \ldots, p_k\}) \rightarrow \text{Mod}(S', \{p_0, \ldots, p_k\})$ be the induced homomorphism. Then the following sequence is exact:*

$$1 \rightarrow \langle T_\beta \rangle \rightarrow \text{Mod}(S, \{p_1, \ldots, p_k\}) \overset{Cap}{\rightarrow} \text{Mod}(S', \{p_0, \ldots, p_k\}) \rightarrow 1.$$

One might also wonder about the case where a boundary component of S' is capped by a (unmarked) disk. The kernel in that case is isomorphic to the fundamental group of the unit tangent bundle of S'; see Section 4.2.

3.6.3 THE CUTTING HOMOMORPHISM

The next geometric operation we consider is the following. Let α be an essential simple closed curve in a surface S. We can delete α from S in order to obtain a surface $S - \alpha$ that has two more punctures than S does. For example, if S has no boundary, then $S - \alpha$ can be identified with the interior of the surface obtained by cutting S along α.

Let a denote the isotopy class of α and let $\mathrm{Mod}(S, a)$ denote the stabilizer in $\mathrm{Mod}(S)$ of a. We would like to show that there is a well-defined homomorphism $\mathrm{Mod}(S, a) \to \mathrm{Mod}(S - \alpha)$. There is an obvious map: given $f \in \mathrm{Mod}(S, a)$, choose a representative ϕ that fixes α. The homeomorphism ϕ restricts to a homeomorphism of $S - \alpha$ and hence gives an element of $\mathrm{Mod}(S - \alpha)$. In order to show that this map $\zeta : \mathrm{Mod}(S, a) \to \mathrm{Mod}(S - \alpha)$ is well defined, we need to show that if two homeomorphisms of the pair (S, α) are homotopic as homeomorphisms of S, then they are homotopic through homeomorphisms that fix α. We now show that this is indeed the case.

Proposition 3.20 (The cutting homomorphism) *Let S be a closed surface with finitely many marked points. Let $\alpha_1, \ldots, \alpha_n$ be a collection of pairwise disjoint, homotopically distinct essential simple closed curves in S. There is a well-defined homomorphism*

$$\zeta : \mathrm{Mod}(S, \{[\alpha_1], \ldots, [\alpha_n]\}) \to \mathrm{Mod}(S - \cup \alpha_i)$$

with kernel $\langle T_{\alpha_1}, \ldots, T_{\alpha_n} \rangle$.

Proof. It is clear that the map ζ defined above is a homomorphism as long as it is well defined. Thus, we only need to show that ζ is well defined.

Let N be an open regular neighborhood of $\cup \alpha_i$. The inclusion $S - N \to S$ induces a homomorphism $\eta_1 : \mathrm{Mod}(S - N) \to \mathrm{Mod}(S)$. The map η_1 surjects onto $\mathrm{Mod}(S, \{[\alpha_1], \ldots, [\alpha_n]\})$, and by Theorem 3.18 its kernel K_1 is generated by elements $T_{\alpha_i^+} T_{\alpha_i^-}^{-1}$, where α_i^+ and α_i^- are the two boundary components of N that are isotopic to α_i in S.

Let $\overline{S - N}$ denote the surface obtained from $S - N$ by capping each boundary component with a punctured disk. The surface $\overline{S - N}$ is naturally homeomorphic to $S - \cup \alpha_i$, and thus there is a canonical isomorphism $\tau : \mathrm{Mod}(\overline{S - N}) \to \mathrm{Mod}(S - \cup \alpha_i)$.

By Theorem 3.18, the kernel of the homomorphism $\eta_2 : \mathrm{Mod}(S - N) \to \mathrm{Mod}(\overline{S - N})$ is the group K_2 generated by the $T_{\alpha_i^+}$ and $T_{\alpha_i^-}$.

We consider the following diagram.

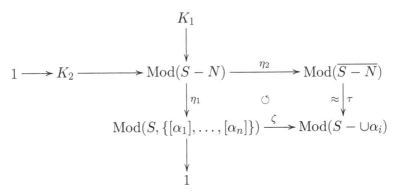

Since $K_1 < K_2$, it follows that $\tau \circ \eta_2 \circ \eta_1^{-1}$ is well defined. But this composition is nothing other than the map ζ defined above, and so we are done.
$\square$

3.6.4 COMPUTATIONS OF MAPPING CLASS GROUPS VIA CAPPING

We can use Proposition 3.19 to determine the mapping class groups of some surfaces with boundary.

Let P denote a *pair of pants*, that is, a compact surface of genus 0 with three boundary components (and no marked points). Recall from Proposition 2.3 that $\mathrm{PMod}(S_{0,3}) = 1$. Starting from this fact and applying Proposition 3.19 three times, we obtain the isomorphism

$$\mathrm{Mod}(P) \approx \mathbb{Z}^3.$$

Let S_1^1 denote a torus minus an open disk. We will show that

$$\mathrm{Mod}(S_1^1) \approx \widetilde{\mathrm{SL}(2,\mathbb{Z})},$$

where $\widetilde{\mathrm{SL}(2,\mathbb{Z})}$ denotes the universal central extension of $\mathrm{SL}(2,\mathbb{Z})$. We will need the following group presentations (see [195, Section 1.5]):

$$\mathrm{SL}(2,\mathbb{Z}) \approx \langle a, b \,|\, aba = bab, \, (ab)^6 = 1 \rangle$$
$$\widetilde{\mathrm{SL}(2,\mathbb{Z})} \approx \langle a, b \,|\, aba = bab \rangle.$$

From these presentations one sees that there is a surjective homomorphism $\widetilde{\mathrm{SL}(2,\mathbb{Z})} \to \mathrm{SL}(2,\mathbb{Z})$ sending a to a and b to b with kernel $\langle (ab)^6 \rangle \approx \mathbb{Z}$. There are also homomorphisms $\widetilde{\mathrm{SL}(2,\mathbb{Z})} \to \mathrm{Mod}(S_1^1)$ and $\mathrm{SL}(2,\mathbb{Z}) \to \mathrm{Mod}(S_{1,1})$, where in each case the generators a and b map to the Dehn twists about the latitude and longitude curves. These maps fit into the following diagram of exact sequences, where each square commutes:

$$
\begin{array}{ccccccccc}
1 & \longrightarrow & \mathbb{Z} & \longrightarrow & \widetilde{\mathrm{SL}(2,\mathbb{Z})} & \longrightarrow & \mathrm{SL}(2,\mathbb{Z}) & \longrightarrow & 1 \\
& & \Big\downarrow{\approx} & & \Big\downarrow & & \Big\downarrow{\approx} & & \\
1 & \longrightarrow & \mathbb{Z} & \longrightarrow & \mathrm{Mod}(S_1^1) & \xrightarrow{\;\mathcal{C}ap\;} & \mathrm{Mod}(S_{1,1}) & \longrightarrow & 1
\end{array}
$$

The desired isomorphism follows from the five lemma.

We mention that the group $\widetilde{\mathrm{SL}(2,\mathbb{Z})}$ is also isomorphic to the braid group on three strands (see Chapter 9) and the fundamental group of the complement of the trefoil knot in S^3, as well as the local fundamental group of the ordinary cusp singularity, that is, the fundamental group of the complement in $\mathbb{C}^2$ of the affine curve $x^2 = y^3$.

Chapter Four

Generating the Mapping Class Group

Is there a way to generate all (homotopy classes of) homeomorphisms of a surface by compositions of simple-to-understand homeomorphisms? We have already seen that $\mathrm{Mod}(T^2)$ is generated by the Dehn twists about the latitude and longitude curves. Our next main goal will be to prove the following result.

THEOREM 4.1 (Dehn–Lickorish theorem) *For $g \geq 0$, the mapping class group $\mathrm{Mod}(S_g)$ is generated by finitely many Dehn twists about nonseparating simple closed curves.*

Theorem 4.1 can be likened to the theorem that for each $n \geq 2$ the group $\mathrm{SL}(n, \mathbb{Z})$ can be generated by finitely many elementary matrices. As with the linear case, Theorem 4.1 is fundamental to our understanding of $\mathrm{Mod}(S_g)$.

In 1938 Dehn proved that $\mathrm{Mod}(S_g)$ is generated by $2g(g-1)$ Dehn twists [51]. Mumford, building on Dehn's work, showed in 1967 that only Dehn twists about nonseparating curves were needed [164]. In 1964 Lickorish, apparently unaware of Dehn's work, gave an independent proof that $\mathrm{Mod}(S_g)$ is generated by the Dehn twists about the $3g-1$ nonseparating curves shown in Figure 4.5 below [131].

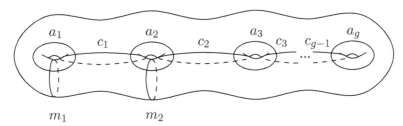

Figure 4.1 Dehn twists about these $2g + 1$ simple closed curves generate $\mathrm{Mod}(S_g)$.

In 1979 Humphries [101] proved the surprising theorem that the twists about the $2g + 1$ curves in Figure 4.1 suffice to generate $\mathrm{Mod}(S_g)$. These generators are often called the *Humphries generators*. Humphries further showed that any set of Dehn twist generators for $\mathrm{Mod}(S_g)$ must have at least $2g + 1$ elements; see Section 6.3 for a proof of this fact.

Punctures and pure mapping class groups. Theorem 4.1 is simply not true for surfaces with multiple punctures since no composition of Dehn twists can permute the punctures. Let $\mathrm{PMod}(S_{g,n})$ denote the *pure mapping class group* of $S_{g,n}$, which is defined to be the subgroup of $\mathrm{Mod}(S_{g,n})$ consisting of elements that fix each puncture individually. The action of $\mathrm{Mod}(S_{g,n})$ on the punctures of $S_{g,n}$ gives us a short exact sequence

$$1 \to \mathrm{PMod}(S_{g,n}) \to \mathrm{Mod}(S_{g,n}) \to \Sigma_n \to 1,$$

where Σ_n is the permutation group on the n punctures. We will show for any surface $S_{g,n}$ that $\mathrm{PMod}(S_{g,n})$ is finitely generated by Dehn twists (see Theorems 4.9 and 4.11). We will give a finite generating set for the full group $\mathrm{Mod}(S_{g,n})$ in Section 4.4.4.

In the case $n = 1$, we have $\mathrm{PMod}(S_{g,1}) = \mathrm{Mod}(S_{g,1})$. If we place a marked point at the rightmost point of S_g in Figure 4.1, we obtain a collection of curves in $S_{g,1}$. A slight modification of our proof of Theorem 4.1 will show that the corresponding Dehn twists form a generating set for $\mathrm{Mod}(S_{g,1})$.

Outline of the proof of Theorem 4.1. In proving Theorem 4.1, we will actually need to prove a more general statement. Precisely, we will prove that $\mathrm{PMod}(S_{g,n})$ is generated by finitely many Dehn twists about nonseparating simple closed curves for any $g \geq 1$ and $n \geq 0$ (Theorem 4.11 below).

We begin by giving a brief outline of the weaker statement that $\mathrm{PMod}(S_{g,n})$ is generated by the (infinite) collection of all Dehn twists about nonseparating simple closed curves. We do this in order to motivate two important tools: the complex of curves and the Birman exact sequence. Each of these tools is of independent interest and is introduced before the proof of Theorem 4.1.

The argument is a double induction on g and n with base case $S_{1,1}$.

Step 1: Induction on genus. Suppose $g \geq 2$ and let $f \in \mathrm{PMod}(S_{g,n})$. Let a be an arbitrary isotopy class of nonseparating simple closed curves in $S_{g,n}$. We want to show that there is a product h of Dehn twists about nonseparating curves in $S_{g,n}$ that takes $f(a)$ to a. For if this is the case, then we can regard hf as an element of the mapping class group of $S_{g-1,n+2}$, the surface obtained from $S_{g,n}$ by deleting a representative of a. Then we can apply induction on genus.

If we are fortunate enough that $i(a, f(a)) = 1$, then Proposition 3.12 gives that $T_{f(a)} T_a$ takes $f(a)$ to a, and we are done. In the general case, we just need to show that there is a sequence of isotopy classes of simple closed curves $a = c_1, \ldots, c_k = f(a)$ in $S_{g,n}$ so that $i(c_i, c_{i+1}) = 1$. This is exactly the content of Lemma 4.5. In the language of Section 4.1, this lemma is phrased in terms of the connectedness of a particular "modified complex of nonseparating curves."

Step 2: Induction on the number of punctures. Suppose $g \geq 1$ and $n \geq 1$. The inductive step on n reads as follows. There is a natural map $S_{g,n} \to S_{g,n-1}$ where one of the punctures/marked points is "forgotten" and this induces a surjective homomorphism $\mathrm{PMod}(S_{g,n}) \to \mathrm{PMod}(S_{g,n-1})$. Elements of the kernel come from "pushing" the nth puncture around the surface, and the Birman exact sequence (Theorem 4.6) identifies the kernel with $\pi_1(S_{g,n-1})$. We also show that generators for $\pi_1(S_{g,n-1})$ correspond to products of Dehn twists about nonseparating simple closed curves; see Fact 4.7. In other words the difference between $\mathrm{PMod}(S_{g,n})$ and $\mathrm{PMod}(S_{g,n-1})$ is (finitely) generated by products of Dehn twists about nonseparating curves, and so this completes the inductive step on the number of punctures.

We give the details of the proof of Theorem 4.11 in Section 4.3.

The word problem. Aside from his seminal work on the mapping class group, another of Max Dehn's highly influential contributions to mathematics is the idea of the word problem for a finitely generated group Γ. The *word problem* for Γ asks for an algorithm that takes as input any finite product w of elements from a fixed generating set for Γ (and their inverses) and as output tells whether or not w represents the identity element of Γ. It is a difficult result of Adian from the 1950s that there are finitely presented groups Γ with an unsolvable word problem; that is, no such algorithm for Γ as above exists. It is not difficult to prove that the (un)solvability of the word problem for a given group does not depend on the generating set.

Now consider $\mathrm{Mod}(S)$ with an explicit finite generating set, say for example the Humphries generators (see below). Suppose we are given any finite product w of these generators. We can choose a collection $\mathcal{C}$ of curves and arcs that fill S, and we can apply each generator in w to each curve and arc of $\mathcal{C}$. We can then use the bigon criterion and the Alexander method to determine whether the element of $\mathrm{Mod}(S)$ is trivial or not. Thus $\mathrm{Mod}(S)$ has a solvable word problem.

THEOREM 4.2 *Let* $S = S_{g,n}$. *The group* $\mathrm{Mod}(S)$ *has a solvable word problem.*

4.1 THE COMPLEX OF CURVES

The *complex of curves* $\mathcal{C}(S)$, defined by Harvey [88], is an abstract simplicial complex associated to a surface S. Its 1-skeleton is given by the following data.

> *Vertices.* There is one vertex of $\mathcal{C}(S)$ for each isotopy class of essential simple closed curves in S.
>
> *Edges.* There is an edge between any two vertices of $\mathcal{C}(S)$ corresponding to isotopy classes a and b with $i(a, b) = 0$.

More generally, $\mathcal{C}(S)$ has a k-simplex for each $(k + 1)$-tuple of vertices where each pair of corresponding isotopy classes has geometric intersection number zero. In other words, $\mathcal{C}(S)$ is a *flag complex*, which means that $k + 1$ vertices span a k-simplex of $\mathcal{C}(S)$ if and only if they are pairwise-connected by edges.[1] While we make use only of the 1-skeleton of $\mathcal{C}(S)$, the higher-dimensional simplices are useful in a number of applications (see, e.g., [107]).

Note that, as far as the complex of curves is concerned, a puncture has the same effect as a boundary component (simple closed curves that are homotopic to either a puncture or a boundary component are inessential). Therefore, we will deal only with punctured surfaces.

4.1.1 CONNECTIVITY OF THE COMPLEX OF CURVES

The following theorem, first stated by Harvey, was essentially proved by Lickorish (Figure 4.2 is his) [131]. Lickorish used it in the same way we will: to show that $\mathrm{Mod}(S)$ is finitely generated.

THEOREM 4.3 *If* $3g + n \geq 5$, *then* $\mathcal{C}(S_{g,n})$ *is connected.*

The hypothesis of Theorem 4.3 is equivalent to the condition that $\mathcal{C}(S_{g,n})$ has edges. In particular, Theorem 4.3 holds for every surface $S_{g,n}$ except when $g = 0$ and $n \leq 4$, or $g = 1$ and $n \leq 1$. We will discuss these sporadic cases below.

Theorem 4.3 can be rephrased as stating that for any two isotopy classes a and b of simple closed curves in $S_{g,n}$, there is a sequence of isotopy classes

$$a = c_1, \ldots, c_k = b$$

so that $i(c_i, c_{i+1}) = 0$.

---

[1] In other words, every nonsimplex contains a nonedge.

Proof. Suppose we are given two vertices $a, b \in \mathcal{C}(S_{g,n})$; thus a and b are isotopy classes of simple closed curves in $S_{g,n}$. We must find a sequence $a = c_1, \ldots, c_k = b$ with $i(c_i, c_{i+1}) = 0$. We induct on $i(a, b)$.

If $i(a, b) = 0$, then there is nothing to prove. If $i(a, b) = 1$, then we can find representatives α and β that intersect in precisely one point. A closed regular neighborhood of $\alpha \cup \beta$ is a torus with one boundary component. Denote by c the isotopy class of this boundary component. If c were not essential, that would mean that either $S_{g,n} \approx S_{1,1}$ or $S_{g,n} \approx T^2$, which violates the condition $3g + n \geq 5$. Therefore, a, c, b gives the desired path in $\mathcal{C}(S_{g,n})$.

For the inductive step we assume that $i(a, b) \geq 2$ and that any two simple closed curves with intersection number strictly less than $i(a, b)$ correspond to vertices that are connected by a path in $\mathcal{C}(S_{g,n})$. We now prove the inductive step by giving a recipe for finding an isotopy class c with both $i(c, a)$ and $i(c, b)$ less than $i(a, b)$.

Let α and β be simple closed curves in minimal position representing a and b. We consider two points of their intersection that are consecutive along β. We orient α and β so that it makes sense to talk about the index of an intersection point of α and β, be it $+1$ or -1.

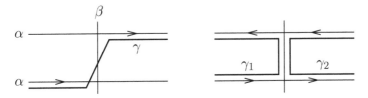

Figure 4.2 The surgered curves in the proof of Theorem 4.3.

If the two intersection points have the same index, then c can be chosen to be the class of γ shown in bold on the left-hand side of Figure 4.2 (outside the figure, γ follows along α). We see that γ is essential since $|\alpha \cap \gamma| = 1$. We emphasize that we construct γ so that, outside the local picture indicated in the figure, γ always lies just to the right of α; in particular, γ can be chosen so that it intersects β fewer times than α does (it "skips" one of the two intersections in the figure).

If the two intersection points have opposite indices, consider the (distinct) simple closed curves γ_1 and γ_2 shown in bold on the right-hand side of Figure 4.2. Neither γ_1 nor γ_2 can be null homotopic since that would mean that α and β were not in minimal position. If both γ_1 and γ_2 are homotopic to a puncture, it follows that α bounds a twice-punctured disk on one of its sides (the side containing γ_1 and γ_2). In this case there are similarly defined curves γ_3 and γ_4 on the other side of α. Again, neither γ_3 nor γ_4 can be null homotopic. Also, it cannot be that both γ_3 and γ_4 are peripheral because that would imply that $S_{g,n} \approx S_{0,4}$, violating the condition $3g + n \geq 5$. Thus we can choose c to be the class of either γ_3 or γ_4.

By construction, it is evident that $i(c, b) < i(a, b)$ and $i(c, a) < i(a, b)$ (in fact, $i(a, c)$ is either 0 or 1). By our inductive hypothesis, the vertices a and c are connected by a path in $\mathcal{C}(S_{g,n})$, and the vertices b and c are connected by a path. The concatenation of these paths is a path between the vertices a and b.   □

We point the reader to Ivanov's survey [107, Section 3.2], where he gives a beautiful alternative proof of Theorem 4.3 using Morse–Cerf theory. The key idea is that two simple closed curves that are level sets of the same Morse function are necessarily disjoint.

Sporadic cases and the Farey complex. In the cases of S^2, $S_{0,1}$, $S_{0,2}$, and $S_{0,3}$, the complex of curves is empty, and in the cases of T^2, $S_{1,1}$, and $S_{0,4}$ it a countable disjoint union of points. If we alter the definition of $\mathcal{C}(S)$ by assigning an edge to each pair of distinct vertices that realizes the minimal possible geometric intersection in the given surface, then the disconnected complexes become connected. In each of the latter three cases above, $\mathcal{C}(S)$ is isomorphic to the Farey complex, which is the ideal triangulation of $\mathbb{H}^2$ indicated in Figure 4.3.

Figure 4.3 The Farey complex.

The more classical description of the Farey complex is as follows. It is the flag complex where vertices correspond to cyclic subgroups of $\mathbb{Z}^2$, and two vertices span an edge if the corresponding primitive vectors span $\mathbb{Z}^2$.

4.1.2 THE COMPLEX OF NONSEPARATING CURVES

Let $\mathcal{N}(S)$ denote the subcomplex of $\mathcal{C}(S)$ spanned by vertices corresponding to nonseparating simple closed curves. This subcomplex is called the *complex of nonseparating curves*. This is an intermediate complex between the complex of curves and the modified complex of nonseparating curves $\widehat{\mathcal{N}}(S)$ (defined below), which is the complex that will actually be used in the proof of Theorem 4.1.

THEOREM 4.4 *If $g \geq 2$, then $\mathcal{N}(S_{g,n})$ is connected.*

Proof. We first prove the theorem for $g \geq 2$ and $n \leq 1$ and then use induction on n to obtain the rest of the cases. So let S be either S_g or $S_{g,1}$. If a and b are arbitrary isotopy classes of simple closed nonseparating simple closed curves in S, then by Theorem 4.3 there is a sequence of isotopy classes $a = c_1, \ldots, c_n = b$ with $i(c_i, c_{i+1}) = 0$.

We will alter the sequence $\{c_i\}$ so that it consists of isotopy classes of nonseparating simple closed curves. Suppose c_i is separating. Let γ_i be a simple closed curve representing c_i and let S' and S'' be the two components of $S_{g,n} - \gamma_i$. By the assumption that $g \geq 2$ and $n \leq 1$, both S' and S'' have positive genus. If c_{i-1} and c_{i+1} have representatives that lie in different subsurfaces, then $i(c_{i-1}, c_{i+1}) = 0$ and we can simply remove c_i from the sequence. If c_{i-1} and c_{i+1} have representatives that both lie in S', then we replace c_i with the isotopy class of a nonseparating simple closed curve in S''. We repeat the above process until each c_i is nonseparating, at which point we have obtained the desired path in $\mathcal{N}(S)$. This proves the theorem in the case $n \leq 1$.

For the induction on n we assume $n \geq 2$ and proceed as above. The only possible problem is that it might happen that representatives of c_{i-1} and c_{i+1} lie on S' and S'' has genus 0. But then S' has genus $g \geq 2$ and has fewer punctures than the original surface S, so by induction we can find a path in $\mathcal{N}(S')$ between the vertices corresponding to c_{i-1} and c_{i+1}, and we replace c_i by the corresponding sequence of isotopy classes of curves in S. $\square$

Theorem 4.4 is not true for any surface of genus 1. Indeed, the map $S_{1,n} \to T^2$ obtained by filling in the n punctures induces a surjective simplicial map $\mathcal{N}(S_{1,n}) \to \mathcal{C}(T^2)$, where the simplicial structure on $\mathcal{C}(T^2)$ is the original simplicial structure, which is disconnected.

4.1.3 A MODIFIED COMPLEX OF NONSEPARATING CURVES

Let $\widehat{\mathcal{N}}(S)$ denote the 1-dimensional simplicial complex whose vertices are isotopy classes of nonseparating simple closed curves in the surface S and whose edges correspond to pairs of isotopy classes a, b with $i(a, b) = 1$.

LEMMA 4.5 *If $g \geq 2$ and $n \geq 0$, then the complex $\widehat{\mathcal{N}}(S_{g,n})$ is connected.*

Proof. Let a and b be two isotopy classes of simple closed curves in $S_{g,n}$. By Theorem 4.4, there is a sequence of isotopy classes $a = c_1, \ldots, c_k = b$ representing vertices of $\widehat{\mathcal{N}}(S_{g,n})$ with $i(c_i, c_{i+1}) = 0$. By the change of coordinates principle, for each i one can find an isotopy class d_i of non-separating simple closed curves with $i(c_i, d_i) = i(d_i, c_{i+1}) = 1$. The sequence $a = c_1, d_1, c_2, \ldots, c_{k-1}, d_{k-1}, c_k = b$ represents the desired path in $\widehat{\mathcal{N}}(S_{g,n})$. $\square$

The conclusion of Lemma 4.5 also holds for any $S_{1,n}$ with $n \geq 0$. This can be proved by induction. The base cases are T^2 and $S_{1,1}$, where $\widehat{\mathcal{N}}(T^2) \approx \widehat{\mathcal{N}}(S_{1,1})$ is the 1-skeleton of the Farey complex. The inductive step on n is similar to the inductive step on punctures in the proof of Theorem 4.4.

4.2 THE BIRMAN EXACT SEQUENCE

As mentioned above, the proof of Theorem 4.1 will be a double induction on genus and the number of punctures. The Birman exact sequence will provide the inductive step for the number of punctures. More generally, it is a basic tool in the study of mapping class groups.

4.2.1 THE POINT-PUSHING MAP, THE FORGETFUL MAP, AND THE BIRMAN EXACT SEQUENCE

Let S be any surface, possibly with punctures (but no marked points) and let (S, x) denote the surface obtained from S by marking a point x in the interior of S. There is a natural homomorphism

$$\mathcal{F}orget : \mathrm{Mod}(S, x) \to \mathrm{Mod}(S)$$

called the *forgetful map*. This map is realized by forgetting that the point x is marked. The forgetful map is clearly surjective: given any homeomorphism of S, we can modify it by isotopy so that it fixes x.

The group $\mathrm{Mod}(S, x)$ is isomorphic to the subgroup G of $\mathrm{Mod}(S - x)$ preserving the puncture coming from x. The forgetful map can be interpreted as the map $G \to \mathrm{Mod}(S)$ obtained by "filling in" the puncture x. In other words, $\mathcal{F}orget$ is the map induced by the inclusion $S - x \to S$.

We would like to describe the kernel of $\mathcal{F}orget$. Let $f \in \mathrm{Mod}(S, x)$ be an element of the kernel of $\mathcal{F}orget$ and let ϕ be a homeomorphism representing f. We can think of ϕ as a homeomorphism $\overline{\phi}$ of S. Since $\mathcal{F}orget(f) = 1$, there is an isotopy from $\overline{\phi}$ to the identity map of S. During this isotopy, the image of the point x traces out a loop α in S based at x. What we will show is that by pushing x along α^{-1} we can recover $f \in \mathrm{Mod}(S, x)$.

Now to make the idea of pushing more precise. Let α be a loop in S based at x. We can think of $\alpha : [0, 1] \to S$ as an "isotopy of points" from x to itself, and this isotopy can be extended to an isotopy of the whole surface S (this is the 0-dimensional version of Proposition 1.11). Let ϕ_α be the homeomorphism of S obtained at the end of the isotopy. By regarding ϕ_α as a homeomorphism of (S, x), and then taking its isotopy class, we obtain a mapping class $\mathcal{P}ush(\alpha) \in \mathrm{Mod}(S, x)$. The way we think of $\mathcal{P}ush(\alpha)$ informally is that we place our finger on x and push x along α, dragging the rest of the surface along as we go.

What one would like of course is for the mapping class $\mathcal{P}ush(\alpha)$ to be well defined, that is, not to depend on the choice of the isotopy extension. One would also want $\mathcal{P}ush(\alpha)$ to not be dependent on the choice of α within its homotopy class. In other words, one hopes to have a well-defined *push map*[2]

$$\mathcal{P}ush : \pi_1(S, x) \to \mathrm{Mod}(S, x).$$

It turns out that this is indeed the case. But it is not obvious at all. To begin with, there is no way in general to extend a homotopy of a loop to a homotopy of a surface (rather, only isotopies can be extended). More to the point, what if we modify α by a homotopy that passes the loop over the marked point x? There is certainly no obvious way to show that the corresponding homeomorphisms of the marked surface (S, x) are homotopic.

The Birman exact sequence gives that the point-pushing map is indeed well defined and that its image is exactly the kernel of the forgetful map.

THEOREM 4.6 (Birman exact sequence) *Let S be a surface with $\chi(S) <$ 0, possibly with punctures and/or boundary. Let (S, x) be the surface obtained from S by marking a point x in the interior of S. Then the following*

---

[2]Birman's original terminology was "spin map."

sequence is exact:

$$1 \longrightarrow \pi_1(S, x) \overset{\mathcal{P}ush}{\longrightarrow} \mathrm{Mod}(S, x) \overset{\mathcal{F}orget}{\longrightarrow} \mathrm{Mod}(S) \longrightarrow 1.$$

Once we know that $\mathcal{P}ush$ is well defined, it follows immediately from the definitions that its image is contained in the kernel of the map $\mathcal{F}orget$ and that it surjects onto the kernel of $\mathcal{F}orget$. Also, it is easy to see that $\mathcal{P}ush$ is injective for $\chi(S) < 0$. Indeed, any representative $\mathcal{P}ush(\alpha) \in \mathrm{Mod}(S, x)$ can be thought of as a map of pairs $(S, x) \to (S, x)$ whose induced automorphism of $\pi_1(S, x)$ is the inner automorphism I_α. Since $\pi_1(S)$ is centerless, we have that I_α is nontrivial whenever α is. Thus if α is nontrivial, then the homeomorphism $\phi_\alpha : (S, x) \to (S, x)$ defined above is not homotopic to the identity as a map of pairs, from which it is immediate that $\mathcal{P}ush(\alpha)$ is nontrivial as an element of $\mathrm{Mod}(S, x)$. In summary, the entire content of Theorem 4.6 is that $\mathcal{P}ush$ is well defined.

We remark that Theorem 4.6 still holds if we replace Mod with $\mathrm{Mod}^\pm$, the extended mapping class group (see Chapter 8).

Also, we can take the restriction of the sequence to any subgroup of $\mathrm{Mod}(S, x)$. The most commonly used restriction is to $\mathrm{PMod}(S, x)$. In this case, $\mathrm{Mod}(S)$ should be replaced with $\mathrm{PMod}(S)$. We can rephrase the Birman exact sequence in this case as follows:

$$1 \to \pi_1(S_{g,n}) \to \mathrm{PMod}(S_{g,n+1}) \to \mathrm{PMod}(S_{g,n}) \to 1.$$

We will show in Section 5.5 that the Birman exact sequence does not split.

A small technical point. Since products in $\mathrm{Mod}(S, x)$ are usually written right to left and products in $\pi_1(S, x)$ are usually written left to right, we should define the map $\pi_1(S, x) \to \mathrm{Mod}(S, x)$ by sending α to the map that pushes x along α^{-1}, not α (otherwise we would obtain an anti-homomorphism instead of a homomorphism). This issue will not play a role in this book.

4.2.2 PUSH MAPS ALONG LOOPS IN TERMS OF DEHN TWISTS

For a simple loop α in S based at the point x, we can give an explicit representative of $\mathcal{P}ush(\alpha)$ as follows. Identify a neighborhood of α with the annulus $S^1 \times [0, 2]$. We orient $S^1 \times [0, 2]$ via the standard orientations on S^1 and $[0, 2]$. Say the marked point x is at the point $(0, 1)$ in this annulus. There is an isotopy of the annulus given by

$$F((\theta, r), t) = \begin{cases} (\theta + 2\pi rt, r) & 0 \le r \le 1, \\ (\theta + 2\pi(2 - r)t, r) & 1 \le r \le 2. \end{cases}$$

We can extend F by the identity to get an isotopy of S. When we restrict F to $\{x\} \times [0, 1]$, we get

$$F((0, 1), t) = (2\pi t, 1).$$

In other words, the isotopy F pushes x around the core of the annulus. Also, the homeomorphism ϕ of (S, x) induced by F at $t = 1$ is a product of two Dehn twists. More precisely, identifying the boundary curve $S^1 \times \{0\}$ of the annulus as a simple closed curve a in (S, x) and identifying $S^1 \times \{2\}$ as a curve b in (S, x), we have that ϕ is (isotopic to) $T_a T_b^{-1}$. A smooth representative of $\mathcal{P}ush(\alpha)$ is shown in Figure 4.4. We summarize this discussion as follows.

Fact 4.7 *Let α be a simple loop in a surface S representing an element of $\pi_1(S, x)$. Then*

$$\mathcal{P}ush([\alpha]) = T_a T_b^{-1},$$

where a and b are the isotopy classes of the simple closed curves in (S, x) obtained by pushing α off itself to the left and right, respectively. The isotopy classes a and b are nonseparating in (S, x) if and only if α is nonseparating in S.

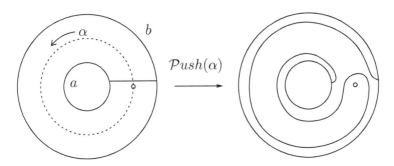

Figure 4.4 The point-pushing map $\mathcal{P}ush$ from the Birman exact sequence.

Naturality. We record the following naturality property for the point-pushing map.

Fact 4.8 *For any $h \in \mathrm{Mod}(S, x)$ and any $\alpha \in \pi_1(S, x)$, we have*

$$\mathcal{P}ush(h_*(\alpha)) = h\mathcal{P}ush(\alpha)h^{-1}.$$

Fact 4.8 follows immediately from the definitions.

4.2.3 THE PROOF

We now give the proof of the existence of the Birman exact sequence.

Proof of Theorem 4.6. There is a fiber bundle

$$\mathrm{Homeo}^+(S, x) \to \mathrm{Homeo}^+(S) \xrightarrow{\mathcal{E}} S \qquad (4.1)$$

with total space $\mathrm{Homeo}^+(S)$, with base space S (i.e., the configuration space of a single point in S), and with fiber the subgroup of $\mathrm{Homeo}^+(S)$ consisting of elements that fix the point x (technically, we should allow only homeomorphisms that fix ∂S pointwise, but this does not affect the proof). The map $\mathcal{E}$ is evaluation at the point x.

We now explain why $\mathcal{E} : \mathrm{Homeo}^+(S) \to S$ is a fiber bundle, that is, why $\mathrm{Homeo}^+(S)$ is locally homeomorphic to a product of an open set U of S with $\mathrm{Homeo}^+(S, x)$ so that the restriction of $\mathcal{E}$ is projection to the first factor. Let U be some open neighborhood of x in S that is homeomorphic to a disk. Given $u \in U$, we can choose a $\phi_u \in \mathrm{Homeo}^+(U)$ so that $\phi_u(x) = u$ and so that ϕ_u varies continuously as a function of u. We have a homeomorphism $U \times \mathrm{Homeo}^+(S, x) \to \mathcal{E}^{-1}(U)$ given by

$$(u, \psi) \mapsto \phi_u \circ \psi.$$

The inverse map is given by $\psi \mapsto (\psi(x), \phi_{\psi(x)}^{-1} \circ \psi)$. For any other point $y \in S$, we can choose a homeomorphism ξ of S taking x to y. Then there is a homeomorphism $\mathcal{E}^{-1}(U) \to \mathcal{E}^{-1}(\xi(U))$ given by $\psi \mapsto \xi \circ \psi$, and so we have verified the fiber bundle property.

The theorem now follows from the long exact sequence of homotopy groups associated to the above fiber bundle. The relevant part of the sequence is the following.

$$\cdots \to \pi_1(\mathrm{Homeo}^+(S)) \to \pi_1(S) \to \pi_0(\mathrm{Homeo}^+(S, x))$$
$$\to \pi_0(\mathrm{Homeo}^+(S)) \to \pi_0(S) \to \cdots .$$

By Theorem 1.14 the group $\pi_1(\mathrm{Homeo}^+(S))$ is trivial, and of course $\pi_0(S)$ is trivial. The remaining terms are isomorphic to the terms of the Birman exact sequence.

Finally, the maps given by the long exact sequence of homotopy groups are exactly the point-pushing map $\mathcal{P}ush$ and the forgetful map $\mathcal{F}orget$. $\square$

There is a version of Theorem 4.6 where one forgets multiple punctures instead of a single version; see Chapter 9. However, in most cases, one can simply apply Theorem 4.6 iteratively in order to forget one puncture at a time.

Surfaces with $\chi(S) \geq 0$. In the proof of Theorem 4.6, we used the assumption that $\chi(S) < 0$ in order to say that $\pi_1(\mathrm{Homeo}^+(S)) = 1$. But we can still use the long exact sequence coming from the fiber bundle (4.1) for other surfaces. For instance, for the torus T^2 we have $\pi_1(\mathrm{Homeo}^+(T^2)) \approx \pi_1(T^2) \approx \mathbb{Z}^2$, and the relevant part of the short exact sequence becomes

$$\cdots \to \mathbb{Z}^2 \xrightarrow{\mathrm{id}} \mathbb{Z}^2 \xrightarrow{0} \mathrm{Mod}(S_{1,1}) \to \mathrm{Mod}(T^2) \to 1 \to \cdots .$$

This gives another proof that $\mathrm{Mod}(S_{1,1}) \approx \mathrm{Mod}(T^2)$.

4.2.4 GENERATING $\mathrm{Mod}(S_{0,n})$

Let $S_{0,n}$ be a sphere with n punctures. As per Section 2.2, $\mathrm{PMod}(S_{0,n}) = 1$ for $n \leq 3$. To understand the situation for more punctures, we can apply the Birman exact sequence:

$$1 \to \pi_1(S_{0,3}) \to \mathrm{PMod}(S_{0,4}) \to \mathrm{PMod}(S_{0,3}) \to 1.$$

Since $\pi_1(S_{0,3}) \approx F_2$, we obtain that $\mathrm{PMod}(S_{0,4}) \approx F_2$. Moreover, the Birman exact sequence gives geometric meaning to this algebraic statement: elements of $\pi_1(S_{0,3})$ represented by simple loops map to Dehn twists in $\mathrm{PMod}(S_{0,4})$, and so the standard generating set for $\pi_1(S_{0,3})$ gives a generating set for $\mathrm{PMod}(S_{0,4})$ consisting of two Dehn twists about simple closed curves with geometric intersection number 2.

We can increase the number of punctures using the Birman exact sequence:

$$1 \to \pi_1(S_{0,4}) \to \mathrm{PMod}(S_{0,5}) \to \mathrm{PMod}(S_{0,4}) \to 1.$$

Since $\pi_1(S_{0,4}) \approx F_3$ and $\mathrm{PMod}(S_{0,4}) \approx F_2$, we obtain $\mathrm{PMod}(S_{0,5}) \approx F_2 \ltimes F_3$. Inductively, we see that $\mathrm{PMod}(S_{0,n})$ is an iterated extension of free groups. Applying Fact 4.7, plus the fact that $\pi_1(S_{0,n})$ is generated by simple loops, we find the following.

THEOREM 4.9 *For $n \geq 0$, the group $\mathrm{PMod}(S_{0,n})$ is generated by finitely many Dehn twists.*

To generate all of $\mathrm{Mod}(S_{0,n})$, we again apply the following exact sequence:

$$1 \to \mathrm{PMod}(S_{0,n}) \to \mathrm{Mod}(S_{0,n}) \to \Sigma_n \to 1.$$

It follows that a generating set for $\mathrm{Mod}(S_{0,n})$ is obtained from a generating set for $\mathrm{PMod}(S_{0,n})$ by adding lifts of generators for Σ_n. We know that Σ_n

is generated by transpositions. A simple lift of a transposition is a half-twist, defined in Chapter 9.

4.2.5 CAPPING THE BOUNDARY

By souping up the proof of the Birman exact sequence, we can give another perspective on the boundary capping sequence (Proposition 3.19) that unifies it with the Birman exact sequence.

Let S° be a surface with nonempty boundary and let $\widehat{S}$ be the surface obtained from S° by capping some component β of ∂S° with a disk. Let p be some point in the interior of this disk. As in Proposition 3.19, we have a short exact sequence

$$1 \to \langle T_\beta \rangle \to \mathrm{Mod}(S^\circ) \overset{Cap}{\to} \mathrm{Mod}(\widehat{S}, p) \to 1. \qquad (4.2)$$

Note that $\langle T_\beta \rangle$ is central in $\mathrm{Mod}(S^\circ)$ since any element of $\mathrm{Mod}(S^\circ)$ has a representative that is the identity in a neighborhood of ∂S°.

We now give our second proof of Proposition 3.19 using the notation from the sequence (4.2).

Second proof of Proposition 3.19. The proof has two steps. Step 1 is to identify $\mathrm{Mod}(S^\circ)$ with a different group and to reinterpret the capping map in the new context, and Step 2 is to apply the method of proof of the Birman exact sequence to the corresponding fiber bundle.

Step 1. Let (p, v) be a point of the unit tangent bundle $UT(\widehat{S})$ that lies in the fiber above p. Let $\mathrm{Diff}^+(\widehat{S}, (p, v))$ denote the group of orientation-preserving diffeomorphisms of $\widehat{S}$ fixing (p, v). The resulting mapping class group, denoted $\mathrm{Mod}(\widehat{S}, (p, v))$, is defined as $\pi_0(\mathrm{Diff}^+(\widehat{S}, (p, v)))$. We claim that there is an isomorphism

$$\mathrm{Mod}(S^\circ) \approx \mathrm{Mod}(\widehat{S}, (p, v)).$$

To prove this isomorphism we first identify $\mathrm{Mod}(S^\circ)$ with $\pi_0(\mathrm{Diff}^+(\widehat{S}, D))$, where D is the boundary capping disk, and diffeomorphisms are taken to fix D pointwise. This identification can be realized by simply removing the interior of D. There is a fiber bundle

$$\mathrm{Diff}^+(\widehat{S}, D) \to \mathrm{Diff}^+(\widehat{S}, (p, v)) \to \mathrm{Emb}^+((D, \widehat{S}), (p, v)),$$

where $\mathrm{Emb}^+((D, \widehat{S}), (p, v))$ is the space of smooth, orientation-preserving embeddings of D into $\widehat{S}$ taking some fixed unit tangent vector in D to the tangent vector (p, v). As in the proof of the Birman exact sequence, we obtain a long exact sequence of homotopy groups that contains the sequence

$$\cdots \to \pi_1(\mathrm{Emb}^+((D, \widehat{S}), (p, v))) \to \pi_0(\mathrm{Diff}^+(\widehat{S}, D))$$
$$\to \pi_0(\mathrm{Diff}^+(\widehat{S}, (p, v))) \to \pi_0(\mathrm{Emb}^+((D, \widehat{S}), (p, v))) \to \cdots .$$

Since D is contractible, the space $\mathrm{Emb}^+((D, \widehat{S}), (p, v)))$ is contractible, and so we obtain the claimed isomorphism $\mathrm{Mod}(\widehat{S}, (p, v)) \approx \mathrm{Mod}(S^\circ)$ (see [107, Theorem 2.6D] and [45]).

The projection map $(p, v) \mapsto p$ induces a map $\mathrm{Mod}(\widehat{S}, (p, v)) \to \mathrm{Mod}(\widehat{S}, p)$ that makes the following diagram commute:

$$
\begin{array}{ccc}
\mathrm{Mod}(\widehat{S}, (p, v)) & & \\
{\scriptstyle \approx} \downarrow & \searrow & \\
\mathrm{Mod}(S^\circ) & \xrightarrow{\ Cap\ } & \mathrm{Mod}(\widehat{S}, p)
\end{array}
$$

Thus we have succeeded in writing the map Cap in terms of $\mathrm{Mod}(\widehat{S}, (p, v))$.

Step 2. We have another fiber bundle

$$\mathrm{Diff}^+(\widehat{S}, (p, v)) \to \mathrm{Diff}^+(\widehat{S}, p) \to UT_p(\widehat{S}),$$

where the second map is the evaluation map onto the fiber over p of the unit tangent bundle of $\widehat{S}$. As in the proof of the Birman exact sequence, we obtain a long exact sequence, part of which is

$$\cdots \to \pi_1(\mathrm{Diff}^+(\widehat{S}, p)) \to \pi_1(UT_p(\widehat{S})) \to \pi_0(\mathrm{Diff}^+(\widehat{S}, (p, v)))$$
$$\to \pi_0(\mathrm{Diff}^+(\widehat{S}, p)) \to \pi_0(UT_p(\widehat{S})) \to \cdots .$$

These terms exactly give the desired short exact sequence.   □

Not only is the last proof similar to the proof of the Birman exact sequence, but both proofs can actually be combined to give the following diagram, which encapsulates the two points of view. In the diagram all sequences are exact and all squares commute.

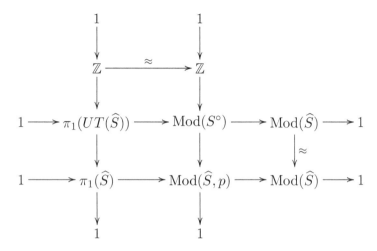

To get the middle row directly, one can consider the fiber bundle

$$\mathrm{Diff}^+(\widehat{S}, (p, v)) \to \mathrm{Diff}^+(\widehat{S}) \to UT(\widehat{S}).$$

4.3 PROOF OF FINITE GENERATION

To show that $\mathrm{Mod}(S)$ is finitely generated we consider its action on complex $\widehat{\mathcal{N}}(S)$. Note that $\mathrm{Mod}(S)$ indeed acts on $\widehat{\mathcal{N}}(S)$ since homeomorphisms take nonseparating simple closed curves to nonseparating simple closed curves and homeomorphisms preserve geometric intersection number. It is a basic principle from geometric group theory that if a group G acts cellularly on a connected cell complex X and if D is a subcomplex of X whose G-translates cover X, then G is generated by the set $\{g \in G : gD \cap D \neq \emptyset\}$ (this idea will be echoed in our proof of Theorem 8.2 below). The next lemma is a specialized version of this fact designed specifically so that we can apply it to the action of $\mathrm{Mod}(S)$ on $\widehat{\mathcal{N}}(S)$.

Lemma 4.10 *Suppose that a group G acts by simplicial automorphisms on a connected, 1-dimensional simplicial complex X. Suppose that G acts transitively on the vertices of X and that it also acts transitively on pairs of vertices of X that are connected by an edge. Let v and w be two vertices of X that are connected by an edge and choose $h \in G$ so that $h(w) = v$. Then the group G is generated by the element h together with the stabilizer of v in G.*

Proof. Let $g \in G$. We would like to show that g is contained in the subgroup

$H < G$ generated by the stabilizer of v together with the element h. Since X is connected, there is a sequence of vertices

$$v = v_0, \ldots, v_k = g(v)$$

where adjacent vertices are connected by an edge. Since G acts transitively on the vertices of X, we can choose elements g_i of G so that $g_i(v) = v_i$. We take g_0 to be the identity and g_k to be g. We will prove by induction that $g_i \in H$. The base case $g_0 \in H$ clearly holds. Now assume that $g_i \in H$. We must prove that $g_{i+1} \in H$.

Applying the element g_i^{-1} to the edge between $v_i = g_i(v)$ and $v_{i+1} = g_{i+1}(v)$, we obtain the edge between v and $g_i^{-1}g_{i+1}(v)$. Since G acts transitively on ordered pairs of vertices of X that are connected by an edge, there is an element $r \in G$ that takes the pair $(v, g_i^{-1}g_{i+1}(v))$ to the pair (v, w). In particular, r lies in the stabilizer of v and $rg_i^{-1}g_{i+1}(v) = w$. We then have that $hrg_i^{-1}g_{i+1}(v) = v$, which means that $hrg_i^{-1}g_{i+1}$ lies in the stabilizer of v. In particular, $hrg_i^{-1}g_{i+1} \in H$. Since h and r lie in H by the definition of H and since g_i lies in H by induction, we have that g_{i+1} lies in H. In particular, $g_k = g$ lies in H, which is what we wanted to show. $\qquad \square$

We are now ready to prove the following theorem, which contains Theorem 4.1 as the special case $n = 0$.

THEOREM 4.11 *Let $S_{g,n}$ be a surface of genus $g \geq 1$ with $n \geq 0$ punctures. Then the group $\mathrm{PMod}(S_{g,n})$ is finitely generated by Dehn twists about nonseparating simple closed curves in $S_{g,n}$.*

Recall that we already showed that $\mathrm{PMod}(S_{0,n})$ is finitely generated by Dehn twists for $n \geq 0$ (Theorem 4.9).

Proof. We will use double induction on genus and the number of punctures of S, with base cases $T^2 = S_{1,0}$ and $S_{1,1}$.

We start with the inductive step on the number of punctures. Let $g \geq 1$ and let $n \geq 0$. Assuming that $\mathrm{PMod}(S_{g,n})$ is generated by finitely many Dehn twists about nonseparating simple closed curves $\{\alpha_i\}$ in $S_{g,n}$, we will show that $\mathrm{PMod}(S_{g,n+1})$ is generated by finitely many Dehn twists about nonseparating curves in $S_{g,n+1}$. We may assume that $(g, n) \neq (1, 0)$ since we know that $\mathrm{Mod}(S_{1,1}) \approx \mathrm{Mod}(T^2)$ is generated by Dehn twists about nonseparating simple closed curves.

We have the Birman exact sequence

$$1 \to \pi_1(S_{g,n}) \to \mathrm{PMod}(S_{g,n+1}) \to \mathrm{PMod}(S_{g,n}) \to 1.$$

Since $g \geq 1$, we have that $\pi_1(S_{g,n})$ is generated by the classes of finitely many simple nonseparating loops. By Fact 4.7, the image of each of these loops is a product of two Dehn twists about nonseparating simple closed curves. We begin building a generating set for $\mathrm{PMod}(S_{g,n+1})$ by taking each of these Dehn twists individually. In order to complete the generating set, it remains to choose a lift to $\mathrm{PMod}(S_{g,n+1})$ of each Dehn twist generator T_{α_i} of $\mathrm{PMod}(S_{g,n})$. But given the nonseparating simple curve α_i in $S_{g,n}$, there exists a nonseparating curve in $S_{g,n+1}$ that maps to α_i under the forgetful map $S_{g,n+1} \to S_{g,n}$. Thus the Dehn twist T_{α_i} in $\mathrm{PMod}(S_{g,n})$ has a preimage in $\mathrm{PMod}(S_{g,n+1})$ that is a Dehn twist about a nonseparating simple closed curve in $S_{g,n+1}$. This completes the inductive step on the number of punctures.

Since we know that $\mathrm{Mod}(T^2)$ and $\mathrm{Mod}(S_{1,1})$ are each generated by two Dehn twists about nonseparating simple closed curves (Section 2.2), it follows from the inductive step on the number of punctures that, for any $n \geq 0$, the group $\mathrm{PMod}(S_{1,n})$ is generated by finitely many Dehn twists about nonseparating simple closed curves.

We now attack the inductive step on the genus g. Let $g \geq 2$ and assume that $\mathrm{PMod}(S_{g-1,n})$ is finitely generated by Dehn twists about nonseparating simple closed curves for any $n \geq 0$. Since $\widehat{\mathcal{N}}(S_g)$ is connected (Lemma 4.5) and since by the change of coordinates principle $\mathrm{Mod}(S_g)$ acts transitively on ordered pairs of isotopy classes of simple closed curves with geometric intersection number 1, we may apply Lemma 4.10 to the case of the $\mathrm{Mod}(S_g)$ action on $\widehat{\mathcal{N}}(S_g)$.

Let a be an arbitrary isotopy class of nonseparating simple closed curves in S_g and let b be an isotopy class with $i(a,b) = 1$. Let $\mathrm{Mod}(S_g, a)$ denote the stabilizer in $\mathrm{Mod}(S_g)$ of a. By Proposition 3.12, we have $T_b T_a(b) = a$. Thus, by Lemma 4.10, $\mathrm{Mod}(S_g)$ is generated by $\mathrm{Mod}(S_g, a)$ together with T_a and T_b. Thus it suffices to show that $\mathrm{Mod}(S_g, a)$ is finitely generated by Dehn twists about nonseparating simple closed curves.

Let $\mathrm{Mod}(S_g, \vec{a})$ be the subgroup of $\mathrm{Mod}(S_g, a)$ consisting of elements that preserve the orientation of a. We have the short exact sequence

$$1 \to \mathrm{Mod}(S_g, \vec{a}) \to \mathrm{Mod}(S_g, a) \to \mathbb{Z}/2\mathbb{Z} \to 1.$$

Since $T_b T_a^2 T_b$ switches the orientation of a (use change of coordinates), it represents the nontrivial coset of $\mathrm{Mod}(S_g, \vec{a})$ in $\mathrm{Mod}(S_g, a)$. Thus it remains to show that $\mathrm{Mod}(S_g, \vec{a})$ is finitely generated by Dehn twists about nonseparating simple closed curves in S_g.

By Proposition 3.20 we have a short exact sequence

$$1 \to \langle T_a \rangle \to \mathrm{Mod}(S_g, \vec{a}) \to \mathrm{PMod}(S_g - \alpha) \to 1,$$

where $S_g - \alpha$ is the surface obtained from S_g by deleting a representative α of a. The surface $S_g - \alpha$ is homeomorphic to $S_{g-1,2}$. By our inductive hypothesis, $\mathrm{PMod}(S_g - \alpha)$ is generated by finitely many Dehn twists about nonseparating simple closed curves. Since each such Dehn twist has a preimage in $\mathrm{Mod}(S_g, \vec{a})$ that is also a Dehn twist about a nonseparating simple closed curve, it follows that $\mathrm{Mod}(S_g, \vec{a})$ is generated by finitely many Dehn twists about nonseparating curves, and we are done. $\qquad\square$

4.4 EXPLICIT SETS OF GENERATORS

The goal of this section is to find an explicit finite set of Dehn twist generators for $\mathrm{Mod}(S)$. Our strategy for accomplishing this is to sharpen our proof that $\mathrm{Mod}(S)$ is generated by finitely many Dehn twists. More specifically, we choose a candidate set of generators and check that each step of the proof of finite generation can be achieved by using our candidate set.

4.4.1 THE CHAIN RELATION

In the very last step of our proof of Theorem 4.13 below, we will require the following relation between Dehn twists. Recall that a sequence of isotopy classes $c_1, \ldots, c_k$ in a surface S is called a *chain* if $i(c_i, c_{i+1}) = 1$ for all i and $i(c_i, c_j) = 0$ for $|i - j| > 1$.

Proposition 4.12 (Chain relation) *Let $k \geq 0$ and let $c_1, \cdots, c_k$ be a chain of curves in a surface S. If we take representatives for the c_i that are in minimal position and then take a closed regular neighborhood of their union, then the boundary of this neighborhood consists of one or two simple closed curves, depending on whether k is even or odd. Denote the isotopy classes of these boundary curves by d in the even case and by d_1 and d_2 in the odd case. Then the following relations hold in $\mathrm{Mod}(S)$:*

$$
\begin{aligned}
(T_{c_1} \cdots T_{c_k})^{2k+2} &= T_d & k \text{ even,} \\
(T_{c_1} \cdots T_{c_k})^{k+1} &= T_{d_1} T_{d_2} & k \text{ odd.}
\end{aligned}
$$

In each case the relation in Proposition 4.12 is called a *chain relation*, or a *k-chain relation*. The chain relation can be proved via the Alexander method. In Chapter 9, we will derive the chain relations as consequences of relations in the braid group.

The *2-chain relation* is a well-known example of the chain relation. In this case, the relation says that if $i(a, b) = 1$, then

$$(T_a T_b)^6 = T_d,$$

where d is the boundary of a regular neighborhood of $a \cup b$. If a and b lie in T^2 or $S_{1,1}$, then T_d is trivial, and we have the relation $(T_a T_b)^6 = 1$. Via the isomorphism of Theorem 2.5, this is simply the relation

$$\left(\begin{pmatrix} 1 & 1 \\ 0 & 1 \end{pmatrix} \begin{pmatrix} 1 & 0 \\ -1 & 1 \end{pmatrix} \right)^6 = 1$$

in $SL(2, \mathbb{Z})$.

There is another version of the chain relation that is sometimes useful. In the above notation, this other version reads

$$(T_{c_1}^2 T_{c_2} \cdots T_{c_k})^{2k} = T_d \quad \text{and} \quad (T_{c_1}^2 T_{c_2} \cdots T_{c_k})^k = T_{d_1} T_{d_2},$$

for k even and odd, respectively.

Dehn twists have roots. A surprising consequence of the last relation is that the Dehn twist about a nonseparating simple closed curve has a nontrivial root in $\mathrm{Mod}(S_g)$ when $g \geq 2$. If we consider a chain of simple closed curves $c_1, \ldots, c_{2g-1}$ in S_g, then the two boundary components of a regular neighborhood of $\cup c_i$ are nonseparating simple closed curves in the same isotopy class d, so we have

$$(T_{c_1}^2 T_{c_2} \cdots T_{c_{2g-1}})^{2g-1} = T_d^2.$$

Thus, since T_d commutes with each T_{c_i}, we have

$$[(T_{c_1}^2 T_{c_2} \cdots T_{c_{2g-1}})^{1-g} T_d]^{2g-1} = T_d.$$

McCullough–Rajeevsarathy proved that $2g-1$ is actually the largest order of a root of T_d for any $g \geq 2$ [147]. It is not difficult to see that Dehn twists about separating simple closed curves have roots: for example, if we imagine fixing the subsurface of S_g to one side of a separating curve d and twisting the other side by an angle π, then we get a square root of T_d. A more formal way to do this is to use the first chain relation with a chain of even length.

4.4.2 THE LICKORISH GENERATORS

Our eventual goal is to show that the Humphries generating set (see the beginning of the chapter) is indeed a generating set for $\mathrm{Mod}(S_g)$. As a first step we show that the Dehn twists about the $3g - 1$ simple closed curves indicated in Figure 4.5 generate $\mathrm{Mod}(S_g)$. This specific generating set was first found by Lickorish, and so we call these Dehn twists the *Lickorish generators* [131].

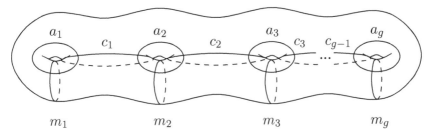

Figure 4.5 The Lickorish generating set for $\mathrm{Mod}(S)$.

THEOREM 4.13 (Lickorish generators) *For $g \geq 1$, the Dehn twists about the isotopy classes*

$$a_1, \ldots, a_g, m_1, \ldots, m_g, c_1, \ldots, c_{g-1}$$

shown in Figure 4.5 generate $\mathrm{Mod}(S_g)$.

In the proof of Theorem 4.13 we refer to the Dehn twists in the statement of the theorem as Lickorish twists, so as not to confuse the issue that we will be proving that they are indeed generators for $\mathrm{Mod}(S)$.

Proof. We proceed by induction on g. Since the Lickorish twists for the torus $T^2 \approx S_1$ are the standard generators for $\mathrm{Mod}(T^2)$, the theorem is true for the case of $g = 1$, and we may assume that $g \geq 2$.

We again apply Lemma 4.10 to the action of $\mathrm{Mod}(S_g)$ on the 1-dimensional simplicial complex $\widehat{\mathcal{N}}(S_g)$ from Section 4.1. By Lemma 3.12, we have $T_{a_1} T_{m_1} T_{a_1}(m_1) = a_1$. Thus by Lemma 4.10, it suffices to show that $\mathrm{Mod}(S_g, m_1)$, the stabilizer in $\mathrm{Mod}(S_g)$ of m_1, lies in the group generated by Lickorish twists.

If $\mathrm{Mod}(S_g, \vec{m}_1)$ is the subgroup of $\mathrm{Mod}(S_g)$ consisting of elements that preserve the orientation of m_1, then we have

$$1 \rightarrow \mathrm{Mod}(S_g, \vec{m}_1) \rightarrow \mathrm{Mod}(S_g, m_1) \rightarrow \mathbb{Z}/2\mathbb{Z} \rightarrow 1.$$

Since the product of Lickorish twists $T_{a_1} T_{m_1}^2 T_{a_1}$ reverses the orientation of m_1, it suffices to show that $\mathrm{Mod}(S_g, \vec{m}_1)$ lies in the group generated by the Lickorish twists.

By Proposition 3.20, we have the following exact sequence:

$$1 \rightarrow \langle T_{m_1} \rangle \rightarrow \mathrm{Mod}(S_g, \vec{m}_1) \rightarrow \mathrm{PMod}(S_{m_1}) \rightarrow 1,$$

where $S_{m_1} \approx S_{g-1,2}$ is the surface obtained by deleting a representative of m_1 from S_g (this is perhaps a slight abuse of notation since we usually write

S_{m_1} to mean the surface obtained from a surface S by cutting along a curve m_1). Since T_{m_1} is a Lickorish twist, it is enough to show that $\mathrm{PMod}(S_{m_1})$ is generated by the images of the Lickorish twists.

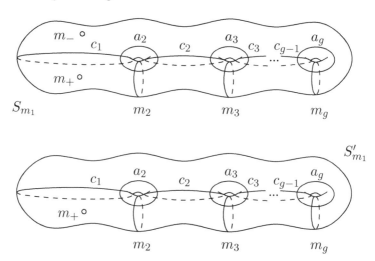

Figure 4.6 The images of the curves from Figure 4.5 in S_{m_1} and S'_{m_1}.

We apply the Birman exact sequence (Theorem 4.6) twice. Let S'_{m_1} denote the surface obtained from S_{m_1} by forgetting the first puncture m_- and let S''_{m_1} be the surface obtained from S'_{m_1} by forgetting the second puncture m_+. We then have the following maps of exact sequences where each square commutes:

$$
\begin{array}{ccccccccc}
1 & \longrightarrow & \pi_1(S'_{m_1}, m_-) & \xrightarrow{\mathit{Push}} & \mathrm{PMod}(S_{m_1}) & \longrightarrow & \mathrm{Mod}(S'_{m_1}) & \longrightarrow & 1 \\
 & & \Big\downarrow{\scriptstyle\approx} & & \Big\downarrow{\scriptstyle\approx} & & \Big\downarrow{\scriptstyle\approx} & & \\
1 & \longrightarrow & \pi_1(S_{g-1,1}) & \longrightarrow & \mathrm{PMod}(S_{g-1,2}) & \longrightarrow & \mathrm{Mod}(S_{g-1,1}) & \longrightarrow & 1
\end{array}
$$
(4.3)

and

$$
\begin{array}{ccccccccc}
1 & \longrightarrow & \pi_1(S''_{m_1}, m_+) & \xrightarrow{\mathit{Push'}} & \mathrm{Mod}(S'_{m_1}) & \longrightarrow & \mathrm{Mod}(S''_{m_1}) & \longrightarrow & 1 \\
 & & \Big\downarrow{\scriptstyle\approx} & & \Big\downarrow{\scriptstyle\approx} & & \Big\downarrow{\scriptstyle\approx} & & \\
1 & \longrightarrow & \pi_1(S_{g-1}) & \longrightarrow & \mathrm{Mod}(S_{g-1,1}) & \longrightarrow & \mathrm{Mod}(S_{g-1}) & \longrightarrow & 1.
\end{array}
$$
(4.4)

In the discussion below, we use the notation S_{m_1}, S'_{m_1}, and S''_{m_1} instead

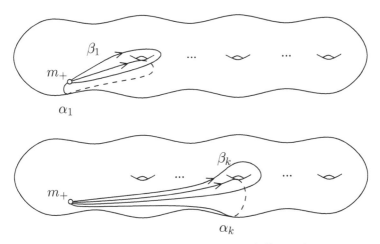

Figure 4.7 Standard generators for $\pi_1(S''_{m_1}, m_+)$

of the simpler notations $S_{g-1,2}$, $S_{g-1,1}$, and S_{g-1} in order to emphasize the point that each of these surfaces comes with fixed maps $S_{m_1} \to S'_{m_1} \to S''_{m_1}$. In particular, there is no choice for the images of the Lickorish twists in $\mathrm{Mod}(S'_{m_1})$ and $\mathrm{Mod}(S''_{m_1})$.

We start with sequence (4.4). The goal is to show that $\mathrm{Mod}(S'_{m_1})$ is generated by the images of the Lickorish twists in $\mathrm{Mod}(S'_{m_1})$; that is, we want to show that $\mathrm{Mod}(S'_{m_1})$ is generated by the Dehn twists about the simple closed curves shown at the bottom of Figure 4.6. By induction, $\mathrm{Mod}(S''_{m_1}) \approx \mathrm{Mod}(S_{g-1})$ is generated by the Dehn twists about the images of these curves in $S''_{m_1} \approx S_{g-1}$, and so by the exact sequence (4.4), it suffices to show that each element of $\mathcal{P}ush'(\pi_1(S''_{m_1}))$ is a product of the Dehn twists given at the bottom of Figure 4.6.

Standard generators for $\pi_1(S''_{m_1}) \approx \pi_1(S_{g-1})$ are shown in Figure 4.7. The mapping class $\mathcal{P}ush'(\alpha_1)$ is equal to the product $T_{c_1}T_{m_2}^{-1}$ (refer to Figure 4.6), so this element is a product of Lickorish twists.

We now explain how to write $\mathcal{P}ush'(\beta_1)$ as a product of Lickorish twists. Using Lemma 3.12, we see that

$$T_{m_2}T_{a_2}(\alpha_1) = \beta_1.$$

Thus, by Fact 4.8, $\mathcal{P}ush'(\beta_1)$ is conjugate to $\mathcal{P}ush'(\alpha_1)$ by a product of Lickorish twists and hence itself is a product of Lickorish twists.

Repeating this conjugation trick, we see that the image of each standard generator for $\pi_1(S''_{m_1})$ under $\mathcal{P}ush'$ is a product of the images of the Lickorish twists in $\mathrm{Mod}(S'_{m_1})$. The required formulas are

$$(T_{c_i}^{-1} T_{a_{i+1}}^{-1})(T_{a_i}^{-1} T_{c_i}^{-1})(\beta_{i-1}) = \beta_i,$$
$$T_{a_{i+1}}^{-1} T_{m_{i+1}}^{-1}(\beta_i) = \alpha_i.$$

We remark that the Lickorish twists seem to be exactly designed for completing this step.

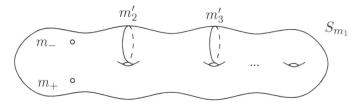

Figure 4.8 The Dehn twists $T_{m'_2}, \ldots, T_{m'_{g-1}}$ are all products of Lickorish twists.

Turning to sequence (4.3), it now remains to show that $\mathcal{P}ush(\pi_1(S'_{m_1}, m_-))$ lies in the group generated by the Dehn twists about the simple closed curves shown at the top of Figure 4.6. The proof is essentially the same as the previous argument. To facilitate the argument, it is helpful to notice that each $T_{m'_i}$ is a product of Lickorish twists where the $m'_2, \ldots, m'_{g-1}$ are the isotopy classes shown in Figure 4.8. This follows from the chain relation

$$(T_{m_g} T_{a_g} T_{c_{g-1}} T_{a_{g-1}} T_{c_{g-2}} \cdots T_{a_{k+1}} T_{c_k})^{2(g-k+1)} = T_{m_k} T_{m'_k}.$$

This completes the proof.   □

4.4.3 THE HUMPHRIES GENERATORS

We can now give Humphries' proof that the Humphries generators do indeed form a generating set for $\mathrm{Mod}(S_g)$.

THEOREM 4.14 (Humphries generators) *Let* $g \geq 2$. *Then the group* $\mathrm{Mod}(S_g)$ *is generated by the Dehn twists about the* $2g + 1$ *isotopy classes of nonseparating simple closed curves*

$$a_1, \ldots, a_g, c_1, \ldots, c_{g-1}, m_1, m_2$$

shown in Figure 4.5.

In Proposition 6.5 below we show that Theorem 4.14 is sharp in the sense that, for $g \geq 2$, any generating set for $\mathrm{Mod}(S_g)$ consisting only of Dehn twists must have at least $2g + 1$ elements.

Proof of Theorem 4.14. By Theorem 4.13 it suffices to show that the Lickorish twists $T_{m_3}, \ldots, T_{m_g}$ can each be written in terms of the other Lickorish twists.

For any $1 \leq i \leq g - 2$, we will find a product h of Dehn twists about the a_i, c_i, and m_{i+1} that takes m_i to m_{i+2}. It will then follow from Fact 3.7 in Section 3.3 that

$$T_{m_{i+2}} = h_i T_{m_i} h_i^{-1},$$

and the theorem will be proved.

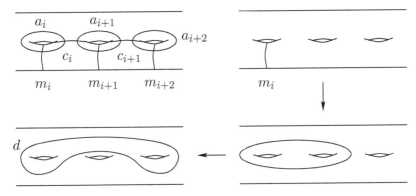

Figure 4.9 Taking m_i to m_{i+2}.

The top left of Figure 4.9 shows the simple closed curves we will use. In the top right of the figure we see m_i. The bottom right shows $T_{m_{i+1}} T_{a_{i+1}} T_{c_i} T_{a_i}(m_i)$, and the bottom left shows the image d of the latter under the product

$$T_{c_{i+1}} T_{a_{i+1}} T_{a_{i+2}} T_{c_{i+1}}.$$

Note that the last curve is symmetric with respect to the ith and $(i+2)$nd holes. It follows that we can use a similar product of Dehn twists h' in order to take d to m_{i+2}. Since h used m_{i+1} and no other m_j, it follows that h' will use m_{i+1} and no other m_j. This completes the proof. □

4.4.4 SURFACES WITH PUNCTURES AND BOUNDARY

Given the Humphries generators for the mapping class group of a closed surface, we can use the Birman exact sequence to find a finite set of generators for the mapping class group of any surface $S_{g,n}$ of genus $g \geq 0$ with $n \geq 0$ punctures.

The $2g + n$ twists about the simple closed curves indicated in Figure 4.10

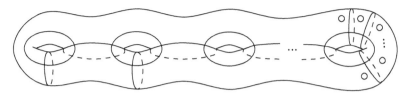

Figure 4.10 Twists about these simple closed curves generate $\mathrm{PMod}(S_{g,n})$.

give a generating set for $\mathrm{PMod}(S_{g,n})$ when $n > 0$. The argument in the last step of Theorem 4.13, that is, the argument that the images of $\mathcal{P}ush$ and $\mathcal{P}ush'$ lie in the group generated by the Lickorish twists, applies in this case to show that the given set of Dehn twists generates $\mathrm{PMod}(S_{g,n})$.

To obtain a generating set for all of $\mathrm{Mod}(S_{g,n})$, we can take a generating set for $\mathrm{PMod}(S_{g,n})$ together with a set of elements of $\mathrm{Mod}(S_{g,n})$ that project to a generating set for the symmetric group Σ_n. One standard generating set for Σ_n consists of $n-1$ transpositions. The most natural elements of $\mathrm{Mod}(S_{g,n})$ that map to transpositions in Σ_n are the half-twists discussed in Chapter 9. We thus have the following corollary of Theorems 4.9 and 4.11.

Corollary 4.15 *For any $g, n \geq 0$, the group $\mathrm{Mod}(S_{g,n})$ is generated by a finite number of Dehn twists and half-twists.*

Finally, let S be a compact surface with boundary (and no marked points). Recall that the elements of $\mathrm{Mod}(S)$ do not permute the boundary components of S. By Proposition 3.19, we see that $\mathrm{Mod}(S)$ is generated by Dehn twists about nonseparating simple closed curves if each Dehn twist about a boundary curve is a product of Dehn twists about nonseparating simple closed curves. It turns out that for $g \geq 2$ this is possible. Consider the simple closed curves shown in Figure 4.11. A special case of the star relation from Section 5.2 gives that

$$(T_{c_1} T_{c_2} T_{c_3} T_b)^3 T_{d_1}^{-1} T_{d_2}^{-1}$$

is equal to the Dehn twist about the boundary curve d.

We thus have the following.

Corollary 4.16 *Let S be any surface of genus $g \geq 2$. The group $\mathrm{PMod}(S)$ is generated by finitely many Dehn twists about nonseparating simple closed curves in S.*

In particular, for any surface S with punctures and/or boundary, $\mathrm{PMod}(S)$ is generated by the Dehn twists about the simple closed curves

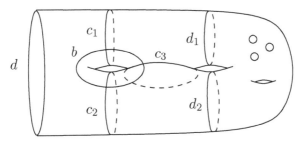

Figure 4.11 Writing the Dehn twist about the boundary in terms of Dehn twists about non-separating curves.

shown in Figure 4.10 (in the figure, one can interpret the small circles as either boundary components or as punctures).

On the other hand, for a genus 1 surface S with more than one boundary component, $\mathrm{Mod}(S)$ is not generated by Dehn twists about nonseparating curves. In this case, there is a generating set consisting of finitely many Dehn twists about nonseparating curves and $b - 1$ Dehn twists about boundary curves, where b is the number of boundary components. It follows from the computation of $H_1(\mathrm{Mod}(S); \mathbb{Z})$ (Section 5.1 below) that all $b - 1$ Dehn twists are needed.

Chapter Five

Presentations and Low-dimensional Homology

Having found a finite set of generators for the mapping class group, we now begin to focus on relations. Indeed, one of our main goals in this chapter is to give a finite presentation for $\mathrm{Mod}(S)$. In doing so, we will see some beautiful topological ideas, as well as some useful techniques from geometric group theory.

The relations in a group G are intimately related to the first and second homology groups of G. Recall that the homology groups of G are defined to be the homology groups of any $K(G, 1)$-space. The first and second homology groups have direct, group-theoretic interpretations. For example, $H_1(G; \mathbb{Z})$ is just the abelianization of G. Also, Hopf's formula, given below, gives an explicit expression for $H_2(G; \mathbb{Z})$ in terms of the generators and relators for G. In this chapter we will give explicit computations of the first and second homology groups of the mapping class group.

5.1 THE LANTERN RELATION AND $H_1(\mathrm{Mod}(S); \mathbb{Z})$

In the late 1970s D. Johnson discovered a remarkable relation among Dehn twists. He called it the lantern relation since his diagram for the relation was "lanternlike" [51, 115]. In the 1990s N. V. Ivanov pointed out that Dehn, in his original paper on mapping class groups from 1938, had already discovered the lantern relation. The existence of this relation has a number of important implications for the structure of mapping class groups. As a first example, we will use the lantern relation to show that $\mathrm{Mod}(S)$ has trivial abelianization for most S.

5.1.1 THE LANTERN RELATION

The *lantern relation* is a relation in $\mathrm{Mod}(S)$ between seven Dehn twists, all lying on a subsurface of S homeomorphic to S_0^4, a sphere with four boundary components.

Proposition 5.1 (Lantern relation) *Let x, y, z, b_1, b_2, b_3, and b_4 be simple closed curves in a surface S that are arranged as the curves shown in*

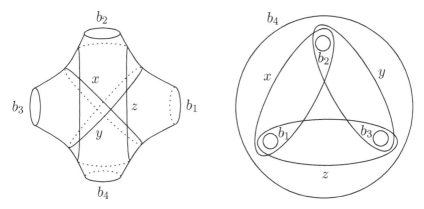

Figure 5.1 Two views of the lantern relation in S_0^4.

Figure 5.1. Precisely, this means that there is an orientation-preserving em-bedding $S_0^4 \hookrightarrow S$ and that each of the above seven curves is the image of the curve with the same name in Figure 5.1. In $\mathrm{Mod}(S)$ we have the relation

$$T_x T_y T_z = T_{b_1} T_{b_2} T_{b_3} T_{b_4}.$$

Proof. As discussed in Section 3.1, any embedding of a compact surface S' into a surface S induces a homomorphism $\mathrm{Mod}(S') \to \mathrm{Mod}(S)$. Since re-lations are preserved by homomorphisms, it suffices to check that the stated relation holds in $\mathrm{Mod}(S_0^4)$.

To check the relation in $\mathrm{Mod}(S_0^4)$, we cut S_0^4 into a disk using three arcs and apply the Alexander method (actually, two arcs would suffice). The computation is carried out in Figure 5.2.

For the computation, it is important to keep track of three conventions: Dehn twists are to the left, the simple closed curves x, y, and z are config-ured clockwise on the surface, and the relation is written using functional notation (i.e., elements on the right are applied first). □

Any surface S with $\chi(S) \leq -2$ contains an essential subsurface S' home-omorphic to S_0^4. Indeed, if x and y are any two simple closed curves in S with $i(x, y) = 2$ and $\hat{i}(x, y) = 0$, then S' can be taken to be any closed regular neighborhood of $x \cup y$. To see this, one can use the fact that if α and β are any two simple closed curves in S, and N is any regular neighborhood of $\alpha \cup \beta$, then $|\chi(N)| = |\alpha \cap \beta|$. As such, we see that the lantern relation occurs in any such S.

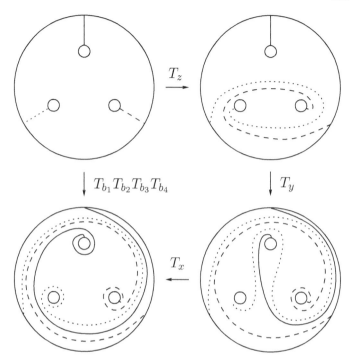

Figure 5.2 Proof of the lantern relation. The simple closed curves x, y, and z are shown in
          Figure 5.1.

The lantern relation implies another relation that is simpler, yet still interesting, namely,

$$T_x T_y T_z = T_y T_z T_x = T_z T_x T_y.$$

This relation follows easily from the lantern relation plus the relation that
each T_{b_i} commutes with each of T_x, T_y, and T_z. We can contrast this result
with Theorem 3.14, which states that there are no relations between Dehn
twists T_a and T_b with $i(a,b) = 2$. Note that $T_x T_y T_z$ is not equal to $T_z T_y T_x$.

The lantern relation via the push map. There is another way to derive the
lantern relation that makes it much less mysterious. Let P be a pair of pants,
that is, a sphere with three boundary components. Embed P in the plane
and label the outer boundary component x and the inner components b_1
and b_2. We obtain an element of $\mathrm{Mod}(P)$ by pushing b_1 around b_2, without
ever turning b_1 (think about a "do-si-do"). From the Alexander method and

Figure 5.3 we see that this map is equal to

$$T_x T_{b_1}^{-1} T_{b_2}^{-1}.$$

More formally, this push map is an element of the image of the homomorphism $\pi_1(UT(A)) \to \mathrm{Mod}(P)$, where A is the annulus obtained by capping b_1 by a closed disk (see Section 4.2).

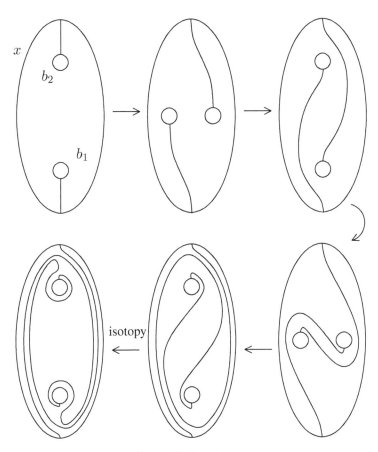

Figure 5.3 A push map.

Let S_0^4 be a sphere with four boundary components. We have the following easy-to-see relation in $\pi_1(UT(P)) < \mathrm{Mod}(S_0^4)$, depicted on the left-hand side of Figure 5.4: pushing b_2 around b_3 and then pushing b_2 around b_1 is the same as pushing b_2 around both b_3 and b_1. In other words, using the simple closed curves shown on the right-hand side of Figure 5.4, we have

$$(T_x T_{b_1}^{-1} T_{b_2}^{-1})(T_y T_{b_2}^{-1} T_{b_3}^{-1}) = T_{b_4} T_{b_2}^{-1} T_z^{-1}.$$

Since the T_{b_i} are central in this group, we can rewrite this as

$$T_x T_y T_z = T_{b_1} T_{b_2} T_{b_3} T_{b_4}.$$

And this is exactly the lantern relation.

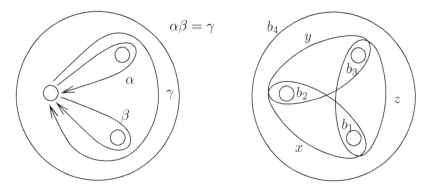

Figure 5.4 A new view of the lantern relation.

5.1.2 FIRST HOMOLOGY OF THE MAPPING CLASS GROUP

It is a basic fact from algebraic topology that, for any path-connected space X, the group $H_1(X; \mathbb{Z})$ is isomorphic to the abelianization of $\pi_1(X)$. Since the homology of a group G is defined as the homology of any $K(G, 1)$, we have that the first homology group of G with integer coefficients is

$$H_1(G; \mathbb{Z}) \approx \frac{G}{[G, G]} \approx G^{\mathrm{ab}},$$

where $[G, G]$ is the commutator subgroup of G and G^{ab} is the abelianization of G.

THEOREM 5.2 *For $g \geq 3$, the group $H_1(\mathrm{Mod}(S_g), \mathbb{Z})$ is trivial. More generally, for any surface S with genus at least 3, we have that $H_1(\mathrm{PMod}(S); \mathbb{Z})$ is trivial.*

In other words, if the genus of S is at least 3, then the group $\mathrm{PMod}(S)$ is equal to its commutator subgroup, or equivalently, $\mathrm{PMod}(S)^{\mathrm{ab}}$ is trivial. A group with this property is called *perfect*. As we will see below, the statement of Theorem 5.2 is false for $g \in \{1, 2\}$.

The following proof is due to Harer [83].

Proof. Let S be a surface whose genus is at least 3. Since Dehn twists about nonseparating simple closed curves are all conjugate (Fact 3.7), it follows that each of them maps to the same element under the natural quotient homomorphism $\mathrm{Mod}(S) \to H_1(\mathrm{Mod}(S); \mathbb{Z})$. Call this element h. Because $\mathrm{Mod}(S)$ is generated by Dehn twists about nonseparating simple closed curves (Corollary 4.16), it follows that $H_1(\mathrm{Mod}(S); \mathbb{Z})$ is generated by h.

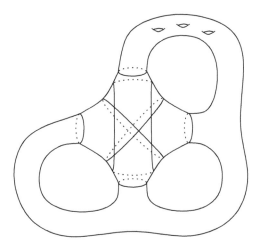

Figure 5.5 A copy of a sphere with four boundary components in a higher-genus surface, which gives rise to a lantern relation between seven nonseparating simple closed curves.

We now claim h is trivial. Since the genus of S is at least 3, it is possible to embed S_0^4 in S so that each of the seven simple closed curves in S_0^4 involved in the lantern relation is nonseparating; see Figure 5.5. The image of this lantern relation under the homomorphism $\mathrm{Mod}(S) \to H_1(\mathrm{Mod}(S); \mathbb{Z})$ gives the relation $h^4 = h^3$, from which we deduce that h is trivial, giving the theorem.   □

The search for the right relation.   Mumford was the first to attack the problem of finding the abelianization of $\mathrm{Mod}(S_g)$. He proved that $H_1(\mathrm{Mod}(S_g); \mathbb{Z})$ is a quotient of $\mathbb{Z}/10\mathbb{Z}$ for $g \geq 2$ [165]. In his paper, he punctuated his result with a question-exclamation mark, ?!, an annotation used in chess for a dubious move. As above, once you know that $\mathrm{Mod}(S_g)$ is generated by Dehn twists about nonseparating simple closed curves, it is a matter of using relations between Dehn twists to determine the abelianization. Mumford used the 3-chain relation $(T_a T_b T_c)^4 = T_d T_e$, hence his result. Birman noticed that one could use a different relation to show that the abelianization of $\mathrm{Mod}(S_g)$ is a quotient of $\mathbb{Z}/2\mathbb{Z}$ for $g \geq 3$ [21, 22]. Powell then produced a product of 15 nonseparating Dehn twists that equals the identity on $\mathrm{Mod}(S_g)$ for $g \geq 3$, finally proving Theorem 5.2 [181]. Later, Harer [83] noticed that the lantern relation can be used to give a simple proof, as above.

For $n > 1$, the group $\mathrm{Mod}(S_{g,n})$ is not perfect: if we take the sign of the induced permutation on the punctures (or marked points), we get a surjective homomorphism from $\mathrm{Mod}(S_{g,n})$ to the abelian group $\mathbb{Z}/2\mathbb{Z}$.

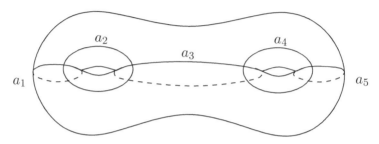

Figure 5.6 The Dehn twists about these simple closed curves generate $\mathrm{Mod}(S_2)$.

5.1.3 LOW-GENUS CASES

In order to determine $H_1(\mathrm{Mod}(S); \mathbb{Z})$ when S is a surface of genus 1 or 2, we work directly from the known presentations of these groups.

Genus 2. The group $\mathrm{Mod}(S_2)$ has the following presentation, due to Birman–Hilden. In the presentation, we use a_i to denote the Dehn twist about the simple closed curve a_i shown in Figure 5.6.

$$\mathrm{Mod}(S_2) = \langle a_1, a_2, a_3, a_4, a_5 \mid [a_i, a_j] = 1 \qquad |i - j| > 1,$$
$$a_i a_{i+1} a_i = a_{i+1} a_i a_{i+1},$$
$$(a_1 a_2 a_3)^4 = a_5^2,$$
$$[(a_5 a_4 a_3 a_2 a_1 a_1 a_2 a_3 a_4 a_5), a_1] = 1,$$
$$(a_5 a_4 a_3 a_2 a_1 a_1 a_2 a_3 a_4 a_5)^2 = 1 \rangle.$$

The first relation is simply disjointness, the second is the braid relation, and the third is a special case of the 3-chain relation (the two simple closed curves forming the boundary of the 3-chain are isotopic). The element $a_5 a_4 a_3 a_2 a_1 a_1 a_2 a_3 a_4 a_5$ appearing in the last two relations is exactly the hyperelliptic involution. We give the Birman–Hilden proof of this presentation in Chapter 9, and we give a brief discussion of the hyperelliptic relations later in this section.

To get a presentation for $\mathrm{Mod}(S_2)^{ab}$, we simply add the relations that all generators commute. This makes the first and fourth relations redundant. The braid relations then tell us that all the a_i represent the same element a in the abelianization. The next relation becomes $a^{12} = a^2$, or $a^{10} = 1$, and the last relation becomes $a^{20} = 1$, which is redundant. Thus $\mathrm{Mod}(S_2)^{ab}$ is a cyclic group of order 10, as proved by Birman–Hilden [26].

It turns out that for any surface $S_{2,n}$ of genus 2 with $n \geq 0$ punctures, we have $H_1(\mathrm{Mod}(S_{2,n}); \mathbb{Z}) \approx \mathbb{Z}/10\mathbb{Z}$; see [124].

Genus 1. Similarly, we can find that $H_1(\mathrm{Mod}(T^2); \mathbb{Z}) \approx \mathbb{Z}/12\mathbb{Z}$ using the classical presentation:

$$\mathrm{Mod}(T^2) \approx \mathrm{SL}(2, \mathbb{Z}) \approx \langle a, b \mid aba = bab, (ab)^6 = 1 \rangle.$$

In $\mathrm{Mod}(T^2)$, the elements a and b are Dehn twists about simple closed curves that intersect once. The relations are the braid relation and the 2-chain relation.

In the genus 1 case, adding punctures does not change the first homology of $\mathrm{Mod}(S)$, but adding boundary does. If S is a genus 1 surface with no boundary, then $H_1(\mathrm{Mod}(S); \mathbb{Z}) \approx \mathbb{Z}/12\mathbb{Z}$, and if S is a genus 1 surface with b boundary components, then $H_1(\mathrm{Mod}(S); \mathbb{Z}) \approx \mathbb{Z}^b$; again, see [124]. Combining the last statement with Proposition 3.19, we see that the mapping class group of a genus 1 surface with multiple boundary components is not generated by Dehn twists about nonseparating simple closed curves (cf. Section 4.4.4).

Genus zero. By again considering presentations, we see that if $S_{0,n}$ is a sphere with n punctures, then $H_1(\mathrm{Mod}(S_{0,n}); \mathbb{Z})$ is isomorphic to a cyclic group of order $2(n-1)$ or $n-1$, depending on whether n is even or odd, respectively. The presentation for $\mathrm{Mod}(S_{0,n})$ is

$$\mathrm{Mod}(S_{0,n}) = \langle \sigma_1, \ldots, \sigma_{n-1} \mid [\sigma_i, \sigma_j] = 1 \qquad |i - j| > 1,$$
$$\sigma_i \sigma_{i+1} \sigma_i = \sigma_{i+1} \sigma_i \sigma_{i+1},$$
$$(\sigma_1 \cdots \sigma_{n-1})^n = 1,$$
$$(\sigma_1 \cdots \sigma_{n-1} \sigma_{n-1} \cdots \sigma_1) = 1 \rangle.$$

One can arrive at this presentation from a presentation for the braid groups; the σ_i correspond to half-twists. See Chapter 9.

5.1.4 THE HYPERELLIPTIC RELATIONS

In our presentation of $\mathrm{Mod}(S_2)$ above we encountered a new, seemingly complicated relation. Here we generalize this relation to higher-genus surfaces, and in Chapter 9 we give a geometric explanation for this relation.

Let $c_1, \ldots, c_{2g+1}$ be a chain of isotopy classes of simple closed curves in the closed surface S_g; that is, $i(c_i, c_{i+1}) = 1$ and $i(c_i, c_j) = 0$ when $|i - j| > 1$. There is only one such chain in S_g up to homeomorphism (this follows from the fact that there is one $2g$-chain in S_g up to homeomorphism, as in Section 1.3). The product

$$T_{c_{2g+1}} \cdots T_{c_1} T_{c_1} \cdots T_{c_{2g+1}}$$

is a hyperelliptic involution (*the* hyperelliptic involution when g is equal to 1 or 2).

Thus we have the following *hyperelliptic relations* in $\mathrm{Mod}(S_g)$:

$$(T_{c_{2g+1}} \cdots T_{c_1} T_{c_1} \cdots T_{c_{2g+1}})^2 = 1,$$
$$[T_{c_{2g+1}} \cdots T_{c_1} T_{c_1} \cdots T_{c_{2g+1}}, T_{c_{2g+1}}] = 1.$$

A strange fact. If we rewrite the first hyperelliptic relation, we see that there is a product of $4g + 1$ Dehn twists that equals the inverse of one Dehn twist. In other words, a right Dehn twist is a product of left Dehn twists. This, plus the Dehn–Lickorish theorem, gives us the following surprising fact (pointed out to us by Luis Paris):

> *Every element of* $\mathrm{Mod}(S_g)$ *is a product of left (positive) Dehn twists.*

5.2 PRESENTATIONS FOR THE MAPPING CLASS GROUP

We have already seen several relations between Dehn twists. In particular, we have the disjointness relation (Fact 3.9), the braid relation, the chain relation, the lantern relation, and the hyperelliptic relation. We will see that these relations suffice to give a finite presentation for $\mathrm{Mod}(S_g)$.

5.2.1 WAJNRYB'S PRESENTATION

Finite presentations for the mapping class groups of closed surfaces of genus 1 and 2 were discussed in Section 5.1.2. McCool gave the first algorithm for finding a finite presentation for the mapping class group of a higher-genus surface [145]. His techniques are algebraic in nature; no explicit presentation has been derived from this algorithm.

Hatcher and Thurston made a breakthrough by finding a topologically fla-vored algorithm for constructing an explicit finite presentation for $\mathrm{Mod}(S)$.

The algorithm was carried out by Harer, who produced a finite but unwieldy presentation [83]. Wajnryb used these ideas to derive the following explicit presentation, which is considered to be the standard presentation for $\mathrm{Mod}(S)$ [29, 209]. The exact form of the presentation given here is taken from a survey paper by Birman [25]. In the statement, we use functional notation as usual (elements applied right to left).

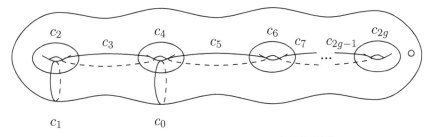

Figure 5.7 The Humphries generators for $\mathrm{Mod}(S)$.

THEOREM 5.3 (Wajnryb's finite presentation) *Let S be either a closed surface or a surface with one boundary component and genus $g \geq 3$. Let a_i denote the Humphries generator T_{c_i}, where c_i is as shown in Figure 5.7. The mapping class group $\mathrm{Mod}(S)$ has a presentation where the generators are $a_0, \ldots, a_{2g}$, and the relations are as follows.*

1. *Disjointness relations*

$$a_i a_j = a_j a_i \quad \text{if} \quad i(c_i, c_j) = 0$$

2. *Braid relations*

$$a_i a_j a_i = a_j a_i a_j \quad \text{if} \quad i(c_i, c_j) = 1$$

3. *3-chain relation*

$$(a_1 a_2 a_3)^4 = a_0 b_0,$$

   where

$$b_0 = (a_4 a_3 a_2 a_1 a_1 a_2 a_3 a_4) a_0 (a_4 a_3 a_2 a_1 a_1 a_2 a_3 a_4)^{-1}$$

4. *Lantern relation*

$$a_0 b_2 b_1 = a_1 a_3 a_5 b_3,$$

where

$$b_1 = (a_4a_5a_3a_4)^{-1}a_0(a_4a_5a_3a_4)$$
$$b_2 = (a_2a_3a_1a_2)^{-1}b_1(a_2a_3a_1a_2)$$
$$b_3 = (a_6a_5a_4a_3a_2ua_1^{-1}a_2^{-1}a_3^{-1}a_4^{-1})a_0(a_6a_5a_4a_3a_2ua_1^{-1}a_2^{-1}a_3^{-1}a_4^{-1})^{-1}$$

and where

$$u = (a_6a_5)^{-1}b_1(a_6a_5)$$

5. *Hyperelliptic relation (S closed)*

$$(a_{2g}\cdots a_1a_1\cdots a_{2g})d = d(a_{2g}\cdots a_1a_1\cdots a_{2g}),$$

where d is any word in the generating set that, under the previous relations, is equivalent to the Dehn twist about the simple closed curve d in Figure 5.8.

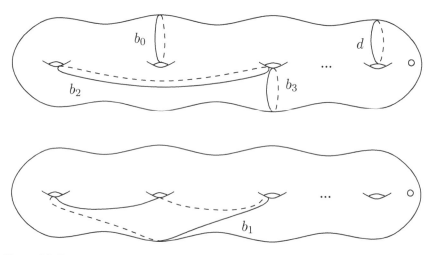

Figure 5.8 Extra elements used in the relations for Wajnryb's presentation for $\mathrm{Mod}(S)$. We have labeled the simple closed curves by the corresponding elements of $\mathrm{Mod}(S)$.

In the statement, we mean that the hyperelliptic relation is only needed (and it is only true) for closed surfaces. The reason for the term "hyperelliptic relation" is that the product $d(a_{2g}\cdots a_1a_1\cdots a_{2g})d$ is a hyperelliptic involution.

Strictly speaking, Theorem 5.3 does not give a formal presentation of $\mathrm{Mod}(S_g)$ since we have not given the element d in terms of the generators, so we take care of that now. If we rephrase things, we need to write the

Dehn twist d as a product of the generators a_i in the mapping class group of the surface with one boundary component. Let $n_1, \ldots, n_g$ be the Dehn twists about the simple closed curves shown in Figure 5.9. Note that n_1, n_2, and n_g are the same as the Dehn twists a_1, b_0, and d from Theorem 5.3. Similarly to Section 4.4.3, we can inductively write the n_i in terms of the Humphries generators. We start with $n_1 = a_1$ and $n_2 = b_0$. Then we have

$$n_{i+2} = w_i n_i w_i^{-1},$$

where

$$w_i = (a_{2i+4} a_{2i+3} a_{2i+2} n_{i+1})(a_{2i+1} a_{2i} a_{2i+2} a_{2i+1})$$
$$(a_{2i+3} a_{2i+2} a_{2i+4} a_{2i+3})(n_{i+1} a_{2i+2} a_{2i+1} a_{2i}).$$

Finally, set $d = n_g$.

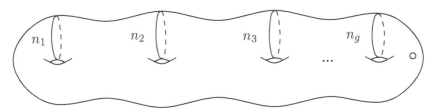

Figure 5.9 Extra elements used in the relations for Wajnryb's presentation for $\mathrm{Mod}(S)$. We have labeled the simple closed curves by the corresponding elements of $\mathrm{Mod}(S)$.

A presentation of the mapping class group of a surface with more than one boundary component can be obtained by applying the Birman exact sequence. Also, a presentation for $\mathrm{Mod}(S_{g,1})$ can be obtained by combining Wajnryb's presentation with Proposition 3.19.

The effect of relations on homology. Harer notes that if we take the abstract group with the Humphries generators and the first two sets of relations in the Wajnryb presentation, then we have a group (an Artin group) whose first homology is $\mathbb{Z}$. We see from our proof of Theorem 5.2 that if we next add in the lantern relation, the resulting group has trivial first homology. At this point, our abstract group has trivial second homology, yet Harer proved that $H_2(\mathrm{Mod}(S_g); \mathbb{Z}) \approx \mathbb{Z}$ (Theorem 5.8 below). Adding in the 3-chain relation corrects this.

The algebrogeometric approach. Years before McCool's result, Baily and Deligne–Mumford gave different compactifications of $\mathcal{M}(S_g)$, the moduli space of Riemann surfaces homeomorphic to S_g, showing that $\mathcal{M}(S_g)$ is a quasiprojective variety [10, 52]. We will prove in Theorem 6.9 below that $\mathrm{Mod}(S_g)$ has a finite-index subgroup Γ that is torsion-free, from which it follows that $\mathcal{M}(S_g)$ has a finite cover (corresponding to Γ) which is a manifold, and so a smooth quasiprojective variety. Lojasiewicz had also shown that any smooth quasiprojective variety has the homotopy type of a finite complex; in particular, its fundamental group is finitely presented. We conclude that Γ, hence $\mathrm{Mod}(S_g)$, is finitely presented. However, this approach does not give an algorithm for finding an explicit finite presentation.

5.2.2 THE CUT SYSTEM COMPLEX

We now very briefly outline the strategy used to derive the presentation in Theorem 5.3. In Section 5.3 below, we will give a complete proof that $\mathrm{Mod}(S_g)$ is finitely presented, although we will not derive an explicit presentation.

The cut system complex. Hatcher–Thurston [89] defined a 2-dimensional CW-complex $X(S_g)$, called the *cut system complex*, as follows. Vertices of $X(S_g)$ correspond to cut systems in S_g, that is, (unordered) sets $\{c_1, \dots, c_g\}$ where

1. each c_i is the isotopy class of a nonseparating simple closed curve γ_i in S_g,

2. $i(c_i, c_j) = 0$ for all i and j, and

3. $S_g - \cup \gamma_i$ is connected.

An example of a vertex in $X(S_g)$ is given by the set of isotopy classes $\{a_1, \dots, a_g\}$ shown in Figure 5.10. Vertices represented by $\{a_i\}$ and $\{b_i\}$ are connected by an edge in $X(S_g)$ if (up to renumbering) $a_i = b_i$ for $2 \leq i \leq g$ and $i(a_1, b_1) = 1$.

Just as the edges of $X(S_g)$ are defined by certain topological configurations of curves, so are the 2-cells of $X(S_g)$. For example, we glue in a triangle to the 1-skeleton of $X(S_g)$ for each triple of vertices that are pairwise-connected by edges. For example, in Figure 5.10, the vertices $v_a = \{a, a_2, \dots, a_g\}$, $v_b = \{b, a_2, \dots, a_g\}$, and $v_c = \{c, a_2, \dots, a_g\}$ span a triangle in $X(S_g)$. The complex $X(S_g)$ also has squares and pentagons; we refer the reader to the paper [89] for the details.

Hatcher–Thurston give a beautiful Morse–Cerf-theoretic proof that $X(S_g)$ is simply connected. Later Hatcher–Lochak–Schneps gave an alternate proof for a closely related complex [92], and Wajnryb gave a combinatorial proof of simple connectivity for the original complex [210].

The mapping class group action. In general, when a group G acts cocompactly on a simply connected complex X with finitely presented vertex stabilizers and finitely generated edge stabilizers, the group G is finitely presented (see Proposition 5.6 below). For each orbit of vertices of X, there are relations in G coming from the relations in those vertex stabilizers; for each orbit of edges of X, there are relations in G coming from the generators of those edge stabilizers (the relations identify elements of the two vertex stabilizers); and finally there is one relation in G for each orbit of 2-cells in X. See the paper by Ken Brown for details [37].

Since the complex $X(S_g)$ is defined by topological rules, it follows that $\mathrm{Mod}(S_g)$ acts on $X(S_g)$. Using the change of coordinates principle, it is not hard to see that the action is cocompact; indeed, there is a single orbit of vertices and a single orbit of edges. Now, the stabilizer in $\mathrm{Mod}(S_g)$ of a vertex of $X(S_g)$ is closely related to a braid group. This is because if we cut S_g along the simple closed curves corresponding to a vertex of S_g, the result is a sphere with $2g$ boundary components; cf. Chapter 9. Therefore, the presentation for a vertex stabilizer can be derived from known presentations of braid groups or mapping class groups of genus 0 surfaces. Generating sets for edge stabilizers are obtained similarly.

Wajnryb's calculation. To give a flavor of the calculation used to get Wajnryb's actual presentation, we explain how the braid relation comes up in his analysis of the action of $\mathrm{Mod}(S_g)$ on $X(S_g)$. Of course, to verify the braid relation in $\mathrm{Mod}(S_g)$ is not difficult (see Proposition 3.11). The point here is that, by the general theory, a full set of relations for $\mathrm{Mod}(S_g)$ is obtained by identifying elements of different cell stabilizers. We will realize the braid relation as one such relation.

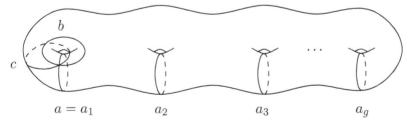

Figure 5.10 The simple closed curves a_i give a vertex of the cut system complex, and the simple closed curves a, b, and c, along with $a_2, \ldots a_g$, give a triangle of the complex.

In what follows, we abuse notation, denoting a simple closed curve and its associated Dehn twist by the same symbol.

Let v_a be the vertex of $X(S_g)$ corresponding to the cut system $\{a_i\}$ given in Figure 5.10. We will make use of two particular elements of the stabilizer G_{v_a} of v_a, namely, the Dehn twist a and the element $s = ba^2b$, where b is the Dehn twist about the simple closed curve shown in Figure 5.10.

Let e_{ab} be the edge of $X(S_g)$ spanned by the vertices v_a and v_b defined above. One element of the stabilizer $G_{e_{ab}}$ of e_{ab} is $r = aba$. Since r interchanges the vertices of e_{ab}, it follows that r^2 is an element of G_{v_a}. In particular, it is the element $sa^2 \in G_{v_a}$. So we obtain the following relation (relation P10 in [210, Theorem 31]):

$$r^2 = sa^2.$$

We now focus on the stabilizer of a 2-cell, namely, the triangle t_{abc} spanned by v_a, v_b, and v_c. The element ar does not stabilize v_a or e_{ab}, but it does stabilize t_{abc}, inducing an order 3 rotation of t_{abc}. Thus $(ar)^3$ is an element of G_{v_a}, and again one can write it as a word in the elements $s, a \in G_{v_a}$, namely, $(asa)^2$. So we have the following relation (relation P11 in [210, Theorem 31]):

$$(ar)^3 = (asa)^2.$$

We can rewrite this last relation using the relation $r^2 = sa^2$ and the trivial relations $aa^{-1} = 1$ and $bb^{-1} = 1$.

$$(ar)^3 = (asa)^2$$
$$\implies (ar)^3 = a(sa^2)sa$$
$$\implies (ar)^3 = ar^2sa$$

Replacing r with aba and s with ba^2b, we find

$$a^2ba^3ba^3ba = a^2ba^2baba^2ba$$
$$\implies (a^2ba^2)aba(a^2ba) = (a^2ba^2)bab(a^2ba)$$
$$\implies aba = bab$$

Thus we see the braid relation arising from the action of $\mathrm{Mod}(S_g)$ on $X(S_g)$; it comes from two relations one gets by flipping edges and by rotating triangles. Deriving the complete presentation of $\mathrm{Mod}(S_g)$ given in Theorem 5.3 is quite involved; we refer the reader to Wajnryb's paper [210] for details.

It is straightforward to carry out this procedure in the case of the torus. The complex $X(T^2)$ is the Farey complex (see Section 4.1), and, in fact, the relations $r^2 = sa^2$ and $aba = bab$ already discussed suffice to present the group $\text{Mod}(T^2) \approx \text{SL}(2, \mathbb{Z})$.

5.2.3 THE GERVAIS PRESENTATION

While Wajnryb's presentation (Theorem 5.3) is the most well known and classical presentation of $\text{Mod}(S)$, there are several other useful ones. We now present one due to Gervais [71]. Some of the features of this presentation are: it is fairly easy to write down explicitly, it works for the pure mapping class group of any surface with boundary, and all of the relations are described on uniformly small subsurfaces (tori with at most three boundary components). Gervais's derivation of this presentation is accomplished by starting from Wajnryb's presentation and simplifying the relations there. The same is true for the beautiful presentation due to Matsumoto [143], which is phrased in terms of Artin groups and which we do not discuss here. Hirose gives a direct derivation of the Gervais presentation [94].

The Gervais presentation uses one new relation which we have not seen before.

The star relation. Consider the torus S_1^3 with three boundary components d_1, d_2, and d_3. Let c_1, c_2, c_3, and b be isotopy classes of simple closed curves configured as in Figure 5.11. Note that S_1^3 is homeomorphic to a closed regular neighborhood of $c_1 \cup c_2 \cup c_3 \cup b$ (really the union of four representatives).

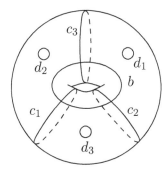

Figure 5.11 The simple closed curves used in the star relation.

Gervais gives the following relation [71]. If c_1, c_2, c_3, b, d_1, d_2, and d_3 are the isotopy classes of simple closed curves in S_1^3 given in Figure 5.11, then we have

$$(T_{c_1} T_{c_2} T_{c_3} T_b)^3 = T_{d_1} T_{d_2} T_{d_3}.$$

As with the lantern relation, this relation can be checked with the Alexander method. We call b the *central curve* of the star relation. For any embedding $S_1^3 \hookrightarrow S$ into a surface S, the image of the star relation under the induced homomorphism $\mathrm{Mod}(S_1^3) \to \mathrm{Mod}(S)$ of course gives a relation (between the images of the above curves) in $\mathrm{Mod}(S)$.

Suppose that S_1^3 is embedded in S in such a way that the isotopy classes c_1 and c_2 are equal but distinct from c_3. This happens when the image of d_3 under the embedding is the trivial isotopy class and the images of d_1 and d_2 are nontrivial. In this case, the star relation becomes

$$(T_{c_1}^2 T_{c_3} T_b)^3 = T_{d_1} T_{d_2}.$$

We call this a *degenerate star relation*. We will not need to consider star relations with $c_1 = c_2 = c_3$. We note that the degenerate star relation is the same as one of the 3-chain relations given in Section 4.4.

Recall that we used the star relation in Section 4.4.4 to prove Corollary 4.16.

The Gervais presentation. Let S be a compact surface of genus g with n boundary components. We begin by giving the generating set for the Gervais presentation of $\mathrm{Mod}(S)$. Each of the generators is a Dehn twist, and so it suffices to list the corresponding simple closed curves. The curves are shown in Figure 5.12, where we have drawn S as a torus with $g-1$ handles attached and n disks removed.

We start at the top of the figure. There is one simple closed curve b which will form the central curve for all of our star relations. There are $2(g-1) + n$ simple closed curves $\{c_i\}$ with $i(b, c_i) = 1$. There are $2(g-1)$ simple closed curves corresponding to the latitudes and longitudes of the $g-1$ handles attached to the central torus. We also include the n boundary components. Finally, for each ordered pair of distinct curves (c_i, c_j), there is a simple closed curve $c_{i,j}$ that lies in a neighborhood of $c_i \cup c_j \cup b$ and that lies in the clockwise direction from c_i along b (note that each $c_{i,i+1}$ has already appeared on the list). The curves $c_{i,j}$ are depicted at the bottom of Figure 5.12; there are $(2g - 2 + n)(2g - 3 + n)$ of these curves.

THEOREM 5.4 (Gervais' finite presentation) *Let S be a surface of genus g with n boundary components. The group $\mathrm{Mod}(S)$ has a presentation with one Dehn twist generator for each simple closed curve shown in Figure 5.12 and with the following relations.*

    1. All disjointness relations between generators

    2. All braid relations between generators

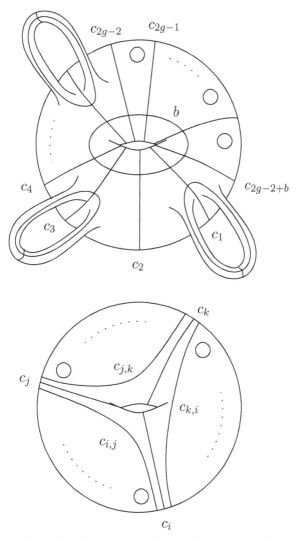

Figure 5.12 The generators for the Gervais presentation.

3. *All star relations between generators, including the degenerate ones,*
   where b is the central curve.

From Theorem 5.4 it is straightforward to write down the presentation explicitly by listing the generators and relations. For the first two kinds of relations, one needs to find all pairs of generators that are disjoint or that have intersection number 1. The degenerate star relations are given by triples $\{c_i, c_i, c_j\}$, where $c_i \neq c_j$, and the other star relations are given by triples of distinct c_i-curves.

By Proposition 3.19, one can get a presentation for the case of a surface with punctures by setting each generator corresponding to a Dehn twist about a boundary curve to be trivial.

5.3 PROOF OF FINITE PRESENTABILITY

We now give a proof that $\mathrm{Mod}(S)$ is finitely presented. While it is possible to give a proof analogous to our proof of finite generation, we instead choose to introduce a new technique. As a result, we obtain a new proof of finite generation.

The strategy, suggested by Andrew Putman, is to show that the arc complex $\mathcal{A}(S)$ is contractible and use the action of $\mathrm{Mod}(S)$ on $\mathcal{A}(S)$ to build a $K(\mathrm{Mod}(S), 1)$ with finite 2-skeleton. It immediately follows that $\mathrm{Mod}(S)$ is finitely presented. While this is a simple proof of finite presentability, we do not know what explicit finite presentation comes out of this approach.

5.3.1 THE ARC COMPLEX

Let S be a compact surface that either has nonempty boundary or has at least one marked point. We define the arc complex $\mathcal{A}(S)$ as the abstract simplicial flag complex described by the following data (cf. Section 4.1).

> *Vertices.* There is one vertex for each free isotopy class of essential simple proper arcs in S.
>
> *Edges.* Vertices are connected by an edge if the corresponding free isotopy classes have disjoint representatives.

If we take a surface S with nonempty boundary and cap one or more boundary components with a once-marked disk, then $\mathcal{A}(S)$ is naturally isomorphic to the arc complex for the capped surface. So in this sense there is no difference between marked points and boundary components in defining the arc complex. When we consider the action of the mapping class group on the arc complex, marked points are more natural than boundary components since Dehn twists about boundary components act trivially on the arc complex.

As a first example, the arc complex of the torus with one boundary component is the Farey complex (see Section 4.1).

The most fundamental fact about the arc complex is the following theorem due to Harer [83].

THEOREM 5.5 *Let S be any compact surface with finitely many marked points. If $\mathcal{A}(S)$ is nonempty, then it is contractible.*

The elegant proof we present is due to Hatcher [90]. A number of other mathematicians made various contributions to the circle of ideas surrounding this theorem, including Thurston, Bowditch–Epstein, Mumford, Mosher, and Penner.

For the proof, recall that the simplicial star of a vertex v in a simplicial complex is the union of closed simplices containing v. The simplicial star of a vertex is contractible.

Proof. We choose some base vertex v of $\mathcal{A}(S)$. To prove that $\mathcal{A}(S)$ is contractible, we will define a flow of $\mathcal{A}(S)$ onto the simplicial star of v.

An arbitrary point p in the simplex of $\mathcal{A}(S)$ spanned by vertices $v_1, \ldots, v_n$ is given by barycentric coordinates, that is, a formal sum $\sum c_i v_i$ where $\sum c_i = 1$ and $c_i \geq 0$ for all i. Let α be a fixed representative of v. We can realize p in S as follows: first realize the v_i as disjoint arcs in S, each in minimal position with α, and then thicken each v_i-arc to a band which is declared to have width c_i.

By an isotopy, we make the intersection of the arc representing v with the union of these bands equal to a closed interval disjoint from ∂S, as on the left-hand side of Figure 5.13. (In the figure we have shown α with its endpoint at a boundary component. If instead its endpoint is at a marked point/puncture, then the boundary component, depicted as a horizontal line at the bottom of the figure, is not in the picture.) Let $\theta = \sum c_i i(v_i, v)$ denote the thickness of this union of bands.

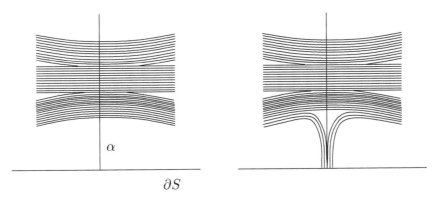

Figure 5.13 The Hatcher flow on $\mathcal{A}(S)$.

The flow is defined as follows. At time t, we push a total band width of $t\theta$ in some prechosen direction along the arc α (see the right-hand side of Figure 5.13). The picture gives barycentric coordinates for some new point in $\mathcal{A}(S)$. At time 1, all of the bands are disjoint from the arc α, and we are in the star of v.

It is not difficult to check that the flow is continuous and well defined on the intersections of simplices. This completes the proof of the theorem.   □

5.3.2 FINITE PRESENTABILITY VIA GROUP ACTIONS ON COMPLEXES

The group $\mathrm{Mod}(S)$ acts by simplicial automorphisms on the contractible simplicial complex $\mathcal{A}(S)$. In order to use this action to analyze $\mathrm{Mod}(S)$, we need to apply some geometric group theory.

The following theorem is adapted from Scott–Wall [191]. In the statement of the theorem, we say that a group G acts on a CW-complex X *without rotations* if, whenever an element $g \in G$ fixes a cell $\sigma \subset X$, then g fixes σ pointwise. Any action of a group on a CW-complex can be turned into an action without rotations by barycentrically subdividing the complex. The benefit of an action without rotations is that the quotient has a natural CW-complex structure coming from the structure of the original complex.

Proposition 5.6 *Let G be a group acting on a contractible CW-complex X without rotations. Suppose that each of the following conditions holds.*

  1. The quotient X/G is finite.

  2. Each vertex stabilizer is finitely presented.

  3. Each edge stabilizer is finitely generated.

Then G is finitely presented.

Proof. Let U be any $K(G, 1)$-complex. Consider the contractible complex $\widetilde{U} \times X$. Since the action of G on $\widetilde{U}$ is free, the diagonal action of G on $\widetilde{U} \times X$ is free. Therefore, as $\widetilde{U} \times X$ is contractible, $(\widetilde{U} \times X)/G$ is another $K(G, 1)$-complex. This construction of a $K(G, 1)$ from a group action on a complex is called the *Borel construction*.

We will show that $(\widetilde{U} \times X)/G$ has the homotopy type of a complex with finite 2-skeleton. Consider the projection

$$(\widetilde{U} \times X)/G \to X/G.$$

If v is a vertex of X with stabilizer G_v in G, then $(\widetilde{U} \times v)/G_v$ is a $K(G_v, 1)$-complex. Moreover, this space maps injectively to $(\widetilde{U} \times X)/G$ and is the preimage of $[v] \in X/G$. In other words, over each vertex of X/G there is in $(\widetilde{U} \times X)/G$ a $K(\pi, 1)$ corresponding to that vertex stabilizer. Similarly, lying over each higher-dimensional open cell is the product of a $K(\pi, 1)$-complex for that cell stabilizer with that open cell.

As a result, we see that $(\widetilde{U} \times X)/G$ has the structure of a complex of spaces, with each vertex space a $K(G_v, 1)$ for a vertex stabilizer G_v and each edge space a $K(G_e, 1)$ for an edge stabilizer G_e. That is, the space $(\widetilde{U} \times X)/G$ is obtained inductively as follows: we start with the disjoint union of the $K(G_v, 1)$-spaces; then, we take the $K(G_e, 1)$-spaces, cross them with intervals, and glue them to the $K(G_v, 1)$-spaces via any map in the unique homotopy class of maps determined by the inclusion $G_e \hookrightarrow G_v$. This process is repeated inductively (and analogously) on higher-dimensional skeletons.

We make the following observation: if each space in the complex of spaces is replaced with another space to which it is homotopy-equivalent (i.e., another $K(\pi, 1)$-space), the homotopy type of the resulting complex does not change. In other words, the homotopy colimit is well defined [91, Proposition 4G.1].

Since the stabilizer G_v of each vertex v is assumed to be finitely presented, each $K(G_v, 1)$-space can be chosen to have a finite 2-skeleton. Since the stabilizer of each edge e is assumed to be finitely generated, each $K(G_e, 1)$-space can be chosen to have a finite 1-skeleton. For the stabilizer G_f of each 2-cell f, the $K(G_f)$-space can be chosen to have a finite 0-skeleton, since for any group H, there is a $K(H, 1)$ with a single vertex).

There are three ways that 2-cells arise in the complex of spaces $(\widetilde{U} \times X)/G$: via 2-cells of $K(G_v, 1)$-spaces, 1-cells of $K(G_e, 1)$-spaces, and 0-cells of $K(G_f, 1)$-spaces. As discussed above, each of these spaces can be chosen to have finite 2-skeleton, 1-skeleton and 0-skeleton, respectively. Since the quotient X/G is finite, the resulting complex of spaces has finitely many 2-cells. Thus we have created a $K(G, 1)$ with a finite 2-skeleton, and so G is finitely presented. $\quad\square$

We remark that the proof of Proposition 5.6 can be slightly modified to work in the case where X is assumed only to be simply connected as opposed to contractible. Actually, the complex of curves $\mathcal{C}(S)$ is simply connected (but not contractible) for most S; see [87, Theorem 3.5] and [108, Theorem 1.3]. The reason we use the arc complex in our application of Proposition 5.6 is simply because it is easier to prove that $\mathcal{A}(S)$ is contractible than it is to prove that $\mathcal{C}(S)$ is simply connected.

5.3.3 PROOF THAT THE MAPPING CLASS GROUP IS FINITELY PRESENTED

We are now ready to prove the following theorem.

THEOREM 5.7 *If S is a compact surface with finitely many marked points, then the group* $\mathrm{Mod}(S)$ *is finitely presented.*

Proof. We first reduce the problem to the case of $S_{g,n}$ with $n > 0$ marked points. Suppose we can prove the theorem in this case. We now explain how to deduce the theorem in the case that S has nonempty boundary and then in the case where S is closed.

Let S be a compact surface with $n > 0$ boundary components and assume that S is not the disk D^2. Also assume by induction that, for any compact surface with $n-1$ boundary components, the mapping class group is finitely presented. We recall Proposition 3.19, which states that if S^* is the surface obtained from a surface S by capping a boundary component β with a once-marked disk, then the following sequence is exact:

$$1 \to \langle T_\beta \rangle \to \mathrm{Mod}(S) \overset{Cap}{\to} \mathrm{Mod}^\star(S^*) \to 1,$$

where $\mathrm{Mod}^\star(S^*)$ is the subgroup of $\mathrm{Mod}(S^*)$ consisting of elements that fix the marked point coming from the capping operation. By the inductive hypothesis, we have that $\mathrm{Mod}(S^*)$ is finitely presented. Since $\mathrm{Mod}^\star(S^*)$ has finite index in $\mathrm{Mod}(S^*)$, it is also finitely presented. Since the extension of a finitely presented group by a finitely presented group is finitely presented, it follows from Proposition 3.19 that $\mathrm{Mod}(S)$ is finitely presented.

A similar argument to the above, using the Birman exact sequence, shows that $\mathrm{Mod}(S_{g,0})$ is finitely presented if $\mathrm{Mod}(S_{g,1})$ is since the quotient of a finitely presented group by a finitely generated group is finitely presented.

We have thus reduced the proof to showing that $\mathrm{Mod}(S_{g,n})$ is finitely presented when $n > 0$. We may assume that $(g, n) \neq (0, 1)$ because we already know $\mathrm{Mod}(S_{0,1}) = 1$. Since a group is finitely presented if and only if any of its finite-index subgroups are finitely presented, it suffices to prove that $\mathrm{PMod}(S_{g,n})$ is finitely presented. We make the inductive hypothesis that $\mathrm{PMod}(S_{g',n'})$ is finitely presented when $g' < g$ or when $g' = g$ and $n' < n$.

We would like to apply Proposition 5.6. By Theorem 5.5, the arc complex $\mathcal{A}(S_{g,n})$ is contractible. Therefore its barycentric subdivision $\mathcal{A}'(S_{g,n})$, on which $\mathrm{PMod}(S_{g,n})$ acts without rotations, is also contractible. Note that vertices of $\mathcal{A}'(S_{g,n})$ correspond to simplices of $\mathcal{A}(S_{g,n})$. It follows from the change of coordinates principle that the quotient of $\mathcal{A}'(S_{g,n})$ by $\mathrm{PMod}(S_{g,n})$ is finite.

Now let v be a vertex of $\mathcal{A}'(S_{g,n})$ and let G_v be its stabilizer in $\mathrm{PMod}(S_{g,n})$. In order to apply Proposition 5.6, we need to show that G_v is finitely presented.

As above, v corresponds to a simplex of $\mathcal{A}(S_{g,n})$, that is, the isotopy class of a collection of disjoint simple proper arcs α_i in $S_{g,n}$. If we cut $S_{g,n}$ along the α_i, we obtain a (possibly disconnected) compact surface with boundary S_α, possibly with marked points in its interior. We may pass from the cut surface S_α to a surface with marked points but no boundary by collapsing each boundary component to a marked point (or, what will have the same effect, capping each boundary component with a once-marked disk). Denote the connected components of the resulting surface by R_i. Each R_i has marked points coming from the marked points of $S_{g,n}$ and/or marked points coming from $\cup\alpha_i$. Note that each $\mathrm{PMod}(R_i)$ falls under the inductive hypothesis.

Let G_v^0 denote the subgroup of G_v consisting of elements that fix each isotopy class $[\alpha_i]$ with orientation. Note that these elements necessarily fix the R_i as well. Since G_v^0 has finite index in G_v, it suffices to show that G_v^0 is finitely presented. There is a map

$$\eta : G_v^0 \to \prod \mathrm{PMod}(R_i).$$

To see that η is a well-defined homomorphism, one needs the fact that if two homeomorphisms of $S_{g,n}$ fixing $\cup\alpha_i$ are homotopic, then they are homotopic through homeomorphisms that fix $\cup\alpha_i$ (cf. Section 3.6).

The map η is also surjective. Indeed, given any element of $\prod \mathrm{PMod}(R_i)$, one can choose a representative homeomorphism that is the identity in a neighborhood of the marked points, and then one can lift this to a representative of an element of G_v^0 that is the identity on a neighborhood of the union of the marked point with the α_i. It follows from Proposition 3.19 that the kernel of η is generated by the Dehn twists about the components of the boundary of the cut surface S_α. Since each $\mathrm{PMod}(R_i)$ is finitely presented, their product is as well. As the kernel of η is finitely generated and its cokernel is finitely presented, it follows that G_v^0 is finitely presented, which is what we wanted to show.

Two vertices of $\mathcal{A}'(S_{g,n})$ are connected by an edge if and only if the corresponding simplices of $\mathcal{A}(S_{g,n})$ share a containment relation (i.e., one is contained in the other). It follows that the stabilizer of an edge in $\mathcal{A}'(S_{g,n})$ is a finite-index subgroup of the larger of the two stabilizers of its vertices. Thus edge stabilizers are finitely presented, and in particular they are finitely generated.

We thus have that $\mathrm{Mod}(S_{g,n})$ acts on the contractible simplicial complex $\mathcal{A}(S)$ without rotations, with finitely presented vertex stabilizers and finitely generated edge stabilizers. Applying Proposition 5.6 to this action gives that $\mathrm{Mod}(S_{g,n})$ is finitely presented. $\qquad\square$

5.4 HOPF'S FORMULA AND $H_2(\mathrm{Mod}(S);\mathbb{Z})$

In Section 5.1.2 we computed $H_1(\mathrm{Mod}(S);\mathbb{Z})$. In this section we compute $H_2(\mathrm{Mod}(S);\mathbb{Z})$. As with first homology, the second homology is a basic invariant of a group G. For example, if $H_2(G;\mathbb{Z})$ is infinitely generated, then G has no finite presentation. The precise connection between $H_2(G;\mathbb{Z})$ and presentations for G is made explicit by Hopf's formula below. Later we will see that $H_2(G;\mathbb{Z})$ is related to $H^2(G;\mathbb{Z})$, which in turn classifies cyclic central extensions of G.

THEOREM 5.8 (Harer) *Let* $g \geq 4$. *Let* S_g^1 *denote a compact surface of genus* g *with one boundary component. Then we have the following isomorphisms:*

$$
\begin{array}{rll}
(i) & H_2(\mathrm{Mod}(S_g);\mathbb{Z}) & \approx \quad \mathbb{Z} \\
(ii) & H_2(\mathrm{Mod}(S_g^1);\mathbb{Z}) & \approx \quad \mathbb{Z} \\
(iii) & H_2(\mathrm{Mod}(S_{g,1});\mathbb{Z}) & \approx \quad \mathbb{Z}^2.
\end{array}
$$

In general, if S is a surface of genus $g \geq 4$ with b boundary components and p punctures, then $H_2(\mathrm{Mod}(S);\mathbb{Z}) \approx \mathbb{Z}^{p+1}$; see [83, 126]. Harer also proved that $H_3(\mathrm{Mod}(S_g);\mathbb{Q}) = 0$ for $g \geq 3$ [85] and $H_4(\mathrm{Mod}(S_g);\mathbb{Q}) \approx \mathbb{Q}^2$ for $g \geq 10$ [82]. The groups $H_k(\mathrm{Mod}(S_g))$ have not been computed for $k \geq 5$, although it is known that $H_k(\mathrm{Mod}(S_g))$ does not depend on g for g large [86].

Harer proved Theorem 5.8 by reducing to the case where S has boundary and using the action of $\mathrm{Mod}(S)$ on the arc complex associated to S. Pitsch gave a completely different proof of the upper bound in Theorem 5.8. That is, he showed that $H_2(\mathrm{Mod}(S_g^1);\mathbb{Z})$ is a quotient of $\mathbb{Z}$. He realized that one can actually apply Hopf's formula to Wajnryb's explicit presentation of $\mathrm{Mod}(S)$. In this section we present what is essentially Pitsch's proof from [179], together with the variations on his argument that are required for the cases of S_g and $S_{g,1}$.

5.4.1 THE HOPF FORMULA

Let G be any group with a finite presentation $G = \langle F | R \rangle$. The group G can also be thought of as F/K, where K is the normal subgroup generated by the *relators*, namely, the elements of R. The classical *Hopf formula* states that

$$
H_2(G;\mathbb{Z}) \approx \frac{K \cap [F,F]}{[K,F]}.
$$

So elements of $H_2(G; \mathbb{Z})$ are cosets represented by relators in G—that is, elements of K—that are products of commutators in F. Given a relator k, we think of any conjugate relator fkf^{-1} as being redundant, and that is why we take the quotient by $[K, F]$. See Brown's book [38, Theorem 5.3] for a proof of Hopf's formula.

The group $(K \cap [F, F])/[K, F]$ is a subgroup of the abelian group $K/[K, F]$. Therefore, as K is normally generated by the finitely many elements of R, the group $K/[K, F]$ is an abelian group generated by the cosets represented by the finitely many elements of R. Hopf's formula thus implies that any element of $H_2(G; \mathbb{Z})$ can be represented (nonuniquely) as a product $\prod r_i^{n_i}$, where $R = \{r_1, \ldots r_N\}$ and $n_i \in \mathbb{Z}$.

5.4.2 THE HOPF FORMULA APPLIED TO THE WAJNRYB PRESENTATION

We start with the case of S_g^1 with $g \geq 4$. We will use Wajnryb's presentation for $\mathrm{Mod}(S_g^1)$, in particular, using the notation from Theorem 5.3. Pitsch's idea is to plug this presentation into Hopf's formula.

We can rewrite each relation from Theorem 5.3 so that we get a word in the generators for $\mathrm{Mod}(S_g^1)$ that is equal to the identity element of $\mathrm{Mod}(S_g^1)$, that is, a relator. We do this by moving all generators to the left-hand side of each relation. We will use the following notation for the relators:

(i)	Disjointness relators	$[a_i, a_j]$	denoted $D_{i,j}$
(ii)	Braid relators	$a_i a_j a_i (a_j a_i a_j)^{-1}$	denoted $B_{i,j}$
(iii)	3-chain relator	$(a_1 a_2 a_3)^4 (a_0 b_0)^{-1}$	denoted C
(iv)	Lantern relator	$(a_0 b_2 b_1)(a_1 a_3 a_5 b_3)^{-1}$	denoted L

In the first two relators, only certain pairs (i, j) are allowed, as governed by the statement of Theorem 5.3. We will not need the precise forms of the relators here—that is, we will not write out the b_i in terms of the a_i—but rather we will only need the number of times, with sign, each a_i appears in each relator. We will give these numbers as needed, though the reader can easily read them off from Theorem 5.3.

Let F be the free group generated by the a_i and let K denote the subgroup of F normally generated by the above relators. As in the above discussion, any element x of the abelian group $K/[K, F]$ is a coset represented by an element of the form

$$x = \left(\prod D_{i,j}^{n_{i,j}} \right) \left(\prod_{i=1}^{2g-1} B_{i,i+1}^{n_i} \right) B_{0,4}^{n_0} C^{n_C} L^{n_L}, \tag{5.1}$$

where the exponents are integers. In the remainder of the proof, we will ignore the distinction between the coset given by such an element of $K/[K, F]$ and the actual element of $K/[K, F]$.

According to Hopf's formula, $H_2(\text{Mod}(S_g^1); \mathbb{Z})$ is isomorphic to the subgroup $(K \cap [F, F])/[K, F]$ of $K/[K, F]$. So which elements of $K/[K, F]$ given by (5.1) are also elements of $[F, F]/[K, F]$? One obvious condition is that the exponent sum of each a_i must be zero. Actually, we will show that, up to multiples, there is at most one element of the form (5.1) that satisfies this condition.

5.4.3 COMMUTING RELATORS

We begin by analyzing the simplest relators, namely, the commuting relators $D_{i,j}$. We will show that each represents the trivial element of $H_2(\text{Mod}(S_g^1); \mathbb{Z})$, and hence these terms can be ignored in (5.1). Choose some particular $D_{i,j} = [a_i, a_j]$. As an element of $F = \langle a_i \rangle$, this word certainly lives in $K \cap [F, F]$, where K is the normal subgroup of F generated by the relators. Our goal is to show that it also lies in $[F, K]$.

In general, if g and h are two commuting elements of $\text{Mod}(S_g^1)$, then $[g, h]$ is an element of $K \cap [F, F]$. Let $\{g, h\}$ denote the corresponding element (coset) in $H_2(\text{Mod}(S_g^1); \mathbb{Z})$.

If g is an element of $\text{Mod}(S_g^1)$ that commutes with the elements h and k of $\text{Mod}(S_g^1)$, then

$$\{g, hk\} = \{g, h\} + \{g, k\} \tag{5.2}$$

in $H_2(\text{Mod}(S_g^1); \mathbb{Z})$. This follows from the fact that, for any three elements x, y, and z in the free group F, we have

$$[x, yz] = [x, y][x, z]^y.$$

We have also used the fact that conjugation "does nothing" in the quotient $(K \cap [F, F])/[K, F]$. It is also easy to check that

$$\{g, h^{-1}\} = -\{g, h\}. \tag{5.3}$$

Lemma 5.9 *Let $g \geq 4$. If a and b are disjoint nonseparating simple closed curves in S_g^1, then $\{T_a, T_b\} = 0$ in $H_2(\text{Mod}(S_g^1); \mathbb{Z})$.*

Proof. We cut S_g^1 along a and obtain a compact surface S' of genus $g - 1$ with three boundary components. The simple closed curve b can be thought of as a simple closed curve on S', and so the Dehn twist T_b can be thought of as an element of $\text{Mod}(S')$. Since $g \geq 4$, we have $g - 1 \geq 3$, so by Theorem 5.2 $\text{Mod}(S')$ has trivial abelianization; that is, it is perfect. We can thus write T_b as a product of commutators $T_b = \prod [x_i, y_i]$, where each x_i and y_i is an element of $\mathrm{Mod}(S')$ and so commutes with T_a.

Using (5.2) and (5.3), we then obtain

$$
\begin{aligned}
\{T_a, T_b\} &= \{T_a, \textstyle\prod_i [x_i, y_i]\} \\[2mm]
&= \sum_i \{T_a, [x_i, y_i]\} \\[2mm]
&= \sum_i [\{T_a, x_i\} + \{T_a, y_i\} - \{T_a, x_i\} - \{T_a, y_i\}] \\[2mm]
&= 0.
\end{aligned}
$$

$\square$

Lemma 5.9 has a topological interpretation. Let $[T^2] \in H_2(T^2; \mathbb{Z}) \approx H_2(\mathbb{Z}^2; \mathbb{Z}) \approx \mathbb{Z}$ denote the fundamental class. Two commuting Dehn twists $g, h \in \mathrm{Mod}(S)$ determine an inclusion $\mathbb{Z}^2 \to \mathrm{Mod}(S)$. This homomorphism determines (up to homotopy) a based map η from the classifying space $K(\mathbb{Z}^2, 1) \approx T^2$ to the classifying space $K(\mathrm{Mod}(S), 1)$. Let $i_* : H_2(\mathbb{Z}^2; \mathbb{Z}) \to H_2(\mathrm{Mod}(S); \mathbb{Z})$ be the induced homomorphism. Lemma 5.9 says precisely that i_* is the zero map.

It follows immediately from Lemma 5.9 that each $D_{i,j}$ represents the trivial element of $H_2(\mathrm{Mod}(S_g^1); \mathbb{Z})$. From this fact and (5.1), we now have that any element x of $(K \cap [F, F])/[K, F]$ has the form

$$
x = \left(\prod_{i=1}^{2g-1} B_{i,j}^{n_i} \right) B_{0,4}^{n_0} C^{n_C} L^{n_L}. \tag{5.4}
$$

5.4.4 COMPLETING THE PROOF

Let $x \in (K \cap [F, F])/[K, F]$. We have shown that x has the form given in (5.4). We will now use the exponent sum condition for elements of $[F, F]$ to reduce the possibilities for x further.

Each relator on the right-hand side of (5.4) is a product of the generators $\{a_i : 1 \le i \le 2g\}$ of F. In order that x lie in $[F, F]$ it must be that the exponent sum of each a_i occurring in x is 0. The only relator involving a_{2g} is $B_{2g-1, 2g}$, in which a_{2g} has exponent sum 1. Thus in the word $B_{2g-1, 2g}^{n_{2g-1}}$ the total exponent sum of a_{2g} is n_{2g-1}. Since no other relator contains a_{2g} and since the exponent sum of a_{2g} in x is 0, it follows that n_{2g-1} is 0. We can thus delete the relator $B_{2g-1, 2g}$ from the expression (5.4) for x.

Now note that the only relator left on the right-hand side of (5.4) involving a_{2g-1} is $B_{2g-2,2g-1}$. By the same argument as above we conclude that $n_{2g-2} = 0$. Continuing in this way, we obtain that $n_i = 0$ for each $i \geq 6$; we stop at $B_{5,6}$ because both a_5 and a_6 appear in other (nonbraid) relators.

Since a_6 appears in $B_{5,6}$ with a total exponent of -1 and since the only other relator in which a_6 appears is L, where it has an exponent sum of 0, it follows that $n_{5,6} = 0$.

At this point we have shown that any element $x \in (K \cap [F,F])/[K,F]$ has the form

$$x = B_{0,4}^{n_0} B_{1,2}^{n_1} B_{2,3}^{n_2} B_{3,4}^{n_3} B_{4,5}^{n_4} C^{n_C} L^{n_L}.$$

The power of the preceding arguments is that, for arbitrary $g \geq 4$, we have reduced the problem to understanding just seven relators and that these relators involve only the generators $a_0, \ldots, a_5$.

Again, in order to get an element of $(K \cap [F,F])/[K,F]$, the exponent sums of each of the six generators $a_0, \ldots, a_5$ must be zero. Since, for example, a_5 occurs in $B_{4,5}$ with exponent sum -1, and in L with exponent sum -1, the fact that the total exponent of a_5 must be 0 gives the equation $-n_4 - n_L = 0$. Continuing in this way, setting each of the exponent sums of $a_0, \ldots, a_5$ equal to 0, we obtain the following system of equations.

$$\begin{pmatrix} 1 & 0 & 0 & 0 & 0 & -2 & 2 \\ 0 & 1 & 0 & 0 & 0 & 4 & -1 \\ 0 & -1 & 1 & 0 & 0 & 4 & 0 \\ 0 & 0 & -1 & 1 & 0 & 4 & -1 \\ -1 & 0 & 0 & -1 & 1 & 0 & 0 \\ 0 & 0 & 0 & 0 & -1 & 0 & -1 \end{pmatrix} \begin{pmatrix} n_0 \\ n_1 \\ n_2 \\ n_3 \\ n_4 \\ n_C \\ n_L \end{pmatrix} = \begin{pmatrix} 0 \\ 0 \\ 0 \\ 0 \\ 0 \\ 0 \end{pmatrix}.$$

An elementary calculation gives that the above matrix has rank 6, and so the linear mapping $\mathbb{Z}^7 \to \mathbb{Z}^6$ has a 1-dimensional kernel. So there is at most one element (up to multiples) that satisfies the given linear equations. A quick check gives that all solutions are simply integral multiples of the vector $(-18, 6, 2, 8, -10, 1, 10)$. It follows that the only possibilities for the arbitrary element $x \in K \cap [F,F]/[F,K] \approx H_2(\mathrm{Mod}(S_g^1); \mathbb{Z})$ are integral powers of the element

$$x_0 = B_{0,4}^{-18} B_{1,2}^6 B_{2,3}^2 B_{3,4}^8 B_{4,5}^{-10} C L^{10}.$$

In other words, x_0 generates $H_2(\mathrm{Mod}(S_g^1); \mathbb{Z})$. Note that we still do not know whether or not x_0 is trivial in $H_2(\mathrm{Mod}(S_g^1); \mathbb{Z})$. We will prove below, by a completely different line of argument, that x_0 has infinite order.

5.4.5 PITSCH'S PROOF FOR CLOSED SURFACES

To extend Pitsch's proof to the case of a closed surface S_g ($g \geq 4$), we only need to show that the hyperelliptic relation from Wajnryb's presentation does not contribute to $H_2(\text{Mod}(S_g); \mathbb{Z})$. The argument, due to Korkmaz–Stipcisz [126] is similar to the proof that the disjointness relations do not contribute.

Recall that the hyperelliptic relation is

$$[a_{2g} \cdots a_1 a_1 \cdots a_{2g}, d] = 1.$$

One would like to directly apply the proof of Lemma 5.9. However, if we cut S_g along a representative of d, the hyperelliptic involution $a_{2g} \cdots a_1 a_1 \cdots a_{2g}$ does not induce an element of the pure mapping class group of the cut surface (it switches the two sides of d). Therefore, we cannot write d as a product of commutators of elements that commute with d.

We must therefore proceed with a different argument. Our first claim is that if a and b are isotopy classes of simple closed curves in S_g with $i(a, b) = 1$, then $\{T_a, (T_a T_b T_a)^2\} = 0$ in $H_2(\text{Mod}(S_g); \mathbb{Z})$. We proceed in three steps. Throughout, we apply the formula (5.2) without mention.

Step 1. The classes $\{T_a, (T_a T_b T_a)^2\}$ and $\{T_b, (T_a T_b T_a)^2\}$ are equal.

Let r be an element of $\text{Mod}(S_g)$ that interchanges a and b. We have

$$\{T_a, (T_a T_b T_a)^2\} = \{r T_a r^{-1}, r(T_a T_b T_a)^2 r^{-1}\}$$
$$= \{T_b, (T_b T_a T_b)^2\}$$
$$= \{T_b, (T_a T_b T_a)^2\}.$$

Step 2. The class $2\{T_a, (T_a T_b T_a)^2\}$ is trivial.

The braid relation gives that $(T_a T_b T_a)^4 = (T_a T_b)^6$, and the 2-chain relation gives that this product is equal to the Dehn twist about the simple closed curve c which is the boundary of a regular neighborhood of minimal position representatives of a and b. We then have

$$2\{T_a, (T_a T_b T_a)^2\} = \{T_a, (T_a T_b T_a)^4\} = \{T_a, T_c\} = 0,$$

where in the last step we have applied Lemma 5.9.

Step 3. The class $3\{T_a, (T_a T_b T_a)^2\}$ is trivial.

To prove this equality, we apply step 1, which gives

$$
\begin{aligned}
3\{T_a, (T_a T_b T_a)^2\} &= \{T_a, (T_a T_b T_a)^2\} + \{T_a, (T_a T_b T_a)^2\} + \{T_a, (T_a T_b T_a)^2\} \\
&= \{T_a, (T_a T_b T_a)^2\} + \{T_b, (T_a T_b T_a)^2\} + \{T_a, (T_a T_b T_a)^2\} \\
&= \{T_a T_b T_a, (T_a T_b T_a)^2\} \\
&= 0.
\end{aligned}
$$

Steps 2 and 3 immediately imply the claim. We can now show that the hyperelliptic relator contributes zero to $H_2(\mathrm{Mod}(S_g); \mathbb{Z})$. In the calculation, we use the identity $\{x, y\} = \{x, x^j y x^k\}$, which follows from formula (5.2) and the fact that $\{x, x\} = 0$. Denote the product $a_{2g-1} \cdots a_1 a_1 \cdots a_{2g-1}$ by A. If a_{2g} represents the Dehn twist $T_{c_{2g}}$, one can check that $A(c_{2g}) = d^2(c_{2g})$, and so $A a_{2g} A^{-1} = d^2 a_{2g} d^{-2}$. We therefore have

$$
\begin{aligned}
\{d, a_{2g} \cdots a_1 a_1 \cdots a_{2g}\} &= \{d, a_{2g} A a_{2g}\} \\
&= \{d, a_{2g} A a_{2g} A^{-1}\} \\
&= \{d, a_{2g} d^2 a_{2g} d^{-2}\} \\
&= \{d, d a_{2g} d^2 a_{2g} d\} \\
&= \{d, (d a_{2g} d)^2\} \\
&= 0.
\end{aligned}
$$

Here the last equality follows from the claim. This completes the proof.

5.5 THE EULER CLASS

In Section 5.4 we proved two of the upper bounds for Theorem 5.8. That is, we showed that $H_2(\mathrm{Mod}(S); \mathbb{Z})$ is cyclic when $S = S_g$ or $S = S_g^1$. In Section 5.6.3 we will use homological algebra to show that $H_2(\mathrm{Mod}(S_{g,1}); \mathbb{Z})$ is generated by at most two elements.

In this section we explicitly construct an infinite-order element of $H^2(S_{g,1}; \mathbb{Z})$ called the Euler class. This will be used, together with the universal coefficients theorem, to provide one of the lower bounds for Theorem 5.8.

The Euler class is not just some element of a cohomology group; it is the most basic and fundamental invariant of surface bundles.

5.5.1 COCYCLES FROM CENTRAL EXTENSIONS

We first recall how central extensions of a group give rise to 2-dimensional cohomology classes. For a more detailed explanation, see, for example, [38, Section IV.3]. Let

$$1 \to A \to E \to G \to 1 \tag{5.5}$$

be a central extension of the group G; in other words, A is central in E and the sequence (5.5) is exact. Note that A is abelian since it lies in the center of E.

If the extension (5.5) is split, then since A is central, it follows that $E \approx A \times G$. Even if E does not split, we still have a (noncanonical) bijection $\phi : A \times G \to E$ obtained by simply picking any set-theoretic section ψ of the map $E \to G$. Moreover, there exists a function $f : G \times G \to A$, called a *factor set*, so that

$$\phi(a_1, g_1)\phi(a_2, g_2) = \phi(a_1 a_2 f(g_1, g_2), g_1 g_2).$$

The factor set f measures the failure of the section ψ to be a homomorphism, or equivalently, the failure of ϕ to be an isomorphism.

While ϕ, and hence f, depended on the choice of section ψ, one can check that f does represent a well-defined element ξ of $H^2(G; A)$. That is, the element ξ depends only on the extension (5.5) and not on any of the choices. The sequence (5.5) splits precisely when the cohomology class ξ is trivial.

5.5.2 THE CLASSICAL EULER CLASS

Before we construct the Euler class in $H^2(\mathrm{Mod}(S_{g,1}); \mathbb{Z})$, we recall the classical Euler class, which is an element of $H^2(\mathrm{Homeo}^+(S^1); \mathbb{Z})$.

Consider the covering $\mathbb{R} \to S^1$ given by the quotient of $\mathbb{R}$ by the group $\mathbb{Z}$ generated by the translation $t(x) = x + 1$. The set of all lifts of an element $\psi \in \mathrm{Homeo}^+(S^1)$ to $\mathrm{Homeo}^+(\mathbb{R})$ is precisely the set of elements of the form $\widetilde{\psi} \circ t^m$ for $m \in \mathbb{Z}$, where $\widetilde{\psi}$ is any fixed lift of ψ. Let $\widetilde{\mathrm{Homeo}}^+(S^1)$ denote the group of all lifts of all elements of $\mathrm{Homeo}^+(S^1)$. In other words, $\widetilde{\mathrm{Homeo}}^+(S^1)$ is the subgroup of $\mathrm{Homeo}^+(\mathbb{R})$ consisting of those homeomorphisms that commute with t, that is, the group of periodic homeomorphisms of period 1. We thus have an exact sequence

$$1 \to \mathbb{Z} \to \widetilde{\mathrm{Homeo}}^+(S^1) \xrightarrow{\pi} \mathrm{Homeo}^+(S^1) \to 1, \tag{5.6}$$

where $\mathbb{Z}$ is generated by t and is thus central. We know that the sequence

(5.6) does not split since $\text{Homeo}^+(S^1)$ has torsion while $\widetilde{\text{Homeo}}^+(S^1)$, being a subgroup of $\text{Homeo}^+(\mathbb{R})$, is torsion-free. As explained above, it follows that the short exact sequence (5.6) gives rise to a nontrivial element of $H^2(\text{Homeo}^+(S^1); \mathbb{Z})$. This element is called the *Euler class*. This Euler class is the most important invariant in the study of circle bundles.

5.5.3 THE EULER CLASS FOR THE MAPPING CLASS GROUP

Let $g \geq 2$. We will show that there is a torsion-free group $\widetilde{\text{Mod}}(S_{g,1})$ and a central extension

$$1 \to \mathbb{Z} \to \widetilde{\text{Mod}}(S_{g,1}) \to \text{Mod}(S_{g,1}) \to 1. \qquad (5.7)$$

Since $\text{Mod}(S_{g,1})$ contains torsion, it follows that the short exact sequence (5.7) does not split, and so we thus obtain a nontrivial element of $H^2(\text{Mod}(S_{g,1}); \mathbb{Z})$ called the Euler class.

We now give two different constructions of the Euler class; that is, we give two derivations of the short exact sequence (5.7) defining the Euler class. The first comes directly from the classical Euler class.

5.5.4 THE EULER CLASS VIA LIFTED MAPPING CLASSES

In Section 8.2 (cf. Theorem 8.7) we will prove that an element of $\text{Mod}(S_{g,1})$ gives rise to a homeomorphism of the circle at infinity in hyperbolic space as follows. Assume that $g \geq 2$ and regard the puncture of $S_{g,1}$ as a marked point p. If we choose a hyperbolic metric on the closed surface S_g, its universal cover is isometric to $\mathbb{H}^2$. Let $\tilde{p}$ be some distinguished lift of p to $\mathbb{H}^2$.

We can represent any $f \in \text{Mod}(S_{g,1})$ by a homeomorphism $\phi : S_g \to S_g$ such that $\phi(p) = p$. There is a unique lift of ϕ to a homeomorphism $\tilde{\phi} : \mathbb{H}^2 \to \mathbb{H}^2$ such that $\tilde{\phi}(\tilde{p}) = \tilde{p}$. In Section 8.2, we will prove that $\tilde{\phi}$ is a $\pi_1(S_g)$-equivariant quasi-isometry of $\mathbb{H}^2$ and that $\tilde{\phi}$ can be extended in a unique way to a homeomorphism

$$\tilde{\phi} \cup \partial\tilde{\phi} : \mathbb{H}^2 \cup \partial\mathbb{H}^2 \to \mathbb{H}^2 \cup \partial\mathbb{H}^2$$

of the closed unit disk. Restricting to $\partial\mathbb{H}^2 \approx S^1$, we obtain an element $\partial\tilde{\phi} \in \text{Homeo}^+(S^1)$. Since S_g is compact, homotopies of S_g move points by a uniformly bounded amount, and so $\partial\tilde{\phi}$ does not depend on the choice of representative ϕ.

We thus have a well-defined map

$$\text{Mod}(S_{g,1}) \hookrightarrow \text{Homeo}^+(S^1).$$

This map is clearly a homomorphism. It is injective because if $\partial\widetilde{\phi}$ fixes each $\gamma_\infty^{\pm 1} \in \partial\mathbb{H}^2$ (using the notation from Section 8.2), it follows that $\phi_\star$ fixes each $\gamma \in \pi_1(S_g)$, and then, since S_g is a $K(G, 1)$-space, it follows that ϕ is homotopic (hence isotopic) to the identity. The construction of the map $\mathrm{Mod}(S_{g,1}) \to \mathrm{Homeo}^+(S^1)$ is due to Nielsen; he used this as a starting point for his analysis and classification of mapping classes.

We finally define the group $\widetilde{\mathrm{Mod}}(S_{g,1})$ as the pullback of $\mathrm{Mod}(S_{g,1})$ to $\widetilde{\mathrm{Homeo}}^+(S^1)$:

$$1 \to \mathbb{Z} \to \widetilde{\mathrm{Mod}}(S_{g,1}) \to \mathrm{Mod}(S_{g,1}) \to 1. \tag{5.8}$$

Thus $\widetilde{\mathrm{Mod}}(S_{g,1})$ is the subgroup of elements of $\widetilde{\mathrm{Homeo}}^+(S^1)$ that project into $\mathrm{Mod}(S_{g,1})$. Because the kernel $\mathbb{Z}$ is central in $\widetilde{\mathrm{Homeo}}^+(S^1)$, it is central in $\widetilde{\mathrm{Mod}}(S_{g,1})$. As above, the central extension (5.8) has an associated cocycle, giving an element $e \in H^2(\mathrm{Mod}(S_{g,1}); \mathbb{Z})$. The element e is called the Euler class for $\mathrm{Mod}(S_{g,1})$.

The group $\widetilde{\mathrm{Mod}}(S_{g,1})$ is torsion-free because it is a subgroup of $\widetilde{\mathrm{Homeo}}^+(S^1)$, which we already noted was torsion-free. On the other hand, $\mathrm{Mod}(S_{g,1})$ has nontrivial torsion (e.g., take any rotation fixing the marked point). As above, it follows that (5.8) does not split, so e is nontrivial. We will later see that e has infinite order in $H^2(\mathrm{Mod}(S_{g,1}); \mathbb{Z})$.

Note that the Euler class for $\mathrm{Mod}(S_{g,1})$ is the pullback of the classical Euler class under the map on cohomology induced by the inclusion $\mathrm{Mod}(S_{g,1}) \to \mathrm{Homeo}^+(S^1)$.

5.5.5 THE RESTRICTION OF THE EULER CLASS TO THE POINT-PUSHING SUBGROUP

We will next evaluate the Euler class $e \in H^2(\mathrm{Mod}(S_{g,1}); \mathbb{Z})$ on a concrete 2-cycle, namely, the one coming from the point-pushing subgroup. We will do this by constructing an easy-to-evaluate cohomology class and by proving that this class equals the Euler class.

Let $g \geq 2$. Recall from Section 4.2 that the point-pushing map is an injective homomorphism $\mathcal{P}ush : \pi_1(S_g) \hookrightarrow \mathrm{Mod}(S_{g,1})$. We can thus pull back the Euler class $e \in H^2(\mathrm{Mod}(S_{g,1}); \mathbb{Z})$ to an element $\mathcal{P}ush^*(e) \in H^2(\pi_1(S_g); \mathbb{Z}) \approx \mathbb{Z}$. Let $\widetilde{\pi_1(S_g)}$ denote the pullback of the subgroup $\pi_1(S_g) < \mathrm{Homeo}^+(S^1)$ to $\widetilde{\mathrm{Homeo}}^+(S^1)$. We have that $\mathcal{P}ush^*(e)$ is the cocycle associated to the following central extension:

$$1 \to \mathbb{Z} \to \widetilde{\pi_1(S_g)} \to \pi_1(S_g) \to 1.$$

Another way to obtain an element of $H^2(\pi_1(S_g); \mathbb{Z})$ is by considering the unit tangent bundle $S^1 \to UT(S_g) \to S_g$. Since S_g is aspherical, the long exact sequence associated to this fiber bundle gives a short exact sequence

$$1 \to \mathbb{Z} \to \pi_1(UT(S_g)) \to \pi_1(S_g) \to 1.$$

This is a central extension, and so it has an associated cocycle $e' \in H^2(S_g; \mathbb{Z})$. We claim that e' is nontrivial. If e were trivial, then there would be a splitting $\pi_1(S_g) \to \pi_1(UT(S_g))$ and hence a section of $UT(S_g) \to S_g$. The latter would give a nonvanishing vector field on S_g, which is prohibited by the Poincaré–Bendixon theorem (for $g \geq 2$). We thus have that e' is nontrivial. In fact, this argument gives that e' has infinite order in $H^2(S_g; \mathbb{Z})$. Indeed, the extension given by ke' is

$$1 \to k\mathbb{Z} \to \pi_1(UT(S_g)) \to \pi_1(S_g) \to 1.$$

If this extension were trivial for some $k \neq 0$, we would again have a nonvanishing vector field on S_g.

Proposition 5.10 *The elements* $\mathcal{P}ush^*(e)$ *and* e' *of* $H^2(\pi_1(S_g); \mathbb{Z})$ *are equal.*

Proposition 5.10 implies that the evaluation of the pullback via $\mathcal{P}ush^*$ of the Euler class for $\mathrm{Mod}(S_{g,1})$ on the fundamental class of $\pi_1(S_g)$ is the Euler number of the unit tangent bundle, which is equal to $2 - 2g$ (the Euler number of the tangent bundle to a Riemannian manifold is always equal to the Euler characteristic of the manifold). In particular, we have that the Euler class for $\mathrm{Mod}(S_{g,1})$ is nontrivial even when restricted to the point-pushing subgroup.

Proof. By the five lemma it suffices to exhibit a homomorphism $\pi_1(UT(S_g)) \to \widetilde{\pi_1(S_g)}$ that makes the following diagram commutative:

$$
\begin{array}{ccccccccc}
1 & \longrightarrow & \mathbb{Z} & \longrightarrow & \pi_1(UT(S_g)) & \longrightarrow & \pi_1(S_g) & \longrightarrow & 1 \\
& & \downarrow & & \downarrow & & \downarrow & & \\
1 & \longrightarrow & \mathbb{Z} & \longrightarrow & \widetilde{\pi_1(S_g)} & \longrightarrow & \pi_1(S_g) & \longrightarrow & 1
\end{array}
$$

The key is the following claim.

> *Claim*: The image of $\pi_1(S_g)$ in $\mathrm{Homeo}^+(S^1)$ given by the composition $\pi_1(S_g) \to \mathrm{Mod}(S_{g,1}) \to \mathrm{Homeo}^+(S^1)$ coincides

with the image of the composition $\pi_1(S_g) \to \mathrm{Isom}^+(\mathbb{H}^2) \to \mathrm{Homeo}^+(S^1)$ obtained by identifying $\pi_1(S_g)$ with the group of deck transformations of the covering $\mathbb{H}^2 \to S_g$.

Proof of claim. For $\alpha \in \pi_1(S_g, p)$, we have that $\mathcal{P}ush(\alpha)$ acts by conjugation on $\pi_1(S_g, p)$, and so the lift of any representative of $\mathcal{P}ush(\alpha)$ fixing $\widetilde{p}$ sends $\gamma \cdot \widetilde{p}$ to $(\alpha\gamma\alpha^{-1}) \cdot \widetilde{p}$ for all $\gamma \in \pi_1(S_g, p)$. On the other hand, the deck transformation corresponding to α sends $\gamma \cdot \widetilde{p}$ to $(\alpha\gamma) \cdot \widetilde{p}$. We can modify this deck transformation by pushing each point $(\alpha\gamma) \cdot \widetilde{p}$ along the unique lift of α^{-1} starting at that point. This induces an isotopy of $\mathbb{H}^2$ moving points a uniformly bounded amount and hence does not change the corresponding element of $\mathrm{Homeo}^+(S^1)$. At the end of this isotopy, each point $(\alpha\gamma) \cdot \widetilde{p}$ gets sent to $(\alpha\gamma\alpha^{-1}) \cdot \widetilde{p}$. Since the lift of $\mathcal{P}ush(\alpha)$ and the (modified) deck transformation corresponding to α agree on the orbit of $\widetilde{p}$, they induce the same element of $\mathrm{Homeo}^+(S^1)$. $\square$

Now let $\widehat{\alpha}$ be an element of $\pi_1(UT(S_g))$. In order to construct the associated element $\widetilde{\psi} \in \widetilde{\mathrm{Homeo}}^+(S^1)$, we need two ingredients:

1. a homeomorphism $\psi \in \mathrm{Homeo}^+(S^1)$, and

2. a path τ in S^1 from some basepoint $x_0 \in S^1$ to $\psi(x_0)$.

Indeed, if $\widehat{x}_0$ is some fixed lift of x_0 to $\mathbb{R}$, and $\widehat{\tau}$ is the unique lift of the path τ starting at x_0, then we can take $\widetilde{\psi}$ to be the unique element of $\widetilde{\mathrm{Homeo}}^+(\mathbb{R})$ that lifts ψ and takes $\widehat{x}_0$ to the endpoint of $\widehat{\tau}$.

After constructing $\widetilde{\psi}$, we will then need to check that it actually lies in $\widetilde{\pi_1(S_g)}$.

As in Section 4.2, the element $\widehat{\alpha} \in \pi_1(UT(S_g))$ gives an element $f_{\widehat{\alpha}} \in \mathrm{Mod}(S_g, (p, v))$, the group of isotopy classes of diffeomorphisms of S_g fixing the point-vector pair (p, v). The mapping class $f_{\widehat{\alpha}}$ is the class of a diffeomorphism $\phi_{\widehat{\alpha}}$ obtained at the end of a smooth isotopy of S_g pushing (p, v) along $\widehat{\alpha}$. By taking the unique lift $\widetilde{\phi}_{\widehat{\alpha}}$ of $\phi_{\widehat{\alpha}}$ to $\mathrm{Homeo}^+(\mathbb{H}^2)$ that fixes the point $\widetilde{p}$, we obtain a well-defined homeomorphism $\overline{f}_{\widehat{\alpha}} \in \mathrm{Homeo}^+(S^1)$ as before. For example, in the case that $\widehat{\alpha}$ is the central element of $\pi_1(UT(S_g))$, the lift of $\phi_{\widehat{\alpha}}$ simply rotates a neighborhood of each lift of p, and the induced element of $\mathrm{Homeo}^+(S^1)$ is trivial.

The homeomorphism $\overline{f}_{\widehat{\alpha}}$ is the desired element of $\mathrm{Homeo}^+(S^1)$. It remains to construct the path τ in S^1 from some fixed basepoint x_0 to $\overline{f}_{\widehat{\alpha}}(x_0)$.

If we forget the datum of the vector v and remember only the point p, then $f_{\widehat{\alpha}}$ also represents $\mathcal{P}ush(\alpha)$, where $\alpha \in \pi_1(S_g)$ is the image of $\widehat{\alpha}$ under the forgetful map $\pi_1(UT(S_g)) \to \pi_1(S_g)$. Thus it follows from the claim that as an element of $\mathrm{Homeo}^+(S^1)$ the mapping class $f_{\widehat{\alpha}}$ agrees with the deck transformation corresponding to α.

Let $(\widetilde{p}, \widetilde{v})$ be a fixed lift of (p, v) to $UT(\mathbb{H}^2)$. Let x_0 be the point of $\partial \mathbb{H}^2 \approx S^1$ to which $(\widetilde{p}, \widetilde{v})$ points. Because $f_{\widehat{\alpha}}$ agrees with the deck transformation α and since deck transformations are isometries, the lifted map $\widetilde{\phi}_{\widehat{\alpha}}$ takes $(\widetilde{p}, \widetilde{v})$ to an element of $UT(\mathbb{H}^2)$ that points to $\overline{f}_{\widehat{\alpha}}(x_0)$.

Recall that $\phi_{\widehat{\alpha}}$ is a diffeomorphism obtained at the end of a smooth isotopy of S_g. Thus $\widetilde{\phi}_{\widehat{\alpha}}$ is a diffeomorphism obtained at the end of a smooth isotopy of $\mathbb{H}^2$. At each point in time during the isotopy of $\mathbb{H}^2$, the pair $(\widetilde{p}, \widetilde{v})$ has a well-defined image, which in turn points to some point on $\partial \mathbb{H}^2$. Thus the isotopy of $\mathbb{H}^2$ coming from $\widehat{\alpha}$ determines a path $\tau_{\widehat{\alpha}}$ in $\partial \mathbb{H}^2 \approx S^1$. Again, at the end of the isotopy, the image of $(\widetilde{p}, \widetilde{v})$ points to the image of x_0, and so $\tau_{\widehat{\alpha}}$ satisfies the desired properties.

We have thus obtained the desired element of $\widetilde{\mathrm{Homeo}}^+(S^1)$. Since the claim implies that $f_{\widehat{\alpha}}$ agrees with a deck transformation, we have in fact constructed an element of $\widetilde{\pi_1(S_g)}$. It follows easily from the above discussion that the resulting map $\pi_1(UT(S_g)) \to \widetilde{\pi_1(S_g)}$ is well defined and that it satisfies the desired commutativity, and we are done. $\qquad\square$

5.5.6 THE EULER CLASS VIA CAPPING THE BOUNDARY

We now give a different construction of the group $\widetilde{\mathrm{Mod}}(S_{g,1})$ and hence a different derivation of the Euler class for $\mathrm{Mod}(S_{g,1})$. Let S_g^1 be the genus g surface with one boundary component. Recall from Proposition 3.19 that there is a short exact sequence

$$1 \to \mathbb{Z} \to \mathrm{Mod}(S_g^1) \to \mathrm{Mod}(S_{g,1}) \to 1 \tag{5.9}$$

where the kernel $\mathbb{Z}$ is generated by the Dehn twist about the boundary of S_g^1 and is thus central. Since the extension is central, it gives an element $e'' \in H^2(\mathrm{Mod}(S_{g,1}); \mathbb{Z})$. Corollary 7.3 gives that $\mathrm{Mod}(S_g^1)$ is torsion-free, and so e'' is nontrivial.

We will show below that $H^2(\mathrm{Mod}(S_{g,1}); \mathbb{Z}) \approx \mathbb{Z}^2$. And we will show that this group is generated by the Euler class and the Meyer signature cocycle. We will also show that the Meyer signature cocycle evaluates trivially on the subgroup $\pi_1(S_g)$ of $\mathrm{Mod}(S_{g,1})$. Thus, to show that e'' is the Euler class, it suffices to check that these two classes agree on the point pushing subgroup $\pi_1(S_g)$. As in Section 4.2, the central extension (5.9) restricts to the central extension:

$$1 \to \mathbb{Z} \to \pi_1(UT(S_g)) \to \pi_1(S_g) \to 1.$$

We thus deduce from Proposition 5.10 that e'' is again the Euler class.

5.5.7 THE BIRMAN EXACT SEQUENCE DOES NOT SPLIT

Let $g \geq 2$. The Birman exact sequence (Theorem 4.6) is

$$1 \to \pi_1(S_g) \to \mathrm{Mod}(S_{g,1}) \to \mathrm{Mod}(S_g) \to 1.$$

Above, we described an embedding $\mathrm{Mod}(S_{g,1}) \to \mathrm{Homeo}^+(S^1)$. Since finite subgroups of $\mathrm{Homeo}^+(S^1)$ are cyclic, it follows that the same is true for $\mathrm{Mod}(S_{g,1})$. It is easy to find finite subgroups of $\mathrm{Mod}(S_g)$ that are not cyclic (e.g., the dihedral group of order $2g$), and so we have the following.

Corollary 5.11 *Let $g \geq 2$. The Birman exact sequence*

$$1 \to \pi_1(S_g) \to \mathrm{Mod}(S_{g,1}) \to \mathrm{Mod}(S_g) \to 1$$

does not split.

5.6 SURFACE BUNDLES AND THE MEYER SIGNATURE COCYCLE

Our next goal is to construct a nontrivial element σ of $H^2(\mathrm{Mod}(S_g); \mathbb{Z})$. We will prove in Section 5.6.3 that σ pulls back to an element of $H^2(\mathrm{Mod}(S_{g,1}); \mathbb{Z})$ that is not a power of the Euler class e. The cocycle σ, called the *Meyer signature cocycle*, is defined using the theory of surface bundles over surfaces.

We will use some homological algebra to show that the Meyer signature cocycle gives rise to nontrivial elements of $H_2(\mathrm{Mod}(S_g))$, $H_2(\mathrm{Mod}(S_g^1))$, and $H_2(\mathrm{Mod}(S_{g,1}))$, and to then complete the proof of Theorem 5.8.

In order to define the Meyer signature cocycle properly, we must clarify the connection between the mapping class group and the theory of surface bundles, so this is where we start.

5.6.1 SURFACE BUNDLES

The basic problem in the theory of surface bundles is to classify, for fixed (Hausdorff, paracompact) base space B, all isomorphism classes of bundles

$$S_g \to E \to B.$$

Recall that a *bundle isomorphism* is a fiberwise homeomorphism of total spaces covering the identity map. Below, we will explain how to reduce the S_g-bundle classification problem to a problem about $\mathrm{Mod}(S_g)$, at least for $g \geq 2$. Before doing this, we first recall some general facts about classifying spaces.

Classifying spaces. Suppose that G is a topological group acting freely, continuously, and properly on a contractible space X. The quotient space X/G is called a *classifying space* for G, and is denoted by BG. Any such space X is denoted by EG. When G is discrete, BG is just a $K(G,1)$-space. For any G, there exists an EG [159]. The space BG is unique up to homotopy equivalence. In fact, any homomorphism of groups $G \to H$ that is a homotopy equivalence induces a homotopy equivalence $BG \to BH$. We will see below the usefulness of BG in the classification of G-bundles and related bundles.

Surface bundles and homeomorphisms. We claim that there is a bijective correspondence:

$$\left\{ \begin{array}{c} \text{Isomorphism classes of} \\ \text{oriented } S_g\text{-bundles over } B \end{array} \right\} \longleftrightarrow \left\{ \begin{array}{c} \text{Homotopy classes of} \\ \text{maps } B \to \mathrm{BHomeo}^+(S_g) \end{array} \right\}$$

This bijection is realized concretely in the following way. The group $\mathrm{Homeo}^+(S_g)$ acts freely and properly discontinuously on the product $\mathrm{EHomeo}^+(S_g) \times S_g$ via the diagonal action. Let E denote the quotient. The projection $\mathrm{EHomeo}^+(S_g) \times S_g \to \mathrm{EHomeo}^+(S_g)$ induces a fiber bundle ζ:

$$S_g \to E \to \mathrm{BHomeo}^+(S_g).$$

The bundle ζ has the universal property that any S_g-bundle over any space B is the pullback of ζ via a continuous map (the *classifying map*) $f : B \to \mathrm{BHomeo}^+(S_g)$. Homotopic classifying maps give isomorphic bundles. Conversely, any bundle induces such a map f. Hence our claim.

Because of this correspondence, the bundle ζ is called the *universal S_g-bundle*. We thus see that $\mathrm{BHomeo}^+(S_g)$ plays the same role for surface bundles as the (infinite) Grassmann manifolds $\mathrm{BSO}(n)$ play for vector bundles.

Surface bundles and the mapping class group. Consider the fiber bundle

$$\mathrm{Homeo}_0(S_g) \to \mathrm{Homeo}^+(S_g) \xrightarrow{\pi} \mathrm{Mod}(S_g).$$

By Theorem 1.14, the fiber is contractible. By the long exact sequence in homotopy, π induces an isomorphism of all homotopy groups. Whitehead's theorem implies that the homomorphism π is a homotopy equivalence, where $\mathrm{Mod}(S_g)$ has the discrete topology. The theory of classifying spaces implies that π induces a homotopy equivalence of classifying spaces $\mathrm{BHomeo}^+(S_g) \to \mathrm{BMod}(S_g)$. In particular, we have the following key fact.

Proposition 5.12 *Suppose* $g \geq 2$. *The classifying space* $\mathrm{BHomeo}^+(S_g)$ *is a* $K(\mathrm{Mod}(S_g), 1)$-*space.*

A continuous map $f : B \to K(\mathrm{Mod}(S_g), 1)$ induces a representation $f_* : \pi_1(B) \to \mathrm{Mod}(S_g)$. Two such representations ρ_1, ρ_2 are called *conjugate* if there exists an $h \in \mathrm{Mod}(S_g)$ so that

$$\rho_1(\gamma) = h\rho_2(\gamma)h^{-1}$$

for all $\gamma \in \pi_1(B)$. Basic algebraic topology gives that the map f is determined up to free homotopy by the conjugacy class of the representation f_* and that every representation is induced by some continuous map. In other words, there is a bijection between free homotopy classes of maps $f : B \to K(\mathrm{Mod}(S_g), 1)$ and conjugacy classes of representations $\pi_1(B) \to \mathrm{Mod}(S_g)$. This bijection, together with Proposition 5.12, gives the following bijective correspondence.

$$\left\{ \begin{array}{c} \text{Isomorphism classes} \\ \text{of oriented } S_g\text{-bundles} \\ \text{over } B \end{array} \right\} \longleftrightarrow \left\{ \begin{array}{c} \text{Conjugacy classes} \\ \text{of representations} \\ \rho : \pi_1(B) \to \mathrm{Mod}(S_g) \end{array} \right\}$$

The simplest (but already interesting) instance of this fact is that isomorphism classes of S_g-bundles over S^1 are in bijection with conjugacy classes of elements in $\mathrm{Mod}(S_g)$. A more remarkable consequence is that, given any group extension

$$1 \to \pi_1(S_g) \to G \to Q \to 1, \tag{5.10}$$

there exist topological spaces (indeed closed manifolds) E and B and a fibration $S_g \to E \to B$ inducing the given group extension (apply the Dehn–Nielsen–Baer theorem from Chapter 8). Why is this surprising? Well, if we are given a representation $\rho : \pi_1(B) \to \mathrm{Homeo}^+(S_g)$, it is easy to see how to build a bundle $S_g \to E \to B$ with monodromy $\pi \circ \rho : \pi_1(B) \to \mathrm{Mod}(S_g)$: just take the quotient of $S_g \times \tilde{B}$ by the obvious $\pi_1(B)$-action. However, the data specified by the group extension (5.10) determines only a representation $\rho : \pi_1(B) \to \mathrm{Mod}(S_g)$. That is, elements of the monodromy are specified only up to isotopy, so it is not at all clear how to use this data to build a well-defined S_g-bundle. In fact, Morita has constructed examples where the monodromy $\rho : \pi_1(B) \to \mathrm{Mod}(S_g)$ does not lift to a representation $\widetilde{\rho} : \pi_1(B) \to \mathrm{Homeo}^+(S_g)$ [160]. Yet the bijection above gives a fiber bundle $S_g \to E \to B$ with B and E closed manifolds that has monodromy ρ.

The above discussion should clarify why the problem of classifying conjugacy classes of representations of various groups into $\mathrm{Mod}(S_g)$ is an important problem.

Cohomology. Another corollary of Proposition 5.12 is that

$$H^*(\mathrm{BHomeo}^+(S_g); \mathbb{Z}) \approx H^*(\mathrm{Mod}(S_g); \mathbb{Z}).$$

This isomorphism is one of the main reasons that we care about the cohomology of $\mathrm{Mod}(S_g)$. It is the reason we think of elements of $H^*(\mathrm{Mod}(S_g); \mathbb{Z})$ as characteristic classes of surface bundles, as we now explain.

Suppose one wants to associate to every S_g-bundle a (say integral) cohomology class on the base of that bundle so that this association is *natural*, that is, it is preserved under pullbacks. By applying this to the universal S_g-bundle ζ, we see that each such cohomology class must be the pullback of some element of $H^*(\mathrm{BHomeo}^+(S_g); \mathbb{Z}) \approx H^*(\mathrm{Mod}(S_g); \mathbb{Z})$. In this sense the classes in $H^*(\mathrm{Mod}(S_g); \mathbb{Z})$ are universal. This is why they are called characteristic classes of surface bundles.

We have already seen that $H_1(\mathrm{Mod}(S_g); \mathbb{Z}) = 0$ if $g \geq 3$ (Theorem 5.2). It follows from the universal coefficients theorem that $H^1(\mathrm{Mod}(S_g); \mathbb{Z}) = 0$. Thus there are no natural 1-dimensional cohomology invariants for these S_g-bundles. In Section 5.4 we proved for $g \geq 4$ that $H_2(\mathrm{Mod}(S_g); \mathbb{Z})$ is cyclic, so that there is at most one natural 2-dimensional invariant. This is the Meyer signature cocycle constructed below.

Remark on the smooth case. Every aspect of the discussion above holds with the smooth category replacing the topological category. Here we replace $\mathrm{BHomeo}^+(S_g)$ with $\mathrm{BDiff}^+(S_g)$, and so on. The key fact is the theorem of Earle–Eells [53] (see also [73]) that the topological group $\mathrm{Diff}_0(S_g)$ is contractible for $g \geq 2$. Following the exact lines of the discussion above, this gives a bijective correspondence between isomorphism classes of *smooth* S_g-bundles over a fixed base space B and conjugacy classes of representations $\rho : \pi_1(B) \to \mathrm{Mod}(S_g)$.

5.6.2 DEFINITION OF THE MEYER SIGNATURE COCYCLE

We are now ready to describe the construction of a nonzero element $\sigma \in H^2(\mathrm{Mod}(S_g); \mathbb{Z})$: the Meyer signature cocycle. Below we will prove that σ pulls back to a nontrivial class both in $H^2(\mathrm{Mod}(S_{g,1}); \mathbb{Z})$ and in $H^2(\mathrm{Mod}(S_g^1); \mathbb{Z})$.

For any closed 4-manifold M, there is a skew-symmetric pairing

$$
\begin{array}{ccccc}
H^2(M; \mathbb{Z}) & \otimes & H^2(M; \mathbb{Z}) & \to & \mathbb{Z} \\
a & \otimes & b & \mapsto & \langle a \cup b, [M] \rangle
\end{array}
$$

given by taking the cup product of two classes and evaluating the result on the fundamental class of M. The signature of the resulting quadratic form is called the *signature* of M, denoted by $\mathrm{sig}(M)$.

We can use signature to give a 2-cochain

$$
\sigma \in C^2(\mathrm{BHomeo}^+(S_g); \mathbb{Z}) \approx \mathrm{Hom}(C_2(\mathrm{BHomeo}^+(S_g); \mathbb{Z}), \mathbb{Z})
$$

as follows. Suppose we are given a chain $c \in C_2(\mathrm{BHomeo}^+(S_g); \mathbb{Z})$. It follows from general facts about 2-chains in topological spaces that c can be represented by a map $f : S_h \to \mathrm{BHomeo}^+(S_g)$, where S_h is a closed surface of genus $h \geq 0$. We then let $\sigma \in C^2(\mathrm{BHomeo}^+(S_g); \mathbb{Z})$ be defined by

$$
\sigma(f) = \mathrm{sig}(f^*\zeta),
$$

where, as above, ζ denotes the universal S_g-bundle over $\mathrm{BHomeo}^+(S_g)$.

It follows from the work of Meyer that σ is a well-defined 2-cocycle [156]. One key ingredient in this is the fact that the signature of a fiber bundle depends only on the action of the fundamental group of the base on the homology of the fiber; another is the Novikov additivity of signature.

It is not easy to prove that the cocycle σ is a nonzero element of $H^2(\mathrm{BHomeo}^+(S_g); \mathbb{Z})$. The hard part is finding a good way to compute signature in terms of the monodromy data. Kodaira, and later Atiyah (see [7]), found a surface bundle over a surface with nonzero signature. This construction can be used to give such a bundle with fiber S_g for any $g \geq 4$. It follows that the signature cocycle $\sigma \in H^2(\mathrm{BHomeo}^+(S_g); \mathbb{Z}) \approx H^2(\mathrm{Mod}(S_g); \mathbb{Z})$ is nonzero. Indeed, this kind of argument can be used to prove that σ has infinite order in $H^2(\mathrm{Mod}(S_g); \mathbb{Z})$.

5.6.3 MATCHING UPPER AND LOWER BOUNDS ON $H_2(\mathrm{Mod}(S); \mathbb{Z})$

In Section 5.4 we used Hopf's formula to give an upper bound on the number of generators of the group $H_2(\mathrm{Mod}(S); \mathbb{Z})$, where S is either S_g or S_g^1

and where $g \geq 4$. So far we have constructed two nontrivial elements of $H^2(\mathrm{Mod}(S); \mathbb{Z})$, the Euler class and the Meyer signature cocycle. We will now use homological algebra to complete the proof of Theorem 5.8.

The universal coefficients theorem and $H_2(\mathrm{Mod}(S); \mathbb{Z})$. Let S be a surface of genus at least 3. In what follows we assume that all homology and cohomology groups have $\mathbb{Z}$-coefficients. The universal coefficients theorem gives the following short exact sequence:

$$1 \to \mathrm{Ext}(H_1(\mathrm{Mod}(S)), \mathbb{Z}) \to H^2(\mathrm{Mod}(S)) \atop \to \mathrm{Hom}(H_2(\mathrm{Mod}(S)), \mathbb{Z}) \to 1. \quad (5.11)$$

Since $H_1(\mathrm{Mod}(S); \mathbb{Z}) = 0$ (Theorem 5.2), the Ext term in (5.11) is trivial. Thus

$$H^2(\mathrm{Mod}(S); \mathbb{Z}) \approx \mathrm{Hom}(H_2(\mathrm{Mod}(S); \mathbb{Z}), \mathbb{Z}).$$

In other words, we have

$$H^2(\mathrm{Mod}(S); \mathbb{Z}) \approx H_2(\mathrm{Mod}(S); \mathbb{Z})/\mathrm{torsion}.$$

Proof that $H_2(\mathrm{Mod}(S_g); \mathbb{Z}) \approx \mathbb{Z}$. In Section 5.4, we proved that the group $H_2(\mathrm{Mod}(S_g); \mathbb{Z})$ is cyclic. Since the Meyer signature cocycle is an infinite-order element of $H^2(\mathrm{Mod}(S_g); \mathbb{Z})$ and since $H^2(\mathrm{Mod}(S_g); \mathbb{Z}) \approx H_2(\mathrm{Mod}(S_g); \mathbb{Z})/\mathrm{torsion}$, we have that

$$H_2(\mathrm{Mod}(S_g); \mathbb{Z}) \approx \mathbb{Z},$$

as stated in Theorem 5.8. Thus we see that, up to multiples, signature is the only 2-dimensional isomorphism invariant for S_g-bundles.

A five-term exact sequence for homology groups. We now introduce a tool that will help us compute $H_2(\mathrm{Mod}(S_g^1))$ and $H_2(\mathrm{Mod}(S_{g,1}))$.

Given any short exact sequence of groups

$$1 \to K \to G \to Q \to 1,$$

there is a five-term exact sequence of homology groups

$$H_2(G) \to H_2(Q) \to H_1(K)_Q \to H_1(G) \to H_1(Q) \to 0$$

where all coefficient groups are $\mathbb{Z}$ and $H_1(K)_Q$ is the *set of coinvariants* of the action of Q by conjugation on $H_1(K; \mathbb{Z})$, that is, the quotient of $H_1(K; \mathbb{Z})$ by all elements $x - q \cdot x$, where $x \in H_1(K; \mathbb{Z})$ and $q \in Q$.

The existence of this five-term exact sequence is a consequence of the Hopf formula (see [38, p. 47]).

Proof that $H_2(\mathrm{Mod}(S_g^1); \mathbb{Z}) \approx \mathbb{Z}$. We saw in Section 5.4 that the group $H_2(\mathrm{Mod}(S_g^1); \mathbb{Z})$ is cyclic. Our aim is to prove that it is isomorphic to $\mathbb{Z}$.

If we apply the five-term exact sequence for homology groups to the short exact sequence

$$1 \to \pi_1(UT(S_g)) \to \mathrm{Mod}(S_g^1) \to \mathrm{Mod}(S_g) \to 1,$$

we obtain the sequence

$$H_2(\mathrm{Mod}(S_g^1)) \to H_2(\mathrm{Mod}(S_g)) \to H_1(\pi_1(UT(S_g)))_{\mathrm{Mod}(S_g)}$$

$$\to H_1(\mathrm{Mod}(S_g^1)) \to H_1(\mathrm{Mod}(S_g)) \to 0,$$

or, by Theorem 5.2,

$$H_2(\mathrm{Mod}(S_g^1)) \to \mathbb{Z} \to H_1(\pi_1(UT(S_g)))_{\mathrm{Mod}(S_g)} \to 0 \to 0 \to 0.$$

We now determine the set of coinvariants in this sequence.

Claim: $H_1(\pi_1(UT(S_g)))_{\mathrm{Mod}(S_g)} \approx \mathbb{Z}/(2g-2)\mathbb{Z}$.

Proof of claim. We start with the presentation

$$\pi_1(UT(S_g)) = \left\langle a_1, b_1, \ldots, a_g, b_g, z \mid \prod_{i=1}^{g} [a_i, b_i] = z^{2-2g}, \ z \text{ central} \right\rangle,$$

where z is the generator of the S^1-fiber; see [190, p. 435]. It follows that

$$H_1(UT(S_g); \mathbb{Z}) \approx \mathbb{Z}^{2g} \oplus \mathbb{Z}/(2g-2)\mathbb{Z} \approx H_1(S_g; \mathbb{Z}) \oplus \mathbb{Z}/(2g-2)\mathbb{Z}.$$

What is more, the action of $\mathrm{Mod}(S_g)$ on $H_1(UT(S_g); \mathbb{Z})$ is given by the standard action of $\mathrm{Mod}(S_g)$ on $H_1(S_g; \mathbb{Z})$ together with the trivial action on $\mathbb{Z}/(2g-2)\mathbb{Z}$. Thus we have

$$H_1(\pi_1(UT(S_g)))_{\mathrm{Mod}(S_g)} \approx H_1(S_g; \mathbb{Z})_{\mathrm{Mod}(S_g)} \oplus \mathbb{Z}/(2g-2)\mathbb{Z},$$

and so it now remains to show that the set of coinvariants $H_1(S_g; \mathbb{Z})_{\mathrm{Mod}(S_g)}$ is trivial.

By the change of coordinates principle and Proposition 6.2, $\mathrm{Mod}(S_g)$ identifies all primitive elements of $H_1(S_g; \mathbb{Z})$. In particular, each primitive element is identified with its inverse. Thus $H_1(S_g; \mathbb{Z})_{\mathrm{Mod}(S_g)}$ is a quotient of $\mathbb{Z}/2\mathbb{Z}$. On the other hand, one can find in $H_1(S_g; \mathbb{Z})$ three primitive elements that sum to zero. It follows that $H_1(S_g; \mathbb{Z})_{\mathrm{Mod}(S_g)}$ is trivial. $\qquad\square$

Our five-term sequence is now reduced to

$$H_2(\mathrm{Mod}(S_g^1)) \to \mathbb{Z} \to \mathbb{Z}/(2g-2)\mathbb{Z} \to 0.$$

It follows that the kernel of the map $\mathbb{Z} \to H_1(\pi_1(UT(S_g)))_{\mathrm{Mod}(S_g)}$ is isomorphic to $\mathbb{Z}$. By exactness of the sequence, we see that $H_2(\mathrm{Mod}(S_g^1); \mathbb{Z})$ contains an infinite cyclic subgroup. On the other hand, we already showed that $H_2(\mathrm{Mod}(S_g^1); \mathbb{Z})$ is a quotient of $\mathbb{Z}$, and so it follows that $H_2(\mathrm{Mod}(S_g^1); \mathbb{Z}) \approx \mathbb{Z}$, as desired.

Actually, we have proven a little more. We have shown that there is an exact sequence

$$H_2(\mathrm{Mod}(S_g^1)) \longrightarrow H_2(\mathrm{Mod}(S_g)) \longrightarrow H_1(\pi_1(UT(S_g)))_{\mathrm{Mod}(S_g)} \longrightarrow 0$$

$$\wr\wr \qquad\qquad\qquad \wr\wr \qquad\qquad\qquad \wr\wr$$

$$\mathbb{Z} \qquad\qquad\qquad \mathbb{Z} \qquad\qquad\qquad \mathbb{Z}/(2g-2)\mathbb{Z}.$$

So we see that the map from $H_2(\mathrm{Mod}(S_g^1)) \approx \mathbb{Z}$ to $H_2(\mathrm{Mod}(S_g)) \approx \mathbb{Z}$ is multiplication by $2g-2$.

Proof that $H_2(\mathrm{Mod}(S_{g,1}); \mathbb{Z}) \approx \mathbb{Z}^2$. We start by showing that the group $H_2(\mathrm{Mod}(S_{g,1}); \mathbb{Z})$ is generated by at most two elements. Recall from Proposition 3.19 that we have a short exact sequence

$$1 \to \langle T_a \rangle \to \mathrm{Mod}(S_g^1) \to \mathrm{Mod}(S_{g,1}) \to 1,$$

where a is the isotopy class of the boundary component of S_g^1. The associated five-term exact sequence of homology groups is

$$H_2(\mathrm{Mod}(S_g^1)) \to H_2(\mathrm{Mod}(S_{g,1})) \to H_1(\langle T_a \rangle)_{\mathrm{Mod}(S_{g,1})}$$

$$\to H_1(\mathrm{Mod}(S_g^1)) \to H_1(\mathrm{Mod}(S_{g,1})) \to 0.$$

We just proved that $H_2(\mathrm{Mod}(S_g^1)) \approx \mathbb{Z}$. Also by Theorem 5.2, the groups $H_1(\mathrm{Mod}(S_g^1))$ and $H_1(\mathrm{Mod}(S_{g,1}))$ are trivial. Finally, since $\langle T_a \rangle$ is central in $\mathrm{Mod}(S_g^1)$, the set of coinvariants $H_1(\langle T_a \rangle)_{\mathrm{Mod}(S_{g,1})}$ is isomorphic to $\mathbb{Z}$. We can thus rewrite the five-term exact sequence as

$$\mathbb{Z} \to H_2(\mathrm{Mod}(S_{g,1})) \to \mathbb{Z} \to 0 \to 0 \to 0.$$

It follows that $H_2(\mathrm{Mod}(S_{g,1}); \mathbb{Z})$ is a quotient of $\mathbb{Z}^2$, as desired.

We obtain one element $e^\star$ of $H_2(S_{g,1}; \mathbb{Z})$ by passing the Euler class $e \in H^2(\text{Mod}(S_{g,1}); \mathbb{Z})$ through the universal coefficients theorem as above.

We obtain another element of $H_2(S_{g,1}; \mathbb{Z})$ from the Meyer signature cocycle $\sigma \in H^2(\text{Mod}(S_g); \mathbb{Z})$ as follows. The universal coefficients theorem identifies σ with an element $\sigma^\star$ of $H_2(\text{Mod}(S_g); \mathbb{Z})$. Then, we consider the Birman exact sequence

$$1 \to \pi_1(S_g) \to \text{Mod}(S_{g,1}) \to \text{Mod}(S_g) \to 1.$$

The associated five-term exact sequence in homology is

$$H_2(\text{Mod}(S_{g,1})) \to H_2(\text{Mod}(S_g)) \to H_1(S_g)_{\text{Mod}(S_g)}$$

$$\to H_1(\text{Mod}(S_{g,1})) \to H_1(\text{Mod}(S_g)) \to 0.$$

We showed above that $H_1(S_g)_{\text{Mod}(S_g)}$ is trivial, and so the map $H_2(\text{Mod}(S_{g,1})) \to H_2(\text{Mod}(S_g))$ is surjective. Thus (abusing notation) there is an element $\sigma^\star \in H_2(\text{Mod}(S_{g,1}))$ mapping to $\sigma^\star \in H_2(\text{Mod}(S_g))$. Applying the universal coefficients theorem one more time, we obtain an element $\sigma \in H^2(\text{Mod}(S_{g,1}))$.

We now show that $e^\star$ and $\sigma^\star$ are distinct elements of $H_2(\text{Mod}(S_{g,1}); \mathbb{Z})$, even up to multiples. By the universal coefficients theorem, it suffices to show that e and σ are distinct elements of $H^2(\text{Mod}(S_{g,1}); \mathbb{Z})$.

By Proposition 5.10, the Euler class e evaluates nontrivially on the 2-cycle given by the fundamental class of the point-pushing subgroup $\pi_1(S_g)$. On the other hand, since $\pi_1(S_g)$ is the kernel of the map $\text{Mod}(S_{g,1}) \to \text{Mod}(S_g)$ (Theorem 4.6), we have that the fundamental class of $\pi_1(S_g)$ pushes forward to zero in $H_2(\text{Mod}(S_g))$. As the signature cocycle $\sigma \in H^2(\text{Mod}(S_{g,1}))$ is the pullback of $\sigma \in H^2(\text{Mod}(S_g); \mathbb{Z})$, it follows that $\sigma \in H^2(\text{Mod}(S_{g,1}))$ evaluates trivially on the fundamental class of $\pi_1(S_g)$. We thus have that $H^2(\text{Mod}(S_{g,1}); \mathbb{Z}) \approx \mathbb{Z}^2$ and hence

$$H_2(\text{Mod}(S_{g,1}); \mathbb{Z}) \approx \mathbb{Z}^2.$$

This completes the proof of Theorem 5.8.

Chapter Six

The Symplectic Representation and the Torelli Group

One of the fundamental aspects of $\mathrm{Mod}(S_g)$ is its action on $H_1(S_g; \mathbb{Z})$. The representation $\Psi : \mathrm{Mod}(S_g) \to \mathrm{Aut}(H_1(S_g; \mathbb{Z}))$ is like a first linear approximation to $\mathrm{Mod}(S_g)$, and we can try to transfer our knowledge of the linear group $\mathrm{Aut}(H_1(S_g; \mathbb{Z}))$ to the group $\mathrm{Mod}(S_g)$.

As we show in Section 6.1, the algebraic intersection number on $H_1(S_g; \mathbb{R})$ gives this vector space a symplectic structure. This symplectic structure is preserved by the image of Ψ, and so Ψ can be thought of as a representation

$$\Psi : \mathrm{Mod}(S_g) \to \mathrm{Sp}(2g, \mathbb{Z})$$

into the integral symplectic group. The homomorphism Ψ is called the *symplectic representation* of $\mathrm{Mod}(S_g)$. The bulk of this chapter is an exposition of the basic properties and applications of Ψ. A sample application is that $\mathrm{Mod}(S_g)$ has a torsion-free subgroup of finite index (Theorem 6.9).

The representation Ψ has a large kernel, called the Torelli group $\mathcal{I}(S_g)$, which can be thought of as the "nonlinear" part of $\mathrm{Mod}(S_g)$. We conclude this chapter with an introduction to the study of $\mathcal{I}(S_g)$, which is an important topic in its own right.

6.1 ALGEBRAIC INTERSECTION NUMBER AS A SYMPLECTIC FORM

In order to understand the symplectic representation $\Psi : \mathrm{Mod}(S_g) \to \mathrm{Sp}(2g, \mathbb{Z})$, one of course needs to know the basic facts about symplectic linear transformations. After describing these, we show how $H_1(S_g; \mathbb{R})$ comes equipped with a natural symplectic structure. This structure relates in a natural way to simple closed curves in S_g.

6.1.1 SYMPLECTIC VECTOR SPACES AND SYMPLECTIC MATRICES

Let $g \geq 1$ be an integer and let $\{x_1, y_1, \ldots, x_g, y_g\}$ be a basis for the vector space $\mathbb{R}^{2g}$. Denote the dual vector space of $\mathbb{R}^{2g}$ by $(\mathbb{R}^{2g})^\star$. The *standard symplectic form* on $\mathbb{R}^{2g}$ is the 2-form

$$\omega = \sum_{i=1}^{g} dx_i \wedge dy_i.$$

Given two vectors $v = (v_1, w_1, \ldots, v_g, w_g)$ and $v' = (v'_1, w'_1, \ldots, v'_g, w'_g)$ in $\mathbb{R}^{2g}$, we compute

$$\omega(v, v') = \sum_{i=1}^{g} (v_i w'_i - v'_i w_i).$$

The 2-form ω is a nondegenerate, alternating bilinear form on $\mathbb{R}^{2g}$. In fact, it is the unique such form up to change of basis of $\mathbb{R}^{2g}$. The vector space $\mathbb{R}^{2g}$ equipped with such a form is called a *real symplectic vector space*.

The *linear symplectic group* $\mathrm{Sp}(2g, \mathbb{R})$ is defined to be the group of linear transformations of $\mathbb{R}^{2g}$ that preserve the standard symplectic form ω. In terms of matrices:

$$\mathrm{Sp}(2g, \mathbb{R}) = \{A \in \mathrm{GL}(2g, \mathbb{R}) : A^\star \omega = \omega\},$$

or in other words,

$$\mathrm{Sp}(2g, \mathbb{R}) = \{A \in \mathrm{GL}(2g, \mathbb{R}) : A^T J A = J\},$$

where J is the $2g \times 2g$ matrix

$$J = \begin{pmatrix}
0 & 1 & 0 & 0 & \cdots & 0 & 0 \\
-1 & 0 & 0 & 0 & \cdots & 0 & 0 \\
0 & 0 & 0 & 1 & \cdots & 0 & 0 \\
0 & 0 & -1 & 0 & \cdots & 0 & 0 \\
\vdots & \vdots & \vdots & \vdots & \ddots & \vdots & \vdots \\
0 & 0 & 0 & 0 & \cdots & 0 & 1 \\
0 & 0 & 0 & 0 & \cdots & -1 & 0
\end{pmatrix}.$$

The *integral symplectic group* $\mathrm{Sp}(2g, \mathbb{Z})$ is defined as

$$\mathrm{Sp}(2g, \mathbb{Z}) = \mathrm{Sp}(2g, \mathbb{R}) \cap \mathrm{GL}(2g, \mathbb{Z}).$$

It is straightforward to check the following facts using basic linear algebra (see, e.g., [148], Lemmas 1.14, 2.19, and 2.20):

1. $\det(A) = 1$ for each $A \in \mathrm{Sp}(2g, \mathbb{R})$.

2. $\mathrm{Sp}(2g, \mathbb{R}) \cap \mathrm{O}(2g, \mathbb{R}) = \mathrm{U}(g)$.

3. λ is an eigenvalue of $A \in \mathrm{Sp}(2g, \mathbb{R})$ if and only if λ^{-1} is. This follows from the fact that A^{-1} and A^T are similar (i.e., conjugate).

We also remark that in the case $g = 1$ we have

$$\mathrm{Sp}(2, \mathbb{R}) = \mathrm{SL}(2, \mathbb{R}) \quad \text{and} \quad \mathrm{Sp}(2, \mathbb{Z}) = \mathrm{SL}(2, \mathbb{Z}).$$

Elementary symplectic matrices. There are symplectic analogues of the elementary matrices for $\mathrm{SL}(n, \mathbb{Z})$. Let σ be the permutation of $\{1, \ldots, 2g\}$ that transposes $2i$ and $2i - 1$ for each $1 \leq i \leq g$. The *elementary symplectic matrices* are the (finitely many) matrices of the form

$$SE_{ij} = \begin{cases} I_{2g} + e_{ij} & \text{if } i = \sigma(j), \\ I_{2g} + e_{ij} - (-1)^{i+j} e_{\sigma(j)\sigma(i)} & \text{otherwise,} \end{cases}$$

where $i \neq j$ and e_{ij} is the matrix with a 1 in the (i, j)-entry and 0s elsewhere. The following result is classical [154, Hilfssatz 2.1].

THEOREM 6.1 $\mathrm{Sp}(2g, \mathbb{Z})$ *is generated by elementary symplectic matrices.*

The Burkhardt generators. In 1890 Burkhardt [40] gave the following generating set for $\mathrm{Sp}(4, \mathbb{Z})$. Below, when we refer to a *factor*, we mean a subgroup of $\mathbb{Z}^{2g}$ spanned by some pair $\{x_i, y_i\}$.

Transvection:

$$(x_1, y_1, x_2, y_2) \mapsto (x_1 + y_1, y_1, x_2, y_2)$$

Factor rotation:

$$(x_1, y_1, x_2, y_2) \mapsto (y_1, -x_1, x_2, y_2)$$

Factor mix:

$$(x_1, y_1, x_2, y_2) \mapsto (x_1 - y_2, y_1, x_2 - y_1, y_2)$$

Factor swap:

$$(x_1, y_1, x_2, y_2) \mapsto (x_2, y_2, x_1, y_1).$$

For $g > 2$, if one adds for each $1 \leq i \leq g$ the factor swap exchanging the adjacent factors $\{x_i, y_i\} \leftrightarrow \{x_{i+1}, y_{i+1}\}$, one can derive the finite generating set given in Theorem 6.1. Thus Burkhardt's elements give a generating set for $\mathrm{Sp}(2g, \mathbb{Z})$. Below we will consider an infinite generating set for $\mathrm{Sp}(2g, \mathbb{Z})$, namely, the set of all transvections.

6.1.2 $H_1(S_g; \mathbb{Z})$ AS A SYMPLECTIC VECTOR SPACE

In what follows we will use $[c]$ to denote the homology class corresponding to an oriented simple closed curve c. Consider the *ordered* basis

$$\{[a_1], [b_1], \ldots, [a_g], [b_g]\}$$

for $H_1(S_g; \mathbb{R}) \approx \mathbb{R}^{2g}$ shown in Figure 6.1. The algebraic intersection number

$$\hat{i} : H_1(S_g; \mathbb{Z}) \wedge H_1(S_g; \mathbb{Z}) \longrightarrow \mathbb{Z}$$

extends uniquely to a nondegenerate, alternating bilinear map

$$\hat{i} : H_1(S_g; \mathbb{R}) \wedge H_1(S_g; \mathbb{R}) \longrightarrow \mathbb{R}.$$

If $[a_i]^\star$ and $[b_i]^\star$ denote the vectors in $H_1(S_g; \mathbb{R})^\star$ dual to $[a_i]$ and $[b_i]$, respectively, then

$$\hat{i} = \sum_{i=1}^{g} [a_i]^\star \wedge [b_i]^\star \in \wedge^2 \left(H_1(S_g; \mathbb{R})^\star \right).$$

With this structure the pair $(H_1(S_g; \mathbb{R}), \hat{i})$ is a symplectic vector space.

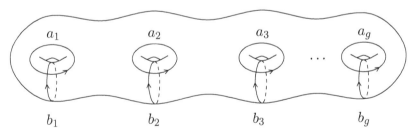

Figure 6.1 The standard geometric symplectic basis for $H_1(S_g; \mathbb{Z})$.

It is an important observation that there is a collection of oriented simple closed curves $\{c_1, \ldots, c_{2g}\}$ in S_g so that the homology classes $\{[c_i]\}$ form a symplectic basis for $H_1(S_g; \mathbb{Z})$ and $i(c_i, c_j) = \hat{i}([c_i], [c_j])$ for all i, j. Such a collection of curves will be called a *geometric symplectic basis* for $H_1(S_g; \mathbb{Z})$.

6.2 THE EUCLIDEAN ALGORITHM FOR SIMPLE CLOSED CURVES

In order to effectively make use of the symplectic structure on $H_1(S_g; \mathbb{Z})$, we will need to strengthen the dictionary between the algebraic and topological aspects of $H_1(S_g; \mathbb{Z})$. As a start, we answer the question: when can an element of $v \in H_1(S_g; \mathbb{Z})$ be represented by an oriented simple closed curve? Of course, if $v \neq 0$, then such a curve must be nonseparating.

Recall that $v \in H_1(S_g; \mathbb{Z}) \approx \mathbb{Z}^{2g}$ is *primitive* if $v \neq nw$ for any $w \in H_1(S_g; \mathbb{Z})$ and any integer $n \geq 2$.

Proposition 6.2 *Let $g \geq 1$. A nonzero element of $H_1(S_g; \mathbb{Z})$ is represented by an oriented simple closed curve if and only if it is primitive.*

Our proof of Proposition 6.2, adapted from Meeks–Patrusky [153], is a topological incarnation of the Euclidean algorithm. We recall the classical Euclidean algorithm for finding the greatest common divisor of two nonnegative integers. Given a pair of nonnegative integers $\{p, q\}$ with $0 < p \leq q$, we subtract p from q to obtain a new set $\{p, q - p\}$ with $\gcd(p, q - q) = \gcd(p, q)$. If we start with the two natural numbers m and n and repeat this process iteratively, then the theorem is that we will eventually arrive at the pair $\{\gcd(m, n), 0\}$.

Proof of Proposition 6.2. Let $\{a_i, b_i\}$ be a geometric symplectic basis shown in Figure 6.1, as well as the corresponding basis $\{[a_i], [b_i]\}$ for $H_1(S_g; \mathbb{Z})$.

The statement of the proposition for the torus is exactly that of Proposition 1.5. Thus we can assume that $g \geq 2$.

One direction of the proposition is simple. By the change of coordinates principle, for any nonseparating oriented simple closed curve γ in S_g, there exists $\phi \in \mathrm{Homeo}^+(S)$ with $\phi(\gamma) = a_1$. Thus the homology class $[\gamma] \in H_1(S_g; \mathbb{Z})$ is part of some basis for $\mathbb{Z}^{2g}$ and is therefore primitive.

The interesting direction of the proposition is to start with a primitive homology class $x \in H_1(S_g; \mathbb{Z})$ and to show that x is represented by a simple closed curve.

Say that with respect to the above basis we have

$$x = (v_1, w_1, \ldots, v_g, w_g).$$

Without loss of generality we may assume that each v_i and w_i is nonnegative, for if not, we can simply switch the orientations of some a_i and b_i so that this condition holds.

For each $1 \leq i \leq g$, take a closed regular neighborhood N_i of $a_i \cup b_i$. We can take the N_i to be disjoint. Each N_i is homeomorphic to a torus with one boundary component. Note that since the proposition is true for the torus, it is also true for a torus with one boundary component. Thus for each i there is an oriented nonseparating simple closed curve γ_i in N_i so that

$$\gcd(v_i, w_i)[\gamma_i] = v_i[a_i] + w_i[b_i] \in H_1(S_g; \mathbb{Z}).$$

We can thus represent x by $\sum \gcd(v_i, w_i)$ pairwise disjoint oriented simple closed curves contained in $\cup N_i$. Our goal is to combine these together to form a single curve.

The following key observation is a consequence of the change of coordinates principle.

Observation: Given any two disjoint, oriented, nonhomologous, nonseparating simple closed curves α and β in S_g, there is an arc joining the left side of α to the left side of β.

Using this observation, we can perform a topological Euclidean algorithm on the $\sum \gcd(v_i, w_i)$ curves above. By this we mean the following. Let $N_{1,2}$ be a closed subsurface of S_g that contains N_1 and N_2 and is disjoint from the other N_i. We can take $N_{1,2}$ to be a surface of genus 2 with one boundary component. As above, we have $\gcd(v_1, w_1)$ parallel copies of γ_1 and $\gcd(v_2, w_2)$ parallel copies of γ_2 in $N_{1,2}$ that together represent $(v_1, w_1, v_2, w_2, 0, \ldots, 0) \in H_1(S_g; \mathbb{Z})$.

By the observation, we can surger the leftmost curve copy of γ_1 with the leftmost curve in γ_2 as in Figure 6.2. Since the surgery adds two parallel arcs with opposite orientations, the homology class of the collection of curves is unchanged. We can repeat this process until we run out of copies of γ_1 or γ_2. We then again have two collections of parallel curves. If $\gcd(v_1, w_1) \geq \gcd(v_2, w_2)$, then the two collections have $\gcd(v_1, w_1) - \gcd(v_2, w_2)$ and $\gcd(v_2, w_2)$ oriented curves, respectively. If we repeat this process in $N_{1,2}$, we will end up, as in the Euclidean algorithm, with

$$\gcd(\gcd(v_1, w_1), \gcd(v_2, w_2)) = \gcd(v_1, w_1, v_2, w_2)$$

parallel oriented simple closed curves in $N_{1,2}$ that together represent the element $(v_1, w_1, v_2, w_2, 0, \ldots, 0)$ of $H_1(S_g; \mathbb{Z})$. Moreover, by our choice of $N_{1,2}$, these curves are disjoint from the γ_i with $i \geq 3$.

We continue the process inductively. Let $N_{1,2,3}$ be a closed surface of genus 3 that contains N_1, N_2, and N_3 and is disjoint from the other N_i. We can apply the above process to the $\gcd(v_1, w_1, v_2, w_2)$ curves obtained in the previous step and $\gcd(v_3, w_3)$ parallel copies of γ_3 in N_3. If we do this, we will find $\gcd(v_1, w_1, v_2, w_2, v_3, w_3)$ parallel oriented simple closed curves in the class $(v_1, w_1, v_2, w_2, v_3, w_3, 0, \ldots, 0) \in H_1(S_g; \mathbb{Z})$.

By induction on genus we find, at the end, $\gcd(v_1, w_1, \ldots, v_g, w_g)$ paral-

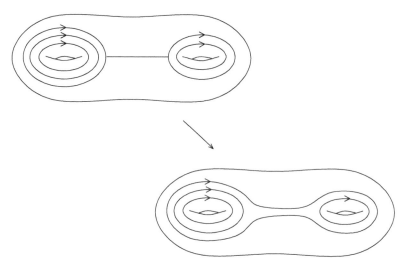

Figure 6.2 Surgering two oriented simple closed curves along an arc.

lel oriented simple closed curves in S_g representing x. Since x is primitive, we have that $\gcd(v_1, w_1, \ldots, v_g, w_g) = 1$, and so we actually have a single oriented simple closed curve in S_g, as desired.   □

Note that since the inclusion maps $S_{g,1} \to S_g$ and $S_g^1 \to S_g$ induce isomorphisms $H_1(S_{g,1}; \mathbb{Z}) \to H_1(S_g; \mathbb{Z})$ and $H_1(S_g^1; \mathbb{Z}) \to H_1(S_g; \mathbb{Z})$, Proposition 6.2 implies the analogous statement for surfaces with either one puncture or one boundary component.

6.3 MAPPING CLASSES AS SYMPLECTIC AUTOMORPHISMS

Any $\phi \in \mathrm{Homeo}^+(S_g)$ induces an automorphism $\phi_* : H_1(S_g; \mathbb{Z}) \to H_1(S_g; \mathbb{Z})$. As homotopic homeomorphisms $\phi \sim \psi$ induce the same map $\phi_* = \psi_*$, there is a representation

$$\Psi_0 : \mathrm{Mod}(S_g) \to \mathrm{Aut}(H_1(S_g; \mathbb{Z})) \approx \mathrm{Aut}(\mathbb{Z}^{2g}) \approx \mathrm{GL}(2g, \mathbb{Z}).$$

The rightmost isomorphism comes from choosing a basis for $H_1(S_g; \mathbb{Z})$. Our goal in this section is to understand the basic properties of Ψ_0, and in particular to compute its image.

Since each element of $\mathrm{Mod}(S_g)$ is represented by an orientation-preserving homeomorphism of S_g, it follows that the image of Ψ_0 lies in $\mathrm{SL}(2g, \mathbb{R})$. Since each $f \in \mathrm{Mod}(S_g)$ preserves the lattice $H_1(S_g; \mathbb{Z})$ inside $H_1(S_g; \mathbb{R})$, it follows that $\Psi_0(\mathrm{Mod}(S_g)) \subseteq \mathrm{SL}(2g, \mathbb{Z})$. Since $\mathrm{Mod}(S_g)$

preserves the nondegenerate, alternating bilinear form $\hat{i}$, it follows that $\Psi_0(\mathrm{Mod}(S_g)) \subset \mathrm{Sp}(2g, \mathbb{R})$. Together these observations give that

$$\Psi_0(\mathrm{Mod}(S_g)) \subset \mathrm{Sp}(2g, \mathbb{Z}).$$

Thus Ψ_0 is better regarded as a representation

$$\Psi : \mathrm{Mod}(S_g) \to \mathrm{Sp}(2g, \mathbb{Z}).$$

The representation Ψ is called the *symplectic representation* of $\mathrm{Mod}(S_g)$.

As we already said, the inclusions $S_{g,1} \to S_g$ and $S_g^1 \to S_g$ induce isomorphisms $H_1(S_{g,1}; \mathbb{Z}) \to H_1(S_g; \mathbb{Z})$ and $H_1(S_g^1; \mathbb{Z}) \to H_1(S_g; \mathbb{Z})$. Therefore, the above discussion applies to give representations

$$\Psi : \mathrm{Mod}(S_{g,1}) \to \mathrm{Sp}(2g, \mathbb{Z}) \quad \text{and} \quad \Psi : \mathrm{Mod}(S_g^1) \to \mathrm{Sp}(2g, \mathbb{Z}).$$

6.3.1 THE ACTION OF A DEHN TWIST ON HOMOLOGY

A first step in understanding Ψ is to compute what it does to Dehn twists. We have the following formula.

Proposition 6.3 *Let a and b be isotopy classes of oriented simple closed curves in S_g. For any $k \geq 0$, we have*

$$\Psi(T_b^k)([a]) = [a] + k \cdot \hat{i}(a, b)[b].$$

Proof. First we treat the case where b is separating. By the change of coordinates principle there is a geometric symplectic basis $\{a_i, b_i\}$ with $i(a_i, b) = i(b_i, b) = 0$ for all i. The proposition follows immediately in this case.

Now assume that b is nonseparating. By change of coordinates there is a geometric symplectic basis $\{a_i, b_i\}$ for $H_1(S_g; \mathbb{Z})$ with $b_1 = b$. It is straightforward to check that the action of T_b^k on $H_1(S_g; \mathbb{Z})$, written with respect to the basis $\{a_i, b_i\}$ is given by

$$\Psi(T_b^k)([c]) = [T_b^k(c)] = \begin{cases} [a_1] + k[b_1] & c = a_1, \\ [c] & c \in \{b_1, a_2, b_2, \ldots, a_g, b_g\}. \end{cases}$$

Now let a be the isotopy class of an arbitrary oriented simple closed curve in S_g. The $[a_1]$-coefficient of $[a]$ in the basis $\{[a_i], [b_i]\}$ is $\hat{i}(a, b)$. By the linearity of $\Psi(T_b^k)$, the proposition follows. □

We caution the reader that if $[c] = [a] + [b] \in H_1(S; \mathbb{Z})$, then

$$\Psi(T_c) \neq \Psi(T_a T_b)$$

in general. It is true, though, that

$$\Psi(T_a) = \Psi(T_{a'}) \iff [a] = [a'],$$

as can be seen from Proposition 6.3. Another consequence of Proposition 6.3 is that if $[a] = 0$, then $\Psi(T_a)$ is trivial.

6.3.2 SURJECTIVITY OF THE SYMPLECTIC REPRESENTATION: THREE PROOFS

It is natural to ask whether every automorphism of $H_1(S_g; \mathbb{Z})$ preserving algebraic intersection number can be realized by some homeomorphism. In other words, is $\Psi : \mathrm{Mod}(S_g) \to \mathrm{Sp}(2g, \mathbb{Z})$ surjective? The first proof one might think of would be to realize each elementary symplectic matrix as the action of some element of $\mathrm{Mod}(S_g)$; since these matrices generate $\mathrm{Sp}(2g, \mathbb{Z})$, surjectivity of Ψ would follow. While some elementary symplectic matrices are the images of a Dehn twist, others are not, and it is not obvious how to prove these lie in the image of Ψ. Nevertheless, Ψ is indeed surjective.

Theorem 6.4 *The representation* $\Psi : \mathrm{Mod}(S_g) \to \mathrm{Sp}(2g, \mathbb{Z})$ *is surjective for* $g \geq 1$.

We give three conceptually distinct proofs of Theorem 6.4 as each demonstrates a different useful concept. The first proof presupposes the Burkhardt generating set for $\mathrm{Sp}(2g, \mathbb{Z})$ and finds particular elements of $\mathrm{Mod}(S_g)$ mapping to those elementary matrices. The second and third proofs offer a more "bare hands" approach, for example, using the Euclidean algorithm from Proposition 6.2.

When S is either $S_{g,1}$ or S_g^1, there is a commutative diagram

$$
\begin{array}{ccc}
\mathrm{Mod}(S) & \longrightarrow & \mathrm{Sp}(2g, \mathbb{Z}) \\
\downarrow & & \downarrow{\scriptstyle \approx} \\
\mathrm{Mod}(S_g) & \longrightarrow & \mathrm{Sp}(2g, \mathbb{Z})
\end{array}
$$

and so $\mathrm{Mod}(S_{g,1})$ and $\mathrm{Mod}(S_g^1)$ both surject onto $\mathrm{Sp}(2g, \mathbb{Z})$ as well.

Theorem 6.4 follows immediately in the case $g = 1$ from the isomorphism $\mathrm{Mod}(T^2) \approx \mathrm{SL}(2, \mathbb{Z}) = \mathrm{Sp}(2, \mathbb{Z})$ given in Theorem 2.5. Hence in what follows we can assume $g \geq 2$.

First proof of Theorem 6.4. The finite generating set for $\mathrm{Sp}(2g, \mathbb{Z})$ given by Burkhardt has four types of generators: one transvection, one factor rotation,

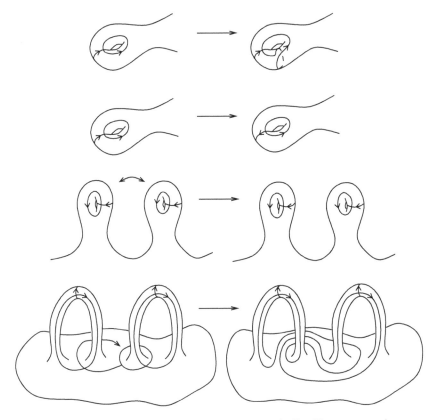

Figure 6.3 Realizing the Burkhardt generators geometrically. From top to bottom: a transvection, a factor rotation, a factor swap, and a factor mix.

one factor mix, and $g - 1$ factor swaps. Let $\{a_i, b_i\}_{i=1}^g$ be oriented simple closed curves in S_g forming a geometric symplectic basis for $H_1(S_g; \mathbb{Z})$ (see Figure 6.1). We show that each of Burkhardt's generators, hence all of $\mathrm{Sp}(2g, \mathbb{Z})$, lies in $\Psi(\mathrm{Mod}(S_g))$. Figure 6.3 illustrates the proof that follows.

By Proposition 6.3, $\Psi(T_{b_1})$ is the transvection generator.

We obtain Burkhardt's factor rotation generator as follows. Let N be a closed regular neighborhood of $a_1 \cup b_1$ in S_g. The subsurface N is homeomorphic to a torus with one boundary component. Think of N as a square with sides identified and an open disk removed from the center. Consider the homeomorphism of N obtained by rotating the boundary of the square by $\pi/2$ and leaving the boundary of N fixed. Extending by the identity map gives a homeomorphism of S_g, hence a mapping class $h \in \mathrm{Mod}(S_g)$ called a *handle rotation*. This handle rotation represents a mapping class which equals the product of Dehn twists: $T_{b_1} T_{a_1} T_{b_1}$. A direct check gives that $\Psi(h)$ is Burkhardt's factor rotation generator.

We next realize Burkhardt's factor mix generator by a mapping class. Consider a closed annular neighborhood of b_1 and push the left-hand boundary component of this annulus along a path in the surface that intersects a_2 once (from the left of a_2) and misses the other curves in the geometric symplectic basis; see the third diagram in Figure 6.3. The resulting mapping class h is called a *handle mix*. We can also describe h as the mapping class obtained by cutting S_g along b_1, pushing one of the new boundary components through the (a_2, b_2)-handle as in Figure 6.3, and then regluing. Alternatively, h is a product of three commuting Dehn twists:

$$h = T_{b_1}^{-1} T_{b_2}^{-1} T_c,$$

where c is a simple closed curve in the homology class $[b_2] - [b_1]$. Compare the handle mix h with our push map description of the lantern relation in Section 5.1. Another direct check gives that $\Psi(h)$ is Burkhardt's factor mix generator.

Finally, we have Burkhardt's $g - 1$ factor swaps. These are obtained as the Ψ-images of handle swaps. The ith *handle swap* h_i for $1 \leq i \leq g - 1$ is easily visualized (see Figure 6.3), but we can also write it as a product of Dehn twists:

$$h_i = (T_{a_{i+1}} T_{b_{i+1}} T_{d_i} T_{a_i} T_{b_i})^3,$$

where d_i is a simple closed curve in the homology class $[a_{i+1}] + [b_i]$. $\square$

We point out that all of the symplectic elementary matrices SE_{ij} are, up to change of basis, equivalent to Burkhardt's transvection generators and factor mix generators for $\mathrm{Sp}(2g, \mathbb{Z})$. Therefore, up to change of coordinates, the proof of Theorem 6.4 shows how to realize the symplectic elementary matrices as Dehn twists and handle mixes.

In the first proof of Theorem 6.4 it was not essential for us to write down explicit products of Dehn twists realizing each Burkhardt generator. In fact, it was not even necessary to say which particular mapping classes descend to those generators. The next proof exploits this idea.

Second proof of Theorem 6.4. Say that $\{a_i, b_i\}$ are oriented simple closed curves in S_g that form a geometric symplectic basis. Let $A \in \mathrm{Sp}(2g, \mathbb{Z})$ and say that we can find a geometric symplectic basis representing $\{A([a_i]), A([b_i])\}$. That is, suppose we can find a geometric symplectic basis $\{a'_i, b'_i\}$ so that $[a'_i] = A([a_i])$, $[b'_i] = A([b_i])$.

If we cut S_g along the union of the a_i and b_i, we get a sphere with g "square" boundary components. Of course each boundary component is a topological circle, but each circle is divided into four segments according to which points came from the left side of a_1, the right side of a_1, the left side of b_1, and the right side of b_1. Similarly, if we cut S_g along the a_i' and b_i', we also get a sphere with g square boundary components.

Choose any homeomorphism ϕ from the first sphere to the second. We can choose ϕ so that it not only takes the ith square to the ith square but also takes the a_i sides to the a_i' sides (with orientation) and the b_i sides to the b_i' sides. Since ϕ respects the required identifications, it follows that ϕ extends to a homeomorphism $S_g \to S_g$. By construction, the action of ϕ on $H_1(S_g; \mathbb{Z})$ is exactly given by A.

Thus to prove the theorem it suffices to show that the image of $\{[a_i], [b_i]\}$ under each of the Burkhardt generators can be realized by a geometric symplectic basis. For the transvection this is easy, and for the permutation generators, namely, the factor rotation and the factor swap, this is essentially obvious. It remains to consider the factor mix

$$([a_1], [b_1], [a_2], [b_2]) \mapsto ([a_1] - [b_2], [b_1], [a_2] - [b_1], [b_2]).$$

But it is easy to realize this basis geometrically (see Figure 6.3 for the solution), and so we are done.   □

We can use the idea from the second proof of Theorem 6.4 to give a proof that does not presuppose that we already know an explicit generating set for $\mathrm{Sp}(2g, \mathbb{Z})$.

Third proof of Theorem 6.4. Let $A \in \mathrm{Sp}(2g, \mathbb{Z})$ be given. Let $\{a_i, b_i\}$ be oriented simple closed curves in S_g that form a geometric symplectic basis. Since $A \in \mathrm{GL}(2g, \mathbb{Z})$, the image vector $A([a_1]) \in H_1(S_g; \mathbb{Z})$ is primitive. By Proposition 6.2 there is an oriented simple closed curve a_1' representing the homology class $A([a_1])$.

Since the vector $A([b_1])$ is primitive, we can represent it by an oriented simple closed curve. Since $\mathrm{Sp}(2g, \mathbb{Z})$ preserves algebraic intersection number, this simple closed curve will necessarily have algebraic intersection $+1$ with a_1'. But we want something better: we want to find a simple closed curve b_1' that represents $A([b_1])$ and has geometric intersection number 1 with a_1'.

We proceed as follows. Choose any geometric symplectic basis $\{a_i'', b_i''\}$ for $H_1(S_g; \mathbb{Z})$, where $a_1'' = a_1'$. The curve b_1'' is the only curve in $\{a_i'', b_i''\}$ that intersects $a_1' = a_1''$, and it intersects it once. We can write $A([b_1])$ uniquely in terms of the basis $\{[a_i''], [b_i'']\}$. Since $\hat{i}(A([a_1]), A([b_1])) = \hat{i}([a_1], [b_1]) = 1$, it follows that the coefficient of b_1'' in this sum is exactly $+1$. This sum gives a nonsimple (and not necessarily connected) representative β of $A([b_1])$. The good news is that β intersects a_1' exactly once.

The strategy now is to convert β into a connected simple closed curve without changing its homology class or its geometric intersection number with a_1'. By "resolving" intersections, we immediately turn β into a disjoint union β' of simple closed curves such that $[\beta'] = A([b_1]) \in H_1(S_g, \mathbb{Z})$. Note that β' has exactly one component simple closed curve that intersects a_1' since the intersection of β with a_1' is 1.

To change β' into a connected simple closed curve without changing its homology class, we apply a slight variation of the Euclidean algorithm for curves from Proposition 6.2. One just needs to notice that, given any two oriented simple closed curves in S_g that are disjoint from a_1' or perhaps an oriented simple closed curve disjoint from a_1' and one that intersects a_1' once, there is an arc that connects the left side of the first curve/arc to the left side of the second and that is disjoint from a_1'. (The reader might prefer to translate this statement into the context of the surface with two boundary components obtained by cutting S_g along a_1'.) Given this fact, we can proceed exactly as in the proof of Proposition 6.2 in order to obtain an oriented simple closed curve b_1' that represents $A([b_1])$ and that intersects a_1' once.

At this point, one can repeat the process to obtain a geometric symplectic basis $\{a_i', b_i'\}$ for $H_1(S_g; \mathbb{Z})$ that represents $\{A([a_i]), A([b_i])\}$. As in the second proof of Theorem 6.4, the result follows. $\square$

6.3.3 MINIMALITY OF THE HUMPHRIES GENERATING SET

The surjectivity of the symplectic representation $\Psi : \mathrm{Mod}(S_g) \rightarrow \mathrm{Sp}(2g, \mathbb{Z})$ can be applied to prove that $\mathrm{Mod}(S_g)$ cannot be generated by fewer than $2g + 1$ Dehn twists. Before proving this, we need a bit of setup.

A *transvection* in $\mathrm{Sp}(2g, \mathbb{Z})$ is an element of $\mathrm{Sp}(2g, \mathbb{Z})$ whose fixed set in $\mathbb{R}^{2g}$ has codimension 1. We claim that each transvection in $\mathrm{Sp}(2g, \mathbb{Z})$ is the image under Ψ of some power of a Dehn twist in $\mathrm{Mod}(S_g)$. Indeed, let $v \in \mathbb{Z}^{2g}$ be any primitive vector that is not fixed by a given transvection A and choose some symplectic basis $\{v, w, x_2, y_2, \ldots, x_g, y_g\}$ for $\mathbb{Z}^{2g}$. Since A preserves the symplectic form restricted to $\mathbb{Z}^{2g}$, it follows that $A(v) = v + kw$ for some $k \in \mathbb{Z}$. By Proposition 6.3, we have $A = \Psi(T_b^k)$, where b is any oriented simple closed curve in S_g with $[b] = w \in H_1(S_g; \mathbb{Z})$.

It follows from the fact that $\mathrm{Mod}(S_g)$ is generated by Dehn twists (Theorem 4.1) that Theorem 6.4 is equivalent to the fact that $\mathrm{Sp}(2g, \mathbb{Z})$ is generated by transvections. That is, we can give another proof of Theorem 6.4 by showing that transvections generate $\mathrm{Sp}(2g, \mathbb{Z})$. Or we can use Theorem 6.4 to deduce the fact that transvections generate $\mathrm{Sp}(2g, \mathbb{Z})$.

If v is a primitive vector in $\mathbb{Z}^{2g}$, we denote by τ_v the corresponding transvection in $\mathrm{Sp}(2g, \mathbb{Z})$, by which we mean that $\tau_v = \Psi(T_c)$, where $[c] = \pm v$. We call an element of $\mathrm{Sp}(2g, \mathbb{Z}/m\mathbb{Z})$ a *transvection* if it is the image of a transvection under the reduction $\mathrm{Sp}(2g, \mathbb{Z}) \to \mathrm{Sp}(2g, \mathbb{Z}/m\mathbb{Z})$. The following proposition (and its proof) are due to Humphries [101].

Proposition 6.5 Let $g \geq 2$. The group $\mathrm{Sp}(2g, \mathbb{Z}/2\mathbb{Z})$ cannot be generated by fewer than $2g + 1$ transvections.

Note that since $\mathrm{Sp}(2, \mathbb{Z}) = \mathrm{SL}(2, \mathbb{Z})$, the conclusion of Proposition 6.5 does not hold for $g = 1$.

Proof. First note that the fixed set of a nontrivial transvection in $\mathrm{Sp}(2g, \mathbb{Z}/2\mathbb{Z})$ has codimension 1 in $(\mathbb{Z}/2\mathbb{Z})^{2g}$. Given a set of transvections, the intersection of their fixed sets is the fixed set for the entire group that they generate. Clearly, there is no nontrivial element of $(\mathbb{Z}/2\mathbb{Z})^{2g}$ fixed by the whole group $\mathrm{Sp}(2g, \mathbb{Z}/2\mathbb{Z})$. It follows that any generating set for $\mathrm{Sp}(2g, \mathbb{Z}/2\mathbb{Z})$ consisting entirely of transvections must have at least $2g$ elements, corresponding to linearly independent vectors.

It remains to show that $\mathrm{Sp}(2g, \mathbb{Z}/2\mathbb{Z})$ cannot be generated by transvections corresponding to $2g$ linearly independent elements of $(\mathbb{Z}/2\mathbb{Z})^{2g}$.

Let $v_1, \ldots, v_{2g}$ be linearly independent elements of $(\mathbb{Z}/2\mathbb{Z})^{2g}$. Note that each nontrivial element of $(\mathbb{Z}/2\mathbb{Z})^{2g}$ is primitive, and in particular that the v_i form a basis for $(\mathbb{Z}/2\mathbb{Z})^{2g}$ (this basis is not necessarily symplectic). We would like to show that the τ_{v_i} do not generate $\mathrm{Sp}(2g, \mathbb{Z}/2\mathbb{Z})$.

We construct a graph G with one vertex for each v_i and an edge between each pair of vertices $\{v_i, v_j\}$ that pair nontrivially (mod 2) under the symplectic form on $(\mathbb{Z}/2\mathbb{Z})^{2g}$ induced by that on $\mathbb{Z}^{2g}$.

To any vector $w \in (\mathbb{Z}/2\mathbb{Z})^{2g}$, we associate a subgraph $G(w)$ of G as follows: if $w = \sum c_i v_i$, where $c_i \in \mathbb{Z}/2\mathbb{Z}$, then $G(w)$ is defined to be the full subgraph of G spanned by the vertices of G corresponding to those v_i with $c_i \neq 0$.

We now argue that, for any transvection τ_{v_i} and any $w \in (\mathbb{Z}/2\mathbb{Z})^{2g}$, the mod 2 Euler characteristics of $G(w)$ and of $G(\tau_{v_i}(w))$ are the same. If the symplectic pairing of v_i with w is 0, then $\tau_{v_i}(w) = w$, and there is nothing to show. If the symplectic pairing of v_i with w is 1, then by Proposition 6.3 $G(\tau_{v_i}(w))$ is obtained from $G(w)$ as follows: first we "add modulo 2" the v_i-vertex of G (i.e., add it if it is not there, delete it if it is); then, so as to preserve the property of being a full subgraph, we add modulo 2 the edges connecting the v_i-vertex to the other vertices of $G(w)$. The first operation changes the Euler characteristic by 1. Since the symplectic pairing of v_i with w is 1 (modulo 2), the second operation changes the Euler characteristic by 1. Thus the mod 2 Euler characteristics of $G(w)$ and $G(\tau_{v_i}(w))$ are the same.

Since $\mathrm{Sp}(2g, \mathbb{Z}/2\mathbb{Z})$ acts transitively on the nontrivial vectors of $(\mathbb{Z}/2\mathbb{Z})^{2g}$, it now suffices to show that there exist nontrivial vectors in

$(\mathbb{Z}/2\mathbb{Z})^{2g}$ whose associated subgraphs have different mod 2 Euler characteristics. Observe that $G(v_1)$ is a single vertex and so has Euler characteristic
equal to 1. If G is not a complete graph, then we can find two vertices, v_i
and v_j, that are not connected by an edge in G, and so $G(v_i + v_j)$ is the
union of two vertices, which has mod 2 Euler characteristic equal to 0, and
we are done in this case. If G is a complete graph, then (since $g \geq 2$) the
graph $G(v_1 + v_2 + v_3)$ is a triangle, which also has Euler characteristic 0.
This completes the proof.   □

Since the symplectic representation $\Psi : \mathrm{Mod}(S_g) \to \mathrm{Sp}(2g, \mathbb{Z})$ is surjective (Theorem 6.4), Proposition 6.5 implies the following.

Corollary 6.6 *Let* $g \geq 2$. *Any generating set for* $\mathrm{Mod}(S_g)$ *consisting entirely of Dehn twists must have cardinality at least* $2g + 1$. *In particular, the
Humphries generating set for* $\mathrm{Mod}(S_g)$ *is minimal among such generating
sets.*

6.4 CONGRUENCE SUBGROUPS, TORSION-FREE SUBGROUPS, AND RESIDUAL FINITENESS

In this section we define the congruence subgroups $\mathrm{Mod}(S_g)[m]$ of
$\mathrm{Mod}(S_g)$ for $m \geq 2$. We will then use these groups to prove two important
algebraic properties of the group $\mathrm{Mod}(S_g)$: it has a torsion-free subgroup of
finite index, and it is residually finite. We will approach these results via the
corresponding theorems in the classical, linear case of $\mathrm{Sp}(2g, \mathbb{Z})$ by using
the symplectic representation.

6.4.1 CONGRUENCE SUBGROUPS OF $\mathrm{Sp}(2g, \mathbb{Z})$

Let $m \geq 2$. The *level* m *congruence subgroup* $\mathrm{Sp}(2g, \mathbb{Z})[m]$ of $\mathrm{Sp}(2g, \mathbb{Z})$
is defined to be the kernel of the reduction homomorphism:

$$\mathrm{Sp}(2g, \mathbb{Z})[m] = \ker \left(\mathrm{Sp}(2g, \mathbb{Z}) \to \mathrm{Sp}(2g, \mathbb{Z}/m\mathbb{Z}) \right).$$

When studying the topology of a space with infinite fundamental group Γ,
it is quite useful to have a torsion-free subgroup of Γ of finite index. For
example, if an orbifold X has orbifold fundamental group Γ and Γ has a
torsion-free subgroup of finite index, then we can sometimes conclude that
X is finitely covered by a manifold; indeed, we will apply this principle
later in this book (Section 12.3).

Proposition 6.7 *Let* $g \geq 1$. *The congruence subgroup* $\mathrm{Sp}(2g, \mathbb{Z})[m]$ *is
torsion-free for* $m \geq 3$.

Note that $\mathrm{Sp}(2g, \mathbb{Z})[2]$ is not torsion-free; consider, for example, the element $-I_{2g}$.

Proof. Since $\mathrm{Sp}(2g, \mathbb{Z})[m] \subset \mathrm{Sp}(2g, \mathbb{Z})[n]$ whenever n is a divisor of m, we can assume that $m = p^a$, where either $p = 2$ and $a > 1$ or p is an odd prime and $a = 1$.

Let $h \in \mathrm{Sp}(2g, \mathbb{Z})[m]$ be nontrivial and let $k \geq 1$ be any positive integer. We must show that $h^k \neq I_{2g}$. Since $h \in \mathrm{Sp}(2g, \mathbb{Z})[m]$, we can write

$$h = I_{2g} + p^d T,$$

where $d \geq a$ and where T is a $2g \times 2g$ matrix with the property that at least one of its entries is not divisible by p. Replacing h by a positive power of h if necessary, we can assume that k is prime. Consider the following two cases.

Case 1: $p = k$. By the binomial theorem,

$$h^k = (I_{2g} + p^d T)^k \equiv I_{2g} + k(p^d T) \equiv I_{2g} + p^{d+1} T \not\equiv I_{2g} \mod p^{d+2}.$$

Note that the first congruence uses $m \neq 2$.

Case 2: $p \neq k$. Note that

$$(p^d T)^2 = p^{2d} T^2 \equiv 0 \mod p^{d+1}.$$

Using this fact, the binomial theorem, and the assumption that k is prime (so $p \nmid k$), it follows that

$$h^k = (I_{2g} + p^d T)^k \equiv I_{2g} + k(p^d T) \not\equiv I_{2g} \mod p^{d+1},$$

as desired. $\qquad\qquad\qquad\qquad\qquad\qquad\qquad\qquad\qquad\qquad\qquad\qquad\qquad\square$

Replacing $\mathrm{Sp}(2g, \mathbb{Z})[m]$ by $\mathrm{SL}(n, \mathbb{Z})[m]$ in the proof of Proposition 6.7 gives that the stronger result that the congruence subgroup $\mathrm{SL}(n, \mathbb{Z})[m]$ is torsion-free.

6.4.2 CONGRUENCE SUBGROUPS OF $\mathrm{Mod}(S_g)$

Let $g \geq 1$ and let $m \geq 2$. The *level m congruence subgroup* $\mathrm{Mod}(S_g)[m]$ of $\mathrm{Mod}(S_g)$ is defined to be the preimage $\Psi^{-1}(\mathrm{Sp}(2g, \mathbb{Z})[m])$ of the level m congruence subgroup $\mathrm{Sp}(2g, \mathbb{Z})[m]$ under the symplectic representation $\Psi : \mathrm{Mod}(S_g) \to \mathrm{Sp}(2g, \mathbb{Z})$. That is, $\mathrm{Mod}(S_g)[m]$ is the kernel of the composition

$$\mathrm{Mod}(S_g) \overset{\Psi}{\to} \mathrm{Sp}(2g, \mathbb{Z}) \to \mathrm{Sp}(2g, \mathbb{Z}/m\mathbb{Z}).$$

Since $\mathrm{Sp}(2g, \mathbb{Z}/m\mathbb{Z})$ is finite, $\mathrm{Mod}(S_g)[m]$ has finite index in $\mathrm{Mod}(S_g)$.

In order to convert our knowledge about torsion in $\mathrm{Sp}(2g, \mathbb{Z})$ into information about torsion in $\mathrm{Mod}(S_g)$, we will need the following.

THEOREM 6.8 *Let $g \geq 1$. If $f \in \mathrm{Mod}(S_g)$ has finite order and is nontrivial, then $\Psi(f)$ is nontrivial.*

We will prove Theorem 6.8 in Section 7.1.2 as an application of the Lefschetz fixed point theorem. With this theorem in hand, we can now prove the following theorem, first observed by Serre [194].

THEOREM 6.9 *Let $g \geq 1$. The group $\mathrm{Mod}(S_g)[m]$ is torsion-free for $m \geq 3$.*

The hyperelliptic involutions of S_g are finite-order elements of $\mathrm{Mod}(S_g)[2]$ (there are no others!). Thus the assumption $m \geq 3$ in Theorem 6.9 is necessary.

Proof. Suppose $f \in \mathrm{Mod}(S)[m]$ has finite order. Since $\mathrm{Sp}(2g, \mathbb{Z})[m]$ is torsion-free (Proposition 6.7), it follows that $\Psi(f)$ is the identity. In other words, f induces the trivial action on $H_1(S_g; \mathbb{Z})$. By Theorem 6.8, f is the identity. $\square$

6.4.3 RESIDUAL FINITENESS

Residual finiteness is one of the most commonly studied concepts in combinatorial group theory. A group G is *residually finite* if it satisfies any one of the following equivalent properties.

1. For each nontrivial $g \in G$, there exists a finite-index subgroup $H < G$ with $g \notin H$.

2. For each nontrivial $g \in G$, there exists a finite-index normal subgroup $N \triangleleft G$ with $g \notin N$.

3. For each nontrivial $g \in G$, there is a finite quotient $\phi : G \to F$ with $\phi(g) \neq 1$.

4. The intersection of all finite-index subgroups in G is trivial.

5. The intersection of all finite-index normal subgroups in G is trivial.

6. G injects into its *profinite completion*

$$\widehat{G} = \varprojlim G/H,$$

where H ranges over all finite-index normal subgroups of G.

It is elementary to check that these six properties are indeed equivalent. A group is thus residually finite if it is well approximated by its finite quotients. Correspondingly, spaces with residually finite fundamental groups can be understood via their finite covers.

Linear groups. By a *linear group* we mean a group that is isomorphic to a subgroup of $\mathrm{GL}(n, \mathbb{C})$ for some n. It is a famous theorem of Malcev that every finitely generated linear group is residually finite. This is easy to see for $\mathrm{Sp}(2g, \mathbb{Z})$ since the intersection

$$\bigcap_{m \geq 3} \mathrm{SL}(n, \mathbb{Z})[m]$$

is trivial. Indeed, if $A \in \mathrm{SL}(n, \mathbb{Z})$ is any matrix lying in the intersection, then all of its off-diagonal entries must be congruent to 0 (mod m) for all $m \geq 3$. Thus all off-diagonal entries of A must be 0, and so $A = I$. Since subgroups of residually finite groups are residually finite, we have the following.

Proposition 6.10 *For each $n \geq 2$, the group $\mathrm{SL}(n, \mathbb{Z})$ is residually finite. In particular, for $g \geq 1$, the group $\mathrm{Sp}(2g, \mathbb{Z})$ is residually finite.*

Mapping class groups. In analogy with linear groups we have the following.

THEOREM 6.11 *Let S be a compact surface. The group $\mathrm{Mod}(S)$ is residually finite.*

Theorem 6.11 was originally proven by Grossman [74]. The idea of her proof is to first show that $\pi_1(S)$ is *conjugacy separable*: given two nonconjugate elements $x, y \in \pi_1(S)$, there is a homomorphism $\phi : \pi_1(S) \to F$ to a finite group F such that $\phi(x)$ and $\phi(y)$ are not conjugate in F. She then proves that any automorphism of $\pi_1(S)$ that preserves conjugacy classes is inner. The outer automorphism group of any finitely generated group with these two properties is residually finite. Theorem 6.11 then follows from the Dehn–Nielsen–Baer theorem (Theorem 8.1 below) and the fact that residual finiteness is inherited by subgroups. See also [11].

Ivanov outlines the following more direct proof in [105, Section 11.1]. The general idea is to derive residual finiteness of $\mathrm{Mod}(S)$ from residual finiteness properties of finitely generated subrings of $\mathbb{R}$.

Proof of Theorem 6.11. First note that $\mathrm{Mod}(S)$ is a subgroup of $\mathrm{Mod}(S')$ where S' is the surface obtained from S by gluing a genus 1 surface with boundary onto each boundary component of S (see Theorem 3.18). Since any subgroup of a residually finite group is clearly residually finite, it suffices to prove the theorem when $\partial S = \emptyset$, which we now assume.

We assume that $S = S_g$ is hyperbolic; for S^2 the theorem is trivial and for T^2 it is easy.

Let $f \in \mathrm{Mod}(S_g)$ be any nontrivial element. We need to find a homomorphism $\phi : \mathrm{Mod}(S_g) \to F$ to a finite group F so that $\phi(f) \neq 1$. By Theorem 6.8 it is enough to consider two cases: either f acts nontrivially on $H_1(S_g; \mathbb{Z})$ or f has infinite order.

In the first case, this says precisely that the image $\Psi(f)$ under the symplectic representation $\Psi : \mathrm{Mod}(S_g) \to \mathrm{Sp}(2g, \mathbb{Z})$ is nontrivial. Since $\mathrm{Sp}(2g, \mathbb{Z})$ is residually finite (Proposition 6.10), there is a finite quotient $\mathrm{Sp}(2g, \mathbb{Z}) \to F$ to which $\Psi(f)$ projects nontrivially, and so we are clearly done.

Now assume that $f \in \mathrm{Mod}(S_g)$ has infinite order. Choose any hyperbolic metric on S_g. This gives a faithful representation

$$\rho : \pi_1(S_g) \to \mathrm{PSL}(2, \mathbb{R}) \approx \mathrm{Isom}(\mathbb{H}^2).$$

Since $\pi_1(S_g)$ is finitely generated, $\rho(\pi_1(S_g))$ is a subgroup of $\mathrm{PSL}(2, A)$ for some finitely generated subring A of $\mathbb{R}$. Such a ring A is *residually finite*: for each nontrivial $a \in A$ there is a ring homomorphism $\phi : A \to R$ to a finite ring R with $\phi(A) \neq 0$. See [211, Section 4.1] for a proof of this fact.

Now, f acts on the set of oriented isotopy classes of simple closed curves in S_g. Since S_g is compact, each free homotopy class γ of curves on S_g contains a unique geodesic, and the isometry $\rho(\gamma)$ is of hyperbolic type. To each such isotopy class γ we associate the hyperbolic length $\ell(\gamma)$ of this unique geodesic. Denote by $|\mathrm{tr}|(\gamma)$ the absolute value of the trace of $\rho(\alpha)$ for any $\alpha \in \pi_1(S_g)$ freely homotopic to γ; this is well defined since geodesics in free homotopy classes are unique. Since $\rho(\gamma)$ is an isometry of $\mathbb{H}^2$ of hyperbolic type, it can be diagonalized, from which we see that $|\mathrm{tr}|(\gamma) = 2\cosh(\ell(\gamma)/2)$.

Since f has infinite order, the action of f on the simple closed curves in S_g must change the hyperbolic length of some conjugacy class γ (this follows, for example, from Lemma 12.4 and the Alexander method). It follows that

$$|\mathrm{tr}|(\gamma) \neq |\mathrm{tr}|(f(\gamma)).$$

Since the ring A is residually finite, we can find a finite-index subring U of A so that $|\mathrm{tr}|(\gamma)$ and $|\mathrm{tr}|(f(\gamma))$ are not equal in A/U. It follows that γ and $f(\gamma)$ are not equal in $\mathrm{PSL}(2, A/U)$.

The action of $\mathrm{Mod}(S_g)$ on $\pi_1(S_g)$ gives rise to a homomorphism $\sigma : \mathrm{Mod}(S_g) \to \mathrm{Out}(\pi_1(S_g))$ (see Chapter 8). We can thus interpret the action of f on free homotopy classes of oriented closed curves in S_g as an action of $\sigma(f)$ on conjugacy classes in $\pi_1(S_g)$.

As $\mathrm{PSL}(2, A/U)$ is finite, the composition

$$\pi_1(S_g) \to \mathrm{PSL}(2, A) \to \mathrm{PSL}(2, A/U)$$

has a finite-index kernel H'. Since $\pi_1(S_g)$ is finitely generated, H' contains a finite-index *characteristic subgroup* H; that is, H is preserved by every automorphism of $\pi_1(S_g)$. Such an H can be constructed by taking the common intersection of the all subgroups in the (finite) $\mathrm{Aut}(\pi_1(S_g))$-orbit of H'. Since H is characteristic, the quotient homomorphism $\pi_1(S_g) \to \pi_1(S_g)/H$ gives rise to a homomorphism

$$\psi : \mathrm{Out}(\pi_1(S_g)) \to \mathrm{Out}(\pi_1(S_g)/H).$$

By construction, $\gamma \neq \sigma(f)(\gamma)$ in $\pi_1(S_g)/H$. It follows that $\psi \circ \sigma(f) \neq 1$. Since $\pi_1(S_g)/H$ is finite, so is $\mathrm{Out}(\pi_1(S_g)/H)$, and we are done. □

When S is allowed to have finitely many punctures, it is still true that $\mathrm{Mod}(S)$ is residually finite. While the proof of Theorem 6.11 given above does not work verbatim in this case, since there are finitely many free homotopy classes (one for each puncture) that do not contain geodesics, a slight variation of the proof can still be used to give the result in this case.

6.5 THE TORELLI GROUP

In this section we give a brief introduction to the Torelli subgroup $\mathcal{I}(S)$ of $\mathrm{Mod}(S)$. In addition to the beauty of the topic, the study of $\mathcal{I}(S)$ has important connections and applications to 3-manifold theory and algebraic geometry.

There is another good reason to study $\mathcal{I}(S)$. One recurring theme in the area is that questions about $\mathrm{Mod}(S)$ can often be answered in two steps: first for the elements that act nontrivially on $H_1(S; \mathbb{Z})$, and then for the elements that act trivially on $H_1(S; \mathbb{Z})$. Since we understand matrix groups comparatively well, the first type of element is usually vastly easier to analyze. We have already seen several instances of this phenomenon:

1. When we computed in Proposition 2.3 that $\mathrm{Mod}(S_{0,3}) \approx \Sigma_3$, all of the work was in showing that an element that acts trivially on $H_1(S_{0,3}; \mathbb{Z})$, that is an element that fixes the three punctures, is the trivial mapping class.

2. When we proved in Proposition 3.1 that Dehn twists are nontrivial

elements of $\mathrm{Mod}(S)$, we easily dispensed with the case of Dehn twists about nonseparating simple closed curves, using their action on $H_1(S; \mathbb{Z})$. For the case of separating curves, we needed a more subtle argument.

3. When we proved in Theorem 6.11 that $\mathrm{Mod}(S)$ is residually finite, we quickly dealt with the case of elements that act nontrivially on $H_1(S; \mathbb{Z})$; the other elements of $\mathrm{Mod}(S)$ required a much more involved argument.

It is therefore important for us to understand the elements of $\mathrm{Mod}(S_g)$ that act trivially on $H_1(S_g; \mathbb{Z})$. These elements form a normal subgroup $\mathcal{I}(S_g)$ of $\mathrm{Mod}(S_g)$ called the *Torelli group*. We have an exact sequence

$$1 \to \mathcal{I}(S_g) \to \mathrm{Mod}(S_g) \xrightarrow{\Psi} \mathrm{Sp}(2g, \mathbb{Z}) \to 1.$$

The Torelli group $\mathcal{I}(T^2)$ of the torus is trivial; this is simply a restatement of Theorem 2.5. In general, we think of $\mathcal{I}(S_g)$ as encoding the more mysterious structure of $\mathrm{Mod}(S_g)$—it is the part that cannot be seen via the symplectic representation Ψ. The study of $\mathcal{I}(S)$ is also of central importance in understanding the structure of congruence subgroups of $\mathrm{Mod}(S)$; for example, see the recent work of Putman [183].

Torelli groups for other surfaces. When S is a surface of genus g with either one puncture or one boundary component, we also have a naturally defined Torelli group $\mathrm{Mod}(S)$, which is again the kernel of the symplectic representation. For other surfaces S, one can still consider the subgroup of $\mathrm{Mod}(S)$ consisting of elements that act trivially on $H_1(S; \mathbb{Z})$. However, there are other natural choices for the Torelli group in these cases; see Putman's paper [184] for an in-depth discussion.

Homology 3-spheres. One purely topological motivation for studying $\mathcal{I}(S_g)$ is the following connection with *integral homology 3-spheres*, which are 3-manifolds that have the same integral homology as S^3. A *standard handlebody H* is a 3-manifold homeomorphic to a closed regular neighborhood of a graph embedded in a plane in S^3. The complement in S^3 of the interior of H is another handlebody H'. Thus we can think of S^3 as the union of two handlebodies glued along their boundaries by a homeomorphism $\phi : \partial H \to \partial H'$, that is,

$$S^3 \approx H \cup_\phi H'.$$

Note that ∂H and $\partial H'$ are homeomorphic closed surfaces. If ψ is a self-homeomorphism of ∂H, we obtain a new 3-manifold

$$M_\psi = H \cup_{\phi \circ \psi} H'.$$

The manifold M_ψ depends only on the isotopy class of ψ. The homology of M_ψ depends only on $\Psi([\psi]) \in \mathrm{Sp}(2g, \mathbb{Z})$. In particular, if $[\psi]$ lies in the Torelli subgroup of $\mathrm{Mod}(\partial H)$, then M_ψ is a homology 3-sphere. What is more, every homology 3-sphere arises in this way [161, Section 2].

The symplectic action. By Theorem 6.4, each matrix $A \in \mathrm{Sp}(2g, \mathbb{Z})$ is the action of some element $\widetilde{A} \in \mathrm{Mod}(S_g)$. The element $\widetilde{A}$ acts by conjugation on the normal subgroup $\mathcal{I}(S_g)$ in $\mathrm{Mod}(S_g)$. A different choice of $\widetilde{A}$ gives an automorphism of $\mathcal{I}(S_g)$ that differs by conjugation by an element of $\mathcal{I}(S_g)$. We therefore have a representation

$$\rho : \mathrm{Sp}(2g, \mathbb{Z}) \to \mathrm{Out}(\mathcal{I}(S_g)).$$

This representation is quite useful; it pervades the study of $\mathcal{I}(S_g)$. For example, the abelian group $H^*(\mathcal{I}(S_g); \mathbb{Z})$ is an $\mathrm{Sp}(2g, \mathbb{Z})$-module. One can then use the representation theory of symplectic groups to greatly constrain the possibilities for $H^*(\mathcal{I}(S_g); \mathbb{Z})$; see [112]. The representation ρ turns out to be an isomorphism; see [33, 34, 59].

6.5.1 TORELLI GROUPS ARE TORSION-FREE

Theorem 6.8 can be rephrased as a theorem about Torelli groups, giving the following basic fact about $\mathcal{I}(S_g)$.

THEOREM 6.12 *For $g \geq 1$, the group $\mathcal{I}(S_g)$ is torsion-free.*

Similarly, we have that $\mathcal{I}(S_{g,1})$ is torsion-free. We could also say that $\mathcal{I}(S_g^1)$ is torsion-free, where S_g^1 is a surface of genus g with one boundary component, but the entire group $\mathrm{Mod}(S_g^1)$ is already torsion-free (Corollary 7.3).

6.5.2 EXAMPLES OF ELEMENTS

We can write down several explicit examples of elements of $\mathcal{I}(S_g)$.

1. Dehn twists about separating curves. Each Dehn twist about a separating simple closed curve γ in S_g is an element of $\mathcal{I}(S_g)$. This is because there exists a basis for $H_1(S_g; \mathbb{Z})$ where each element is represented by an oriented simple closed curve disjoint from γ. Since T_γ fixes each of these curves, it in particular fixes the corresponding homology classes and is hence an element of $\mathcal{I}(S_g)$.

Another way to see that T_γ is an element of $\mathcal{I}(S_g)$ is to apply Proposition 6.3, which gives that

$$T_\gamma(x) = x + \hat{i}([\gamma], x)[\gamma]$$

for any $x \in H_1(S_g; \mathbb{Z})$. Since γ is separating, we have $[\gamma] = 0$, and so $T_\gamma(x) = x$.

The group generated by Dehn twists about separating simple closed curves is denoted $\mathcal{K}(S_g)$. In the 1970s Birman asked whether $\mathcal{K}(S_g)$ is equal to all of $\mathcal{I}(S_g)$ or at least has finite index in $\mathcal{I}(S_g)$. To prove that this is not the case, one has to find some invariant to tell that an element of $\mathcal{I}(S_g)$ does not lie in $\mathcal{K}(S_g)$. We explain below Johnson's construction of such an invariant.

2. Bounding pair maps. A *bounding pair* in a surface is a pair of disjoint, homologous, nonseparating simple closed curves. A *bounding pair map* is a mapping class of the form

$$T_a T_b^{-1},$$

where a and b form a bounding pair. Since a and b are homologous, Proposition 6.3 gives that the images of T_a and T_b in $\mathrm{Sp}(2g, \mathbb{Z})$ are equal. Thus any bounding pair map is an element of $\mathcal{I}(S_g)$.

We have seen bounding pair maps once before: the kernel of the forgetful map $\mathrm{Mod}(S_{g,1}) \to \mathrm{Mod}(S_g)$ is generated by bounding pair maps. This follows from Theorem 4.6 together with Fact 4.7 and the fact that $\pi_1(S_g)$ is generated by simple nonseparating loops.

3. Fake bounding pair maps. In verifying that a bounding pair map acts trivially on homology, we never used the fact that the curves in the bounding pair were disjoint—just that they were homologous. Thus $T_a T_b^{-1}$ is an element of $\mathcal{I}(S_g)$ whenever a and b are homologous. A special case of this is the mapping class $[T_a, T_c]$, where $\hat{i}(a, c) = 0$. Indeed,

$$T_a T_c T_a^{-1} T_c^{-1} = T_a T_{T_c(a)}^{-1},$$

and, by Proposition 6.3, the simple closed curves a and $T_c(a)$ are homologous.

4. Point pushes and handle pushes. The Birman exact sequence gives us the point-pushing homomorphism

$$\mathcal{P}ush : \pi_1(S_g) \to \mathrm{Mod}(S_{g,1}).$$

Since $\pi_1(S_g)$ is generated by simple loops, and these elements map to bounding pair maps in $\mathrm{Mod}(S_{g,1})$ (Fact 4.7), we have that the entire image $\mathrm{Push}(\pi_1(S_g))$ lies in $\mathcal{I}(S_{g,1})$.

We would like to make an analogous statement for $\mathcal{I}(S_g^1)$. Since the map $S_g^1 \to S_{g,1}$ induces a canonical isomorphism $H_1(S_g^1; \mathbb{Z}) \to H_1(S_{g,1}; \mathbb{Z})$, the boundary-capping homomorphism $\mathrm{Mod}(S_g^1) \to \mathrm{Mod}(S_{g,1})$ induces a surjective homomorphism $\mathcal{I}(S_g^1) \to \mathcal{I}(S_{g,1})$. By Proposition 3.19 and the fact that Dehn twists about separating curves lie in the Torelli group, we obtain a short exact sequence

$$1 \to \mathbb{Z} \to \mathcal{I}(S_g^1) \to \mathcal{I}(S_{g,1}) \to 1,$$

where the kernel $\mathbb{Z}$ is generated by the Dehn twist about the boundary of S_g^1.

Recall from from Section 4.2 that we also have a homomorphism $\pi_1(UT(S_g)) \to \mathrm{Mod}(S_g^1)$ that makes the following diagram commute:

$$\begin{array}{ccc} \pi_1(UT(S_g)) & \longrightarrow & \mathrm{Mod}(S_g^1) \\ \downarrow & & \downarrow \\ \pi_1(S_g) & \longrightarrow & \mathrm{Mod}(S_{g,1}) \end{array}$$

By the commutativity of the diagram, the fact that the image of $\pi_1(S_g)$ in $\mathrm{Mod}(S_{g,1})$ lies in $\mathcal{I}(S_{g,1})$, the fact that the kernel of the map $\pi_1(UT(S_g)) \to \pi_1(S_g)$ maps to $\mathcal{I}(S_g^1)$, and the fact that $\mathcal{I}(S_g^1)$ surjects onto $\mathcal{I}(S_{g,1})$, we obtain that the image of $\pi_1(UT(S_g))$ in $\mathrm{Mod}(S_g^1)$ lies in $\mathcal{I}(S_g^1)$.

The natural inclusion $S_g^1 \to S_{g+1}$ induces an injective homomorphism $\mathrm{Mod}(S_g^1) \to \mathrm{Mod}(S_{g+1})$ that restricts to an injective homomorphism $\mathcal{I}(S_g^1) \to \mathcal{I}(S_{g+1})$. Precomposing with the homomorphism $\pi_1(UT(S_g)) \to \mathcal{I}(S_g^1)$ we obtain an inclusion

$$\pi_1(UT(S_g)) \to \mathcal{I}(S_{g+1}).$$

We think of the elements in the image of this map as *handle pushes*, obtained by pushing the $(g+1)$st handle around the surface S_g.

6.5.3 A BIRMAN EXACT SEQUENCE FOR THE TORELLI GROUP

The above discussion about point pushes and handle pushes gives the following result, which allows us to translate results back and forth between the three groups $\mathcal{I}(S_g)$, $\mathcal{I}(S_{g,1})$, and $\mathcal{I}(S_g^1)$:

Proposition 6.13 *Let* $g \geq 2$. *The forgetful map* $S_{g,1} \to S_g$ *induces a short exact sequence*

$$1 \to \pi_1(S_g) \to \mathcal{I}(S_{g,1}) \to \mathcal{I}(S_g) \to 1,$$

and the boundary-capping map $S_g^1 \to S_g$ *gives a short exact sequence*

$$1 \to \pi_1(UT(S_g)) \to \mathcal{I}(S_g^1) \to \mathcal{I}(S_g) \to 1.$$

6.5.4 THE ACTION ON SIMPLE CLOSED CURVES

Similar to Section 1.3, we can classify the orbits of simple closed curves in S_g up to the action of $\mathcal{I}(S_g)$. While the statement is perhaps not so surprising, the proof is more subtle than the usual change of coordinates principle.

To state the result we need the fact that a separating simple closed curve in S_g (or its isotopy class) induces a splitting of $H_1(S_g; \mathbb{Z})$. By a *splitting* of $H_1(S_g; \mathbb{Z})$ we mean a decomposition as a direct product of subgroups that are orthogonal with respect to skew-symmetric bilinear pairing given by algebraic intersection number $\hat{i}$ on $H_1(S_g; \mathbb{Z})$. A simple closed curve γ that separates S_g into two subsurfaces S' and S'' gives a splitting of $H_1(S_g; \mathbb{Z})$ into the product of the two subgroups $H_1(S'; \mathbb{Z})$ and $H_1(S''; \mathbb{Z})$, each subgroup consisting of those homology classes supported on one side of γ or the other. We say that two isotopy classes of simple closed curves are $\mathcal{I}(S_g)$-equivalent if there is an element of $\mathcal{I}(S_g)$ taking one to the other.

The following theorem, observed by Johnson [110, Section 6], gives that the obvious necessary condition for two simple closed curves on S_g to be $\mathcal{I}(S_g)$-equivalent is also sufficient.

Proposition 6.14 *Let* c *and* c' *be two isotopy classes of simple closed curves in* S_g. *If* c *and* c' *are separating, then they are* $\mathcal{I}(S_g)$-*equivalent if and only if they induce the same splitting of* $H_1(S_g; \mathbb{Z})$. *If* c *and* c' *are nonseparating, then they are* $\mathcal{I}(S_g)$-*equivalent if and only if, up to sign, they represent the same element of* $H_1(S_g; \mathbb{Z})$.

Proof. For both cases, one direction is obvious, and so we only need to prove that the obvious necessary condition for $\mathcal{I}(S_g)$ equivalence is sufficient. Let γ and γ' be representative curves for the isotopy classes c and c'.

Suppose that c and c' are separating. Let S_1 and S_2 be the two embedded subsurfaces of S_g bounded by γ and let S_1' and S_2' be the two embedded subsurfaces bounded by γ'. Up to renumbering, our hypothesis tells us that $H_1(S_1; \mathbb{Z})$ and $H_1(S_1'; \mathbb{Z})$ are equal as subsets of $H_1(S_g; \mathbb{Z})$. Therefore, S_1 and S_1' have the same genus and hence are homeomorphic. Fix a homeomorphic identification of γ with γ' and choose any homeomorphism $\phi_1 : (S_1, \gamma) \to (S'_1, \gamma')$ respecting this identification. By Theorem 6.4 and by the hypothesis, there is a homeomorphism $\psi_1 \in \mathrm{Homeo}^+(S'_1, \gamma')$ so that $\psi_1 \circ \phi_1$ is the identity automorphism of $H_1(S_1; \mathbb{Z}) = H_1(S'_1; \mathbb{Z})$. Here we are invoking our claim that all of the results in Section 6.3 work for surfaces with one boundary component. We similarly choose $\psi_2 \circ \phi_2 : S_2 \to S'_2$. Together, the maps $\psi_1 \circ \phi_1$ and $\psi_2 \circ \phi_2$ induce a homeomorphism of S_g that takes γ to γ' and acts trivially on $H_1(S_g; \mathbb{Z})$.

Now suppose that c and c' are nonseparating. We would like to proceed similarly to the previous case. One difficulty is that we do not have a surjectivity statement for the action of the stabilizer of c in $\mathrm{Mod}(S_g)$ on the homology of the surface obtained by cutting along c. Instead, we proceed as follows.

Let β be any simple closed curve in S_g that intersects γ once. By the argument in the third proof of Theorem 6.4, there is a simple closed curve β' that intersects γ' once and is homologous to β. Let δ be the boundary of a regular neighborhood of $\beta \cup \gamma$ and let δ' be the boundary of a regular neighborhood of $\beta' \cup \gamma'$. Applying the present proposition to the case of separating simple closed curves, there is an element of $\mathcal{I}(S_g)$ taking δ to δ'. Since $\mathcal{I}(S_{1,1})$ is trivial (Theorem 2.5), it follows that this element of $\mathcal{I}(S_g)$ takes c to c', and we are done. $\square$

The statement of Proposition 6.14 can be sharpened in the case of isotopy classes of oriented simple closed curves. Two isotopy classes of oriented nonseparating simple closed curves are $\mathcal{I}(S_g)$-equivalent if and only if they represent the same element of $H_1(S_g; \mathbb{Z})$. Two isotopy classes of oriented separating simple closed curves are $\mathcal{I}(S_g)$-equivalent if and only if they induce the same *ordered splitting* of $H_1(S_g; \mathbb{Z})$, where the ordering of the factors comes from the fact that the curve has well-defined left and right sides.

The statement of Proposition 6.14 (and its proof) apply to the cases of surfaces with either one boundary or one puncture.

6.5.5 GENERATORS FOR THE TORELLI GROUP

Birman and Powell proved that $\mathcal{I}(S_g)$ is generated by the infinite collection of all Dehn twists about separating simple closed curves and all bounding pair maps [23, 180]. The general method they used is as follows.

From relations to generators. Let

$$1 \to K \to E \xrightarrow{\rho} Q \to 1$$

be a short exact sequence of groups. Suppose that E is generated by $\{e_1, \ldots, e_k\}$ and that Q has a presentation with generators $\rho(e_1), \ldots, \rho(e_k)$ and relations $\{r_i = 1\}$, where each r_i is a word in the $\{\rho(e_i)\}$. For each i, let $\widetilde{r}_i$ be the corresponding word in the e_i. As an element of E, each $\widetilde{r}_i$ lies in K. It is easy to check that the $\{\widetilde{r}_i\}$ is a normal generating set for K, that is, the set of all conjugates of all r_i by elements of E generate K.

An infinite generating set for $\mathcal{I}(S_g)$. Birman's idea was to apply the above general fact to the short exact sequence

$$1 \to \mathcal{I}(S_g) \to \mathrm{Mod}(S_g) \to \mathrm{Sp}(2g, \mathbb{Z}) \to 1.$$

Birman determined a finite presentation for $\mathrm{Sp}(2g, \mathbb{Z})$ and made the remark that the relators for $\mathrm{Sp}(2g; \mathbb{Z})$ give rise to generators for $\mathcal{I}(S_g)$. Then her student Powell interpreted each of these generators as products of Dehn twists about separating curves and bounding pair maps, thus proving that $\mathcal{I}(S_g)$ is generated by (infinitely many) such maps.

Putman has recently shown that the same generating set for $\mathcal{I}(S_g)$ can be derived from methods similar to the ones that we used to show that $\mathrm{Mod}(S_g)$ is generated by Dehn twists; see [184].

Whittling down the infinite generating set. Johnson showed that, for $g \geq 3$, the Dehn twists about separating simple closed curves are not needed to generate $\mathcal{I}(S_g)$. In other words, he proved that any such Dehn twist is a product of bounding pair maps. This can be deduced from the lantern relation as shown in Figure 6.4. In the figure, the pairs of simple closed curves (x, b_3), (y, b_1), and (z, b_4) are all bounding pairs, and so, using the fact that the T_{b_i} commute with the Dehn twists about all seven simple closed curves in the picture, the lantern relation $T_x T_y T_z = T_{b_1} T_{b_2} T_{b_3} T_{b_4}$ can be written as the desired relation in $\mathcal{I}(S_g)$:

$$(T_x T_{b_3}^{-1})(T_y T_{b_1}^{-1})(T_z T_{b_4}^{-1}) = T_{b_2}.$$

The genus of a bounding pair map $T_a T_b^{-1}$ is the minimum of the genera of the two components of $S_g - \{a \cup b\}$. It is easy to see that a genus k bounding pair map is a product of k genus 1 bounding pair maps, and so $\mathcal{I}(S_g)$ is generated by genus 1 bounding pair maps. This implies, by the change of coordinates principle, that $\mathcal{I}(S_g)$ is normally generated in $\mathrm{Mod}(S_g)$ by a single genus 1 bounding pair map.

Finite generation. In his clever and beautiful paper [111] Johnson proved the following.

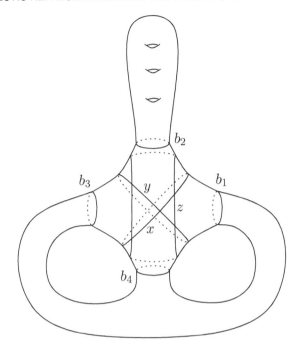

Figure 6.4 A lantern showing how to write the twist about the separating simple closed curve a as a product of bounding pair maps.

THEOREM 6.15 *For $g \geq 3$, the Torelli group $\mathcal{I}(S_g)$ is generated by a finite number of bounding pair maps.*

While $\mathrm{Mod}(S_g)$ can be generated by $2g+1$ Dehn twists that can easily be displayed in one figure, we will see below that any generating set for $\mathcal{I}(S_g)$ must have at least $O(g^3)$ generators (Theorem 6.19). Thus any such generating set for $\mathcal{I}(S_g)$ is not so easy to write in a single figure. This indicates the combinatorial complexity needed to prove Theorem 6.15. What is particularly remarkable is that for $g \geq 3$, Johnson finds a generating set for $\mathcal{I}(S_g)$ with $O(2^g)$ elements (for $g = 20$, he gives over one trillion generators); even naming that many elements in a coherent way is not so trivial!

Johnson's strategy for Theorem 6.15 is as follows. He first produces an explicit list of bounding pair maps in $\mathcal{I}(S_g)$, some of which are genus 1, and shows that the group generated by these is normal in $\mathrm{Mod}(S_g)$. To check normality, it suffices to check that the conjugate of each bounding pair map on the list by each Humphries generator for $\mathrm{Mod}(S_g)$ is a product of bounding pair maps on the list. Since any single genus 1 bounding pair map normally generates $\mathcal{I}(S_g)$ (in $\mathrm{Mod}(S_g)$), this proves the theorem. Of course, the hard part is coming up with the explicit list. The proof of Theorem 6.15 would take us too far afield, but we encourage the reader to read the proof in [111].

While Theorem 6.15 settles the question of finite generation of $\mathcal{I}(S_g)$, we do not have an analogue of the Humphries generating set. In fact, Johnson has conjectured that $\mathcal{I}(S_g)$ has a generating set with $O(g^3)$ elements, as the rank of $H_1(\mathcal{I}(S_g);\mathbb{Q})$ has this order. If this conjecture is true, Johnson's generating set with $O(2^g)$ elements is far from minimal. In the case $g = 3$, Johnson was able to whittle down the cardinality of his generating set for $\mathcal{I}(S_3)$ to 35, which is exactly the rank of $H_1(\mathcal{I}(S_3);\mathbb{Q})$. Johnson conjectures that this should persist in higher genus. However, it is still an open question even to find a generating set for $\mathcal{I}(S_g)$ whose number of elements is polynomial in g.

Two related open questions are: is $\mathcal{I}(S_g)$ finitely presented for $g \geq 3$? is $\mathcal{K}(S_g)$ finitely generated for $g \geq 3$?

Genus 2. In genus 2 the story is quite different. McCullough–Miller showed that $\mathcal{I}(S_2)$ is not finitely generated [146]. Mess sharpened this result by showing that $\mathcal{I}(S_2)$ is an infinitely generated free group, with one Dehn twist generator for each orbit of the action of $\mathcal{I}(S_2)$ on the set of separating simple closed curves in S_2 [155]. Note that there are no bounding pairs in S_2, and so $\mathcal{I}(S_2)$ is generated by Dehn twists; that is, $\mathcal{I}(S_2) = \mathcal{K}(S_2)$.

Nonclosed surfaces. For the surfaces $S_{g,1}$ and S_g^1, it follows from Theorem 6.15 and the Birman exact sequences for $\mathcal{I}(S_g)$ that $\mathcal{I}(S_{g,1})$ is generated by finitely many bounding pair maps and that $\mathcal{I}(S_g^1)$ is generated by finitely many bounding pair maps together with the Dehn twist about the boundary curve of S_g^1.

6.6 THE JOHNSON HOMOMORPHISM

In this section we explain the Johnson homomorphism and some of its applications.

6.6.1 CONSTRUCTION

We now describe another of Johnson's fundamental contributions to our understanding of the Torelli group, the Johnson homomorphism [109]. This is a surjective homomorphism

$$\tau : \mathcal{I}(S_g) \rightarrow \left(\wedge^3 H_1(S_g;\mathbb{Z})\right)/H_1(S_g;\mathbb{Z}),$$

where $\wedge^3 H_1(S_g; \mathbb{Z})$ is the third exterior power of $H_1(S_g; \mathbb{Z})$. The map τ exactly captures the torsion-free part of $H_1(\mathcal{I}(S_g); \mathbb{Z})$ (Theorem 6.19). It is a useful invariant of elements of $\mathcal{I}(S_g)$, as we shall see.

We begin by considering the case of S_g^1, a surface of genus $g \geq 2$ with one boundary component. We do this for simplicity since we can choose a basepoint on ∂S_g^1 and so any element of $\mathrm{Mod}(S_g^1)$ gives an automorphism of $\pi_1(S_g^1)$ as opposed to just an outer automorphism. Also the target of τ in this case is simply $\wedge^3 H_1(S_g; \mathbb{Z})$.

Let $\Gamma = \pi_1(S_g^1)$, which is isomorphic to the free group of rank $2g$. Let Γ' denote the commutator subgroup $[\Gamma, \Gamma]$ of Γ. By definition, $\mathcal{I}(S_g^1)$ is the subgroup of $\mathrm{Mod}(S_g^1)$ that acts trivially on Γ/Γ'. Johnson's key idea is to look at the action of $\mathcal{I}(S_g^1)$ on the quotient of Γ by the next term in its lower central series, namely, $[\Gamma, \Gamma'] = [\Gamma, [\Gamma, \Gamma]]$.

There is a short exact sequence

$$1 \to \Gamma'/[\Gamma, \Gamma'] \to \Gamma/[\Gamma, \Gamma'] \to \Gamma/\Gamma' \to 1,$$

which we rewrite as

$$1 \to N \to E \to H \to 1$$

by simply renaming the groups. The *Johnson homomorphism* is the homomorphism

$$\tau : \mathcal{I}(S_g^1) \to \mathrm{Hom}(H, N)$$

given by

$$\tau(f)(x) = f(e)e^{-1},$$

where e is any lift of $x \in H$ to E. It is straightforward to check that $\tau(f)$ is a well-defined homomorphism and that τ itself is a homomorphism.

In the literature and in applications, $\tau(f)$ is usually thought of as an element of $\wedge^3 H$. This involves a little bit of an algebraic juggle as follows.

1. There is a homomorphism $\psi : \wedge^2 H \to N$ defined as follows. For $a, b \in H$, we take lifts $\tilde{a}$ and $\tilde{b}$ in E and let

$$\psi(a \wedge b) = [\tilde{a}, \tilde{b}] \in N.$$

Now extend ψ linearly. Note that $\mathrm{Sp}(2g, \mathbb{Z})$ acts on both the domain and the range of ψ, and it is not hard to prove that τ is an $\mathrm{Sp}(2g, \mathbb{Z})$-module homomorphism. Using, for example, the classical Witt formula to count dimensions, one can check that ψ is an Spmodule isomorphism. Therefore, $\mathrm{Hom}(H, N)$ is naturally isomorphic to $\mathrm{Hom}(H, \wedge^2 H)$.

2. $\mathrm{Hom}(H, \wedge^2 H)$ is canonically isomorphic to $H^\star \otimes \wedge^2 H$. Using the nondegenerate symplectic form given by algebraic intersection number, we can canonically identify H with its dual $H^\star$. This gives a canonical isomorphism $\mathrm{Hom}(H, \wedge^2 H) \approx H \otimes \wedge^2 H$.

3. There is a natural inclusion of $\wedge^3 H$ into $H \otimes \wedge^2 H$ given by

$$a \wedge b \wedge c \mapsto a \otimes (b \wedge c) + b \otimes (c \wedge a) + c \otimes (a \wedge b),$$

and we will show below that the image of τ is exactly $\wedge^3 H$.

Naturality. The action of $\mathrm{Mod}(S_g^1)$ on $H = H_1(S_g^1; \mathbb{Z})$ induces an action of $\mathrm{Mod}(S_g^1)$ on $\wedge^3 H$. A crucial and easily verified property of τ is the following *naturality property*: for any $f \in \mathcal{I}(S_g^1)$ and $h \in \mathrm{Mod}(S_g^1)$, we have

$$\tau(hfh^{-1}) = h_\star(\tau(f)). \tag{6.1}$$

Closed and once-punctured surfaces. We will compute below that for the isotopy class c of ∂S_g^1, $\tau(T_c) = 0$. It then follows that $\tau : \mathcal{I}(S_g^1) \to \wedge^3 H$ factors through a homomorphism $\tau : \mathcal{I}(S_{g,1}) \to \wedge^3 H$.

For closed surfaces S_g, the Johnson homomorphism is a surjective homomorphism $\tau : \mathcal{I}(S_g) \to \wedge^3 H/H$. The inclusion of H into $\wedge^3 H$ is given by

$$a \mapsto \left(\sum x_i \wedge y_i \right) \wedge a,$$

where x_i and y_i represent a symplectic basis for $H = H_1(S_g; \mathbb{Z})$. The reason that we need to take the quotient $\wedge^3 H/H$ is in order for τ to be well defined on $\mathcal{I}(S_g)$ comes from the fact that $\mathcal{I}(S_g)$ is the quotient of $\mathcal{I}(S_{g,1})$ by the normal subgroup $\pi_1(S_g)$ (Proposition 6.13). In computing the image of a bounding pair map in $\mathcal{I}(S_g)$ under τ, we can think of the quotient by H as accounting for the fact that there is no preferred side of a bounding pair in a closed surface; in S_g^1 the two subsurfaces cut off by a bounding pair are distinguished from each other by whether or not they contain ∂S_g^1.

We can now deduce Corollary 6.17 for the closed surface S_g. The analogues of Theorems 6.18 and 6.19 also hold for closed surfaces; see [111, 112, 113, 114].

An interpretation via mapping tori. The Johnson homomorphism τ can also be defined using topology. Let $f \in \mathcal{I}(S_{g,1})$ and think of $S_{g,1}$ as S_g with a marked point. We wish to come up with an element of $\wedge^3 H$. Let ϕ be a representative of f and consider the mapping torus

$$M_f = \frac{S_g \times [0,1]}{(x,0) \sim (\phi(x),1)}.$$

Since $f \in \mathcal{I}(S_{g,1})$, it follows that $H_1(M_\phi; \mathbb{Z}) \approx H_1(S_g \times S^1; \mathbb{Z})$. The projection map $S_g \times S^1 \to S_g$ induces a projection $H_1(S_g \times S^1; \mathbb{Z}) \to H_1(S_g) \approx \mathbb{Z}^{2g}$. Composing these maps and then precomposing with the abelianization homomorphism $\pi_1(M_\phi) \to H_1(M_\phi; \mathbb{Z})$ gives a homomorphism

$$\pi_1(M_\phi) \to \mathbb{Z}^{2g}.$$

Since T^{2g} is a $K(\mathbb{Z}^{2g}, 1)$, it follows that this homomorphism is induced by a continuous based map of spaces

$$M_\phi \to T^{2g},$$

where T^{2g} is the $2g$-dimensional torus. This map is well defined up to (based) homotopy, and it induces a homomorphism

$$\psi : H_3(M_\phi; \mathbb{Z}) \to H_3(T^{2g}; \mathbb{Z}).$$

Since $H_3(T^{2g}; \mathbb{Z}) \approx \wedge^3(H)$, the image $\psi([M_\phi])$ of the fundamental class of M_ϕ in $H_3(T^{2g}; \mathbb{Z})$ specifies an element of $\wedge^3(H)$. This element is precisely $\tau(f)$. This can be proven by a straightforward algebraic topology argument; see [47].

Another interpretation via mapping tori. There is a different way to use mapping tori in order to obtain a description of $\tau(f)$. Specifically, we will find a homomorphism $\mathcal{I}(S_{g,1}) \to \mathrm{Hom}(\wedge^3 H, \mathbb{Z}) \approx \wedge^3 H$ that agrees with τ.

Let $x \in H$. Represent x by an oriented multicurve μ in $S_{g,1}$. The cylinder $C = \mu \times [0,1]$ lies in $S_{g,1} \times [0,1]$ and hence maps to the mapping torus M_ϕ, where ϕ is a representative of f. The cylinder C is in general not a closed surface. However, since $\phi(\mu)$ is homologous to μ, there is an immersed surface R in $S_{g,1} \times \{0\} \approx S_{g,1}$ with $\mu - \phi(\mu)$ as its boundary. Since $S_{g,1}$ has a marked point, the choice of R is unique. The union $C \cup R$ is a surface Σ_x, and so it represents an element $[\Sigma_x] \in H_2(M_\phi; \mathbb{Z}) \approx H^1(M_\phi; \mathbb{Z})$, this last isomorphism coming from Poincaré duality.

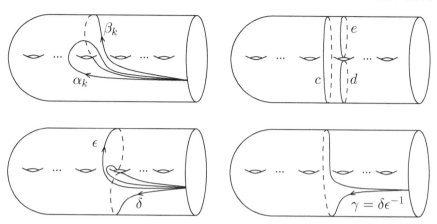

Figure 6.5 The simple closed curves c and d and the elements of $\pi_1(S_g^1)$ used to compute $\tau(T_c)$ and $\tau(T_d T_e^{-1})$.

Given any $x \wedge y \wedge z \in \wedge^3 H$, the cup product

$$[\Sigma_x] \cup [\Sigma_y] \cup [\Sigma_z] \in H^3(M_\phi; \mathbb{Z})$$

can be paired with the fundamental class $[M_\phi]$ to give an element of $\mathbb{Z}$. Equivalently, one can take the triple (algebraic) intersection $\Sigma_x \cap \Sigma_y \cap \Sigma_z$ to obtain this element of $\mathbb{Z}$. We have thus constructed a map $\mathcal{I}(S_{g,1}) \to \mathrm{Hom}(\wedge^3 H, \mathbb{Z}) \approx \wedge^3 H$ that one can check is a homomorphism and that agrees with τ; see [112].

6.6.2 Computing the Image of τ

We now explain how to explicitly calculate τ on certain elements of $\mathcal{I}(S_g^1)$ and compute its image.

The image of a Dehn twist. Let c be the standard separating simple closed curve shown in Figure 6.5. We claim that $\tau(T_c) = 0$.

To prove this claim we begin by taking the generators $\{\alpha_i, \beta_i\}$ for $\pi_1(S_g^1)$ shown in Figure 6.5. Let k be the genus of the subsurface of S_g^1 cut off by c and not containing ∂S_g^1; in Figure 6.5 this is the surface to the left of c. We see that T_c fixes α_i and β_i for $k+1 \le i \le g$. Let γ be the element of $\pi_1(S_g^1)$ shown at the bottom right of Figure 6.5. For $x \in \{\alpha_1, \beta_1, \ldots, \alpha_k, \beta_k\}$, we find that

$$T_c(x) = \gamma x \gamma^{-1},$$

and so

$$T_c(x)x^{-1} = [\gamma, x].$$

But γ is a separating simple closed curve, and so $\gamma \in \Gamma'$, so that $[\gamma, x] \in [\Gamma, \Gamma']$. Thus we have $[\gamma, x] = 0 \in \wedge^2 H$, and so $\tau(T_c) = 0$.

By the change of coordinates principle and the naturality property of τ (formula (6.1)) it follows that $\tau(T_{c'}) = 0$ for any separating simple closed curve c' in S_g^1. We thus have $\mathcal{K}(S_g^1) \leq \ker(\tau)$.

The image of a bounding pair map. As in the case of Dehn twists, in order to understand the image of an arbitrary bounding pair map, it suffices to compute $\tau(T_d T_e^{-1})$ for the standard bounding pair $\{d, e\}$ shown in Figure 6.5.

Let $f = T_d T_e^{-1}$ and suppose that the bounding pair $\{d, e\}$ has genus k; that is, the subsurface of S_g^1 cut off by $d \cup e$ and not containing ∂S_g^1 has genus k. It is straightforward to compute directly the induced action of f on $\pi_1(S_g^1)$ by computing the action on each generator α_i and β_i of $\pi_1(S_g^1)$. Doing this, we obtain

$$
\begin{array}{ll}
f(\alpha_i) = \delta \alpha_i \delta^{-1} \quad i \leq k & f(\beta_{k+1}) = \beta_{k+1} \\
f(\beta_i) = \delta \beta_i \delta^{-1} \quad i \leq k & f(\alpha_i) = \alpha_i \quad i \geq k+1 \\
f(\alpha_{k+1}) = \delta \epsilon^{-1} \alpha_{k+1} & f(\beta_i) = \beta_i \quad i \geq k+1
\end{array}
$$

where δ and ϵ are the elements of $\pi_1(S_g^1)$ shown at the bottom left of Figure 6.5.

From here we can write down the product $f(x)x^{-1}$ for each $x \in \{\alpha_i, \beta_i\}$. Recall that $f(x)x^{-1}$ lies in N and that it corresponds to an element of $\wedge^2 H$. In the calculation we will use the fact that

$$\delta \epsilon^{-1} = \prod_{i=1}^{k} [\alpha_i, \beta_i]$$

and the fact that the homology classes $[\delta]$ and $[\beta_{k+1}]$ are equal. Denoting by $\leftrightarrow$ the correspondence between elements of N and $\wedge^2 H$ via the isomorphism described above, we have

$$f(\alpha_i)\alpha_i^{-1} = [\delta, \alpha_i] \qquad \leftrightarrow \qquad [\beta_{k+1}] \wedge [\alpha_i] \qquad\qquad i \leq k$$

$$f(\beta_i)\beta_i^{-1} = [\delta, \beta_i] \qquad \leftrightarrow \qquad [\beta_{k+1}] \wedge [\beta_i] \qquad\qquad i \leq k$$

$$f(\alpha_{k+1})\alpha_{k+1}^{-1} = \delta\epsilon^{-1} \qquad \leftrightarrow \qquad \sum_{i=1}^{k} [\alpha_i] \wedge [\beta_i]$$

$$f(\beta_{k+1})\beta_{k+1}^{-1} = 1 \qquad \leftrightarrow \qquad 0$$

$$f(\alpha_i)\alpha_i^{-1} = 1 \qquad \leftrightarrow \qquad 0 \qquad\qquad i \geq k+1$$

$$f(\beta_i)\beta_i^{-1} = 1 \qquad \leftrightarrow \qquad 0 \qquad\qquad i \geq k+1$$

This gives that $\tau(f)$, as an element of $H \otimes \wedge^2 H$, is

$$\tau(f) = \sum_{i=1}^{k} ([\beta_i] \otimes ([\beta_{k+1}] \wedge [\alpha_i]) - [\alpha_i] \otimes ([\beta_{k+1}] \wedge [\beta_i]))$$

$$+ [\beta_{k+1}] \otimes \left(\sum_{i=1}^{k} [\alpha_i] \wedge [\beta_i] \right)$$

$$= \sum_{i=1}^{k} ([\alpha_i] \otimes ([\beta_i] \wedge [\beta_{k+1}]) + [\beta_i] \otimes ([\beta_{k+1}] \wedge [\alpha_i])$$

$$+ [\beta_{k+1}] \otimes ([\alpha_i] \wedge [\beta_i]))$$

$$= \left(\sum_{i=1}^{k} [\alpha_i] \wedge [\beta_i] \right) \wedge [\beta_{k+1}].$$

In summary, we have shown that

$$\tau(T_d T_e^{-1}) = \sum_{i=1}^{k} x_i \wedge y_i \wedge z,$$

where $z \in H$ is the homology class $[d] = [e]$ and $x_1, y_1, \ldots, x_k, y_k, z$ form a (degenerate) symplectic basis for the homology of the component of $S_g^1 - (d \cup e)$ not containing the basepoint of $\pi_1(S_g^1)$.

The image of $\mathcal{I}(S_g^1)$. Choosing $k = 1$ in the above computation gives that the wedge product

$$[\alpha_1] \wedge [\beta_1] \wedge [\beta_2]$$

lies in the image of $\tau : \mathcal{I}(S_g^1) \to \wedge^3 H$. We will now use the naturality property (6.1) together with the fact that $\mathrm{Mod}(S_g^1)$ surjects onto $\mathrm{Sp}(2g, \mathbb{Z})$ to show that τ surjects onto $\wedge^3 H$.

Assume that $g \geq 3$. By Theorem 6.4, there is some $f \in \mathrm{Mod}(S_g^1)$ so that f_* maps the pair $([\alpha_1], [\alpha_3])$ to the pair $([\alpha_1] + [\beta_1] - [\beta_3], [\alpha_3] - [\beta_1] + [\beta_3])$ and fixes all other basis elements of H. Then

$$f_*([\alpha_1] \wedge [\beta_1] \wedge [\beta_2]) = [\alpha_1] \wedge [\beta_1] \wedge [\beta_2] - [\beta_1] \wedge [\beta_2] \wedge [\beta_3].$$

Since we have already shown that $[\alpha_1] \wedge [\beta_1] \wedge [\beta_2]$ lies in the image of τ, it follows from the naturality property of τ that $[\beta_1] \wedge [\beta_2] \wedge [\beta_3]$ does as well. Applying factor swaps and factor rotations gives that every wedge product $x \wedge y \wedge z$ is in the image of τ, where $x, y, z \in \{[\alpha_i], [\beta_i]\}$. Since such elements span $\wedge^3 H$, this completes the proof that τ is surjective when $g \geq 3$. We leave the case of $g = 2$ as an exercise.

We have therefore proved the following result of Johnson.

Proposition 6.16 *If $g \geq 2$, then $\tau(\mathcal{I}(S_g^1)) = \wedge^3 H$.*

There is another way to prove the slightly weaker fact that $\tau(\mathcal{I}(S_g^1)) \otimes \mathbb{Q} = \wedge^3 H \otimes \mathbb{Q}$. Let $H_\mathbb{Q} = H \otimes \mathbb{Q}$. Then the vector space $\wedge^3 H_\mathbb{Q}$ decomposes as a direct sum of irreducible $\mathrm{Sp}(2g, \mathbb{Q})$-modules as follows:

$$\wedge^3 H_\mathbb{Q} = \wedge^3 H_\mathbb{Q} / H_\mathbb{Q} \oplus H_\mathbb{Q}.$$

Since these two summands are irreducible and since τ satisfies the naturality property (6.1), we could prove that τ is a surjection onto $\wedge^3 H$ (after tensoring with $\mathbb{Q}$) by finding one element with τ-image in the first summand and one element with τ-image in the second summand and then applying Schur's lemma. Note that $\wedge^3 H$ is a small, nonobviously embedded subspace of $\wedge^2 H \otimes H$. How did Johnson know to prove that the image of τ is contained in this subspace? Well, he knew that the image of τ has to be a direct sum of Sp-invariant subspaces, so after computing a few elements in the image he might have guessed which subspaces would be needed.

6.6.3 SOME APPLICATIONS

The Johnson homomorphism τ is the most important invariant in the study of the Torelli group. Here we give two example applications.

The kernel of τ. As $\mathcal{K}(S_g^1)$ is contained in the kernel of τ and since the image of τ is infinite, and indeed the image of any bounding pair map is infinite, we immediately deduce the following.

Corollary 6.17 *If $g \geq 3$, then $\mathcal{K}(S_g^1)$ has infinite index in $\mathcal{I}(S_g^1)$. In fact, no bounding pair map or any of its nontrivial powers lie in $\mathcal{K}(S_g^1)$.*

Thus, by using a purely algebraically defined "invariant" τ, Johnson deduced a purely topological statement, namely, that no nontrivial power of any bounding pair can be written as a product of Dehn twists of separating curves. Before Johnson's work, Chillingworth had already shown that $\mathcal{K}(S_g) \neq \mathcal{I}(S_g)$ [46]. Johnson actually proved the following much deeper result [113].

THEOREM 6.18 *If $g \geq 3$, then $\ker(\tau) = \mathcal{K}(S_g^1)$.*

In other words, the kernel of τ, which is defined purely algebraically, is simply the group $\mathcal{K}(S_g^1)$, which is defined purely topologically.

The abelianization of the Torelli group. That fact that $\tau : \mathcal{I}(S_g^1) \to \wedge^3 H$ is surjective immediately implies that the abelianization $H_1(\mathcal{I}(S_g^1); \mathbb{Z})$ must contain an isomorphic copy of $\wedge^3 H$. It turns out that τ captures the entire torsion-free part of $H_1(\mathcal{I}(S_g^1); \mathbb{Z})$, but there is more to the story. Johnson proved the following [114].

THEOREM 6.19 *Let $g \geq 2$. Then*

$$H_1(\mathcal{I}(S_g^1); \mathbb{Z}) \approx \wedge^3 H \times (\mathbb{Z}/2\mathbb{Z})^N,$$

where

$$N = \binom{2g}{2} + \binom{2g}{1} + \binom{2g}{0}.$$

The $\wedge^3 H$ in the theorem is exactly what is detected by the Johnson homomorphism. The torsion part is detected by the Birman–Craggs–Johnson homomorphisms, which are defined using the Rochlin invariant, an invariant coming from the theory of 3-manifolds. See Johnson's lovely survey paper [112] for a discussion.

A filtration of the mapping class group. Let S be either S_g or S_g^1 and let $\Gamma = \pi_1(S)$. The symplectic representation of $\mathrm{Mod}(S)$ describes the action of $\mathrm{Mod}(S)$ on $H_1(S; \mathbb{Z}) = \Gamma/\Gamma'$, where $\Gamma' = [\Gamma, \Gamma]$. The kernel of this representation is the Torelli group $\mathcal{I}(S)$. The Johnson homomorphism describes the action of $\mathcal{I}(S)$ on the quotient $\Gamma/[\Gamma', \Gamma]$. By Theorem 6.18, the kernel of this map is $\mathcal{K}(S)$. One would like to continue this line of analysis to $\mathcal{K}(S)$ and beyond.

To this end, we consider the *lower central series* of $\Gamma = \pi_1(S)$, which is the sequence of groups

$$\Gamma = \Gamma_1 \supset \Gamma_2 \supset \cdots$$

defined inductively by

$$\Gamma_1 = \Gamma \quad \text{and} \quad \Gamma_i = [\Gamma, \Gamma_{i-1}].$$

Since each Γ_i in the lower central series is *characteristic*, that is, fixed by $\mathrm{Aut}(\Gamma)$, there is a natural homomorphism $\mathrm{Aut}(\Gamma) \to \mathrm{Aut}(\Gamma/\Gamma_i)$ that descends to a homomorphism

$$\Psi_i : \mathrm{Out}(\Gamma) \to \mathrm{Out}(\Gamma/\Gamma_{i+1}).$$

As explained in Chapter 8, the (outer) action of $\mathrm{Mod}(S)$ on $\pi_1(S)$ gives a homomorphism $\mathrm{Mod}(S) \to \mathrm{Out}(\pi_1(S))$. (The Dehn–Nielsen–Baer theorem says that this map is an isomorphism when S is closed.)

We define the kth Torelli group $\mathcal{I}^k(S)$ to be the kernel of Ψ_k restricted to $\mathrm{Mod}(S)$. We have already seen the following.

$$\mathcal{I}^0(S) = \mathrm{Mod}(S) \qquad \mathcal{I}^1(S) = \mathcal{I}(S) \qquad \mathcal{I}^2(S) = \mathcal{K}(S)$$

It is a theorem of Magnus that the intersection of the Γ_i is trivial [65, 137]. Using this fact, Bass and Lubotzky proved that the intersection of the $\mathcal{I}^k(S)$ is trivial and so the $\mathcal{I}^k(S)$ give a *filtration* of $\mathcal{I}(S)$, that is, a descending sequence of normal subgroups that intersect in the identity [12]. This filtration of $\mathcal{I}(S)$ is called the *Johnson filtration*. In the same way that the Torelli group captures some mysterious aspects of the mapping class group, we can think of the Johnson filtration as probing even more deeply.

Chapter Seven

Torsion

In this chapter we investigate finite subgroups of the mapping class group. After explaining the distinction between finite-order mapping classes and finite-order homeomorphisms, we then turn to the problem of determining what is the maximal order of a finite subgroup of $\mathrm{Mod}(S_g)$. We will show that, for $g \geq 2$, finite subgroups have order at most $84(g-1)$ and cyclic subgroups have order at most $4g+2$. We will also see that there are finitely many conjugacy classes of finite subgroups in $\mathrm{Mod}(S)$. At the end of the chapter, we prove that $\mathrm{Mod}(S_g)$ is generated by finitely many elements of order 2.

7.1 FINITE-ORDER MAPPING CLASSES VERSUS FINITE-ORDER HOMEOMORPHISMS

In this section we will see that problems about finite-order mapping classes can be converted to (easier) problems about finite-order homeomorphisms.

7.1.1 NIELSEN REALIZATION

Assume $g \geq 2$ and suppose that $G < \mathrm{Homeo}^+(S_g)$ is a finite subgroup. It follows from Theorem 6.8 that the natural projection $\mathrm{Homeo}^+(S_g) \to \mathrm{Mod}(S_g)$ restricted to G is injective. That is, any finite subgroup of $\mathrm{Homeo}^+(S_g)$ is isomorphic to a finite subgroup of $\mathrm{Mod}(S_g)$.

What about the converse? Even the case of a single element is interesting. Suppose $f \in \mathrm{Mod}(S)$ has order k and suppose $\phi \in \mathrm{Homeo}^+(S)$ is any representative of f. It follows from the definition of $\mathrm{Mod}(S)$ that ϕ^k is *isotopic* to the identity. The question is whether or not ϕ can be chosen so that ϕ^k is exactly the identity in $\mathrm{Homeo}^+(S)$. The following classical theorem, due to Fenchel and Nielsen, answers this question in the affirmative.

THEOREM 7.1 *Let* $S = S_{g,n}$ *and suppose* $\chi(S) < 0$. *If* $f \in \mathrm{Mod}(S)$ *is an element of finite order* k, *then there is a representative* $\phi \in \mathrm{Homeo}^+(S)$ *so that* ϕ *has order* k. *Further,* ϕ *can be chosen to be an isometry of some hyperbolic metric on* S.

Our proof of Theorem 7.1 relies on basic properties of Teichmüller space, and so we relegate it to Section 13.2. The following theorem of Kerckhoff is a generalization of Theorem 7.1 from finite cyclic groups to arbitrary finite groups [122]. Its proof is much harder than the proof of Theorem 7.1 and is beyond the scope of this book.

THEOREM 7.2 (Nielsen realization theorem) *Let $S = S_{g,n}$ and suppose $\chi(S) < 0$. Suppose $G < \mathrm{Mod}(S)$ is a finite group. Then there exists a finite group $\widetilde{G} < \mathrm{Homeo}^+(S)$ so that the natural projection $\mathrm{Homeo}^+(S) \to \mathrm{Mod}(S)$ restricts to an isomorphism $\widetilde{G} \to G$. Further, $\widetilde{G}$ can be chosen to be a subgroup of isometries of some hyperbolic metric on S.*

In other words, every finite subgroup of $\mathrm{Mod}(S)$ comes from a finite subgroup of $\mathrm{Homeo}^+(S)$.

Mapping class groups of surfaces with boundary are torsion-free. Recall that a *frame* at a point $x \in S$ is a basis for the tangent space at x. If $\partial S \neq \emptyset$, then any isometry that fixes ∂S pointwise must clearly fix each frame at each point of ∂S. Since an isometry of a surface is determined by what it does to a point and a frame, any such isometry is equal to the identity.

When $\partial S \neq \emptyset$, our proof of Theorem 7.1 applies to produce an isometry $\phi \in \mathrm{Homeo}^+(S)$ (not $\mathrm{Homeo}^+(S, \partial S)$) in the free homotopy class of f. Using the fact that Dehn twists about components of ∂S have infinite order, we obtain the following.

Corollary 7.3 *If $\partial S \neq \emptyset$, then $\mathrm{Mod}(S)$ is torsion-free.*

Isometries of the torus. Since $\mathrm{Mod}(T^2) \approx \mathrm{SL}(2, \mathbb{Z})$, torsion in $\mathrm{Mod}(T^2)$ is the same as torsion in $\mathrm{SL}(2, \mathbb{Z})$. The group $\mathrm{SL}(2, \mathbb{Z})$ has eight nontrivial conjugacy classes of finite-order elements. There are elements of 2, 3, 4, and 6 given by the matrices

$$\begin{pmatrix} -1 & 0 \\ 0 & -1 \end{pmatrix}, \begin{pmatrix} 0 & -1 \\ 1 & -1 \end{pmatrix}, \begin{pmatrix} 0 & -1 \\ 1 & 0 \end{pmatrix}, \text{ and } \begin{pmatrix} 0 & 1 \\ -1 & 1 \end{pmatrix}$$

and their inverses. Each of these matrices can be realized as an isometry of either the square torus or the hexagonal torus; compare Section 12.2.

Isometries of punctured spheres. Let $S_{0,n}$ be a sphere with $n \geq 3$ punctures and let $f \in \mathrm{Mod}(S_{0,n})$ be a finite-order element. By Theorem 7.1, there is a hyperbolic metric on $S_{0,n}$ and a representative $\phi \in \mathrm{Homeo}^+(S_{0,n})$

of f so that ϕ acts by isometries. In particular, ϕ is a finite-order homeomorphism. What is more, we can fill in the punctures of $S_{0,n}$ and so regard ϕ as a finite-order homeomorphism of the 2-sphere S^2.

Now, any finite-order homeomorphism f of S^2 is topologically conjugate to an isometry of S^2 in the standard round metric; see, for example, [128, Section 2.2]. When f is a diffeomorphism, one can see this by averaging a metric to obtain an f-invariant metric and then pulling back this f-invariant metric to the round metric, which one can do by the uniformization theorem. The conjugation of f by the uniformizing map will then act by isometries on the round metric on S^2.

Any orientation-preserving isometry of the round metric on S^2 is a rotation. Therefore, up to taking powers, there are exactly three conjugacy classes of finite-order elements of $\mathrm{Mod}(S_{0,n})$ when $n \geq 4$, since there are 0, 1, or 2 punctures on the axis of rotation. When $n = 3$, there are only two nontrivial conjugacy classes since any element of $\mathrm{Mod}(S_{0,3})$ that fixes two punctures must also fix the third.

7.1.2 DETECTING TORSION WITH THE SYMPLECTIC REPRESENTATION

Using Theorem 7.1, we can now prove Theorem 6.8, which states that if $f \in \mathrm{Mod}(S_g)$ has finite order, then its image under $\Psi : \mathrm{Mod}(S_g) \to \mathrm{Sp}(2g, \mathbb{Z})$ is nontrivial.

Proof of Theorem 6.8. For $g = 1$, the theorem follows immediately from Theorem 2.5, so assume $g \geq 2$. By Theorem 7.1, the mapping class f is represented by an element $\phi \in \mathrm{Diff}^+(S_g)$ of order n, where $1 < n < \infty$. Choose any Riemannian metric h on S_g. Average h by taking $h + \phi^* h + \cdots + (\phi^{n-1})^* h$, which is a ϕ-invariant Riemannian metric on S_g. Thus ϕ acts as an isometry in this metric.

Consider any fixed point $x \in S_g$ of ϕ if one exists. Since ϕ is an isometry, it is determined by its derivative $D\phi_x$ at x, which is a 2×2 orthogonal matrix. Since ϕ is orientation-preserving, the matrix $D\phi$ has determinant 1. Since ϕ is nontrivial, $D\phi_x$ is a nontrivial rotation, and so x is an isolated fixed point of ϕ of index 1.

Since ϕ is a continuous map with isolated fixed points, we can apply the Lefschetz fixed point theorem, which says in this case that the sum $M(\phi)$ of the indices of the fixed points of ϕ is equal to the *Lefschetz number* $L(\phi)$, which is defined as

$$L(\phi) = \sum_{i=0}^{2} (-1)^i \mathrm{Trace}(\phi_\star : H_i(S_g; \mathbb{Z}) \to H_i(S_g; \mathbb{Z}))$$
$$= 1 - \mathrm{Trace}(\phi_\star : H_1(S_g; \mathbb{Z}) \to H_1(S_g; \mathbb{Z})) + 1.$$

Since each fixed point of ϕ has index 1, it follows that $M(\phi) \geq 0$, so that $L(\phi) \geq 0$. But since $g \geq 2$, the matrix ϕ_* cannot be the identity, for then its trace would be at least 4, giving $L(\phi) < 0$, a contradiction. Thus $\Psi(f) = \phi_*$ is nontrivial, as desired. $\qquad\square$

7.2 ORBIFOLDS, THE $84(g-1)$ THEOREM, AND THE $4g+2$ THEOREM

By rotating a flat torus X in one circle factor by $2\pi/n$, one obtains an isometry of X of any order n. In contrast, the possible isometries of hyperbolic surfaces are highly constrained. In this section we will prove two theorems along these lines. The first result was proved in 1893 by Hurwitz. It bounds the order of any finite group of hyperbolic isometries of a genus $g \geq 2$ surface.

THEOREM 7.4 ($84(g-1)$ theorem) *If X is a closed hyperbolic surface of genus $g \geq 2$, then*

$$|\operatorname{Isom}^+(X)| \leq 84(g-1).$$

One remarkable aspect of Theorem 7.4 is that the number 84 appears (why 84?) and yet the given bound is sharp in the sense that the $84(g-1)$ bound is realized for infinitely many g; see the discussion below.

The following theorem was proved in 1895 by Wiman [214].

THEOREM 7.5 ($4g+2$ theorem) *Let X be a closed hyperbolic surface of genus $g \geq 2$. Then any element of $\operatorname{Isom}^+(X)$ has order at most $4g+2$.*

The upper bound of Theorem 7.5 is attained for every $g \geq 2$: we simply realize S_g as a regular hyperbolic $(4g+2)$-gon with angle sum 2π and with opposite sides identified, and we consider the rotation by one click.

Combining Theorems 7.2, 7.4, and 7.5 gives the following.

Corollary 7.6 *Let $g \geq 2$. The order of any finite subgroup of $\operatorname{Mod}(S_g)$ is at most $84(g-1)$, and the order of any finite cyclic subgroup of $\operatorname{Mod}(S_g)$ is at most $4g+2$.*

Since Theorem 7.1 is proved in Section 13.2, this book contains a complete proof of the second statement of Corollary 7.6.

7.2.1 THE ISOMETRY GROUP OF A CLOSED HYPERBOLIC SURFACE IS FINITE

A first step toward obtaining upper bounds on the orders of finite subgroups of isometry groups of surfaces is to show that these groups are finite to begin with.

Proposition 7.7 *Let X be a hyperbolic surface homeomorphic to S_g with $g \geq 2$. Then $\mathrm{Isom}(X)$ is finite in any hyperbolic metric.*

Proof. The isometry group of any compact Riemannian manifold is a compact topological group.[1] This follows easily from the Arzela–Ascoli theorem. It therefore suffices to prove that $\mathrm{Isom}(X)$ is discrete or, what is the same thing, that the connected component in $\mathrm{Isom}(X)$ of the identity is trivial. Since the topology on $\mathrm{Isom}(X)$ agrees with the subspace topology inherited from $\mathrm{Homeo}^+(S_g)$, it is enough to prove that $\mathrm{Isom}(X) \cap \mathrm{Homeo}_0(S_g) = \{1\}$.

Suppose that $\phi \in \mathrm{Isom}(X) \cap \mathrm{Homeo}_0(S_g)$. This says precisely that $\phi \in \mathrm{Isom}(X)$ is isotopic to the identity. Then ϕ has a lift to $\mathrm{Isom}(\mathbb{H}^2)$ that is a bounded distance from the identity map of $\mathbb{H}^2$. By the classification of hyperbolic isometries, any such isometry is equal to the identity. Thus ϕ is the identity, as desired.   □

Proposition 7.7 is simply not true for the torus: the standard square torus has infinitely many isometries. Indeed, the isometry group contains a copy of $S^1 \times S^1 \approx T^2$. On the other hand, these isometries all represent the trivial element of $\mathrm{Mod}(T^2)$. In general, if X is any flat torus, then we still have that $\mathrm{Isom}(X)$ is compact. From this it follows that the projection

$$\mathrm{Isom}^+(X) \to \mathrm{Mod}(X) = \mathrm{Mod}(T^2)$$

has finite image.

7.2.2 ORBIFOLDS

As the hypothesis of Theorem 7.4, we are given a closed hyperbolic surface X of genus $g \geq 2$. The basic strategy of the proof of Theorem 7.4 is to study the quotient space

$$Y = X/\mathrm{Isom}^+(X).$$

When $\mathrm{Isom}^+(X)$ acts freely on X, the quotient Y is itself a hyperbolic surface. However, elements of $\mathrm{Isom}^+(X)$ can have fixed points in X, so

---

[1]It is a theorem of Myers–Steenrod that the isometry group of a compact Riemannian manifold is in fact a Lie group, but we will not need this.

that it is not even clear that Y is a manifold (we will prove below that it is). Since $\mathrm{Isom}^+(X)$ is a finite group for $g \geq 2$ (Proposition 7.7), the space Y has a well-defined area given by

$$\mathrm{Area}(Y) = \mathrm{Area}(X)/|\mathrm{Isom}^+(X)|.$$

By the Gauss–Bonnet theorem we have $\mathrm{Area}(X) = 2\pi(2g-2)$. Thus if we find a universal lower bound on $\mathrm{Area}(Y)$, we obtain a universal upper bound on the order of $\mathrm{Isom}^+(X)$. Theorem 7.10 gives that $\mathrm{Area}(Y) \geq \pi/21$, and we will use this to easily prove Theorem 7.4.

In order to prove that $\mathrm{Area}(Y) \geq \pi/21$, we will need to better understand the geometry of quotients of hyperbolic surfaces by (possibly nonfree) actions of finite groups. This is best accomplished via the theory of hyperbolic orbifolds.

A *2-dimensional (orientable) hyperbolic orbifold*[2] is a quotient X/G, where X is an orientable surface with a hyperbolic metric and G is a subgroup of the finite group $\mathrm{Isom}^+(X)$. Our main goal is to find an Euler characteristic for orbifolds, to prove a Gauss–Bonnet theorem for orbifolds, and to use these results to show that there is a universal lower bound of $\pi/21$ for the area of any 2-dimensional orientable hyperbolic orbifold. As explained above, applying this lower bound to the orbifold $Y = X/\mathrm{Isom}^+(X)$ gives the $84(g-1)$ theorem.

Orbifold fundamental group. By the *orbifold fundamental group* of an orbifold X, we mean the deck transformation group of the universal cover $\widetilde{X} \approx \mathbb{H}^2$. Elements of the orbifold fundamental group of X can be represented by loops in X.

Cone points and signature. Let Y be any 2-dimensional hyperbolic orbifold. Any point $y \in Y$ has a neighborhood isometric to the quotient of an open ball in $\mathbb{H}^2$ by a finite group of rotations F_y of $\mathbb{H}^2$. Under this isometry, the point y is mapped to the fixed point of F_y. This follows from the fact that any finite subgroup of $\mathrm{Isom}^+(\mathbb{H}^2)$ is a finite group of rotations fixing some point. If F_y is trivial, then y is called a *regular point* of Y; if F_y is not trivial, then y is called a *cone point of order* $|F_y|$. There are finitely many cone points on a 2-dimensional hyperbolic orbifold.

If X is a 2-dimensional hyperbolic orbifold where the underlying topological surface (the surface obtained by forgetting the extra structure of the cone points) is homeomorphic to S_g and where the cone points have orders

---

[2]What we are referring to as an orbifold is sometimes called a "good orbifold"; see [206, Chapter 13].

$p_1, p_2, \ldots, p_m$, then we define the *signature* of X to be the $(m+1)$-tuple $(g; p_1, p_2, \ldots, p_m)$.

Orbifolds from hyperbolic triangle groups. We can use triangles in $\mathbb{H}^2$ to build examples of 2-dimensional orientable hyperbolic orbifolds as follows. Consider a triangle T in $\mathbb{H}^2$ with angles π/p, π/q, and π/r, where $p, q, r \in \mathbb{N}$ and $1/p + 1/q + 1/r < 1$. Each side of T can be extended to a unique geodesic line in $\mathbb{H}^2$. Let $\Gamma < \mathrm{Isom}(\mathbb{H}^2)$ denote the group generated by the reflections in these three geodesic lines. The elements of Γ that are orientation-preserving form a subgroup Γ_0 of index 2. Note that Γ_0 contains rotations about the vertices of T of orders p, q and r. By the Selberg lemma [193] (or by a direct argument), Γ_0 contains a normal, torsion-free subgroup Γ_1 of finite index. Note that Γ_1 acts properly discontinuously and cocompactly on $\mathbb{H}^2$ since Γ does. Since Γ_1 is torsion-free, it also acts freely, so that $\mathbb{H}^2/\Gamma_1$ is a closed hyperbolic surface. By basic covering space theory this surface admits an isometric action by the finite group Γ_0/Γ_1 with quotient $\mathbb{H}^2/\Gamma_0$. Thus $\mathbb{H}^2/\Gamma_0$ is a 2-dimensional (orientable) hyperbolic orbifold. It has signature $(0; p, q, r)$.

We will see that the combinatorial data of signature is enough to determine the hyperbolic area of a 2-dimensional hyperbolic orbifold. This is essentially the content of the Gauss–Bonnet theorem for orbifolds explained below.

In order to get to that point, we will first need to find an Euler characteristic for orbifolds. This invariant should agree with the classical Euler characteristic when evaluated on surfaces and should be multiplicative with respect to coverings. Of course, the key issue here is to find such a definition that gives a well-defined number; this is not trivial to do since there are many coverings of and many finite group actions on hyperbolic surfaces. In order to give the definition we will use the notion of orbifold covering maps.

Orbifold covering maps. By an *isometry* of a 2-dimensional hyperbolic orbifold X, we mean an isometry of the metric space X. Such an isometry necessarily is an isometry of $X - \{\text{cone points}\}$ thought of as a Riemannian manifold, and it takes cone points to other cone points of the same order.

A map $X \to Y$ between 2-dimensional hyperbolic orbifolds is a *regular d-fold orbifold covering map* if it is a quotient map by an order d group of orientation-preserving isometries of X.

For example, if Z is a hyperbolic surface and $H \lhd G < \mathrm{Isom}^+(Z)$, then the orbifold $X = Z/H$ covers the orbifold $Y = Z/G$ since Y is the quotient of X by G/H:

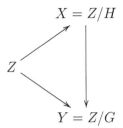

$$X = Z/H$$

$$Z$$

$$Y = Z/G$$

Consider a d-fold orbifold covering $\pi : X \to Y$. The *degree of π at a point x* is the order of the cone point $\pi(x)$ divided by the order of the cone point x. The sum of the degrees of π at the preimages of a given point $y \in Y$ is always equal to d. In other words, if the preimage of a cone point of order p is a collection of cone points in X of orders $q_1, \ldots, q_k$, then

$$\sum_{i=1}^{k} \frac{p}{q_i} = d.$$

One way to see that this equality holds is to notice that π is a true covering map away from the cone points and to consider a regular point close to y.

By summing over all cone points in Y, we have that if X has signature $(h; q_1, \ldots, q_n)$ and Y has signature $(g, p_1, \ldots, p_m)$ and $X \to Y$ is a d-fold cover, then

$$\sum_{i=1}^{n} \frac{1}{q_i} = d \sum_{i=1}^{m} \frac{1}{p_i}. \qquad (7.1)$$

The Riemann–Hurwitz formula. We want to find an orbifold Euler characteristic, that is, a function of the signature of an orbifold that is multiplicative under orbifold covers.

Consider the 2-dimensional hyperbolic orbifold Y with signature $(g; p_1, p_2, \ldots, p_m)$. We think of constructing Y by starting with a closed surface of genus g, removing m open disks, and gluing in "fractions of disks." This leads to the *Riemann–Hurwitz formula*, an Euler characteristic for 2-dimensional orientable hyperbolic orbifolds. We define

$$\chi(Y) = (2 - 2g) - m + \sum_{i=1}^{m} \frac{1}{p_i} \qquad \text{(Riemann–Hurwitz formula)}.$$

First note that $\chi(Y)$ clearly agrees with the classical Euler characteristic when Y has no singular points.

Proposition 7.8 (Multiplicativity of orbifold Euler characteristic) *If* $\pi :$ $X \to Y$ *is a d-fold orbifold cover, then we have*

$$\chi(X) = d\chi(Y).$$

Proof. Denote the signatures of the orbifolds X and Y by $(h; q_1, \ldots, q_n)$ and $(g; p_1, p_2, \ldots, p_m)$, respectively. Let Y° be the complement in Y of disjoint open neighborhoods of the cone points of Y and let $X^\circ = \pi^{-1}(Y^\circ)$. Note that $X - X^\circ$ is an open neighborhood of the cone points in X and that $\pi|_{X^\circ} : X^\circ \to Y^\circ$ is a d-fold covering map of surfaces. We now compute:

$$\chi(X) = (2 - 2h) - n + \sum_{i=1}^{n} \frac{1}{q_i}$$

$$= \chi(X^\circ) + \sum_{i=1}^{n} \frac{1}{q_i}$$

$$= d\chi(Y^\circ) + d\sum_{i=1}^{m} \frac{1}{p_i}$$

$$= d((2 - 2g) - m) + d\sum_{i=1}^{m} \frac{1}{p_i}$$

$$= d\chi(Y).$$

The first and fifth equalities follow from the Riemann–Hurwitz formula. The third equality follows from (7.1) and the multiplicativity of the Euler characteristic for surfaces. The second and fourth equalities follow from the fact that deleting an open disk from a surface reduces the Euler characteristic by 1. This completes the proof. □

The orbifold Gauss–Bonnet formula. The classical Gauss-Bonnet formula for closed hyperbolic surfaces X gives that $\text{Area}(X) = -2\pi\chi(X)$. For an orbifold Y that is the quotient of a hyperbolic surface X by a group G of isometries, the area $\text{Area}(Y)$ is $\text{Area}(X)/|G|$ (this agrees with the area of $Y - \{\text{cone points}\}$, thought of as a Riemannian manifold). With this generalized notion of area and the generalized notion of Euler characteristic χ for orbifolds, the Gauss–Bonnet formula extends to hyperbolic orbifolds.

Proposition 7.9 (Orbifold Gauss–Bonnet formula) *Suppose Y is a 2-dimensional hyperbolic orbifold. If the signature of Y is $(g; p_1, p_2, \ldots, p_m)$,* *then*

$$\text{Area}(Y) = -2\pi\chi(Y)$$

$$= -2\pi\left((2-2g) - \sum_{i=1}^{m}\left(1 - \frac{1}{p_i}\right)\right).$$

Proof. Verifying this formula is easy, given the Gauss–Bonnet theorem for surfaces and the multiplicativity of the orbifold Euler characteristic (Proposition 7.8). Indeed, if $Y = X/G$, we have

$$\text{Area}(Y) = \frac{\text{Area}(X)}{|G|} = -2\pi\frac{\chi(X)}{|G|} = -2\pi\chi(Y).$$

$\square$

The smallest 2-dimensional hyperbolic orbifold. Armed with the orbifold Gauss-Bonnet formula, we are now able to find a lower bound on the area of any 2-dimensional hyperbolic orbifold. As noted at the start of this section, this will give the desired upper bound on the order of $\text{Isom}^+(X)$.

Theorem 7.10 *If Y is any compact 2-dimensional (orientable) hyperbolic orbifold, then $\chi(Y) \le -1/42$. Equivalently, $\text{Area}(Y) \ge \pi/21$. Further, the orbifold with signature $(0; 2, 3, 7)$ is the unique 2-dimensional hyperbolic orbifold with Euler characteristic $-1/42$.*

The fact that $\chi(Y) \le -1/42$ is equivalent to $\text{Area}(Y) \ge \pi/21$ follows immediately from the orbifold Gauss-Bonnet formula (Proposition 7.9). To construct the orbifold with signature $(0; 2, 3, 7)$, simply choose any triangle in $\mathbb{H}^2$ with angles $\pi/2$, $\pi/3$, and $\pi/7$, consider the group Γ generated by the reflections in the unique lines containing its sides, and take the quotient of $\mathbb{H}^2$ by the index 2 subgroup of Γ consisting of orientation-preserving isometries.

Proof. We begin with a simple but useful observation. Any cone point has order at least 2. Thus for each cone point of order p, the corresponding term $1 - \frac{1}{p}$ from the Riemann–Hurwitz formula is at least $1/2$.

Assume that X is a 2-dimensional orientable hyperbolic orbifold with $\chi(X) \ge -1/42$. We will rule out all possibilities for X except the hyperbolic orbifold with signature $(0; 2, 3, 7)$. We accomplish this with a case-by-case analysis, applying the Riemann–Hurwitz formula repeatedly.

We can immediately rule out that X has no cone points since in this case $\chi(X)$ is a negative integer and so is less than $-1/42$. We can also dispense with all orbifolds X of genus greater than 1, since in this case $\chi(X) \leq -2$. Similarly, any orbifold X of genus 1 must have at least one cone point in order to be hyperbolic, and hence $\chi(X) \leq -1/2$. The case that X has genus 0 and more than four cone points can be eliminated since in this case $\chi(X) \leq 2 - 5 \cdot 1/2 = -1/2$.

Consider the case when X is genus 0 with four cone points. If X has signature $(0; 2, 2, 2, 2)$, then $\chi(X) = 0$, contradicting the fact that X is hyperbolic. If any of the four cone points of X has order greater than 2, then

$$\chi(X) \leq 2 - 3 \cdot 1/2 - 2/3 = -1/6 < -1/42.$$

We are now reduced to checking orbifolds X with genus 0 and three cone points. If 3 is the smallest order of a cone point of X, then either $\chi(X) = 2 - 3 \cdot 2/3 = 0$, contradicting the fact that X is hyperbolic, or

$$\chi(X) \leq 2 - 2 \cdot 2/3 - 3/4 = -1/12 < -1/42.$$

Thus we can assume that X has at least one cone point of order 2. We know that X cannot have two cone points of order 2, for otherwise $\chi(X) > 0$. If X has no cone point of order 3, then $\chi(X) \geq 2 - 1/2 - 2 \cdot 3/4 = 0$ (a contradiction) or

$$\chi(X) \leq 2 - 1/2 - 3/4 - 4/5 = -1/20 < -1/42.$$

It now remains to check orbifolds X of signature $(0; 2, 3, p)$. It is easy to check that the smallest p for which $\chi(X) < 0$ is $p = 7$. If $p > 7$, then $\chi(X) < -1/42$. Combining all of the observations above, we see that $\chi(X) < -1/42$ for every hyperbolic orbifold except for the hyperbolic orbifold of signature $(0; 2, 3, 7)$, which has Euler characteristic $-1/42$. $\square$

7.2.3 PROOF OF THE $84(g - 1)$ THEOREM

As explained above, the $84(g - 1)$ theorem follows rather directly from the inequality of Theorem 7.10.

Proof of the $84(g - 1)$ theorem. Let $G = \text{Isom}^+(X)$. By Proposition 7.7, the group G is finite. Thus X/G is a 2-dimensional orientable hyperbolic orbifold. By Theorem 7.10, we have

$$\text{Area}(X/G) \geq \frac{\pi}{21},$$

and by the orbifold Gauss–Bonnet formula, this becomes

$$\frac{2\pi(2g - 2)}{|G|} \geq \frac{\pi}{21},$$

which gives the result. □

7.2.4 PROOF OF THE $4g + 2$ THEOREM

Let X be a closed hyperbolic surface of genus g. In this subsection we prove Wiman's theorem (Theorem 7.5) that every element of $\mathrm{Isom}^+(X)$ has order at most $4g+2$. As explained above, this bound is attained for every $g \geq 1$ by considering the rotations of the $(4g+2)$-gon about its center. The quotient of X by this cyclic group of rotations is a 2-dimensional orientable hyperbolic orbifold of signature $(0; 2, 2g + 1, 4g + 2)$.

Let G be a cyclic subgroup of $\mathrm{Isom}^+(X)$. To prove that $|G| \leq 4g + 2$ we will apply a case-by-case analysis similar to the proof of the $84(g - 1)$ theorem. In order to get a better upper bound than $84(g-1)$ for $|G|$, we will of course have to exploit the fact that the orbifold covering map $X \to Y = X/G$ is cyclic.

Lemma 7.11 *Let $X \to Y$ be an orbifold covering with cyclic covering group $G < \mathrm{Isom}^+(X)$. Suppose that the signature of Y is $(0; p_1, \ldots, p_m)$. Then for any $1 \leq i \leq m$,*

$$\mathrm{lcm}(p_1, \ldots, p_{i-1}, p_{i+1}, \ldots, p_m) = |G|.$$

That is, the least common multiple of the orders of any $m - 1$ cone points is equal to $|G|$.

Proof. The covering group over any 2-dimensional hyperbolic orbifold of genus 0 with m cone points is generated by simple loops that go around any $m - 1$ of the cone points. This is analogous to the fact that fundamental groups of punctured spheres are generated by such loops. A simple loop going around a cone point of order p_i represents an element of order p_i in the covering group. The lemma now follows from the fact that the order of a cyclic group is the least common multiple of the orders of any set of generators. □

Proof of the $4g + 2$ theorem. Let X be a closed hyperbolic surface of genus $g \geq 2$, let $G < \mathrm{Isom}^+(X)$ be a cyclic subgroup, and let $X \to Y = X/G$ be the induced orbifold covering map. Say that the orbifold signature of Y is $(h; p_1, \ldots, p_m)$. Since the orbifold Euler characteristic is multiplicative (Proposition 7.8), we have $\chi(X)/|G| = \chi(Y)$, which we write as

$$\frac{2g - 2}{|G|} = -\chi(Y). \tag{7.2}$$

The proof proceeds as follows. We systematically go through all possibilities for the signature of Y. For each signature that is not $(0; 2, 2g+1, 4g+2)$, we will either show that the signature cannot possibly be the signature of a quotient of X or show that $|G| < 4g + 2$. Sometimes the latter will be accomplished by showing that $-\chi(Y) = (2g - 2)/|G|$ is at least $1/2$ (note that $(2g - 2)/(4g + 2) < 1/2$).

First suppose that $h \geq 1$. By the Riemann–Hurwitz formula, we have

$$-\chi(Y) = 2h - 2 + \sum_{i=1}^{m}\left(1 - \frac{1}{p_i}\right) \geq \sum_{i=1}^{m}\left(1 - \frac{1}{p_i}\right).$$

If $h = 1$, then $m > 0$ (otherwise Y is not hyperbolic), and so $-\chi(Y) \geq 1/2$. If $h \geq 2$, then $2g - 2 \geq 2$, and so $-\chi(Y) \geq 2 > 1/2$. Thus it remains to consider orbifolds of signature $(0; p_1, \ldots, p_m)$, and so we can write

$$-\chi(Y) = -2 + \sum_{i=1}^{m}\left(1 - \frac{1}{p_i}\right). \tag{7.3}$$

Suppose that $m \geq 5$. Again, since $(1 - 1/p_i) \geq 1/2$ for each i, we have $-\chi(Y) \geq 1/2$. It follows easily from the Riemann–Hurwitz formula that a 2-dimensional hyperbolic orbifold of genus 0 must have at least three cone points. Thus we may assume that $m = 3$ or $m = 4$.

First we treat the case $m = 4$. In this case, (7.2) and (7.3) give

$$\frac{2g - 2}{|G|} = 2 - \left(\frac{1}{p_1} + \frac{1}{p_2} + \frac{1}{p_3} + \frac{1}{p_4}\right).$$

Say that $p_1 \leq p_2 \leq p_3 \leq p_4$. If $p_3 \geq 4$, then $p_4 \geq 4$, and we again find $-\chi(Y) \geq 1/2$. If $p_3 = 3$, then $p_1 \leq 3$ and $p_2 \leq 3$, and so $\text{lcm}(p_1, p_2, p_3)$ is equal to 3 or 6. Applying Lemma 7.11 then gives that $|G|$ is equal to 3 or 6. In either case, $|G| < 4g + 2$ since $g \geq 2$. Finally, if $p_3 = 2$, then $p_1 = p_2 = p_3 = 2$, and Lemma 7.11 gives that $|G| = 2$.

It remains to consider orbifolds of signature $(0; p_1, p_2, p_3)$. Now (7.2) and (7.3) give

$$\frac{2g - 2}{|G|} = 1 - \left(\frac{1}{p_1} + \frac{1}{p_2} + \frac{1}{p_3}\right). \tag{7.4}$$

As above, we assume $p_1 \leq p_2 \leq p_3$. We deal with two subcases, according to whether or not p_1 divides p_2.

If p_1 divides p_2, then $\text{lcm}(p_1, p_2) = p_2$, and Lemma 7.11 gives $p_2 = |G|$. Lemma 7.11 also gives $\text{lcm}(p_2, p_3) = p_2$. Since $p_3 \geq p_2$, we have $p_2 = p_3 = |G|$. Substituting $|G|$ for p_2 and p_3 in (7.4) and simplifying, we obtain

$$2g = |G|\left(1 - \frac{1}{p_1}\right).$$

Since $1/2 \leq 1 - 1/p_1 < 1$, it follows that $2g < |G| \leq 4g$.

Finally, we treat orbifolds of signature $(0; p_1, p_2, p_3)$ where p_1 does not divide p_2. If $p_1 \geq 6$, then (7.4) gives $-\chi(Y) \geq 1/2$, and so we may assume $p_1 \leq 5$; in particular, p_1 is either 2, 3, 4, or 5. An elementary case-by-case argument using Lemma 7.11 then gives that $|G| = \operatorname{lcm}(p_1, p_2)$ is equal to p_3 (this means that G has a fixed point at the cone point of order p_3). Substituting $|G|$ for p_3 in (7.4) and simplifying, we obtain

$$2g - 1 = |G| \left(1 - \frac{1}{p_1} - \frac{1}{p_2} \right). \tag{7.5}$$

If $p_1 \geq 4$, then the right-hand side of (7.5) is at least $|G|/2$, and so $|G| \leq 4g - 2$. If $p_1 = 3$, then Lemma 7.11 gives that $p_2 = |G|/3$. Plugging into (7.5) then gives $|G| = 3g + 3$, which is strictly less than $4g + 2$ for $g \geq 2$. Finally, if $p_1 = 2$, then Lemma 7.11 implies that $p_2 = |G|/2$, and we find that $|G| = 4g + 2$. This is exactly the case where the quotient orbifold has signature $(0; 2, 2g + 1, 4g + 2)$, as desired. $\square$

Combined with the results of Section 10.5, our proof of the $4g+2$ theorem really proves a stronger result, namely, that (up to isometry) there is only one hyperbolic structure X on S_g that admits a symmetry of order $4g + 2$, and moreover the corresponding element of $\operatorname{Mod}(S_g)$ is unique up to conjugacy (cf. Theorem 7.14 below).

7.3 REALIZING FINITE GROUPS AS ISOMETRY GROUPS

The $84(g - 1)$ theorem gives a restriction on those finite groups that can act effectively by isometries on some hyperbolic surface of genus $g \geq 2$. One can ask for a sort of converse: can any given group be realized as a group of isometries of some closed hyperbolic surface? If so, what is the smallest genus of such a surface?

THEOREM 7.12 *Let G be any finite group. Then G can be realized as a subgroup of $\operatorname{Mod}(S_g)$ for some $g \geq 2$. In fact, G is a subgroup of $\operatorname{Isom}^+(X)$ for some hyperbolic surface $X \approx S_g$.*

We give two proofs of Theorem 7.12, one using geometric group theory and one using covering spaces.

First proof of Theorem 7.12. Let G be a nontrivial finite group and let Γ be the Cayley graph of G with respect to any generating set. Let S_g be the surface obtained as follows. We start by taking one torus for each vertex of Γ. Then, for each edge of Γ, we perform a connect sum operation on the corresponding tori. The result is a closed surface S_g. Since G is nontrivial, the graph Γ has at least two vertices, and so $g \geq 2$.

The action by G on Γ on the left by automorphisms induces an action of G on S_g by orientation-preserving homeomorphisms. We prove in Theorem 6.8 below that the natural projection $\mathrm{Homeo}^+(S_g) \to \mathrm{Mod}(S_g)$ is faithful when restricted to any finite subgroup. (Alternatively, to see that the action is faithful, we can notice that the action of G on $H_1(S_g; \mathbb{Z})$ is faithful since there is a torus for each vertex of Γ and each torus carries a nontrivial subspace of $H_1(S_g; \mathbb{Z})$.)

As mentioned above, any finite group G of homeomorphisms of S_g, where $g \geq 2$, preserves some hyperbolic metric on S_g: one just averages any metric to obtain a G-invariant metric, uniformizes that metric, and then conjugates the G-action by this uniformizing map to obtain a G-invariant hyperbolic metric. $\square$

We note that it is possible to perturb any G-invariant hyperbolic metric within the space of hyperbolic metrics so that $G = \mathrm{Isom}^+(X)$ for some hyperbolic metric X.

Second proof of Theorem 7.12. Let $S_{0,n+1}$ be a sphere with $n+1$ punctures, where n is the size of some generating set for G. Since $\pi_1(S_{0,n+1})$ is a free group on n letters, it surjects onto G, and so there is a covering map $S' \to S_{0,n+1}$ with covering group G. We can fill in the punctures of S' to get a closed surface S_g on which G acts effectively by homeomorphisms. (An alternative way to obtain that the action of G is effective is to modify the surface S_g by adding extra handles equivariantly; it is then clear that each element of G acts nontrivially on the first homology of the resulting surface.) As in the previous proof, this proves the theorem. $\square$

For a classical treatment of the problem of understanding finite-group actions on surfaces, see [41, Chapter XII].

It is natural to ask how often the bound of $84(g-1)$ in Theorem 7.4 is realized. It is a classical fact that it is realized for infinitely many g and not realized for infinitely many g. One can find infinitely many $g \geq 2$ for which there is a closed genus g hyperbolic surface X with $|\mathrm{Isom}^+(X)| = 84(g-1)$ as follows. Consider the quotient of $\mathbb{H}^2$ by the congruence group $\mathrm{PSL}(2, \mathbb{Z})[7]$ (see Chapter 6 below) and fill in the punctures of the resulting surface. This gives a closed surface admitting a hyperbolic metric. This surface X is known as the *Klein quartic surface*. It has genus 3. A straightforward but detailed analysis gives that

$$|\mathrm{Isom}^+(X)| = 168 = 84(3-1).$$

The group $\mathrm{PSL}(2,\mathbb{Z})[7]$ acts on the Farey complex, and the resulting triangulation on the Klein quartic surface is exactly the fundamental domain for the action of the isometry group. Examples of surfaces in higher genus realizing the $84(g-1)$ bound are obtained by simply taking normal covers of this one. Larsen proved the remarkable result that the frequency of g for which the bound $84(g-1)$ is attained is the same as the frequency of the perfect cubes in the integers [129].

7.4 CONJUGACY CLASSES OF FINITE SUBGROUPS

We have seen above that a finite subgroup of $\mathrm{Homeo}^+(S_g)$ gives rise to an orbifold covering map $X \to Y$, where X is a hyperbolic surface homeomorphic to S_g. If we have two orbifold coverings $X \to Y$ and $X' \to Y'$, where $X, X' \approx S_g$, then a necessary condition for the covering groups to be conjugate in $\mathrm{Homeo}^+(S_g)$ is that Y and Y' have the same signature. However, this is not sufficient, even in the case where Y and Y' have no cone points. Indeed, we also need for the maps from the orbifold fundamental groups of Y and Y' to the deck group to be the same, up to precomposition by an automorphism of the orbifold fundamental group.

By the Riemann–Hurwitz formula, there are finitely many orbifolds that can be covered by a fixed S_g. The fundamental group of each such orbifold has finitely many homomorphisms onto some fixed finite group. Finally, by the orbifold Gauss–Bonnet formula and the fact that area is multiplicative under orbifold covers, the order of the deck transformation group of S_g over a fixed orbifold is completely determined. We thus deduce the following.

Theorem 7.13 *Let $g \geq 2$. There are finitely many conjugacy classes of finite subgroups in $\mathrm{Homeo}^+(S_g)$. In particular, there are finitely many conjugacy classes of finite-order elements in $\mathrm{Homeo}^+(S_g)$.*

If we then quote the Nielsen realization theorem (Theorem 7.2), we obtain the following.

Theorem 7.14 *Let $g \geq 2$. There are finitely many conjugacy classes of finite subgroups in $\mathrm{Mod}(S_g)$. In particular, there are finitely many conjugacy classes of finite-order elements in $\mathrm{Mod}(S_g)$.*

Uniqueness of hyperelliptic involutions. In Chapter 2, we said that the element of $\mathrm{Mod}(S_g)$ obtained by reflecting a regular $(4g+2)$-gon through its center is called a hyperelliptic involution. A more sophisticated definition of a *hyperelliptic involution* is that it is an order 2 element of $\mathrm{Mod}(S_g)$ that acts by $-I$ on $H_1(S_g; \mathbb{Z})$. In what follows we take this new definition of a hyperelliptic involution.

As an illustration of the above criterion for distinguishing conjugacy classes of finite subgroups, we have the following.

Proposition 7.15 *Let $g \geq 1$. Any two hyperelliptic involutions in $\mathrm{Mod}(S_g)$ are conjugate.*

Proof. First note that the quotient orbifold corresponding to a hyperelliptic involution must have genus 0, otherwise the involution permutes handles of S_g and hence does not act by $-I$ on $H_1(S_g; \mathbb{Z})$. By the Riemann–Hurwitz formula, the quotient has $2g+2$ cone points of order 2. The involution is then determined by the homomorphism from this orbifold fundamental group to $\mathbb{Z}/2\mathbb{Z}$. But each generator must map nontrivially to $\mathbb{Z}/2\mathbb{Z}$, for otherwise the cover, which is supposed to be S_g, would have cone points. Therefore, there is only one possible homomorphism and hence one conjugacy class of hyperelliptic involutions. $\square$

The element of $\mathrm{Mod}(S_g)$ obtained by reflecting a $(4g+2)$-gon through its center has order 2, and it acts by $-I$ on $H_1(S_g; \mathbb{Z})$. The element of $\mathrm{Mod}(S_g)$ depicted in Figure 2.3 also has these properties, and so it, too, is a hyperelliptic involution. By Proposition 7.15, these mapping classes are conjugate.

Proposition 7.15 implies that we could alternatively define hyperelliptic involutions as the (homotopy classes of) order 2 homeomorphisms with $2g+2$ fixed points.

Recall from the discussion after Theorem 3.10 that the hyperelliptic involutions in $\mathrm{Mod}(T^2)$ and $\mathrm{Mod}(S_2)$ are central. So in these cases the hyperelliptic involution is not only unique up to conjugacy but is completely unique. For $g \geq 3$, there are infinitely many hyperelliptic involutions in $\mathrm{Mod}(S_g)$.

7.5 GENERATING THE MAPPING CLASS GROUP WITH TORSION

We conclude this chapter with the following curious theorem of Feng Luo [132]. By an *involution* in a group we simply mean any element of order 2.

THEOREM 7.16 *For $g \geq 3$, the group $\mathrm{Mod}(S_g)$ is generated by finitely many involutions.*

Proof. Theorem 4.1 states that $\mathrm{Mod}(S_g)$ is generated by finitely many Dehn twists about nonseparating simple closed curves. So to prove the theorem it suffices to show that every Dehn twist about a nonseparating curve is a product of involutions. By the change of coordinates principle and Fact 3.7, any two twists about nonseparating curves are conjugate, so it suffices to prove that any specific such twist is the product of involutions.

Recall from Section 5.1 that, since $g \geq 3$, we can find a lantern relation

$$T_x T_y T_z = T_a T_b T_c T_d$$

where each of the seven simple closed curves in the relation is nonseparating (cf. Figure 5.5). What is more, we can arrange that each of $x \cup a$, $y \cup b$, and $z \cup c$ is nonseparating.

To prove the theorem, we only need to show that T_d is a product of involutions. Using the fact that each of T_a, T_b, and T_c commutes with each of T_x, T_y, and T_z, we can rewrite the above lantern relation as

$$(T_x T_a^{-1})(T_y T_b^{-1})(T_z T_c^{-1}) = T_d.$$

The theorem is now reduced to showing that if $\{u, v\}$ is a pair of simple closed curves in S_g where $u \cup v$ is nonseparating, then $T_u T_v^{-1}$ is a product of involutions. Indeed, it then follows from the change of coordinates principle that each of $T_x T_a^{-1}$, $T_y T_b^{-1}$, and $T_z T_c^{-1}$ is a product of involutions, and then so is T_d.

Let u and v be curves in S_g as above. We claim that there is an involution $f \in \text{Mod}(S_g)$ interchanging u and v. Indeed, there is an involution of S_g interchanging the simple closed curves s and t in Figure 7.1. Our claim then follows from the change of coordinates principle.

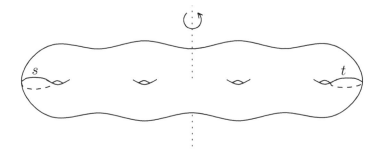

Figure 7.1 Rotation by π is an involution of S_g interchanging s and t.

Since $f(u) = v$ and since $f = f^{-1}$, we can use Fact 3.7 to write

$$T_u T_v^{-1} = T_u (f T_u^{-1} f).$$

By simply changing the parentheses on the right-hand side of the last equation we have

$$T_u T_v^{-1} = (T_u f T_u^{-1}) f.$$

We know that f is an involution by assumption, and so $T_u f T_u^{-1}$ is an involution since it is conjugate to f. Thus $T_u T_v^{-1}$ is a product of two involutions, and we are done.   □

Luo asked if there was a universal bound on the number of torsion elements needed to generate $\mathrm{Mod}(S_g)$. Korkmaz showed that $\mathrm{Mod}(S_g)$ can actually be generated by two torsion elements, which is obviously optimal [125]. Building on work of Brendle–Farb, Kassabov proved that $\mathrm{Mod}(S_g)$ is generated by four involutions when $g \geq 7$ [32, 117]. Now $\mathrm{Mod}(S_g)$ does not have a finite-index cyclic subgroup, so it is not generated by two involutions. The question of whether or not $\mathrm{Mod}(S_g)$ can be generated by three involutions remains open.

Chapter Eight

The Dehn–Nielsen–Baer Theorem

The Dehn–Nielsen–Baer theorem states that $\mathrm{Mod}(S_g)$ is isomorphic to an index 2 subgroup of the group $\mathrm{Out}(\pi_1(S_g))$ of outer automorphisms of $\pi_1(S_g)$. This is a beautiful example of the interplay between topology and algebra in the mapping class group. It relates a purely topological object, $\mathrm{Mod}(S_g)$, to a purely algebraic one, $\mathrm{Out}(\pi_1(S_g))$. Further, these are related via hyperbolic geometry!

8.1 STATEMENT OF THE THEOREM

We begin by defining the objects in the statement of the theorem.

Extended mapping class group. Let S be a surface without boundary. The *extended mapping class group*, denoted $\mathrm{Mod}^{\pm}(S)$, is the group of isotopy classes of all homeomorphisms of S, including the orientation-reversing ones.[1] The group $\mathrm{Mod}(S)$ is an index 2 subgroup of $\mathrm{Mod}^{\pm}(S)$. There is a homomorphism $\mathrm{Mod}^{\pm}(S) \to \mathbb{Z}/2\mathbb{Z}$ which records whether or not an element is orientation-preserving, and we have the short exact sequence

$$1 \to \mathrm{Mod}(S) \to \mathrm{Mod}^{\pm}(S) \to \mathbb{Z}/2\mathbb{Z} \to 1.$$

For any S, there is an order 2 element of $\mathrm{Mod}^{\pm}(S)$ that reverses orientation, and so this sequence is split.

As a first example, we have $\mathrm{Mod}^{\pm}(S^2) \approx \mathbb{Z}/2\mathbb{Z}$. Also, it follows from the fact that $\mathrm{Mod}(T^2) \approx \mathrm{SL}(2, \mathbb{Z})$ (Theorem 2.5) that

$$\mathrm{Mod}^{\pm}(T^2) \approx \mathrm{GL}(2, \mathbb{Z}).$$

---

[1] For the surfaces $S_{0,1}$ and $S_{0,2}$, we must be careful to define $\mathrm{Mod}^{\pm}(S)$ as the group of isotopy classes of homeomorphisms; for these surfaces, every homeomorphism is homotopic to an orientation-preserving homeomorphism.

Similarly, we have

$$\mathrm{Mod}^{\pm}(S_{0,3}) \approx \Sigma_3 \times \mathbb{Z}/2\mathbb{Z}$$
$$\mathrm{Mod}^{\pm}(S_{0,4}) \approx \mathrm{PGL}(2,\mathbb{Z}) \ltimes (\mathbb{Z}/2\mathbb{Z} \times \mathbb{Z}/2\mathbb{Z})$$
$$\mathrm{Mod}^{\pm}(S_{1,1}) \approx \mathrm{GL}(2,\mathbb{Z}).$$

We remark that, the way we have defined things, we do not automatically have a definition of the extended mapping class group for a surface S with boundary since a homeomorphism that is the identity on ∂S is necessarily orientation-preserving.

Outer automorphism groups. For a group G, let $\mathrm{Aut}(G)$ denote the group of automorphisms of G. For any $h \in G$, there is an associated *inner automorphism* $I_h : G \to G$ given by

$$g \mapsto hgh^{-1}$$

for all $g \in G$. For $\Phi \in \mathrm{Aut}(G)$ and $h \in G$, we have

$$\Phi \circ I_h \circ \Phi^{-1} = I_{\Phi(h)}.$$

Thus the inner automorphisms form a normal subgroup of $\mathrm{Aut}(G)$, called the *inner automorphism group* of G, denoted $\mathrm{Inn}(G)$.

The *outer automorphism group* of G is defined as the quotient

$$\mathrm{Out}(G) = \mathrm{Aut}(G)/\mathrm{Inn}(G).$$

In other words, $\mathrm{Out}(G)$ is the group of automorphisms of G considered up to conjugation. Note that while an element of $\mathrm{Out}(G)$ does not act on the set of elements in G, it does act on the set of conjugacy classes of elements in G.

A natural homomorphism. Let S be a surface with $\chi(S) \leq 0$. The universal cover of S is contractible, and so S is a $K(\pi_1(S), 1)$-space. We thus have a correspondence:

$$\left\{ \begin{array}{c} \text{Free homotopy classes of} \\ \text{(unbased) maps } S \to S \end{array} \right\} \longleftrightarrow \left\{ \begin{array}{c} \text{Conjugacy classes of homo-} \\ \text{morphisms } \pi_1(S) \to \pi_1(S) \end{array} \right\}$$

Let $p \in S$. Given a map $\phi : S \to S$ and a path γ from p to $\phi(p)$, we obtain a homomorphism $\phi_* : \pi_1(S, p) \to \pi_1(S, p)$ as follows. For a loop α based at p, we set

$$\phi_*([\alpha]) = [\gamma * \phi(\alpha) * \gamma^{-1}].$$

For fixed ϕ, different choices of γ give rise to maps ϕ_* that differ by conjugation.

If ϕ is a homeomorphism, then it is invertible, and so ϕ_* is an automorphism. It follows that we have a well-defined homomorphism

$$\sigma : \mathrm{Mod}^\pm(S) \to \mathrm{Out}(\pi_1(S))$$

which is injective by the correspondence given above. We have the following remarkable theorem.

THEOREM 8.1 (Dehn–Nielsen–Baer) *Let $g \geq 1$. The homomorphism*

$$\sigma : \mathrm{Mod}^\pm(S_g) \longrightarrow \mathrm{Out}(\pi_1(S_g))$$

is an isomorphism.

As noted above, the proof of Theorem 8.1 reduces to the statement that σ is surjective. The original proof of this is due to Dehn [51], although Nielsen was the first to publish a proof [168]. Baer was the first to prove injectivity.

Note that in the case $g = 1$, the Dehn–Nielsen–Baer theorem recovers the fact that $\mathrm{Mod}^\pm(T^2) \approx \mathrm{GL}(2, \mathbb{Z})$. Note too that the statement of the theorem does not hold when $g = 0$ since

$$\mathrm{Mod}^\pm(S^2) \approx \mathbb{Z}/2\mathbb{Z} \not\approx 1 \approx \mathrm{Out}(\pi_1(S^2)).$$

Action on the fundamental class. The action of $\mathrm{Mod}^\pm(S_g)$ on $H_2(S_g; \mathbb{Z}) \approx \mathbb{Z}$ and the action of $\mathrm{Out}(\pi_1(S_g))$ on $H_2(\pi_1(S_g); \mathbb{Z}) \approx \mathbb{Z}$ are related by the Dehn–Nielsen–Baer theorem in the sense that the following diagram is commutative.

$$
\begin{array}{ccc}
\mathrm{Mod}^\pm(S_g) & \xrightarrow{\ \approx\ } & \mathrm{Out}(\pi_1(S_g)) \\
\downarrow & & \downarrow \\
\mathbb{Z}/2\mathbb{Z} \approx \mathrm{Out}(H_2(S_g; \mathbb{Z})) & \xrightarrow{\ \approx\ } & \mathrm{Out}(H_2(\pi_1(S_g); \mathbb{Z}))
\end{array}
$$

An element of $\mathrm{Mod}^\pm(S_g)$ is orientation-preserving if and only if the induced element of $\mathrm{Out}(H_2(S_g; \mathbb{Z}))$ is trivial. This gives an algebraic characterization of $\mathrm{Mod}(S_g)$ inside $\mathrm{Mod}^\pm(S_g)$: it is the subgroup of $\mathrm{Out}(\pi_1(S_g))$ that acts trivially on $H_2(\pi_1(S_g); \mathbb{Z})$.

The case of punctured surfaces. The Dehn–Nielsen–Baer theorem does not hold as stated for surfaces with punctures. For example, let $S_{0,3}$ be the thrice-punctured sphere. We have $\pi_1(S_{0,3}) \approx F_2$, the free group on two generators. Also, it is a theorem of Nielsen that $\mathrm{Out}(F_2) \approx \mathrm{GL}(2,\mathbb{Z})$; see [133, Proposition 4.5] or [20, Section 5.3]. Thus $\mathrm{Out}(\pi_1(S_{0,3})) \approx \mathrm{GL}(2,\mathbb{Z})$, but, as above, $\mathrm{Mod}^\pm(S_{0,3})$ is isomorphic to the finite group $\Sigma_3 \times \mathbb{Z}/2\mathbb{Z}$.

For punctured surfaces, we will see in Theorem 8.8 below that $\mathrm{Mod}^\pm(S)$ is isomorphic to the subgroup of $\mathrm{Out}(\pi_1(S))$ that preserves the collection of conjugacy classes of elements corresponding to punctures of S (the primitive parabolic elements).

8.2 THE QUASI-ISOMETRY PROOF

Dehn's original proof of the Dehn–Nielsen–Baer theorem uses the notion of quasi-isometry. Again, the goal is to show that each element of $\mathrm{Out}(\pi_1(S_g))$ is induced by an element of $\mathrm{Mod}^\pm(S_g)$. The key step is to show that an element of $\mathrm{Out}(\pi_1(S_g))$, which a priori preserves only algebraic properties/objects, must in fact preserve topological ones. For example, the first step in the proof will be to prove that an element of $\mathrm{Out}(\pi_1(S_g))$ must respect the topological property of whether or not the free homotopy classes of two simple closed curves have geometric intersection number 0. We will prove this by studying the behavior of $\pi_1(S_g)$ "at infinity" in $\mathbb{H}^2$.

8.2.1 METRICS ON $\pi_1(S)$

Let G be a group with a fixed finite generating set S. The *Cayley graph* $\Gamma(G,S)$ for G with respect to S is the abstract graph with a vertex for each element $g \in G$ and an edge between the vertices g and gs if $s \in S$ or $s^{-1} \in S$. The group G acts on $\Gamma(G,S)$ on the left by graph automorphisms.

There is a natural metric on $\Gamma(G,S)$ given by taking each edge to have length 1 and putting the *path metric* on $\Gamma(G,S)$, whereby the distance between two points is the length of the shortest path between them. Restricting this metric to the vertices of $\Gamma(G,S)$ gives a G-invariant metric on G called the *word metric* on G with respect to S. For $g \in G$, the distance $d_S(1,g)$ is called the *word length* of g. By left invariance, for any $g, h \in G$, the distance $d_S(g,h)$ is the word length of $g^{-1}h$.

For a surface S with $\chi(S) < 0$, another way to get a metric on $\pi_1(S)$ is to choose a covering map $\mathbb{H}^2 \to S$ that endows S with a hyperbolic metric (recall that, by "hyperbolic metric," we mean a complete, finite-area Riemannian metric with constant curvature -1). If we fix a basepoint in S, its set of lifts to $\mathbb{H}^2$ are in bijection with elements of $\pi_1(S)$. We can therefore define the *hyperbolic distance* between two elements of $\pi_1(S)$ as the hyperbolic distance between the corresponding lifts.

Clearly, the word metric on $\pi_1(S)$ depends on the choice of generating set, and the hyperbolic metric on $\pi_1(S)$ depends on the choice of covering map. We would like to understand what properties of the metric do not depend on these choices. In short, the answer is that all choices give metrics that look the same, up to a universally bounded stretch, at large scales. This brings us to the notion of quasi-isometry.

8.2.2 QUASI-ISOMETRIES

A function $f\colon X \to Y$ between metric spaces (X, d_X) and (Y, d_Y) is a *quasi-isometric embedding* if there are constants K and C so that

$$\frac{1}{K} d_X(x, x') - C \le d_Y(f(x), f(x')) \le K d_X(x, x') + C$$

for any choice of x and x' in X. We say that f is a *quasi-isometry* if there is a constant D so that the D-neighborhood of $f(X)$ is equal to Y. In this case we say that X and Y are *quasi-isometric*. Quasi-isometry is an equivalence relation on metric spaces.

There is a more symmetric definition of quasi-isometry, as follows. Two metric spaces (X, d_X) and (Y, d_Y) are quasi-isometric if and only if there are maps $f : X \to Y$ and $\overline{f} : Y \to X$ and constants K, C, and D such that

$$d_Y(f(x), f(x')) \le K d_X(x, x') + C \quad d_X(\overline{f}(y), \overline{f}(y')) \le K d_Y(y, y') + C$$

and

$$d_X(\overline{f} \circ f(x), x) \le D \qquad d_Y(f \circ \overline{f}(y), y) \le D$$

for all $x, x' \in X$ and $y, y' \in Y$.

As a first exercise, one can show that given two word metrics on the same finitely generated group G, the identity map $G \to G$ is a quasi-isometry. This fact also follows from the first statement of Theorem 8.2; see Corollary 8.3 below.

8.2.3 THE FUNDAMENTAL OBSERVATION OF GEOMETRIC GROUP THEORY

The following theorem, sometimes called the Milnor–Švarc lemma, is one of the most basic theorems in geometric group theory. It first appeared in the work of Efremovič [55], Švarc [202], and Milnor [158].

Recall that the action of a group G on a topological space X is *properly discontinuous* if, for each compact K in X, the set $\{g \in G : (g \cdot K) \cap K \ne \emptyset\}$ is finite. Let X be some metric space. The space X is *proper* if closed balls in X are compact. A *geodesic* in X is a distance-preserving map of a closed interval into X. Finally, X is a *geodesic metric space* if there exists a geodesic connecting any two points in X.

THEOREM 8.2 (Fundamental oberservation of geometric group theory)
Let X be a proper geodesic metric space and suppose that a group G acts properly discontinuously on X via isometries. If the quotient X/G is compact, then G is finitely generated and G is quasi-isometric to X. More precisely, there is a word metric for G so that, for any point $x_0 \in X$, the map

$$G \to X$$
$$g \mapsto g \cdot x_0$$

is a quasi-isometry.

One example of the phenomenon described in Theorem 8.2 is given by the action by deck transformations of a compact Riemannian manifold on its universal cover.

Proof. Let x_0 be some fixed basepoint of X. Since the action of G on X is properly discontinuous, the metric on X induces a metric on X/G. Indeed, the distance between two points in the quotient is the infimum of the distances between any two of their preimages; the proper discontinuity implies the infimum is a minimum. As X/G is compact, it has finite diameter R. It follows that X is covered by the G-translates of $B = B(x_0, R)$, the ball of radius R about x_0. Let

$$\mathcal{S} = \{g \in G : g \neq 1 \text{ and } g \cdot B \cap B \neq \emptyset\}.$$

By the properness of X and the proper discontinuity of the action of G on X, the set $\mathcal{S}$ is finite.

Let d denote the metric on X. We define

$$\lambda = \max_{s \in \mathcal{S}} d(x_0, s \cdot x_0) \quad \text{and} \quad r = \inf\{d(B, g \cdot B) \mid g \notin \mathcal{S} \cup \{1\}\}.$$

Note that, since the action of G is properly discontinuous and since X is proper, r is actually a minimum.

If $r = 0$, then G is finite, and the theorem is trivial in this case. So we may assume $r > 0$.

Let $g \in G$. As X is geodesic, it is in particular path-connected. Given a path from x_0 to $g \cdot x_0$, we can choose points $x_1, \ldots, x_n = g \cdot x_0$ along this path so that $d(x_i, x_{i+1}) < r$. Since the $\{g \cdot B\}$ cover X, we may choose $g_1, \ldots, g_n \in G$ so that $x_i \in g_i \cdot B$. If we set $g_0 = 1$ and $s_i = g_{i-1}^{-1} g_i$, we have that $s_1 s_2 \cdots s_n = g$. We have

$$d(s_i \cdot B, B) = d(g_{i-1}^{-1} g_i \cdot B, B) = d(g_i \cdot B, g_{i-1} \cdot B) \le d(x_i, x_{i-1}) < r.$$

By the definition of r, we see that $s_i \in \mathcal{S} \cup \{1\}$ for all i. Thus $\mathcal{S}$ generates G, and G is finitely generated.

We will now show that the map $g \mapsto g \cdot x_0$ defines a quasi-isometric embedding $G \to X$, where G is given the word metric associated to $\mathcal{S}$. In other words, we will show that for $g_1, g_2 \in G$ we have

$$\frac{1}{\lambda} d(g_1 \cdot x_0, g_2 \cdot x_0) \le d_{\mathcal{S}}(g_1, g_2) \le \frac{1}{r} d(g_1 \cdot x_0, g_2 \cdot x_0) + 1.$$

Since G acts by isometries on itself and on X, this is equivalent to the statement that

$$\frac{1}{\lambda} d(x_0, g \cdot x_0) \le d_{\mathcal{S}}(1, g) \le \frac{1}{r} d(x_0, g \cdot x_0) + 1$$

for any $g \in G$ (substitute $g_1^{-1} g_2$ for g). In our definition of a quasi-isometric embedding, one can take $K = \max\{\lambda, \frac{1}{r}\}$ and $C = 1$. The constant C cannot be taken to be 0 because, for instance, g could be in the stabilizer of x_0.

The inequality $\frac{1}{\lambda} d(x_0, g \cdot x_0) \le d_{\mathcal{S}}(1, g)$ follows immediately from the triangle inequality, the definitions of $\mathcal{S}$ and λ, and the fact that $s \in \mathcal{S}$ if and only if $s^{-1} \in \mathcal{S}$. Thus "short" paths in G give rise to short paths in X.

We must now show that short paths in X correspond to short paths in G. Precisely, we will prove the inequality $d_{\mathcal{S}}(1, g) \le \frac{1}{r} d(x_0, g \cdot x_0) + 1$. The argument is a souped-up version of the argument that $\mathcal{S}$ generates G. Let $g \in G$. Since X is geodesic, we may find a geodesic of length $d(x_0, g \cdot x_0)$ connecting x_0 to $g \cdot x_0$. Let n be the smallest integer strictly greater than $d(x_0, g \cdot x_0)/r$, so $n \le d(x_0, g \cdot x_0)/r + 1$. We can find points $x_1, \ldots, x_{n-1}, x_n = g \cdot x_0$ in X so that $d(x_i, x_{i+1}) < r$ for $0 \le i \le n - 1$. Since the G-translates of B cover X, we can choose elements $1 = g_0, g_1 \ldots, g_{n-1}, g_n = g$ of G so that $x_i \in g_i \cdot B$. If we set $s_i = g_{i-1}^{-1} g_i$, then $g = s_1 \cdots s_n$. Again, by the definition of r, we have $s_i \in \mathcal{S}$, and so the word length of g is at most n. In summary, we have

$$d(1, g) \le n \le \frac{1}{r} d(x_0, g \cdot x_0) + 1,$$

which is what we wanted to show.

By the definition of R, the R-neighborhood of the image of G is all of X, and so the quasi-isometric embedding $G \to X$ is a quasi-isometry. $\qquad \square$

Applications to Cayley graphs. Any Cayley graph for a finitely generated group is a proper, geodesic metric space. Thus, by considering the action of a group G on an arbitrary Cayley graph for G, we obtain the following fact.

Corollary 8.3 *For any two word metrics on a finitely generated group G, the identity map $G \to G$ is a quasi-isometry.*

The following corollary of Corollary 8.3 represents the first step in our proof of the Dehn–Nielsen–Baer theorem.

Corollary 8.4 *Any automorphism of a finitely generated group is a quasi-isometry.*

By Corollary 8.3, we do not need to specify which word metric we are using in the statement of Corollary 8.4.

Proof. Let $\Phi : G \to G$ be an automorphism of a finitely generated group G and let $\mathcal{S}$ be a finite generating set for G. Since Φ is an automorphism, we have that $\Phi^{-1}(\mathcal{S}) = \{\Phi^{-1}(s) : s \in \mathcal{S}\}$ is a finite generating set for G. What is more, we have

$$d_{\mathcal{S}}(\Phi(g), \Phi(h)) = d_{\Phi^{-1}(\mathcal{S})}(g, h).$$

In other words, the amount word length in G is stretched under the map Φ is equivalent to the amount of stretch word length undergoes when changing the finite generating set. The result now follows immediately from Corollary 8.3. $\qquad\square$

Combining Theorem 8.2 and Corollary 8.3, we have that any two word metrics on $\pi_1(S_g)$ are quasi-isometric, and for $g \geq 2$, each word metric is quasi-isometric to each hyperbolic metric on $\pi_1(S_g)$. What is more, the quasi-isometry in each case is the identity map. In other words, there is only one natural metric on $\pi_1(S_g)$ up to the equivalence relation of quasi-isometry. Thus in our arguments we will be able to switch back and forth between word metrics and hyperbolic metrics. For instance, Corollary 8.4 is proved using the word metric, and then it is applied in the proof of Lemma 8.5, where we use a hyperbolic metric on $\pi_1(S_g)$.

Now that we have a well-defined metric on $\pi_1(S_g)$, we can begin our study of its large-scale behavior.

8.2.4 LINKING AT INFINITY

Let S be a hyperbolic surface. We say that an element of $\pi_1(S)$ is *hyperbolic* if the corresponding deck transformation is a hyperbolic isometry of $\mathbb{H}^2$. Recall that the axis of a hyperbolic element α of $\pi_1(S)$ has a pair of endpoints $\partial\alpha$ lying in $\partial\mathbb{H}^2$. Two hyperbolic elements α, β of $\pi_1(S)$ are *linked at infinity* if $\partial\alpha$ and $\partial\beta$ are linked in $\partial\mathbb{H}^2 \approx S^1$, that is, if the pair $\partial\alpha$ separates the pair $\partial\beta$ (and vice versa).

A priori this notion depends on the choice of hyperbolic metric on S. One can prove that actually the property of being linked at infinity is independent of the choice of metric. For simplicity, though, we will use a fixed covering, so there is no ambiguity.

Lemma 8.5 *Let $g \geq 2$ and let $\mathbb{H}^2 \rightarrow S_g$ be a fixed covering map. Let Φ be an automorphism of $\pi_1(S_g)$ and let γ and δ be nontrivial elements of $\pi_1(S_g)$. Then the elements $\Phi(\gamma)$ and $\Phi(\delta)$ are linked at infinity if and only if γ and δ are linked at infinity.*

Proof of Lemma 8.5. Since S_g is a closed hyperbolic surface, all nontrivial elements are hyperbolic, and so it makes sense to talk about linking at infinity. Because Φ is invertible, it suffices to show that if γ and δ are not linked at infinity, then $\Phi(\gamma)$ and $\Phi(\delta)$ are not linked at infinity. Also, we may assume that γ and δ do not share an axis since having the same axis is equivalent to having equal (nontrivial) powers, and this property is preserved by the automorphism Φ.

By Corollary 8.4, Φ is a quasi-isometry of $\pi_1(S_g)$. Say that with respect to the hyperbolic metric coming from the fixed covering $\mathbb{H}^2 \rightarrow S_g$, the quasi-isometry constants are $K \geq 1$ and $C \geq 0$. Let D be the diameter of some fixed fundamental domain for $\pi_1(S_g)$ in $\mathbb{H}^2$.

Fix some $R > 2DK^2 + 2CK$. Let x_0 some fixed basepoint for $\mathbb{H}^2$ and consider the orbit

$$\mathcal{O}_\gamma = \{\gamma^k \cdot x_0 : k \in \mathbb{Z}\}.$$

Connect the points of $\mathcal{O}_\gamma$ by an infinite piecewise-geodesic path, where each segment of the path connects two points in the orbit of x_0 that lie in adjacent fundamental domains and where the entire path lies in some fixed metric neighborhood of the axis for γ. We can denote such a path by its set of vertices, say $\{\alpha_i\}$.

Since γ and δ are unlinked hyperbolic isometries of $\mathbb{H}^2$, and since the path $\{\alpha_i\}$ lies in a metric neighborhood of the axis for γ, we may choose an $N = N(R)$ so that each point of

$$\mathcal{O}_{\delta^N} = \{(\delta^N)^k \cdot x_0 : k \in \mathbb{Z}, k \neq 0\}$$

has a distance at least $R + D$ from each point of $\mathcal{O}_\gamma$. Note that $\mathcal{O}_{\delta^N}$ is not the entire orbit of x_0 under δ^N since it is missing the point x_0. We can connect the points of $\mathcal{O}_{\delta^N}$ by a piecewise-geodesic path $\{\beta_i\}$ where each β_i is in the orbit of x_0, so that the path $\{\beta_i\}$ stays a hyperbolic distance at least R from the path $\{\alpha_i\}$ and so that consecutive vertices β_i and β_{i+1} lie in adjacent fundamental domains. To find the β_i, we start with any bi-infinite continuous path that connects the points of $\mathcal{O}_{\delta^N}$ and stays outside the $(R + D)$-neighborhood of the path $\{\alpha_i\}$, and we keep track of the fundamental domains through which this path passes.

For both of the paths we just constructed, the length of each geodesic segment is at most $2D$ (any pair of points in adjacent fundamental domains have distance at most $2D$). The vertices of the two paths are identified with particular elements of $\pi_1(S_g)$.

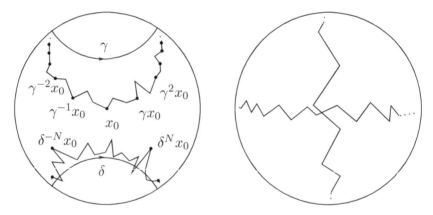

Figure 8.1 Left: the polygonal paths constructed in the proof of Lemma 8.5. Right: polygonal paths that are linked at infinity.

Assume, for the purposes of contradiction, that the hyperbolic isometries $\Phi(\gamma)$ and $\Phi(\delta)$ are linked at infinity. It follows that the polygonal paths $\{\Phi(\alpha_i)\}$ and $\{\Phi(\beta_i)\}$ have to cross. Since Φ is a quasi-isometry with constants K and C, each geodesic segment of $\{\Phi(\alpha_i)\}$ and $\{\Phi(\beta_i)\}$ has length at most $K(2D) + C$. But if these paths cross, two of the geodesic segments themselves must cross—see the right-hand side of Figure 8.1. Now, each segment has at least one endpoint whose distance from the crossing point is less than or equal to $(K(2D) + C)/2$, and so these two endpoints lie at a distance of at most $K(2D) + C$.

What we have now is that there exist elements $\alpha, \beta \in \pi_1(S_g)$ with $d(\alpha, \beta) \geq R$ and $d(\Phi(\alpha), \Phi(\beta)) \leq K(2D) + C$. Since $R > 2DK^2 + 2CK$, we obtain a contradiction with the assumption that Φ is a quasi-isometry with constants K and C. Thus it must be the case that $\Phi(\gamma)$ and $\Phi(\delta)$ are not linked at infinity, and we are done.   □

Sides. In addition to linking, we can also talk about two hyperbolic elements $\alpha, \beta \in \pi_1(S)$ being on the same *side* of a hyperbolic element $\gamma \in \pi_1(S)$. That is, if α and β are unlinked with γ (and do not share an axis with γ), then their axes either lie on the same side of the axis for γ or do not. One can also formulate this notion purely topologically at infinity in terms of the endpoints of the axes on $\partial \mathbb{H}^2$.

Corollary 8.6 *Let $g \geq 2$ and let $\mathbb{H}^2 \to S_g$ be a fixed covering map. Let Φ be an automorphism of $\pi_1(S_g)$. If α, β, and γ are elements of $\pi_1(S_g)$ with distinct axes, then the axes for $\Phi(\alpha)$ and $\Phi(\beta)$ lie on the same side of $\Phi(\gamma)$ if and only if the axes for α and β lie on the same side of γ.*

Proof. The axes for α and β lie on the same side of the axis for γ if and only if there is an element $\delta \in \pi_1(S_g)$ that is linked at infinity with α and β but not with γ. Apply Lemma 8.5.   □

8.2.5 FINISHING THE PROOF

We can now prove the Dehn–Nielsen–Baer theorem.

Proof of the Dehn–Nielsen–Baer theorem. As discussed above, we need only prove that the homomorphism $\sigma : \mathrm{Mod}^{\pm}(S_g) \to \mathrm{Out}(\pi_1(S_g))$ is surjective. Let any $[\Phi] \in \mathrm{Out}(\pi_1(S_g))$ be given and let Φ be a representative automorphism. Also, fix once and for all a covering map $\mathbb{H}^2 \to S_g$.

Let $(c_1, \dots, c_{2g})$ be a chain of isotopy classes of simple closed curves in S_g. As in Section 1.3, this means that $i(c_i, c_{i+1}) = 1$ and $i(c_i, c_j) = 0$ otherwise. For concreteness, we take the curves shown on the top of Figure 8.2. Orient each c_i so that each algebraic intersection number $\hat{i}(c_i, c_{i+1})$ is $+1$.

Recall that free homotopy classes of oriented curves in S_g correspond to conjugacy classes of elements of $\pi_1(S_g)$; each c_i is the conjugacy class of the element γ_i shown on the bottom of Figure 8.2.

Since Φ is an automorphism of $\pi_1(S_g)$, it acts on the set of conjugacy classes of $\pi_1(S_g)$. We claim that $\{\Phi(c_i)\}$ is also a chain of isotopy classes of simple closed curves and that the algebraic intersections $\hat{i}(\Phi(c_i), \Phi(c_{i+1}))$ are all $+1$ or all -1. We prove this claim in four steps:

1. $\Phi(c_i)$ is a simple closed curve for each i.

2. $i(\Phi(c_i), \Phi(c_j)) = 0$ for $|i - j| > 1$.

3. $i(\Phi(c_i), \Phi(c_{i+1})) = 1$ for each i.

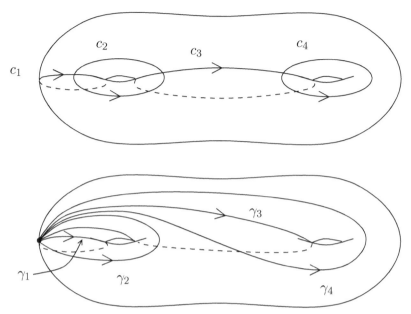

Figure 8.2 A chain on a genus 2 surface (top) and representatives in the fundamental group (bottom).

4. $\hat{i}(\Phi(c_i), \Phi(c_{i+1}))$ does not depend on i.

Each of the four steps will follow from Lemma 8.5. For step 1, recall that a conjugacy class of a primitive element of $\pi_1(S_g)$ has a simple representative if and only if each pair of representatives for the class is not linked at infinity (cf. the proof of Proposition 1.6). Now simply note that, as proved in Lemma 8.5, Φ preserves whether or not axes are linked.

Similarly, for step 2, we use the fact that two conjugacy classes have geometric intersection number 0 if and only if any pair of representatives is unlinked at infinity and this latter property is preserved by Φ. For step 3 we use the following Φ-invariant characterization of when two conjugacy classes have representatives with geometric intersection number 1 (plus Lemma 8.5):

> *Two conjugacy classes a and b have geometric intersection number 1 if and only if for some representative α of a that is linked at infinity with a given representative β of b, the set of representatives of a that are linked at infinity with β is precisely $\{\beta^k \alpha \beta^{-k} : k \in \mathbb{Z}\}$.*

Step 4 is more intricate. It suffices to prove that given three conjugacy classes a, b, and c with $i(a, b) = i(b, c) = 1$ and $i(a, c) = 0$, we can characterize the agreement of the signs of $\hat{i}(a, b)$ with $\hat{i}(b, c)$ in terms of data we know to be preserved by Φ. Let α, β, and γ be any representatives for a, b, and c so that the axes for α and γ are disjoint and the axis for β intersects each of the axes for α and γ once each. Now note the following.

With the above notation, $\hat{i}(a, b)$ has the same sign as $\hat{i}(b, c)$ if and only if the axes for $\alpha\beta\alpha^{-1}$ and $\gamma\beta\gamma^{-1}$ lie on different sides of the axis for β.

Replacing a, b, and c with c_i, c_{i+1}, and c_{i+2}, we apply Lemma 8.5 and Corollary 8.6 to complete step 4, thus proving the claim.

By the change of coordinates principle, more precisely by example 6 in Section 1.3, there is a homeomorphism ϕ that fixes the basepoint of $\pi_1(S_g)$ and satisfies $\phi_*(c_i) = \Phi_*(c_i)$ (with orientation) for each i. Here ϕ_* and Φ_* denote the induced actions on (conjugacy classes of) elements of $\pi_1(S_g)$.

To complete the proof of the theorem, we must now prove that the mapping class $[\phi]$, acting on $\pi_1(S_g)$, induces the outer automorphism $[\Phi]$. For any $\beta \in \pi_1(S_g)$, let I_β denote the inner automorphism of $\pi_1(S_g)$ given by $\gamma \mapsto \beta\gamma\beta^{-1}$. Since the representatives γ_i generate $\pi_1(S_g)$, it suffices to show that there is an inner automorphism I_α of $\pi_1(S_g)$ so that

$$I_\alpha \circ \phi_*^{-1} \circ \Phi(\gamma_i) = \gamma_i$$

for each i.

Note that it is simply not true in general that if an automorphism of a group fixes the conjugacy class of each generator, then it is an inner automorphism. As an example, take the free group on $\{x, y, z\}$ and consider the automorphism given by $x \mapsto yxy^{-1}$, $y \mapsto y$, and $z \mapsto z$.

We will use the fact that the particular representatives γ_i of the c_i shown in Figure 8.2 form a chain in the sense that the lifts of γ_i and γ_{i+1} to $\mathbb{H}^2$ are linked at infinity for each i. This follows from the fact that γ_i and γ_{i+1} are linked on the surface; more precisely, if we take a small closed neighborhood of the basepoint of $\pi_1(S_g)$, γ_i and γ_{i+1} are linked on the boundary of this neighborhood. Arbitrary lifts of c_i and c_{i+1} may or may not be linked at infinity.

Denote $\phi_*^{-1} \circ \Phi$ by F. Since ϕ induces an automorphism of $\pi_1(S_g)$, we see that F still preserves linking at infinity. Again, the goal is to find an element α so that $I_\alpha \circ F$ is the identity automorphism of $\pi_1(S)$.

Since $F(c_i) = c_i$ with orientation for all i, we have in particular $F(c_1) = c_1$, and so $F(\gamma_1) = \alpha_1^{-1}\gamma_1\alpha_1$ for some $\alpha_1 \in \pi_1(S_g)$. Thus

$$I_{\alpha_1} \circ F(\gamma_1) = \gamma_1.$$

We know that $F(c_2) = c_2$, that $I_{\alpha_1} \circ F$ preserves linking at infinity, and that γ_1 and γ_2 are linked. It follows from the characterization of conjugacy classes with geometric intersection number 1 given above that $I_{\alpha_1} \circ F(\gamma_2) = \gamma_1^{-k} \gamma_2 \gamma_1^k$ for some $k \in \mathbb{Z}$. Therefore,

$$I_{\gamma_1^k \alpha_1} \circ F(\gamma_1) = I_{\gamma_1^k} \circ I_{\alpha_1} \circ F(\gamma_1) = \gamma_1$$

and

$$I_{\gamma_1^k \alpha_1} \circ F(\gamma_2) = I_{\gamma_1^k} \circ I_{\alpha_1} \circ F(\gamma_2) = \gamma_2.$$

We can now see inductively that $I_{\gamma_1^k \alpha_1} \circ F(\gamma_i) = \gamma_i$ for each $i \geq 3$, and so $I_{\gamma_1^k \alpha_1}$ is the desired inner automorphism. Indeed, since γ_1 and γ_2 are both fixed by $I_{\gamma_1^k \alpha_1} \circ F$, it follows that each element of $\{\gamma_1^l \gamma_2 \gamma_1^{-l}\}$ is fixed. But since γ_3 is linked with γ_2, it is characterized in $\pi_1(S_g)$ by the properties that it is linked with γ_2 and that its axis in $\mathbb{H}^2$ lies between the axes for $\gamma_2^l \gamma_1 \gamma_2^{-l}$ and $\gamma_2^{l+1} \gamma_1 \gamma_2^{-(l+1)}$ for some particular l. Thus γ_3 is fixed by $I_{\gamma_1^k \alpha_1} \circ F$ and, by induction, each γ_i for $i > 3$ is also fixed (the inductive step for γ_i uses that both γ_{i-1} and γ_{i-2} are fixed). We have thus found the required inner automorphism, and so the proof is complete. $\square$

8.2.6 THE INDUCED HOMEOMORPHISM AT INFINITY

Our proof of the Dehn–Nielsen–Baer theorem suggests an elegant way to think about the automorphism Φ, namely, through an induced action $\partial \Phi$ on $\partial \mathbb{H}^2 \approx S^1$. We now explain this idea.

If γ is an element of $\pi_1(S_g)$, then the forward endpoint of the (oriented) axis of γ in $\mathbb{H}^2$ is identified with a point $\gamma_\infty \in \partial \mathbb{H}^2$. Let

$$\Gamma_\infty = \{\gamma_\infty : \gamma \in \pi_1(S_g)\}.$$

Since the action of $\pi_1(S_g)$ on $\mathbb{H}^2$ is cocompact, the set Γ_∞ is dense in $\partial \mathbb{H}^2$. We define $\partial \Phi : \Gamma_\infty \to \Gamma_\infty$ by

$$\partial \Phi(\gamma_\infty) = (\Phi(\gamma))_\infty.$$

Note that $\partial \Phi$ is well defined on this set because the axes of two elements of $\pi_1(S_g)$ can share an endpoint at infinity only if they share a common power. Since Φ is an automorphism, $\partial \Phi$ is a bijection.

Denote the backward endpoint of the axis for $\gamma \in \pi_1(S_g)$ by $\gamma_{-\infty}$. The set

$$\Gamma_{\pm\infty} = \{(\gamma_\infty, \gamma_{-\infty}) : \gamma \in \pi_1(S_g)\}$$

is dense in $\partial \mathbb{H}^2 \times \partial \mathbb{H}^2$; see [13, Theorem 5.3.8]. This fact was used implicitly in our proof of Corollary 8.6.

The following theorem underlies much of Nielsen's work on surface homeomorphisms. We already used this fact in Section 5.5.

Theorem 8.7 *Let $g \geq 2$. Any automorphism Φ of $\pi_1(S_g)$ induces a homeo-morphism of $\partial\mathbb{H}^2$.*

Proof. It suffices to show that $\partial\Phi$ induces a homeomorphism of Γ_∞. Since Γ_∞ is dense in $\partial\mathbb{H}^2$, there is then a unique extension to a homeomorphism of $\partial\mathbb{H}^2$.

Let $\delta \in \pi_1(S_g)$. Let δ_R denote the set of elements γ of $\pi_1(S_g)$ so that γ_∞ lies to the right of the oriented axis of δ. The sets δ_R are identified with subsets of Γ_∞ via the correspondence $\gamma \leftrightarrow \gamma_\infty$.

By the density of $\Gamma_{\pm\infty}$ in $\partial\mathbb{H}^2 \times \partial\mathbb{H}^2$, the sets $\{\delta_R : \delta \in \pi_1(S_g)\}$ form a basis for the topology of Γ_∞.

To show that $\partial\Phi$ is a homeomorphism, we will show that in fact Φ (hence Φ^{-1}) takes each element of $\{\delta_R\}$ to another such element. More precisely, we will show that $\Phi(\delta_R)$ is equal to either $\Phi(\delta)_R$ or $\Phi(\delta^{-1})_R$.

Indeed, let α be any element of δ_R that is unlinked with δ. Assume for concreteness that $\Phi(\alpha)$ is contained in $\Phi(\delta)_R$. We will show that $\Phi(\delta_R) \subseteq \Phi(\delta)_R$ (if $\Phi(\alpha)$ were contained in $\Phi(\delta^{-1})_R$, the same argument would show that $\Phi(\delta_R) \subseteq \Phi(\delta^{-1})_R$).

First, let β be an element of δ_R that is not linked with δ. Applying Corollary 8.6 to δ, α, and β, we find that $\Phi(\beta) \in \Phi(\delta)_R$.

Now suppose β is an element of δ_R that is linked with δ. By Lemma 8.5, we immediately obtain that $\Phi(\beta)$ is linked with $\Phi(\delta)$. Because the axis for β crosses the axis for δ from left to right, it follows that $\beta^{-1}\delta\beta$ is unlinked with δ and lies in δ_R. By the previous case, $\Phi(\beta^{-1}\delta\beta) = \Phi(\beta)^{-1}\Phi(\delta)\Phi(\beta)$ lies in $\Phi(\delta)_R$. But then it must be that $\Phi(\beta)$ crosses $\Phi(\delta)$ from left to right, which means $\Phi(\beta) \in \Phi(\delta)_R$.

We have thus proven that $\Phi(\delta_R) \subseteq \Phi(\delta)_R$. Since Φ is invertible, we in fact have that $\Phi(\delta_R) = \Phi(\delta)_R$. This completes the proof. $\qquad\square$

In fact, a much more general statement than Theorem 8.7 is true: any quasi-isometry of $\mathbb{H}^2$ induces a homeomorphism of $\partial\mathbb{H}^2$. A related fact is that any $\pi_1(S_g)$-equivariant homeomorphism of $\mathbb{H}^2$ extends to a $\pi_1(S_g)$-equivariant homeomorphism of the closed disk $\mathbb{H}^2 \cup \partial\mathbb{H}^2$. This last fact will be used in the proof of Theorem 14.20.

8.2.7 THE PUNCTURED CASE

There is a version of the Dehn–Nielsen–Baer theorem for punctured surfaces as follows. Let $\text{Out}^\star(\pi_1(S))$ be the subgroup of $\text{Out}(\pi_1(S))$ consisting of elements that preserve the set of conjugacy classes of the simple closed curves surrounding individual punctures. Note that these conjugacy classes are precisely the primitive conjugacy classes that correspond to the parabolic elements of $\mathrm{Isom}(\mathbb{H}^2)$.

THEOREM 8.8 *Let $S = S_{g,p}$ be a hyperbolic surface of genus g with p punctures. Then the natural map*

$$\mathrm{Mod}^{\pm}(S) \to \mathrm{Out}^{\star}(\pi_1(S))$$

is an isomorphism.

The proof of this more general theorem follows the same outline as in the proof of the closed case (Theorem 8.1). We content ourselves to point out the two main differences.

1. In the case $S = S_g$, we knew automatically that any automorphism of $\pi_1(S_g)$ must send hyperbolic elements to hyperbolic elements since all nontrivial elements of $\pi_1(S_g)$ are hyperbolic. If S is not closed, then an arbitrary automorphism of $\pi_1(S)$ can exchange hyperbolic elements with parabolic elements. But the fact that we consider $\mathrm{Out}^{\star}(\pi_1(S))$ instead of $\mathrm{Out}(\pi_1(S))$ in the statement of Theorem 8.8 exactly accounts for this.

2. The map $\pi_1(S) \to \mathbb{H}^2$ given by taking the orbit in $\mathbb{H}^2$ of a single point is not a quasi-isometry. To remedy this, we truncate S by deleting a small neighborhood of each puncture. We can choose the neighborhoods to be small enough so that the preimage in $\mathbb{H}^2$ of the truncated surface is a connected space X. If we endow X with the path metric, then the action of $\pi_1(S)$ on X satisfies the conditions of Theorem 8.2, and so $\pi_1(S)$ is quasi-isometric to X.

   The proof of Lemma 8.5 now proceeds similarly as before. Points are farther in X than they are in $\mathbb{H}^2$, so there is no problem in choosing N so that the sets $\mathcal{O}_\gamma$ and $\mathcal{O}_{\delta N}$ are far apart. Also, there is no obstruction to choosing the paths $\{\alpha_i\}$ and $\{\beta_i\}$. If $\{\Phi(\alpha_i)\}$ and $\{\Phi(\beta_i)\}$ were to cross, we would still have a short path in X between two vertices of the paths, which would give the desired contradiction.

We have already mentioned the theorem of Nielsen that $\mathrm{Out}(F_2) \approx \mathrm{GL}(2,\mathbb{Z})$. Thus we have

$$\mathrm{Out}(F_2) \approx \mathrm{GL}(2,\mathbb{Z}) \approx \mathrm{Mod}^{\pm}(S_{1,1}).$$

In the language of Theorem 8.8, this means that the group $\mathrm{Out}^{\star}(\pi_1(S_{1,1}))$ is the entire group $\mathrm{Out}(\pi_1(S_{1,1}))$. In other words, every element of the outer automorphism group of $F_2 = \langle x, y \rangle$ preserves the conjugacy class $[x, y]$. Thus $\mathrm{GL}(n, \mathbb{Z})$, $\mathrm{Mod}^{\pm}(S)$, and $\mathrm{Out}(F_n)$ can be viewed as three different generalizations of the same group.

Once-punctured versus closed. The Dehn–Nielsen–Baer theorem can be used to relate the group $\mathrm{Mod}(S_g)$ to the group $\mathrm{Mod}(S_{g,1})$, where $S_{g,1}$ is the genus $g \geq 2$ surface with one marked point. This is done by the following isomorphism of exact sequences, where each square is a commutative diagram:

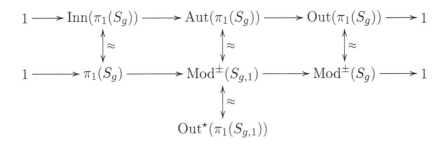

The first row is the usual relationship between the automorphism group and outer automorphism group of any group. The second row is a version of the Birman exact sequence for the extended mapping class group.

The isomorphism $\mathrm{Inn}(\pi_1(S_g)) \approx \pi_1(S_g)$ is equivalent to the statement that $\pi_1(S_g)$ has trivial center, and the isomorphism $\mathrm{Out}(\pi_1(S_g)) \approx \mathrm{Mod}^{\pm}(S_g)$ is the Dehn–Nielsen–Baer theorem. Now there certainly is a map $\mathrm{Mod}^{\pm}(S_{g,1}) \to \mathrm{Aut}(\pi_1(S_g))$ that makes the diagram (as described so far) commutative—simply choose the basepoint for $\pi_1(S_g)$ to be the marked point. The five lemma then tells us that the middle vertical map is an isomorphism from $\mathrm{Mod}^{\pm}(S_{g,1})$ to $\mathrm{Aut}(\pi_1(S_g))$.

Finally, we examine the isomorphism $\mathrm{Out}^{\star}(\pi_1(S_{g,1})) \to \mathrm{Aut}(\pi_1(S_g))$. At first it seems odd to have an outer automorphism group of a surface group be the same as the automorphism group of another surface group. However, given $\phi \in \mathrm{Out}^{\star}(\pi_1(S_{g,1}))$, we get an element of $\mathrm{Aut}(\pi_1(S_g))$ by taking the unique representative automorphism of ϕ that fixes the loop corresponding to the puncture (not just up to conjugacy), and this gives the desired isomorphism.

8.3 TWO OTHER VIEWPOINTS

In this section we provide two other proofs of the Dehn–Nielsen–Baer theorem, one inspired by 3-manifold theory (adapted from [93, Theorem 13.6]) and one using harmonic maps. There are various other proofs, each involving a different kind of mathematics. For example, in Theorem 1.8 of [43] Calegari exploits the relationship between simple closed curves on S_g and HNN extensions of $\pi_1(S_g)$ to give an inductive argument for the Dehn–Nielsen–Baer theorem. Zieschang–Vogt–Coldewey give a combinatorial-group-theoretical proof in [218, Section 5.6], and Seifert gives an elementary covering space argument in [192].

Let S be a surface with $\chi(S) < 0$. Since S is a $K(\pi_1(S), 1)$-space, every outer automorphism of $\pi_1(S)$ is induced by some (unbased) map $S \to S$. By the Whitehead theorem [91, Theorem 4.5] and the fact that $\pi_i(S) = 0$ for $i > 1$, we have that this self-map of S is a homotopy equivalence. Thus, for the surjectivity part of the Dehn–Nielsen–Baer theorem, it suffices to show that every homotopy equivalence of S is homotopic to a homeomorphism of S.

THEOREM 8.9 *If $g \geq 2$, then any homotopy equivalence $\phi : S_g \to S_g$ is homotopic to a homeomorphism.*

We give two approaches to Theorem 8.9 below, one topological and one analytical.

8.3.1 THE TOPOLOGICAL APPROACH: PANTS DECOMPOSITIONS

Recall that a *pair of pants* is a compact surface of genus 0 with three boundary components. Let S be a compact surface with $\chi(S) < 0$. A *pair of pants decomposition* of S, or a *pants decomposition* of S, is a collection of disjoint simple closed curves in S with the property that when we cut S along these curves, we obtain a disjoint union of pairs of pants. Equivalently, a pants decomposition of S is a maximal collection of disjoint, essential simple closed curves in S with the property that no two of these curves are isotopic.

We can easily prove the equivalence of the two definitions of a pants decomposition. First, suppose we have a collection of simple closed curves that cuts S into pairs of pants. We immediately see that every curve is essential since there are no disk components when we cut S. Further, since any simple closed curve on a pair of pants is either homotopic to a point or to a boundary component, it follows that the given collection is maximal. For the other direction, suppose we have a collection of disjoint, nonisotopic essential simple closed curves in S. If the surface obtained from S by cutting along these curves is not a collection of pairs of pants, then it follows from the classification of surfaces and the additivity of Euler characteristic that one component of the cut surface either has positive genus or is a sphere with more than three boundary components. On such a surface there exists an essential simple closed curve that is not homotopic to a boundary component. Thus the original collection of curves was not maximal.

A pair of pants has Euler characteristic -1. If we cut a surface along a collection of disjoint simple closed curves, the cut surface has the same Euler characteristic as the original surface. Thus a pants decomposition of S cuts S into $-\chi(S)$ pairs of pants. It follows that, for a compact surface S of genus g with b boundary components, a pants decomposition for S has

$$\frac{-3\chi(S) - b}{2} = 3g + b - 3$$

curves. Indeed, each pair of pants has three boundary curves and, aside from the curves coming from ∂S, these curves match up in pairs to form curves in S. In particular, a pants decomposition of S_g for $g \geq 2$ has $3g - 3$ curves, cutting S_g into $2g - 2$ pairs of pants.

First proof of Theorem 8.9. We modify ϕ in steps by homotopies until it is a homeomorphism; at each stage, the resulting map will be called ϕ. Choose some pants decomposition $\mathcal{P}$ of S_g consisting of smooth simple closed curves. We first approximate ϕ by a smooth map that is transverse to $\mathcal{P}$. By choosing a close enough approximation we can assume that the approximation is homotopic to ϕ (see [95, p. 124]). By transversality we have that $\phi^{-1}(\mathcal{P})$ is a collection of simple closed curves. If any component of $\phi^{-1}(\mathcal{P})$ is inessential, we can homotope ϕ to remove that component since such a curve bounds a disk, and we can use that disk to define the homotopy.

Since ϕ induces an automorphism on $\pi_1(S_g)$, it takes primitive conjugacy classes in $\pi_1(S_g)$ to primitive conjugacy classes in $\pi_1(S_g)$. Thus the restriction of ϕ to any particular component of $\phi^{-1}(\mathcal{P})$ has degree ± 1 as a map $S^1 \to S^1$. We can therefore homotope ϕ so that it restricts to a homeomorphism on each component of $\phi^{-1}(\mathcal{P})$.

Since ϕ is a homotopy equivalence, it has degree ± 1, and so ϕ is surjective. It follows that $\phi^{-1}(\mathcal{P})$ has at least $3g - 3$ components. If it had more, then two such components would necessarily be isotopic, and the annulus between them would give rise to a homotopy of ϕ reducing the number of components of $\phi^{-1}(\mathcal{P})$.

At this point ϕ is a homeomorphism on each component of $\phi^{-1}(\mathcal{P})$, and ϕ maps each component of $S_g - \phi^{-1}(\mathcal{P})$ to a single component of $S_g - \mathcal{P}$. It therefore suffices to show that if R and R' are pairs of pants and if $\phi : R \to R'$ is a continuous map such that $\phi|_{\partial R}$ is a homeomorphism, then there is a homotopy of ϕ to a homeomorphism $R \to R'$, so that the homotopy restricts to the identity map on ∂R.

Let X be the union of three disjoint arcs in R', one connecting each pair of boundary components. Note that it must be that $R' - (\partial R' \cup X)$ is homeomorphic to a disjoint union of two open disks. Again, we may assume that ϕ is smooth, and so $\phi^{-1}(X)$ is a properly embedded 1-manifold with boundary lying in ∂R. If any component of $\phi^{-1}(X)$ is closed, then it is necessarily null homotopic (since all nonperipheral simple closed curves on a pair of pants are null homotopic), and we may modify ϕ by homotopy to remove this component.

Since $\phi|_{\partial R}$ is assumed to be a homeomorphism, and so it takes distinct boundary components to distinct boundary components, $\phi^{-1}(X)$ consists of exactly three arcs, one for each pair of boundary components of R. We can modify ϕ so that it restricts to a homeomorphism on each component of X. By the Alexander lemma ϕ is homotopic to a homeomorphism. $\qquad\square$

8.3.2 THE ANALYTIC APPROACH: HARMONIC MAPS

We now give an analytic proof of Theorem 8.9. While this proof relies on the machinery of harmonic maps, it is conceptually straightforward.

A *harmonic map* $f : M \to N$ between Riemannian manifolds is one that minimizes the energy functional

$$E(f) = \int_M \|df\|^2.$$

Second proof of Theorem 8.9. We endow S with a hyperbolic metric. It is a theorem of Eells–Sampson that, with respect to this metric, there is a harmonic map h in the homotopy class of ϕ [54]. Since h is a homotopy equivalence, we must have that the degree of h is ± 1. Then by a theorem of Schoen–Yau, any degree one harmonic map between compact surfaces of negative curvature is necessarily a diffeomorphism [189]. $\qquad\square$

Chapter Nine

Braid Groups

In this chapter we give a brief introduction to Artin's classical braid groups B_n. While B_n is just a special kind of mapping class group, namely, that of a multipunctured disk, the study of B_n has its own special flavor. One reason for this is that multipunctured disks can be embedded in the plane, so that elements of B_n lend themselves to specialized kinds of pictorial representations.

9.1 THE BRAID GROUP: THREE PERSPECTIVES

The notion of a mathematical braid is quite natural and classical. For instance, this concept appeared in Gauss's study of knots in the early nineteenth century (see [182]) and in Hurwitz's 1891 paper on Riemann surfaces [102]. The first rigorous definition of the braid group was given by Artin in 1925 [6].

In this section we give three equivalent ways of thinking about the braid group, starting with Artin's classical definition. We will then explain how to go back and forth between the different points of view.

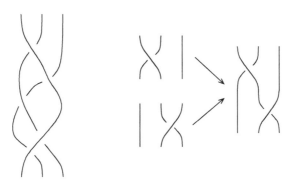

Figure 9.1 Left: a sample 3-braid. Right: the product of two 3-braids.

9.1.1 BRAIDED STRINGS

Let $p_1, \ldots, p_n$ be distinguished points in the complex plane $\mathbb{C}$. A *braid* is a collection of n paths $f_i : [0, 1] \to \mathbb{C} \times [0, 1]$, $1 \leq i \leq n$, called *strands*, and a permutation $\bar{f}$ of $\{1, \ldots, n\}$ such that each of the following holds:

- the strands $f_i([0, 1])$ are disjoint

- $f_i(0) = p_i$

- $f_i(1) = p_{\bar{f}(i)}$

- $f_i(t) \in \mathbb{C} \times \{t\}$.

See the left-hand side of Figure 9.1 for a typical example of a braid on three strands. In our figures, we draw $t = 0$ as the top of the braid.

A braid $(f_1(t), \ldots, f_n(t))$ is determined by its *braid diagram*, which is the picture obtained by projecting the images of the f_i to the plane $\mathbb{R} \times [0, 1]$. In order that this picture carry all of the information, we must indicate which strands are passing over which other strands at the crossings, as in Figure 9.1.

The *braid group on n strands*, denoted B_n, is the group of isotopy classes of braids. The key is that strands are not allowed to cross each other during the isotopy. It also follows from the definitions that an isotopy of braids fixes the set $\{p_i\} \times \{0, 1\}$ and is level-preserving.

The product of the braid $(f_1(t), \ldots f_n(t))$ with the braid $(g_1(t), \ldots, g_n(t))$ is the braid $(h_1(t), \ldots, h_n(t))$, where

$$h_i(t) = \begin{cases} f_i(2t) & 0 \leq t \leq 1/2 \\ g_{\bar{f}(i)}(2t - 1) & 1/2 \leq t \leq 1. \end{cases}$$

In other words, to multiply $f, g \in B_n$ one takes braid representatives for f and g, scales their heights by $1/2$, and then stacks the braid corresponding to f on top of that corresponding to g, thus giving a braid representative for $fg \in B_n$. See the right-hand side of Figure 9.1 for an example of braid multiplication. There we use the typical convention of *not* rescaling the vertical direction (this makes it possible to draw increasingly complicated braids).

The inverse of a given braid is obtained by taking its reflection either through the plane $\mathbb{C} \times \{0\}$ or through the plane $\mathbb{C} \times \{1\}$. See Figure 9.2. Notice that the resulting composition is isotopic to the trivial braid, thus showing that the two braids are indeed inverses.

For $1 \leq i \leq n - 1$, let $\sigma_i \in B_n$ denote the braid whose only crossing is the $(i+1)$st strand passing in front of the ith strand, as shown in Figure 9.3. We claim that the group B_n is generated by elements $\sigma_1, \ldots, \sigma_{n-1}$. The

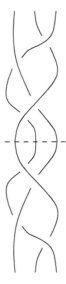

Figure 9.2 A braid (above the dotted line) and its inverse (below the dotted line).

claim follows immediately from the fact that any braid β can be isotoped so that its finitely many crossings occur at different horizontal levels (i.e., different values of t). Reading off the crossings in β from top to bottom then gives β as a product of the σ_i's and their inverses.

We remark that if in the definition of B_n we allow an isotopy between two braids to pass through n-tuples $(f_1(t), \ldots, f_n(t))$ that satisfy all parts of the definition of a braid except the condition $f_i(t) \in \mathbb{C} \times \{t\}$, then the resulting group is the same.

Figure 9.3 A generator σ_i for the braid group: the $(i+1)$st strand passes in front of the ith strand.

9.1.2 FUNDAMENTAL GROUPS OF CONFIGURATION SPACES

Let $C^{ord}(S, n)$ denote the configuration space of n distinct, ordered points in a surface S:

$$C^{ord}(S, n) = S^{\times n} - \text{BigDiag}(S^{\times n}),$$

where $S^{\times n}$ is the n-fold Cartesian product of S and $\text{BigDiag}(S^{\times n})$ is the *big diagonal* of $S^{\times n}$, that is, the subset of $S^{\times n}$ where at least two coordinates are equal. The symmetric group Σ_n acts on $S^{\times n}$ by permuting the coordinates. This action clearly preserves $\text{BigDiag}(S^{\times n})$ and thus induces an action of Σ_n by homeomorphisms on $C^{ord}(S, n)$. Since the action of Σ_n permutes the n coordinates and since these coordinates are always distinct for points in $C^{ord}(S, n)$, we see that this action is free. The quotient space

$$C(S, n) = C^{ord}(S, n)/\Sigma_n$$

is just the configuration space of n distinct, unordered points in S. Since $C(S, n)$ is the quotient of a manifold by a free action (by homeomorphisms) of a finite group, it follows that $C(S, n)$ is a manifold.

It is almost immediate from the definitions that

$$B_n \approx \pi_1(C(\mathbb{C}, n)).$$

Indeed, since each strand of a braid is a map $f_i : I \to \mathbb{C} \times I$ with $f_i(t) \in \mathbb{C} \times \{t\}$, we can think of each f_i as a map $I \to \mathbb{C}$, and this identification gives the isomorphism. Said another way, the intersection of any slice $\mathbb{C} \times \{t\}$ with any braid is a point in $C(\mathbb{C}, n)$, and so the full collection of slices gives an element of $\pi_1(C(\mathbb{C}, n))$. In this way, we can think of a braid $\sigma = (f_1(t), \ldots, f_n(t))$ as tracing out a loop of n-point configurations in $\mathbb{C}$ as t increases from 0 to 1.

The generator σ_i of B_n described above corresponds to the element of $\pi_1(C(\mathbb{C}, n))$ given by the loop of n-point configurations in $\mathbb{C}$ where the ith and $(i + 1)$st points switch places by moving in a clockwise fashion, as indicated in Figure 9.4, and the other $n - 2$ points remain fixed.

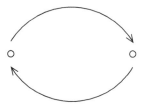

Figure 9.4 A standard generator of B_n in the configuration space model.

The configuration space $C^{ord}(\mathbb{C}, n)$ can also be written as

$$C^{ord}(\mathbb{C}, n) = \mathbb{C}^n - \bigcup_{i<j}\{(z_1, \ldots, z_n) : z_i = z_j\}.$$

Thus $C^{ord}(\mathbb{C}, n)$ is the complement of a *complex hyperplane arrangement*, that is, the complement of a finite union of hyperplanes in $\mathbb{C}^n$. Since

$$B_n = \pi_1(C(\mathbb{C}, n)) = \pi_1(C^{ord}(\mathbb{C}, n))/\Sigma_n,$$

the group B_n is also isomorphic to the fundamental group of the quotient of a complex hyperplane complement by the action of Σ_n.

Since $C^{ord}(\mathbb{C}, n)$ is a space of *ordered* n-tuples, there is a map

$$\psi_n : C^{ord}(\mathbb{C}, n) \to C^{ord}(\mathbb{C}, n-1)$$

defined by forgetting the last point. The fiber $\psi_n^{-1}(x_1, \ldots, x_{n-1})$ is clearly $C^{ord}((\mathbb{C} - (x_1, \ldots, x_{n-1})), 1) \approx \mathbb{C} - (x_1, \ldots, x_{n-1})$. Fadell–Neuwirth [57] proved that

$$\mathbb{C} - (x_1, \ldots, x_{n-1}) \to C^{ord}(\mathbb{C}, n) \to C^{ord}(\mathbb{C}, n-1)$$

is a fibration (what is more, it is a fibration with section). Note that the space $\mathbb{C} - (x_1, \ldots, x_{n-1})$ is aspherical, that is, all of its higher homotopy groups vanish. An application of the homotopy long exact sequence of a fibration gives by an inductive argument that $C^{ord}(\mathbb{C}, n)$ is aspherical for every $n \geq 1$. Since $C(\mathbb{C}, n)$ is finitely covered by $C^{ord}(\mathbb{C}, n)$, all of its higher homotopy groups vanish as well. Thus $C(\mathbb{C}, n)$ is a $K(B_n, 1)$-space.

9.1.3 MAPPING CLASS GROUP OF A PUNCTURED DISK

Finally, we describe B_n as a mapping class group. Let D_n be a closed disk D^2 with n marked points. Then B_n is also isomorphic to the mapping class group of D_n:

$$B_n \approx \text{Mod}(D_n) = \pi_0(\text{Homeo}^+(D_n, \partial D_n)).$$

The isomorphism between $\text{Mod}(D_n)$ and $\pi_1(C(\mathbb{C}, n)) \approx B_n$ can be described as follows. Let ϕ be a homeomorphism of D^2 that leaves invariant the set of n marked points. If we forget that the marked points are distinguished, then ϕ is just a homeomorphism of D^2 fixing ∂D^2 pointwise, so by the Alexander lemma ϕ is isotopic to the identity. Throughout any such isotopy the marked points move around the interior of D^2 (which we identify with $\mathbb{C}$) and return to where they started, thus effecting a loop in $C(\mathbb{C}, n)$. We have thus produced a braid. We will prove in Theorem 9.1 below that this association gives a well-defined homomorphism $B_n \to \mathrm{Mod}(D_n)$ and that this is in fact an isomorphism.

Under the isomorphism $B_n \approx \mathrm{Mod}(D_n)$, each generator σ_i corresponds to the homotopy class of a homeomorphism of D_n that has support a twice-punctured disk and is described on this support by Figure 9.5. We denote such a half-twist as H_α, and we can think of α as either a simple closed curve with two punctures in its interior or a simple proper arc connecting two punctures.

Figure 9.5 A half-twist.

9.1.4 SURFACE BRAID GROUPS AND MAPPING CLASS GROUPS

We have given three different ways of thinking about the braid group. We have already seen that the first two are equivalent. Now we prove that both are equivalent to the third. Specifically, we will prove the isomorphism

$$\pi_1(C(\mathbb{C}, n)) \approx \mathrm{Mod}(D_n).$$

To do this we will require a generalization of the Birman exact sequence that is also due to Birman [24]. In the process we will need to consider the fundamental group $\pi_1(C(S, n))$ for an arbitrary surface S. This group is called the *n-stranded surface braid group of* S.

Let S be a compact surface, perhaps with finitely many punctures but with no marked points. Let $(S, \{x_1, \ldots, x_n\})$ denote S with n marked points $x_1, \ldots, x_n$ in the interior. We are using both punctures and marked points here to distinguish the two, as they will play different roles. As in Section 4.2, there is a forgetful homomorphism $\mathrm{Mod}(S, \{x_1, \ldots, x_n\}) \to \mathrm{Mod}(S)$ given by forgetting that the marked points are marked. As in the proof of Theorem 4.6, there is a fiber bundle

$$\mathrm{Homeo}^+((S, \{x_1, \ldots, x_n\}), \partial S) \to \mathrm{Homeo}^+(S, \partial S) \to C(S^\circ, n)$$

where S° is the interior of S and $\mathrm{Homeo}^+((S, \{x_1, \ldots, x_n\}), \partial S)$ is the group of orientation-preserving homeomorphisms of S that preserve the set $\{x_1, \ldots, x_n\}$ and fix the boundary of S pointwise. As a consequence, we obtain the following generalization of the Birman exact sequence.

THEOREM 9.1 (Birman exact sequence, generalized) *Let S be a surface without marked points and with $\pi_1(\text{Homeo}^+(S, \partial S)) = 1$. The following sequence is exact:*

$$1 \longrightarrow \pi_1(C(S, n)) \xrightarrow{Push} \text{Mod}(S, \{x_1, \ldots, x_n\}) \xrightarrow{Forget} \text{Mod}(S) \longrightarrow 1.$$

Recall that the hypothesis $\pi_1(\text{Homeo}^+(S, \partial S)) = 1$ holds whenever $\chi(S)$ is negative (Theorem 1.14). Also, it follows from the Alexander trick that $\text{Homeo}^+(D^2, \partial D^2)$ is contractible.

Note that we have replaced $C(S^\circ, n)$ with $C(S, n)$ in the statement of Theorem 9.1 since these spaces are homotopy-equivalent.

When $n = 1$, Theorem 9.1 reduces to the usual Birman exact sequence (Theorem 4.6) since $C(S, 1) \approx S$. When $S = D^2$, Theorem 9.1 gives an exact sequence

$$1 \to \pi_1(C(D^2, n)) \to \text{Mod}(D_n) \to \text{Mod}(D^2) \to 1.$$

Since $\text{Mod}(D^2)$ is trivial (Lemma 2.1), and since $\pi_1(C(D^2, n)) \approx \pi_1(C(\mathbb{C}, n)) \approx B_n$, it follows that $B_n \approx \text{Mod}(D_n)$. Note that since $\pi_1(\text{Homeo}^+(\mathbb{C})) \approx \mathbb{Z}$ (see [217]), Theorem 9.1 does *not* give that $B_n \approx \text{Mod}(\mathbb{C} - \{n \text{ points}\})$.

Spherical braid groups. As in the case of Theorem 4.6, the fiber bundle picture still gives us information in the case where $\pi_1(\text{Homeo}^+(S, \partial S))$ is nontrivial. We still have a point-pushing map $\pi_1(C(S, n)) \to \text{Mod}(S, x_1, \ldots, x_n)$, but the kernel of this map is isomorphic to the image of $\pi_1(\text{Homeo}^+(S))$ in $\pi_1(C(S, n))$.

Consider for instance the case $S = S^2$. The group $\pi_1(C(S^2, n))$ is called *the spherical braid group on n strands*. The group $\text{Homeo}^+(S^2)$ has the homotopy type of $\text{SO}(3)$ [197], and so $\pi_1(\text{Homeo}^+(S^2)) \approx \mathbb{Z}/2\mathbb{Z}$. When $n \geq 2$, this group maps nontrivially into $\pi_1(C(S^2, n))$. Combining this with the fact that $\text{Mod}(S^2) = 1$ gives a short exact sequence

$$1 \to \mathbb{Z}/2\mathbb{Z} \to \pi_1(C(S^2, n)) \to \text{Mod}(S_{0,n}) \to 1. \tag{9.1}$$

The nontrivial element of the kernel in (9.1) is given by rotating the n marked points by a 2π twist. In $S^2 \times [0, 1]$ the points trace out n paths, as shown in Figure 9.6 for the case $n = 3$. This is a nontrivial element α of $\pi_1(C(S^2, n))$ because there is no way to untangle the strands in the figure. The image of α in $\text{Mod}(S_{0,n})$ is a Dehn twist about a simple closed curve that surrounds all of the punctures, which is the trivial mapping class. However, the 4π twist α^2 is trivial in $\pi_1(C(S^2, n))$. The fact that the spherical braid α^2 can be unraveled is an example of the belt trick.

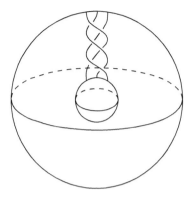

Figure 9.6 The nontrivial element of the kernel $\pi_1(C(S^2, n)) \to \mathrm{Mod}(S^2 - n$ points).

9.2 BASIC ALGEBRAIC STRUCTURE OF THE BRAID GROUP

In this section we investigate some of the basic algebraic properties of B_n.

A finite presentation. In his seminal paper on braid groups, Artin [6] gives the following presentation for B_n.

$$B_n = \langle \sigma_1, \dots, \sigma_{n-1} \mid \begin{array}{ll} \sigma_i\sigma_{i+1}\sigma_i = \sigma_{i+1}\sigma_i\sigma_{i+1} & \text{for all } i, \\ \sigma_i\sigma_j = \sigma_j\sigma_i & \text{for } |i - j| > 1 \end{array} \rangle$$

Here, and in general with braid groups, we use algebraic notation: the element on the left of a word comes first (for the given presentation this does not matter).

We can see in Figure 9.7 that the given relations hold in B_n. Note that the relation $\sigma_i\sigma_{i+1}\sigma_i = \sigma_{i+1}\sigma_i\sigma_{i+1}$ corresponds to the type 3 Reidemeister move from knot theory. In fact, it is possible to derive the above presentation for B_n using the fact that any two planar diagrams for a given knot differ by a finite sequence of Reidemeister moves; see [118, Theorem 1.6]. Another derivation of the presentation is given in [61, Théorème 5].

Computations. It follows from the presentation of B_n that

$$B_1 = 1$$
$$B_2 \approx \mathbb{Z}$$
$$B_3 \approx \widetilde{\mathrm{SL}(2, \mathbb{Z})},$$

Figure 9.7 Relations in the braid group: the *commuting relation* and the *braid relation*.

where $\widetilde{SL(2,\mathbb{Z})}$ is the central extension

$$1 \to \mathbb{Z} \to \widetilde{SL(2,\mathbb{Z})} \to SL(2,\mathbb{Z}) \to 1.$$

The abelianization. It is easy to see from the presentation of B_n that the abelianization of B_n is $\mathbb{Z}$ and that this $\mathbb{Z}$ is generated by the image of any σ_i under the abelianization map $B_n \to \mathbb{Z}$ (cf. Section 5.1). The abelianization map $B_n \to \mathbb{Z}$ is the length homomorphism which counts the signed word length of elements of B_n in terms of the standard generators.

Torsion-freeness. Since D_n is a surface with boundary, Corollary 7.3 implies that B_n is torsion-free for any n. If G is a group with nontrivial torsion, then any $K(G,1)$-space must be infinite-dimensional [91, Proposition 2.45]. Therefore, the fact that B_n is torsion-free also follows from the fact that $C(\mathbb{C},n)$ is a finite-dimensional $K(B_n,1)$.

The center. For $n \geq 3$, the braid group B_n has an infinite cyclic center $Z(B_n)$ generated by

$$z = (\sigma_1 \cdots \sigma_{n-1})^n.$$

Note that $Z(B_2) = \langle \sigma_1 \rangle$. Figure 9.8 demonstrates that z is indeed central.

From the point of view of mapping class groups, z corresponds to the Dehn twist about the boundary of D_n. This Dehn twist commutes with the standard half-twist generators for B_n, and so we again see that z is central.

We now prove that $\langle z \rangle$ is the entire center of B_n. There is a homomorphism $B_n \to \mathrm{Mod}(S_{0,n+1})$ obtained by capping the boundary of D_n with a once-punctured disk. By Proposition 3.19 the kernel of this homomorphism is $\langle z \rangle$. Now any surjective homomorphism between groups takes central elements to central elements. Since $Z(\mathrm{Mod}(S_{0,n+1}))$ is trivial (cf. Section 3.4), it follows that z generates $Z(B_n)$.

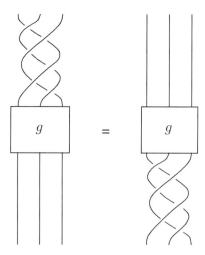

Figure 9.8 Twisting the box by 2π along the vertical axis takes the braid zg to the braid gz.

Braid group modulo center. The previous paragraph gives that the quotient $B_n/Z(B_n)$ is isomorphic to the index n subgroup of $\mathrm{Mod}(S_{0,n+1})$ consisting of elements that fix one distinguished puncture. By taking the distinguished puncture to be the point at infinity, we see that this is the same as the mapping class group of the n-times-punctured plane. One can also derive this description of $B_n/Z(B_n)$ from the long exact sequence used in the proof of Theorem 9.1 and the fact that $\pi_1(\mathrm{Homeo}^+(\mathbb{R}^2)) \approx \mathbb{Z}$.

Roots of central elements. In Section 7.1.1, we classified all finite-order elements of the mapping class group of a multipunctured sphere: they are all conjugate to Euclidean rotations of the sphere. By our above description of $B_n/Z(B_n)$, roots of central elements in B_n correspond to finite-order elements in the subgroup of $\mathrm{Mod}(S_{0,n+1})$ consisting of elements that fix some distinguished puncture. Therefore, up to powers, any root of a central element of B_n is conjugate to one of the elements shown in Figure 9.9. In terms of the generators for B_n, the first root is given by $\sigma_1 \cdots \sigma_{n-1}$, and the second is given by $\sigma_1^2 \sigma_2 \cdots \sigma_{n-1}$.

9.3 THE PURE BRAID GROUP

The *pure braid group* PB_n is the kernel of the homomorphism from B_n to the permutation group Σ_n given by the definition of $\overline{f}$ above:

$$1 \to PB_n \to B_n \to \Sigma_n \to 1.$$

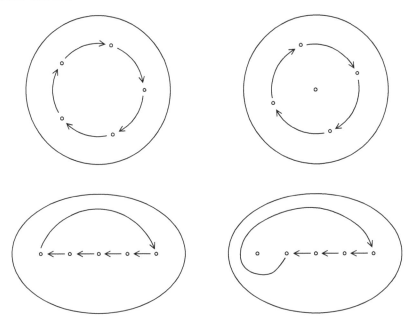

Figure 9.9 Two views of each of the types of roots of central elements.

In other words, a pure braid is a braid where each strand begins and ends at the same point of $\mathbb{C}$. A small variation of Theorem 9.1 gives the following isomorphisms:

$$PB_n \approx \pi_1(C^{ord}(\mathbb{C}, n)) \approx \mathrm{PMod}(D_n).$$

Generators. Artin proved that PB_n is generated by the elements

$$a_{i,j} = (\sigma_{j-1} \cdots \sigma_{i+1})\sigma_i^2 (\sigma_{j-1} \cdots \sigma_{i+1})^{-1}$$

for $1 \leq i < j \leq n$. Since each $a_{i,j}$ is the conjugate of a square of a half-twist, we see that each $a_{i,j}$ is a Dehn twist about a simple closed curve surrounding exactly two punctures. In fact, we can see exactly which simple closed curves. If σ_i^2 corresponds to the Dehn twist T_{c_i}, then $(\sigma_{j-1} \cdots \sigma_{i+1})^{-1}$ corresponds to the mapping class $f = H_{c_{j-1}}^{-1} \cdots H_{c_{i+1}}^{-1}$ and $a_{i,j}$ corresponds to the mapping class $fT_{c_i}f^{-1} = T_{f(c_i)}$ (note that we have passed to functional notation). The effect of f on c_i is shown in Figure 9.10. We see $a_{i,j}$ is the Dehn twist about a simple closed curve that surrounds the ith and jth punctures.

One can derive the above generating set for $PB_n \approx \mathrm{PMod}(D_n)$ from the

Birman exact sequence, as in our proof of Theorem 4.9.

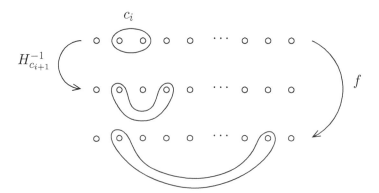

Figure 9.10 Writing the generators of the pure braid group.

The center. The central element z of B_n is also an element of PB_n. For the same reason as before (the Alexander method), z generates the center of PB_n. It is not at all obvious how to write z in terms of the generators $\{a_{i,j}\}$ for PB_n. We claim that

$$z = (a_{1,2}\,a_{1,3}\,\cdots\,a_{1,n})\,\cdots\,(a_{n-2,n-1}\,a_{n-2,n})\,(a_{n-1,n}).$$

We now prove this claim. We think of the product on the right hand side as a product $g_1 g_2 \cdots g_{n-1}$, where

$$g_i = a_{i,i+1}\,a_{i,i+2}\,\cdots\,a_{i,n}.$$

In terms of configuration spaces, g_i is the element obtained by pushing the ith point around the $(i+1)$st point, around the $(i+2)$nd point, and so on, all the way up to the nth point. The orientations of these paths agree, and so this loop in $C(\mathbb{C}, n)$ is isotopic to the loop that pushes the ith point around the last $n - i$ points all at once. In the mapping class group, this push map (see Section 4.2) is equal to the product of two Dehn twists: $T_{d_{i-1}} T_{d_i}^{-1}$ (see Figure 9.11). We then have that the product $g_1 g_2 \cdots g_{n-1}$ is equal to

$$(T_{d_0} T_{d_1}^{-1})(T_{d_1} T_{d_2}^{-1}) \cdots (T_{d_{n-1}} T_{d_n}^{-1}).$$

All terms in this expression cancel except the first, which is the Dehn twist about ∂D_n, and the last, which is trivial (it is the Dehn twist about a simple closed curve with one puncture in its interior). This proves our claim.

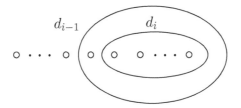

Figure 9.11 The simple closed curves d_i and d_{i-1}.

A finite presentation. Artin's original presentation for PB_n is considerably more complicated than that for B_n. We give here a slightly modified version of his presentation:

$$PB_n \approx \langle a_{i,j} \mid$$

$$[a_{p,q}, a_{r,s}] = 1 \qquad\qquad p < q < r < s$$

$$[a_{p,s}, a_{q,r}] = 1 \qquad\qquad p < q < r < s$$

$$a_{p,r}a_{q,r}a_{p,q} = a_{q,r}a_{p,q}a_{p,r} = a_{p,q}a_{p,r}a_{q,r} \; p < q < r$$

$$[a_{r,s}a_{p,r}a_{r,s}^{-1}, a_{q,s}] = 1 \qquad\qquad p < q < r < s \rangle.$$

Each of the relations in this presentation can be viewed as a type of commutation relation. The four diagrams in Figure 9.12 show the configurations of arcs that appear in the four types of relations (recall that elements are applied left to right).

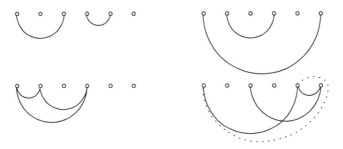

Figure 9.12 Relations for PB_n.

The first two relations are the familiar commutations of Dehn twists about disjoint simple closed curves. The third relation corresponds to the relation $T_x T_y T_z = T_y T_z T_x$, which we discussed in Section 5.1 as a consequence of the lantern relation. By Facts 3.7 and 3.9, we can rephrase the fourth and final relation as follows: if we twist the (p, r)-arc about the (r, s)-arc, the result—namely, the dotted arc in Figure 9.12—is disjoint from the (q, s)-arc.

It is possible to refine the above presentation for PB_n so all of the relations are disjointness relations and lantern relations [141, Theorem 4.10].

A splitting. One important (and nonobvious) fact about PB_n that can be deduced from the above presentation is that PB_n splits as a direct product over its center:

$$PB_n \approx PB_n/Z(PB_n) \times Z(PB_n).$$

To verify that PB_n splits as above, it suffices to show that there is a homomorphism $f : PB_n \to Z(PB_n)$ such that the composition

$$Z(PB_n) \hookrightarrow PB_n \xrightarrow{f} Z(PB_n)$$

is the identity map. We can define such an f by $f(a_{1,2}) = z$ and $f(a_{i,j}) = 1$ otherwise (the choice of $a_{1,2}$ is noncanonical). The map f is a well-defined homomorphism because all of the defining relations for PB_n are commutations. The composition is the identity since $f(z) = z$.

The homomorphism $B_n/Z(B_n) \to \mathrm{Mod}(S_{0,n+1})$ from page 247 identifies $PB_n/Z(PB_n)$ isomorphically with $\mathrm{PMod}(S_{0,n+1})$, and so we have

$$PB_n \approx \mathrm{PMod}(S_{0,n+1}) \times \mathbb{Z}.$$

We can think of the projection $PB_n \to \mathbb{Z}$ geometrically as the map $PB_n \to PB_2$ obtained by forgetting $n - 2$ of the strands.

The abelianization. Another consequence of the fact that all of the defining relations for PB_n are commutations is that the abelianization of PB_n is a free abelian group with one generator for each generator of PB_n. Thus

$$H_1(PB_n; \mathbb{Z}) \approx \mathbb{Z}^{\binom{n}{2}}.$$

A decomposition. Since the pure braid group can be thought of as the pure mapping class group of the n-times-punctured disk, we can apply the Birman exact sequence (Theorem 9.1), which in this context takes the form

$$1 \to F_{n-1} \to PB_n \to PB_{n-1} \to 1.$$

As usual, F_{n-1} denotes the free group on $n - 1$ letters, which is isomorphic to the fundamental group of the disk with $n - 1$ punctures. There is a natural splitting $PB_{n-1} \to PB_n$ obtained by adding an extra strand, and so we see that $PB_n \approx PB_{n-1} \ltimes F_{n-1}$. What is more, by repeating this argument, we see that PB_n is an iterated extension of free groups.

This splitting of PB_n follows from the theorem of Fadell–Neuwirth that the map $C^{ord}(\mathbb{C}, n) \to C^{ord}(\mathbb{C}, n-1)$ is a fiber bundle with section.

9.4 BRAID GROUPS AND SYMMETRIC MAPPING CLASS GROUPS

Besides the relation to mapping class groups of punctured spheres, braid groups arise in the study of the mapping class groups of higher-genus surfaces.

Let S_g^1 be a surface of genus g with one boundary component. We define a homomorphism $\psi : B_n \to \mathrm{Mod}(S_g^1)$ for $n \le 2g+1$ as follows. Choose a chain of simple closed curves $\{\alpha_i\}$ in S_g^1, that is, a collection of simple closed curves satisfying $i(\alpha_i, \alpha_{i+1}) = 1$ for all i and $i(\alpha_i, \alpha_j) = 0$ otherwise. We then define ψ via $\psi(\sigma_i) = T_{\alpha_i}$. By the disjointness relation (Fact 3.9) and the braid relation (Proposition 3.11) for Dehn twists, the map ψ does indeed define a homomorphism. We will prove below that ψ is injective. Even without knowing injectivity, ψ is useful because it allows us to transfer relations from B_n to $\mathrm{Mod}(S_g^1)$.

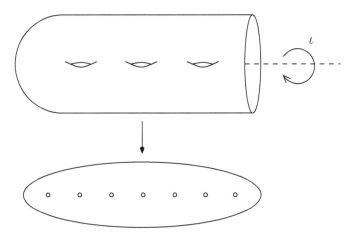

Figure 9.13 The Birman–Hilden double cover

9.4.1 THE BIRMAN–HILDEN THEOREM

Let ι be the order 2 element of $\mathrm{Homeo}^+(S_g^1)$ as shown in Figure 9.13 and let $\mathrm{SHomeo}^+(S_g^1)$ be the centralizer in $\mathrm{Homeo}^+(S_g^1)$ of ι:

$$\mathrm{SHomeo}^+(S_g^1) = C_{\mathrm{Homeo}^+(S_g^1)}(\iota).$$

The group $\mathrm{SHomeo}^+(S_g^1)$ is called the group of orientation-preserving *symmetric homeomorphisms* of S_g^1. The *symmetric mapping class group* is the group

$$\mathrm{SMod}(S_g^1) = \mathrm{SHomeo}^+(S_g^1)/\text{isotopy},$$

that is, the subgroup of $\mathrm{Mod}(S_g^1)$ that is the image of $\mathrm{SHomeo}^+(S_g^1)$.

The homeomorphism ι has $2g + 1$ fixed points in S_g^1. The quotient of S_g^1 by $\langle \iota \rangle$ is a topological disk D_{2g+1} with $2g + 1$ cone points of order 2, with each cone point coming from a fixed point of ι. Since the elements of $\mathrm{SHomeo}^+(S_g^1)$ commute with ι, they descend to homeomorphisms of the quotient disk. Also, by the commutativity, they must preserve the set of $2g + 1$ fixed points of ι, and so there is a homomorphism

$$\mathrm{SHomeo}^+(S_g^1) \to \mathrm{Homeo}^+(D_{2g+1}).$$

This homomorphism is easily seen to be injective. It is actually an isomorphism of topological groups since any element of $\mathrm{Homeo}^+(D_{2g+1})$ can be lifted to $\mathrm{SHomeo}^+(S_g^1)$. We thus have

$$
\begin{aligned}
\mathrm{SHomeo}^+(S_g^1)/\text{symmetric isotopy} \ &= \pi_0(\mathrm{SHomeo}^+(S_g^1)) \\
&\approx \pi_0(\mathrm{Homeo}^+(D_{2g+1})) \\
&= \mathrm{Mod}(D_{2g+1}) \\
&\approx B_{2g+1}.
\end{aligned}
$$

We would like to show that $\mathrm{SMod}(S_g^1) \approx B_{2g+1}$. Since

$$\mathrm{SHomeo}^+(S_g^1)/\text{symmetric isotopy} \approx B_{2g+1},$$

this amounts to showing that if two symmetric homeomorphisms of S_g^1 are isotopic, then they must actually be symmetrically isotopic. Birman–Hilden proved that this is indeed the case [27].

THEOREM 9.2 *Using the above notation,* $\mathrm{SMod}(S_g^1) \approx B_{2g+1}$.

We will give a proof of Theorem 9.2 at the end of the section.

As an illustration of Theorem 9.2, take $g = 1$. The theorem of Birman–Hilden tells us that

$$\mathrm{Mod}(S_1^1) = \mathrm{SMod}(S_1^1) \approx B_3 \approx \mathrm{Mod}(D_3).$$

The Birman–Hilden theorem also holds for surfaces with two (symmetric) boundary components that are interchanged by ι (see top right of Figure 9.15), and so we have

$$\mathrm{SMod}(S_g^2) \approx B_{2g+2}.$$

This implies

$$\mathrm{SMod}(S_g^2)/Z(\mathrm{SMod}(S_g^2)) \approx B_{2g+2}/Z(B_{2g+2}),$$

which in the case $g = 1$ gives

$$\mathrm{PMod}(S_{1,2}) \approx B_4/Z(B_4).$$

From this isomorphism we obtain that $\mathrm{Mod}(S_{1,2}) \approx B_4/Z(B_4) \times \mathbb{Z}/2\mathbb{Z}$, where the last factor is generated by the hyperelliptic involution ι.

Dehn twists and half-twists. Let α be a nonseparating simple closed curve in S_g^1 that is fixed by ι and let N be a neighborhood of α that is fixed by ι. Since $\iota_*([\alpha]) = -[\alpha]$, the restriction of ι to α is a flip. Thus, the restriction of ι to N is a rotation that switches the two boundary components, and $\overline{N} = N/\iota$ is a disk with two cone points of order 2. The Dehn twist T_α commutes with $\iota|_N$ and hence descends to a homeomorphism of $\overline{N}$. The induced homeomorphism of $\overline{N}$ is nothing other than the half-twist about the arc that is the image of α in $\overline{N}$ (the half-twist interchanges the two cone points). See the bottom arrow in Figure 9.14.

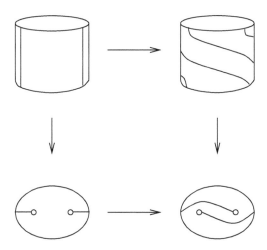

Figure 9.14 The Dehn twist about the core of the annulus covers the half-twist in the disk with two punctures/marked points.

In the other direction, we see that any Dehn twist T_γ in B_{2g+1} either lifts to the square of a Dehn twist, a product of two Dehn twists, or the square root of a Dehn twist, depending on whether the preimage of γ in S_g^1 has two isotopic components, two nonisotopic components, or one component (equivalently, whether γ surrounds two punctures, an even number of punctures greater than 2, or an odd number of punctures greater than 1).

9.4.2 DERIVING RELATIONS IN $\mathrm{Mod}(S_g)$ FROM RELATIONS IN B_n

The connection between the braid relation in $\mathrm{Mod}(S_g^1)$ and the braid relation in the braid group is now apparent. If α and β are the arcs in D_3 shown at the bottom left of Figure 9.15, then the half-twists H_α and H_β satisfy the braid relation in B_3. Via the Birman–Hilden theorem, these half-twists lift to the Dehn twists $T_{\widetilde{\alpha}}$ and $T_{\widetilde{\beta}}$ in $\mathrm{Mod}(S_1^1)$ (see top left of Figure 9.15), which also satisfy the braid relation.

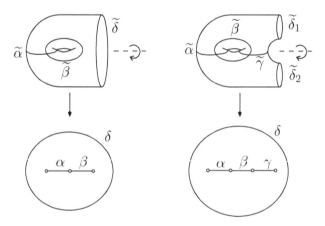

Figure 9.15 The braid relation and the chain relations via the Birman–Hilden theorem.

Our next goal is to explain the chain relations. Recall the relation $(\sigma_1\sigma_2)^3 = z$, where σ_1 and σ_2 are the standard generators for B_3 and z generates the center $Z(B_3)$. Via the isomorphism $B_3 \approx \mathrm{Mod}(D_3)$, this relation becomes $(H_\alpha H_\beta)^3 = T_\delta$, where α, β, and δ are the arcs and the curve in D_3 shown at the bottom left of Figure 9.15. Via the isomorphism $B_3 \approx \mathrm{SMod}(S_1^1)$, the Dehn twist T_δ corresponds to a half-twist about the curve $\widetilde{\delta}$ in S_1^1 shown at the top left of the figure; this mapping class is achieved by holding the boundary fixed and twisting the rest of the surface halfway around. So if we want to get a relation in $\mathrm{SMod}(S_1^1)$ between full Dehn twists, we should consider the relation $(H_\alpha H_\beta)^6 = T_\delta^2$ in $\mathrm{Mod}(D_3)$. In $\mathrm{SMod}(S_g^1)$, this corresponds to the relation $(T_{\widetilde{\alpha}}T_{\widetilde{\beta}})^6 = T_{\widetilde{\delta}}$, where the curves are as shown at the top left of Figure 9.15. This is precisely the 2-

chain relation.

Similarly, the relation $(H_\alpha H_\beta H_\gamma)^4 = T_\delta$ in B_4 (see bottom right of Figure 9.15) corresponds to the relation $(T_{\tilde\alpha} T_{\tilde\beta} T_{\tilde\gamma})^4 = T_{\tilde\delta_1} T_{\tilde\delta_2}$ in $\mathrm{SMod}(S_1^2)$. This is exactly the 3-chain relation. The other k-chain relations are obtained similarly.

Comparing with Figure 9.9 (and the surrounding discussion), we see that the k-chain relation in the mapping class group corresponds to a rotation of order $k + 1$ in the punctured disk.

If instead of using the factorization $(\sigma_1 \cdots \sigma_k)^{k+1}$ of $z \in Z(B_{k+1})$ we use the factorization $(\sigma_1^2 \sigma_2 \cdots \sigma_k)^k = z$, we obtain the alternate chain relations discussed in Section 4.4.

We also mention that the star relation comes from an embedding of the Artin group of type D_4 into the mapping class group of a torus with three boundary components [178].

Closed surfaces. For closed surfaces, the Birman–Hilden theorem takes the form

$$\mathrm{SMod}(S_g)/\langle \iota \rangle \approx \mathrm{Mod}(S_{0,2g+2}),$$

where $S_{0,2g+2}$ is a sphere with $2g + 2$ marked points. Birman–Hilden used this version of their theorem in order to obtain the presentation for $\mathrm{Mod}(S_2)$ given in Section 5.1. Since each standard generator for $\mathrm{Mod}(S_2)$ has a representative in $\mathrm{SHomeo}^+(S_2)$, we have $\mathrm{SMod}(S_2) = \mathrm{Mod}(S_2)$. Thus

$$\mathrm{Mod}(S_2)/\langle [\iota] \rangle = \mathrm{SMod}(S_2)/\langle [\iota] \rangle \approx \mathrm{Mod}(S_{0,6}).$$

Certain relations in $\mathrm{Mod}(S_g)$ can also be interpreted from this point of view. For instance, the hyperelliptic relation in $\mathrm{Mod}(S_g)$ (see Section 5.1) becomes the relation in $\mathrm{Mod}(S_{0,2g+2})$ that pushing a puncture around a simple loop surrounding all of the other punctures is the trivial mapping class; the other side of the loop is a disk.

9.4.3 PROOF OF THE BIRMAN–HILDEN THEOREM

Here we give a new proof of the Birman–Hilden theorem. Our proof is combinatorially flavored, relying on the bigon criterion and the Alexander method. For concreteness, we deal with the case of a closed surface S_g with $g \geq 2$. At the end, we discuss various other surfaces for which the proof applies.

Below, when we say that two symmetric simple closed curves are *symmetrically isotopic*, we mean that they are isotopic through symmetric simple closed curves.

Lemma 9.3 *Let $g \geq 2$ and let α and β be two symmetric nonseparating simple closed curves in S_g. If α and β are isotopic, then they are symmetrically isotopic.*

This lemma is not true for the torus since there exist simple closed curves that are isotopic but pass through different fixed points of ι.

Proof. Let $\overline{\alpha}$ and $\overline{\beta}$ denote the images of α and β in $S_{0,2g+2} \approx S_g/\langle \iota \rangle$. As above, $\overline{\alpha}$ and $\overline{\beta}$ are simple proper arcs in $S_{0,2g+2}$. Any isotopy between these arcs will lift to a symmetric isotopy between α and β.

We can modify α by a symmetric isotopy so that it is transverse to β. We claim that α cannot be disjoint from β. Indeed, for then $\overline{\alpha}$ and $\overline{\beta}$ are disjoint, including endpoints. But such arcs cannot correspond to isotopic curves in S_g. Indeed, any arc $\overline{\gamma}$ that shares an odd number of endpoints with $\overline{\alpha}$ and an even number of endpoints with $\overline{\beta}$ lifts to a simple closed curve γ in S_g with $i(\alpha, \gamma)$ odd and $i(\beta, \gamma)$ even.

Since α is isotopic to β and $\alpha \cap \beta \neq \emptyset$, the bigon criterion gives that α and β form a bigon B. We assume that B is an innermost bigon. As α and β are both fixed by ι, we have that $\iota(B)$ is another innermost bigon in the graph $\alpha \cup \beta$.

Notice that we cannot have $\iota(B) = B$. One way to see this is to note that B lies to one particular side of α and ι takes α to α, reversing its orientation. It follows that the image of B in $S_{0,2g+2}$ is an innermost bigon $\overline{B}$ between $\overline{\alpha}$ and $\overline{\beta}$. What is more, since $\iota(B) \neq B$, there are no fixed points of ι in B and hence no marked points of $S_{0,2g+2}$ in $\overline{B}$.

The bigon $\overline{B}$ can have zero, one, or two of its vertices on marked points of $S_{0,2g+2}$. In the first two cases, we can modify $\overline{\alpha}$ by isotopy in order to remove the bigon, reducing the intersection number of $\overline{\alpha}$ with $\overline{\beta}$. In the last case, since $\overline{B}$ is innermost, we see that $\overline{\alpha} \cup \overline{\beta}$ is a simple loop bounding a disk, and we can push $\overline{\alpha}$ onto $\overline{\beta}$. Removing bigons inductively, we see that $\overline{\alpha}$ is isotopic to $\overline{\beta}$, and this isotopy lifts to a symmetric isotopy between α and β. $\square$

We say that two symmetric homeomorphisms of S_g are *symmetrically isotopic* if they are isotopic through symmetric homeomorphisms, that is, if they lie in the same component of $\mathrm{SHomeo}(S_g)$.

Proposition 9.4 *Let $g \geq 2$ and let $\phi, \psi \in \mathrm{SHomeo}^+(S_g)$. If ϕ and ψ are isotopic, then they are symmetrically isotopic.*

Proof. It suffices to treat the case where ψ is the identity since any symmetric homotopy from $\phi \circ \psi^{-1}$ to identity gives a symmetric homotopy from ϕ to ψ. Let $\overline{\phi}$ denote the induced homeomorphism of $S_{0,2g+2}$.

Let $(\gamma_1, \ldots, \gamma_{2g+1})$ be a chain of nonseparating symmetric simple closed curves in S_g. By assumption, ϕ is isotopic to the identity, and so for each i the curve $\phi(\gamma_i)$ is isotopic to γ_i. By Lemma 9.3, we have that $\phi(\gamma_i)$ is symmetrically isotopic to γ_i for each i. If $\overline{\gamma_i}$ and $\overline{\phi(\gamma_i)}$ are the images in $S_{0,2g+2}$ of γ_i and $\phi(\gamma_i)$, then the last sentence implies that for each i the arc $\overline{\phi(\gamma_i)}$ is isotopic to $\overline{\gamma_i}$. What is more, each such isotopy must fix the endpoints of the arcs throughout and must avoid the marked points of $S_{0,2g+2}$ throughout (if the interior of an arc were to cross a marked point in $S_{0,2g+2}$, then its preimage in S_g would fail to be simple). Applying the Alexander method to the collection of arcs $\overline{\gamma_i}$ in $S_{0,2g+2}$, we conclude that $\overline{\phi}$ is isotopic to the identity. The isotopy induces a symmetric isotopy of ϕ to the identity. □

Proof of the Birman–Hilden theorem. If we compose the natural surjective homomorphism $\mathrm{SHomeo}^+(S_g) \to \mathrm{Homeo}^+(S_{0,2g+2})$ with the projection $\mathrm{Homeo}^+(S_{0,2g+2}) \to \mathrm{Mod}(S_{0,2g+2})$, we obtain a surjective homomorphism $\mathrm{SHomeo}^+(S_g) \to \mathrm{Mod}(S_{0,2g+2})$. By Proposition 9.4, the latter factors through a surjective homomorphism

$$\mathrm{SMod}(S_g) = \pi_0(\mathrm{SHomeo}^+(S_g)) \to \mathrm{Mod}(S_{0,2g+2}).$$

It remains to determine the kernel of this map. Let $f \in \mathrm{SMod}(S_g)$ and let $\phi \in \mathrm{SHomeo}^+(S_g)$ be a symmetric representative. Let $\overline{\phi}$ be the image of ϕ in $\mathrm{Homeo}^+(S_{0,2g+2})$. Since $f \mapsto 1$, we have that $\overline{\phi}$ is isotopic to the identity. This isotopy lifts to an isotopy of ϕ to either the identity or ι. We thus have

$$\mathrm{SMod}(S_g)/\langle[\iota]\rangle \approx \mathrm{Mod}(S_{0,2g+2}),$$

as desired.   □

It is straightforward to generalize our proof of the Birman–Hilden theorem. For instance, in the case of S_g^1, we simply use a chain of $2g$ curves. The quotient of S_g^1 by the hyperelliptic involution $\iota : S_g^1 \to S_g^1$ is a disk with $2g + 1$ marked points. Since ι is not an element of $\mathrm{Homeo}^+(S_g^1, \partial S_g^1)$, it does not represent an element of $\mathrm{SMod}(S_g^1)$, and so we obtain $\mathrm{SMod}(S_g^1) \approx \mathrm{Mod}(D_{2g+1}) \approx B_{2g+1}$, as desired.

PART 2
Teichmüller Space and Moduli Space

Chapter Ten

Teichmüller Space

This chapter introduces another main player in our story: the Teichmüller space $\operatorname{Teich}(S)$ of a surface S. For $g \geq 2$, the space $\operatorname{Teich}(S_g)$ parameterizes all hyperbolic structures on S_g up to isotopy. After defining a topology on $\operatorname{Teich}(S)$, we give a few heuristic arguments for computing its dimension. The length and twist parameters of Fenchel and Nielsen are then introduced in order to prove that $\operatorname{Teich}(S_g)$ is homeomorphic to $\mathbb{R}^{6g-6}$. At the end of the chapter, we prove the $9g - 9$ theorem, which tells us that a hyperbolic structure on S_g is completely determined by the lengths assigned to $9g - 9$ isotopy classes of simple closed curves in S_g.

In Chapter 12, we will prove that $\operatorname{Teich}(S)$ admits a properly discontinuous action of $\operatorname{Mod}(S)$. The quotient $\mathcal{M}(S) = \operatorname{Teich}(S)/\operatorname{Mod}(S)$ is the moduli space of Riemann surfaces. The interplay between properties of $\operatorname{Teich}(S)$, properties of $\operatorname{Mod}(S)$, and properties of this action provide us with information on $\operatorname{Teich}(S)$, $\operatorname{Mod}(S)$, and $\mathcal{M}(S)$. For example, in Chapter 13 we will use the action of $\operatorname{Mod}(S)$ on $\operatorname{Teich}(S)$ to give a classification of elements of $\operatorname{Mod}(S)$.

10.1 DEFINITION OF TEICHMÜLLER SPACE

Let S be a compact surface with finitely many (perhaps zero) points removed from the interior. We assume for now that $\chi(S) < 0$. After some preparation, we will define the Teichmüller space of S to be the set of isotopy classes of hyperbolic structures on S. While implicit in the work of Poincaré, Riemann, and Klein, Teichmüller space was first defined and studied by Fricke, Teichmüller, Fenchel, and Nielsen.

By a *hyperbolic structure* on S we will mean a diffeomorphism $\phi : S \to X$, where X is a surface with a complete, finite-area hyperbolic metric with totally geodesic boundary. We can record the hyperbolic structure $\phi : S \to X$ by the pair (X, ϕ). The diffeomorphism ϕ is referred to as the *marking*, and either X or (X, ϕ) can be referred to as a *marked hyperbolic surface* (depending on whether or not we need to be explicit about the marking).

Two hyperbolic structures $\phi_1 : S \to X_1$ and $\phi_2 : S \to X_2$ on S are *homotopic* if there is an isometry $X_1 \to X_2$ so that the markings $I \circ \phi_1 :$

$S \to X_2$ and $\phi_2 : S \to X_2$ are homotopic. This is to say that the following diagram commutes up to homotopy:

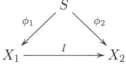

Here homotopies are allowed to move points in the boundary of X_2.

We can then define the *Teichmüller space* of S as the set of homotopy classes of hyperbolic structures on S:

$$\mathrm{Teich}(S) = \{\text{hyperbolic structures on } S\}/\text{homotopy}$$

In slightly different language,

$$\mathrm{Teich}(S) = \{(X, \phi)\}/\sim,$$

where two marked hyperbolic surfaces are equivalent if the hyperbolic structures they define are homotopic.

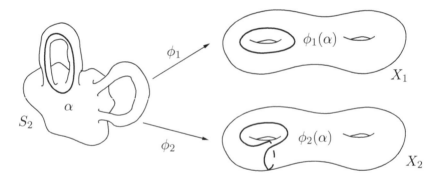

Figure 10.1 The hyperbolic surfaces X_1 and X_2 are isometric, but $\mathfrak{X}_1 = [(X_1, \phi_1)]$ and $\mathfrak{X}_2 = [(X_2, \phi_2)]$ are not the same point of $\mathrm{Teich}(S_2)$ since, for example, the way we have arranged things, $\ell_{\mathfrak{X}_1}(\alpha)$ is not equal to $\ell_{\mathfrak{X}_2}(\alpha)$.

Teichmüller space as a set of metrics. A marking $\phi : S \to X$ of course gives rise to an actual hyperbolic Riemannian metric on S, namely, the pullback of the hyperbolic metric on X. Thus we can also describe Teichmüller space as the set of isotopy classes of hyperbolic metrics on S:

$$\mathrm{Teich}(S) = \mathrm{HypMet}(S)/\mathrm{Diff}_0(S),$$

where the action of $\mathrm{Diff}_0(S)$ on the set of hyperbolic metrics $\mathrm{HypMet}(S)$ is by pullback. While this second definition is in a sense more direct—for instance, there are no auxiliary surfaces required—our first definition will be easier to use in practice.

Length functions. Let S denote the set of isotopy classes of essential simple closed curves in S. The hyperbolic structure on S corresponding to a point $X \in \text{Teich}(S)$ is defined only up to isotopy, but this is exactly enough information to define a *length function*

$$\ell_X : S \to \mathbb{R}_+.$$

If X is the equivalence class of the marked hyperbolic surface (X, ϕ) and c is an isotopy class of simple closed curves in S, then $\ell_X(c)$ is the length of the unique geodesic in X in the isotopy class $\phi(c)$. As we proved in Proposition 1.3, there is a unique such curve in X realizing this minimum.

Understanding points of $\text{Teich}(S)$ via the length functions they define is a useful point of view. See Figure 10.1. Indeed, as we will prove in Section 10.7 below, if $\mathbb{R}^S$ denotes the set of real-valued functions on S, the map $\ell : \text{Teich}(S) \to \mathbb{R}^S$ given by $X \mapsto \ell_X$ is injective. Actually, we will prove something much stronger: an element of $\text{Teich}(S)$ is determined by finitely many coordinates of the length function ℓ.

Change of marking. Given two hyperbolic structures $\phi : S \to X$ and $\psi : S \to Y$ on S, there is a bijective correspondence between $\text{Homeo}(S)$ and $\text{Homeo}(X, Y)$ given by $f \leftrightarrow \psi \circ f \circ \phi^{-1}$. The only canonical homeomorphism $S \to S$ is the identity map. The corresponding canonical homeomorphism $X \to Y$ is the *change of marking map* $\psi \circ \phi^{-1}$.

10.2 TEICHMÜLLER SPACE OF THE TORUS

The Gauss–Bonnet theorem implies that any closed hyperbolic surface X has fixed area $-2\pi\chi(X)$. In contrast, a flat metric on the torus T^2 can have any positive number as its area. Of course, any flat metric on the torus can be multiplied pointwise by a fixed real number so that the area of the resulting metric equals 1. It is thus natural to define the Teichmüller space $\text{Teich}(T^2)$ as the set of isotopy classes of unit-area flat structures on T^2.

The space $\text{Teich}(T^2)$ will serve as a simple example with which we can compute explicitly, in contrast to the case of $\text{Teich}(S_g)$ with $g \geq 2$. As a first example, we have the following.

Proposition 10.1 *There is a natural bijective correspondence*

$$\text{Teich}(T^2) \leftrightarrow \mathbb{H}^2.$$

We will give two proofs of Proposition 10.1, one using the upper half-plane model for $\mathbb{H}^2$ and one using the open unit disk model.

In the first proof of Proposition 10.1 we will describe $\text{Teich}(T^2)$ in terms of lattices. By a *lattice* in $\mathbb{R}^2$, we mean a discrete subgroup Λ of the additive group $\mathbb{R}^2$ with $\Lambda \approx \mathbb{Z}^2$. Equivalently, a discrete subgroup $\Lambda < \mathbb{R}^2$ is a lattice if $\mathbb{R}^2/\Lambda$ is compact.

Note that the $\mathbb{R}$-span of any pair of generators for the group Λ is all of $\mathbb{R}^2$. The *area* of a lattice Λ in $\mathbb{R}^2$ is the Euclidean area of the torus $\mathbb{R}^2/\Lambda$. Any lattice in $\mathbb{R}^2$ is homothetic to a unique unit-area lattice. Recall that a *homothety* of $\mathbb{R}^2$ is a map $z \mapsto \lambda z$ for some $\lambda \in \mathbb{R}_+$.

We say that a lattice in $\mathbb{R}^2$ is *marked* if it comes equipped with an ordered set of two generators. Equivalently, we can say that a lattice in $\mathbb{R}^2$ is marked if it comes with a fixed isomorphism with $\mathbb{Z}^2$.

First proof of Proposition 10.1. We proceed in two steps.

Step 1: $\text{Teich}(T^2) \longleftrightarrow \{\text{marked lattices in } \mathbb{R}^2\}/\sim$, where the equivalence relation is generated by Euclidean isometries and homotheties.

Fix a standard ordered generating set for $\pi_1(T^2)$. The ordered generating set for a marked lattice Λ in $\mathbb{R}^2$ descends to an ordered generating set for $\pi_1(\mathbb{R}^2/\Lambda)$. It is possible to find a diffeomorphism $\phi : T^2 \to \mathbb{R}^2/\Lambda$ that takes the first and second generators of $\pi_1(T^2)$ to the first and second generators for $\pi_1(\mathbb{R}^2/\Lambda)$. We can scale the flat torus $\mathbb{R}^2/\Lambda$ so that it has unit area, and the diffeomorphsm ϕ induces a marking of this unit area flat torus. We have thus obtained a point of $\text{Teich}(T^2)$.

On the other hand, if we start with a point $[(X, \phi)] \in \text{Teich}(T^2)$, where $\phi : T^2 \to X$ is a unit-area flat structure, then the metric universal cover of X is isometric to $\mathbb{R}^2$. The group of deck transformations is a lattice Λ in $\mathbb{R}^2$, and the image under ϕ of the ordered set of generators for $\pi_1(T^2)$ is a marking of Λ.

Step 2: $\mathbb{H}^2 \leftrightarrow \{\text{marked lattices in } \mathbb{R}^2\}/\sim$.

Let Λ be a marked lattice in $\mathbb{C} \approx \mathbb{R}^2$. We can describe Λ by an ordered pair of complex numbers (ν, τ), namely, the ordered set of generators coming from the marking. We think of each of these complex numbers as vectors in the plane. Staying within the same equivalence class of lattices, we can scale and rotate Λ so that ν becomes 1. In other words, Λ is equivalent to the lattice corresponding to $(1, \tau)$. The choice of τ here is not unique since $(1, \tau)$ and $(1, \overline{\tau})$ correspond to equivalent marked lattices (they differ by reflection across the x-axis, which is a Euclidean isometry). Reflecting across the x-axis, we can assume τ lies in the upper half-plane, which we identify with $\mathbb{H}^2$. The map $[\Lambda] \mapsto \tau$ from the set of equivalence classes of marked lattices to $\mathbb{H}^2$ is the desired bijection. $\square$

For the second proof of Proposition 10.1 we need to discuss how to write down real linear maps $\mathbb{R}^2 \to \mathbb{R}^2$ using complex notation. Let $f : \mathbb{R}^2 \to \mathbb{R}^2$ be a linear map. We can represent f by a matrix

$$f = \begin{pmatrix} a & b \\ c & d \end{pmatrix}.$$

Using complex notation and setting $z = x + iy$, we can rewrite f as

$$f(z) = \alpha z + \beta \bar{z},$$

where

$$\alpha = \frac{(a + ic) - i(b + id)}{2} \quad \text{and} \quad \beta = \frac{(a + ic) + i(b + id)}{2}.$$

Indeed, it is straightforward to check that the latter map sends 1 to $a + ic$ and i to $b + id$. We have

$$|\alpha|^2 - |\beta|^2 = ad - bc = \det f.$$

Thus f is a linear isomorphism if and only if $|\alpha| \neq |\beta|$, and in this case f is orientation-preserving if and only if $|\alpha| > |\beta|$.

Second proof of Proposition 10.1. We again prove the proposition in two steps.

Step 1: $\mathrm{Teich}(T^2) \leftrightarrow \{\text{orientation-preserving isomorphisms } \mathbb{R}^2 \to \mathbb{R}^2\}/\sim$, where two linear maps are equivalent if they differ by rotation and/or dilation.

This bijection is essentially a restatement of the bijection in step 1 of the first proof of Proposition 10.1. Indeed, a linear map is exactly given by a marked lattice (the image of the standard basis).

Step 2: $\mathbb{H}^2 \leftrightarrow \{\text{orientation-preserving isomorphisms } \mathbb{R}^2 \to \mathbb{R}^2\}/\sim$.

Let $f : \mathbb{R}^2 \to \mathbb{R}^2$ be an orientation-preserving linear automorphism of $\mathbb{R}^2$. As above, we can write f uniquely in complex notation

$$f(z) = \alpha z + \beta \bar{z}$$

for some $\alpha, \beta \in \mathbb{C}$. Since multiplication by a complex number is the composition of a rotation with a dilation, we can postcompose f by multiplication by α^{-1}, staying in the same equivalence class. We then obtain the linear map

$$z \mapsto z + \mu \bar{z},$$

where $\mu = \beta/\alpha$. Again, since f is orientation-preserving, the complex number μ lies inside the unit disk in $\mathbb{C}$ and hence gives a point of $\mathbb{H}^2$ via the unit disk model. This process is reversible, so we have exhibited the desired bijection.   □

The complex number μ attached to the map f is called the *complex dilatation* of f. We will see in Chapter 11 that μ conveys salient information about the map f in that it records the amount of stretching that f effects on $\mathbb{R}^2$.

There is a third space lurking that is also equivalent to $\text{Teich}(T^2)$ and $\mathbb{H}^2$, namely,

$$\text{SL}(2, \mathbb{R})/\text{SO}(2, \mathbb{R}).$$

Indeed, our description of $\text{Teich}(T^2)$ in step 1 of the second proof of Proposition 10.1 is equivalent to this quotient: given an orientation-preserving linear map, we can scale in order to get an element of $\text{SL}(2, \mathbb{R})$. Then the rotations in $\text{SL}(2, \mathbb{R})$ are exactly the elements of $\text{SO}(2, \mathbb{R})$.

There is a direct way to see the bijection $\text{SL}(2, \mathbb{R})/\text{SO}(2, \mathbb{R}) \leftrightarrow \mathbb{H}^2$: the group $\text{SL}(2, \mathbb{R})$ acts transitively on $\mathbb{H}^2$ with point stabilizers isomorphic to $\text{SO}(2, \mathbb{R})$.

A topology on Teich(T^2). The bijection $\text{Teich}(T^2) \leftrightarrow \mathbb{H}^2$ induces a topology on $\text{Teich}(T^2)$ by declaring the bijection to be a homeomorphism (one can check that the various bijections are compatible). We will see below that this idea generalizes to give a topology on $\text{Teich}(S)$ for arbitrary S.

Sample tori. We explore the dictionary between $\text{Teich}(T^2)$ and $\mathbb{H}^2$ given by our first proof of Proposition 10.1. To start, the points i and $i + 1$ both represent the standard lattice in $\mathbb{R}^2$. However, the point i corresponds to the marking $(1, i)$, whereas the point $i + 1$ corresponds to the marking $(1, i + 1)$, and so these are different points of $\text{Teich}(T^2)$. Viewed as marked tori, both are isometric to the standard square torus, but i corresponds to the square torus with the standard marking, and $i + 1$ corresponds to the square torus where the marking differs from the standard marking by a Dehn twist. Similarly, one can check that the points ni and i/n represent isometric tori but different points in $\text{Teich}(T^2)$ for any $n > 0$. Finally, one can check that, for $\epsilon \in (0, 1)$, the points i and $i + \epsilon$ represent nonisometric tori and hence different points in $\text{Teich}(T^2)$.

10.3 THE ALGEBRAIC TOPOLOGY

There is an alternate characterization of $\mathrm{Teich}(S)$ which gives rise to a natural topology on $\mathrm{Teich}(S)$, called the algebraic topology. To describe this characterization we will use a higher-genus analogue of the description of $\mathrm{Teich}(T^2)$ in terms of marked lattices. We begin with the closed case $S = S_g$ for $g \geq 2$.

Recall that $\mathrm{Isom}^+(\mathbb{H}^2) \approx \mathrm{PSL}(2, \mathbb{R})$ and $\mathrm{Isom}(\mathbb{H}^2) \approx \mathrm{PGL}(2, \mathbb{R})$. A representation $\rho : \pi_1(S_g) \to \mathrm{PSL}(2, \mathbb{R})$ is called *faithful* if it is injective. Such a representation ρ is called *discrete* if $\rho(\pi_1(S_g))$ is discrete in $\mathrm{PSL}(2, \mathbb{R})$.

The group $\mathrm{PGL}(2, \mathbb{R})$ acts on the space $\mathrm{DF}(\pi_1(S_g), \mathrm{PSL}(2, \mathbb{R}))$ of discrete, faithful representations $\rho : \pi_1(S_g) \to \mathrm{PSL}(2, \mathbb{R})$ by conjugation: for each $\gamma \in \pi_1(S_g)$ and each $h \in \mathrm{PGL}(2, \mathbb{R})$, we let

$$(h \cdot \rho)(\gamma) = h\rho(\gamma)h^{-1}.$$

The quotient

$$\mathrm{DF}(\pi_1(S_g), \mathrm{PSL}(2, \mathbb{R}))/\mathrm{PGL}(2, \mathbb{R})$$

is the set of $\mathrm{PGL}(2, \mathbb{R})$ conjugacy classes of discrete, faithful representations of $\pi_1(S_g)$ into $\mathrm{PSL}(2, \mathbb{R})$. The following is an analogue of Proposition 10.1.

Proposition 10.2 *Let $g \geq 2$. There is a natural bijective correspondence:*

$$\mathrm{Teich}(S_g) \leftrightarrow \mathrm{DF}(\pi_1(S_g), \mathrm{PSL}(2, \mathbb{R}))/\mathrm{PGL}(2, \mathbb{R}).$$

To make the analogy between Propositions 10.2 and 10.1 more clear, we can think of an equivalence class of marked lattices in $\mathbb{R}^2$ as a discrete, faithful representation of $\pi_1(T^2) \approx \mathbb{Z}^2$ into $\mathrm{Isom}^+(\mathbb{R}^2)$ up to conjugation by $\mathrm{Isom}(\mathbb{R}^2)$ and homothety. Again, the reason that homothety does not appear in Proposition 10.2 is that the Gauss–Bonnet theorem implies that all hyperbolic structures on S_g have the same area when $g \geq 2$.

Proof. Let $[(X, \phi)] \in \mathrm{Teich}(S_g)$. There is an isometric identification $\eta : \widetilde{X} \to \mathbb{H}^2$, where $\widetilde{X}$ is the metric universal cover of X. The group $\pi_1(X)$ acts isometrically and properly discontinuously on $\widetilde{X}$. The marking ϕ identifies $\pi_1(S_g)$ with $\pi_1(X)$ and hence with the group of deck transformations of $\widetilde{X}$. These identifications give rise to a discrete, faithful representation $\rho : \pi_1(S_g) \to \mathrm{PSL}(2, \mathbb{R})$.

In determining ρ we made several choices: the choice of (X, ϕ) in the class $[(X, \phi)]$, the choice of η, the choice of isomorphism $\phi_*(\pi_1(S_g)) \to$

$\pi_1(X)$, and the choice of identification of $\pi_1(X)$ with the group of deck transformations of $\widetilde{X}$. We claim that none of these choices affect the equivalence class of ρ. For example, the choice of η is unique up to postcomposing by an element of $\mathrm{Isom}(\mathbb{H}^2)$. If we replace η with $\eta \circ \nu$, where $\nu \in \mathrm{Isom}(\mathbb{H}^2) \approx \mathrm{PGL}(2, \mathbb{R})$, then ρ simply becomes $\nu \cdot \rho$. Changing (X, ϕ) within its equivalence class is tantamount to changing ϕ within its homotopy class. But changing ϕ by homotopy does not affect ρ. One way to see this is to observe that if we lift any isotopy of X to $\widetilde{X} \approx \mathbb{H}^2$, then points of $\mathbb{H}^2$ move a uniformly bounded distance and so the induced action on $\partial \mathbb{H}^2$ is trivial. On the other hand, an isometry of $\mathbb{H}^2$ is determined by its action on $\partial \mathbb{H}^2$. Finally, the choices of isomorphisms between $\phi_*(\pi_1(S_g))$, $\pi_1(X)$, and the group of deck transformations are well defined up to conjugation, and so the resulting ρ is well defined up to conjugation.

For the other direction, let $\rho \in \mathrm{DF}(\pi_1(S_g), \mathrm{PSL}(2, \mathbb{R}))$. We claim that ρ is a covering space action on $\mathbb{H}^2$. Since $\rho(\pi_1(S_g))$ is discrete, the action of $\rho(\pi_1(S_g))$ on $\mathbb{H}^2$ is properly discontinuous. Thus to prove the claim we must show that this action is free. If the action of ρ were not free, then the image of ρ would contain a nontrivial elliptic isometry of $\mathbb{H}^2$, that is, a rotation. Since ρ is faithful and $\pi_1(S_g)$ is torsion-free, this elliptic element must have infinite order. This violates the discreteness of ρ. Thus the action of $\rho(\pi_1(S_g))$ on $\mathbb{H}^2$ must be free.

Since ρ is a covering space action, it follows that $X = \mathbb{H}^2 / \rho(\pi_1(S_g))$ has fundamental group $\pi_1(S_g)$. Thus, by the classification of surfaces, X is diffeomorphic to S_g.

We can recover a homomorphism $\rho_* : \pi_1(S_g) \to \pi_1(X)$ from ρ since ρ maps elements of $\pi_1(S_g)$ to covering transformations over X, which in turn correspond to elements of $\pi_1(X)$. Since S_g and X are $K(\pi_1(S_g), 1)$-spaces, it follows that there is a unique homotopy class of homotopy equivalences from S_g to X that realizes the map ρ_*. But any homotopy equivalence $S_g \to X$ is homotopic to a diffeomorphism (Proposition 8.9 plus Theorem 1.13), and this diffeomorphism serves as the desired marking.

Suppose we replace ρ by one of its $\mathrm{PGL}(2, \mathbb{R})$ conjugates, ρ'. The resulting Riemann surface X' is isometric to X, and it follows that the resulting point of $\mathrm{Teich}(S_g)$ is the same.

The two maps described above are inverses of each other, so the proof is complete. $\square$

The topology.   There is a natural topology on the quotient space $\mathrm{DF}(\pi_1(S_g), \mathrm{PSL}(2, \mathbb{R})) / \mathrm{PGL}(2, \mathbb{R})$, which we now describe.

We endow $\pi_1(S_g)$ with the discrete topology and $\mathrm{PSL}(2, \mathbb{R})$ with its usual topology as a Lie group and then give the set $\mathrm{Hom}(\pi_1(S_g), \mathrm{PSL}(2, \mathbb{R}))$

the compact-open topology. There is a more concrete way to describe this topology. Pick a set of $2g$ generators for $\pi_1(S_g)$. Since a homomorphism $\pi_1(S_g) \to \mathrm{PSL}(2,\mathbb{R})$ is determined by where it sends a generating set, there is a natural inclusion of $\mathrm{Hom}(\pi_1(S_g), \mathrm{PSL}(2,\mathbb{R}))$ into the direct product $\mathrm{PSL}(2,\mathbb{R})^{2g}$ of $2g$ copies of $\mathrm{PSL}(2,\mathbb{R})$. We endow $\mathrm{PSL}(2,\mathbb{R})^{2g}$ with the usual Lie group topology. Then the set $\mathrm{Hom}(\pi_1(S_g), \mathrm{PSL}(2,\mathbb{R}))$ inherits the subspace topology. It is straightforward to check that different choices of generating sets for $\pi_1(S_g)$ give rise to equivalent topologies on $\mathrm{Hom}(\pi_1(S_g), \mathrm{PSL}(2,\mathbb{R}))$. It is also not hard to verify that the two topologies on $\mathrm{Hom}(\pi_1(S_g), \mathrm{PSL}(2,\mathbb{R}))$ described above give rise to equivalent topologies.

Since $\mathrm{DF}(\pi_1(S_g), \mathrm{PSL}(2,\mathbb{R}))$ is a subset of $\mathrm{Hom}(\pi_1(S_g), \mathrm{PSL}(2,\mathbb{R}))$, it inherits the subspace topology. Finally, we endow the quotient

$$\mathrm{DF}(\pi_1(S_g), \mathrm{PSL}(2,\mathbb{R}))/\mathrm{PGL}(2,\mathbb{R})$$

with the quotient topology. We then obtain via Proposition 10.2 a topology on $\mathrm{Teich}(S_g)$ called the *algebraic topology* on $\mathrm{Teich}(S_g)$.

In Chapter 11, we will define a metric on $\mathrm{Teich}(S_g)$ called the Teichmüller metric, and we will check that the induced topology on $\mathrm{Teich}(S_g)$ is homeomorphic to the algebraic topology on $\mathrm{Teich}(S_g)$.

Continuity of length functions. Let γ be some fixed element of $\pi_1(S_g)$, where $g \geq 2$. The function $[\rho] \mapsto \mathrm{trace}(\rho(\gamma))$ is a continuous function on $\mathrm{DF}(\pi_1(S_g), \mathrm{PSL}(2,\mathbb{R}))/\mathrm{PGL}(2,\mathbb{R})$. For $\mathcal{X} \in \mathrm{Teich}(S_g)$, denote by $\rho_\mathcal{X}$ some corresponding representation. Since

$$\ell_\mathcal{X}(\gamma) = 2\cosh^{-1}(\mathrm{trace}(\rho_\mathcal{X}(\gamma))/2),$$

we have the following consequence of Proposition 10.2.

Proposition 10.3 *Let S be any hyperbolic surface and let c be an isotopy class of simple closed curves in S. The function $\mathrm{Teich}(S) \to \mathbb{R}$ given by*

$$\mathcal{X} \mapsto \ell_\mathcal{X}(c)$$

is continuous.

Nonclosed surfaces. The procedure just described for obtaining a topology on $\mathrm{Teich}(S)$ is more delicate when S is not closed. The reason is that there are nonhomeomorphic surfaces with the same fundamental group, for example, $\pi_1(S_{0,3}) \approx \pi_1(S_{1,1}) \approx F_2$. In these cases we do not simply consider all discrete faithful representations of $\pi_1(S)$ into $\mathrm{PSL}(2,\mathbb{R})$. Instead, we consider the subset of $\mathrm{DF}(\pi_1(S), \mathrm{PSL}(2, \mathbb{R}))$ consisting of those representations corresponding to complete, finite-area hyperbolic surfaces with geodesic boundary homeomorphic to S. For example, this restricts the conjugacy class in $\pi_1(S)$ corresponding to a loop around a puncture to map to a unipotent element of $\mathrm{PSL}(2, \mathbb{R})$, so that the corresponding isometry of $\mathbb{H}^2$ will be of parabolic type. Indeed, if a loop around a puncture corresponds to a hyperbolic isometry, the hyperbolic structure on the surface will have infinite area.

10.4 TWO DIMENSION COUNTS

In Section 10.6, we will give a formal proof that $\mathrm{Teich}(S_g) \approx \mathbb{R}^{6g-6}$ when $g \geq 2$. Before doing this we first arrive at the correct dimension for $\mathrm{Teich}(S_g)$ via two different heuristic counts. This dimension was first stated by Riemann in his paper on abelian functions [185].

10.4.1 TEICHMÜLLER SPACE AS A REPRESENTATION SPACE

For the first dimension count we use the bijection between $\mathrm{Teich}(S_g)$ and $\mathrm{DF}(\pi_1(S_g), \mathrm{PSL}(2, \mathbb{R}))/\mathrm{PGL}(2, \mathbb{R})$ given in Proposition 10.2.

The Lie group $\mathrm{PGL}(2, \mathbb{R})$ is 3-dimensional and acts on the space $\mathrm{DF}(\pi_1(S_g), \mathrm{PSL}(2, \mathbb{R}))$ with 3-dimensional orbits (this is not hard to check). Thus the dimension of the quotient

$$\mathrm{DF}(\pi_1(S_g), \mathrm{PSL}(2, \mathbb{R}))/\mathrm{PGL}(2, \mathbb{R})$$

can be computed as the dimension of $\mathrm{DF}(\pi_1(S_g), \mathrm{PSL}(2, \mathbb{R}))$ minus 3.

The set $\mathrm{DF}(\pi_1(S_g), \mathrm{PSL}(2, \mathbb{R}))$ is open in $\mathrm{Hom}(\pi_1(S_g), \mathrm{PSL}(2, \mathbb{R}))$ (see [212]), so the dimensions of the two spaces are the same, and it suffices to find the dimension of the latter. Let $\gamma_1, \ldots, \gamma_{2g} \in \pi_1(S_g)$ so that

$$\pi_1(S_g) = \langle \gamma_1, \ldots, \gamma_{2g} \,|\, [\gamma_1, \gamma_2] \cdots [\gamma_{2g-1}, \gamma_{2g}] \rangle.$$

A homomorphism $\rho : \pi_1(S_g) \to \mathrm{PSL}(2, \mathbb{R})$ is determined by choosing the $2g$ images $\rho(\gamma_i) \in \mathrm{PSL}(2, \mathbb{R})$. However, by the relation

$$[\rho(\gamma_1), \rho(\gamma_2)] \cdots [\rho(\gamma_{2g-1}), \rho(\gamma_{2g})] = I,$$

we see that $\rho(\gamma_{2g})$ is completely determined by the other $\rho(\gamma_i)$. This cuts down on 3 degrees of freedom in our choices of the $\rho(\gamma_i)$.

We thus arrive at the following count.

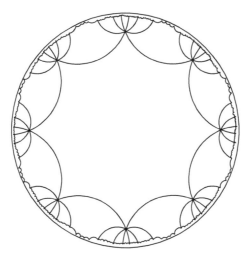

Figure 10.2 The tiling of $\mathbb{H}^2$ by regular octagons.

Dimension count 1: the space of representations

$+6g$: Choose elements $\rho(\gamma_1), \ldots, \rho(\gamma_{2g}) \in \mathrm{PSL}(2, \mathbb{R})$.

-3 : The $\rho(\gamma_i)$ must satisfy one relation.

-3 : Conjugate representations are equivalent.

$= 6g - 6$ total dimensions

10.4.2 TEICHMÜLLER SPACE AS A SPACE OF TILINGS

We define a *hyperbolic S_g-tile* as a geodesic hyperbolic $4g$-gon with the following properties

1. The sum of the interior angles is 2π.

2. Reading clockwise, the edges are labeled

$$\gamma_1, \gamma_2, \gamma_1, \gamma_2, \ldots, \gamma_{2g-1}, \gamma_{2g}, \gamma_{2g-1}, \gamma_{2g}.$$

3. Edges with the same labels have the same hyperbolic length.

If we identify the sides of a hyperbolic S_g-tile according to the labels, we obtain a closed hyperbolic surface of genus g.

We will give a bijection between $\mathrm{Teich}(S_g)$ and equivalence classes of S_g-tiles and use this identification to find the dimension of $\mathrm{Teich}(S_g)$.

Fix a collection of simple loops $\{\gamma_1, \ldots, \gamma_{2g}\}$ in S_g that are based at a common point x, that are disjoint away from x, and that cut S_g into a $4g$-gon whose labels agree with that of a hyperbolic S_g-tile.

Let $[(X, \phi)] \in \text{Teich}(S_g)$. If we consider homotopy classes relative to $\phi(x)$, then each $\phi(\gamma_i)$ has a unique shortest representative δ_i in X. The δ_i are all simple and intersect pairwise only at $\phi(x)$. When we cut X along the δ_i, we obtain a hyperbolic S_g-tile.

Any path lifting of any δ_i to the universal cover $\widetilde{X} \approx \mathbb{H}^2$ is a geodesic segment, and the union of all path lifts of all the δ_i gives a tiling of $\mathbb{H}^2$ by hyperbolic S_g-tiles. See Figures 10.2 and 10.3 for sample tilings of $\mathbb{H}^2$ by hyperbolic S_2-tiles.

In passing from $[(X, \phi)]$ to a hyperbolic S_g-tile, we had to choose a specific marking in the homotopy class $\phi : S_g \to X$. If we modify ϕ by homotopy, then the based geodesics δ_i, and hence the hyperbolic S_g-tile, will change. The resulting geodesics δ_i are not sensitive to the specific homotopy; they only depend on the path that $\phi(x)$ traces out during the homotopy. What is more, the δ_i only depend on the relative homotopy class of this path. In other words, the choice we made in going from $[(X, \phi)]$ to a hyperbolic S_g-tile amounts to a choice of point in $\widetilde{X}$.

To summarize the previous paragraph, there is a map from $\text{Teich}(S_g)$ to the set of hyperbolic S_g-tiles. This map is not well defined, but the ambiguity is exactly accounted for by $\widetilde{X}$, which is 2-dimensional.

Conversely, given a hyperbolic S_g-tile, the space X obtained by identifying the sides in pairs is isometric to a hyperbolic surface that is homeomorphic to S_g. Moreover, the labeling induces an identification of $\pi_1(S_g)$ with $\pi_1(X)$. As in Section 10.3, there is then a diffeomorphism $\phi : S_g \to X$ realizing this isomorphism, which is a marking.

We thus have a bijective correspondence between points of $\text{Teich}(S_g)$ and the set of equivalence classes of hyperbolic S_g-tiles, where two hyperbolic S_g-tiles are equivalent if they differ by marked, orientation-preserving isometry and by "pushing the basepoint." We will count the dimension of the set of these equivalence classes.

In our dimension count, we will use the fact that, given any geodesic polygon in $\mathbb{H}^2$, we can scale the polygon so that its interior angles sum to 2π. This is true because scaling a polygon in $\mathbb{H}^2$ continuously varies the interior angle sum from nearly the Euclidean angle sum (small polygons) to nearly zero (big polygons). When $g \geq 2$, the Euclidean angle sum of a geodesic $4g$-gon is greater than 2π, and so it follows that we can scale any geodesic $4g$-gon so that the angle sum is exactly 2π.

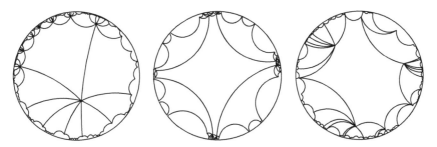

Figure 10.3 Some possible tilings of $\mathbb{H}^2$ coming from hyperbolic structures on S_2. These pictures are sketches of images from the (existing but inactive) web page for the "Teichmüller Navigator" on the Geometry Center's web site.

Dimension count 2: the space of tilings

$+8g$: Choose a set of $4g$ vertices in $\mathbb{H}^2$.

$-2g$: Side lengths must match in pairs.

-1 : Scale so the sum of interior angles is 2π.

-3 : Isometric tilings are equivalent.

-2 : Pushing the base point gives different tilings representing the same point of $\mathrm{Teich}(S_g)$.

$= 6g - 6$ Total dimensions

10.5 THE TEICHMÜLLER SPACE OF A PAIR OF PANTS

Let P denote a pair of pants, that is, a compact surface of genus 0 with three boundary components. Recall from Chapter 8 that a pants decomposition of a surface S is a maximal collection of pairwise disjoint, pairwise non-isotopic, essential simple closed curves in S. When S is given a hyperbolic metric, the curves in a pants decomposition can be represented by geodesics in S.

Decomposing a hyperbolic surface S with totally geodesic boundary along a collection of disjoint geodesics gives another hyperbolic surface S' with totally geodesic boundary. The surface S' has smaller complexity than S in the sense that the number of curves in a pants decomposition for S' is strictly less than the number for S. This cutting procedure thus gives us an inductive method for understanding the hyperbolic structure of a surface. Since the only geodesic simple closed curves on a hyperbolic pair of pants are the three boundary components, the pair of pants serves as our base case for the induction.

So our first goal is to determine $\mathrm{Teich}(P)$ for a pair of pants P. To do this we will reduce the problem to understanding a certain space of right-angled hexagons, as we now explain.

Hyperbolic hexagons. By a *marked hexagon* we will mean a hexagon with one vertex distinguished. Let $\mathcal{H}$ denote the set of equivalence classes of marked right-angled geodesic hexagons in $\mathbb{H}^2$, where two such hexagons are equivalent if there is an orientation-preserving isometry of $\mathbb{H}^2$ taking one hexagon to the other and taking the marked point of the first to the marked point of the second.

Any space of metrics on a surface is a priori infinite-dimensional. It seems difficult to find a precise set of constraints on a metric so that the space of such metrics is finite-dimensional but still not empty. Thus the finite dimensionality of $\text{Teich}(S)$ of any compact surface is quite remarkable. At some point one has to make the jump from an infinite-dimensional space of possibilities to a finite-dimensional one. The following key proposition is precisely where this jump occurs.

Proposition 10.4 *The map $W : \mathcal{H} \to \mathbb{R}^3_+$ defined by taking the lengths of every other side of a hexagon, starting at the marked point and traveling counterclockwise, is a bijection.*

Proof of Proposition 10.4. We will define a two-sided inverse $\mathbb{R}^3_+ \to \mathcal{H}$ to W. That is, given an arbitrary triple $(L_\alpha, L_\beta, L_\gamma) \in \mathbb{R}^3_+$, we will construct a marked right-angled hexagon H that is unique up to marked orientation-preserving isometry and that satisfies $W(H) = (L_\alpha, L_\beta, L_\gamma)$. Throughout, the reader should refer to Figure 10.4.

There is a basic fact from hyperbolic geometry that we will use: given two disjoint geodesics in $\mathbb{H}^2$ with four distinct endpoints at infinity, there is a unique geodesic perpendicular to both.

For any $t > 0$, let α_t and β_t be a pair of geodesics in $\mathbb{H}^2$ a distance t apart and let γ'_t be the unique geodesic segment realizing this distance. Let α'_t and β'_t be geodesics on the same side of γ'_t such that α'_t has a perpendicular intersection with β_t at a distance L_β away from γ'_t and β'_t has a perpendicular intersection with α_t at a distance L_α away from γ'_t. We further require that if γ'_t is oriented from α_t to β_t, then α'_t and β'_t lie to the left of the γ'_t.

There is a value $t_0 > 0$ so that α'_{t_0} and β'_{t_0} share an endpoint on $\partial \mathbb{H}^2$. For $t > t_0$, let γ_t be the unique geodesic segment perpendicular to α'_t and β'_t.

As t varies from t_0 to infinity, the length of γ_t varies continuously and monotonically from zero to infinity, so there is a unique t so that the length of γ_t is exactly L_γ.

Trimming geodesics to segments as necessary and marking the intersection of α_t and β'_t, we obtain a right-angled hexagon that represents the desired point of $\mathcal{H}$.

In the construction just described, we made no choices after the initial choice of α_t and β_t. What is more, up to orientation-preserving isometries of $\mathbb{H}^2$, there is a unique ordered pair of geodesics whose distance from each other is a given positive length. In other words, even the choices of α_t and β_t were unique up to isometries of $\mathbb{H}^2$. It follows that the point of $\mathcal{H}$ we constructed is uniquely defined, and we have indeed given a two-sided inverse of W.   □

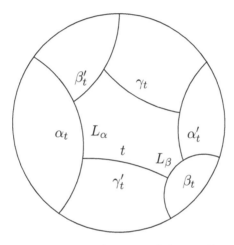

Figure 10.4 The picture for the proof of Proposition 10.4.

Pairs of pants. Having determined $\mathcal{H}$ we can now show that $\mathrm{Teich}(P) \approx \mathbb{R}^3$.

Proposition 10.5 *Let P be a pair of pants with boundary components α_1, α_2, and α_3. The map $\mathrm{Teich}(P) \to \mathbb{R}_+^3$ defined by*

$$\mathcal{X} \mapsto (\ell_{\mathcal{X}}(\alpha_1), \ell_{\mathcal{X}}(\alpha_2), \ell_{\mathcal{X}}(\alpha_3))$$

is a homeomorphism.

Proof. We first establish a bijection between $\mathrm{Teich}(P)$ and $\mathcal{H}$, the set of oriented isometry classes of marked right-angled hyperbolic hexagons.

Let $\mathcal{X} = [(X, \phi)] \in \mathrm{Teich}(P)$, where X is a hyperbolic surface with totally geodesic boundary and $\phi : P \to X$ is a homeomorphism. For each pair of distinct boundary components of X, there is a unique isotopy class of arcs connecting them; let $\delta_{ij} = \delta_{ji}$ denote the geodesic representative of the arc connecting $\phi(\alpha_i)$ and $\phi(\alpha_j)$. By the first variation principle, each of the δ_{ij} is perpendicular to ∂X at both of its endpoints. The closures of the two components of $X - \cup \delta_{ij}$ are hyperbolic right-angled hexagons H_1 and H_2.

An application of Proposition 10.4 gives that H_1 and H_2 are abstractly isometric since the lengths of the δ_{ij} determine the hyperbolic structure on each. Let H be a marked right-angled hexagon in $\mathbb{H}^2$ that is the isometric image of the marked hexagon H_1, where the image of $\delta_{13} \cap \phi(\alpha_1)$ is the marked point and where the images of the $\phi(\alpha_1)$-, $\phi(\alpha_2)$-, and $\phi(\alpha_3)$-edges appear in counterclockwise order. The equivalence class of this hexagon is an element of $\mathcal{H}$.

On the other hand, given an element of $\mathcal{H}$, we realize it as a marked hexagon H in $\mathbb{H}^2$, create a second hexagon H' by reflecting H over the edge lying first in the clockwise direction from the marked point, label the sides as in Figure 10.5, and obtain a hyperbolic pair of pants X by identifying the pairs of sides labeled δ_{12} and δ_{23}. Then, as the marking we take the unique isotopy class of diffeomorphisms $P \to X$ (remember: isotopies are free on the boundary) respecting the labels of the boundary components.

We have thus established a bijection between $\mathrm{Teich}(P)$ and $\mathcal{H}$. Composing with the map W from Proposition 10.4, we obtain a bijection between $\mathrm{Teich}(P)$ and $\mathbb{R}^3_+$. This bijection is a homeomorphism because if two points of $\mathbb{R}^3_+$ are close, then the corresponding right-angled hexagons are nearly isometric and so the corresponding representations $\pi_1(P) \approx F_2 \to \mathrm{PSL}(2, \mathbb{R})$ (defined by identifying two side pairs of a doubled hexagon in $\mathbb{H}^2$) are close in the algebraic topology on $\mathrm{Teich}(P)$. □

Consider the thrice-punctured sphere $S_{0,3}$, which is homeomorphic to the interior of P. An argument similar to that given above shows that $\mathrm{Teich}(S_{0,3})$ is a single point. The reason for this is that we can identify the point(s) of $\mathrm{Teich}(S_{0,3})$ with the space of ideal triangles in $\mathbb{H}^2$ (we can think of an ideal triangle as a hexagon with three degenerate sides); but there is a unique ideal triangle in $\mathbb{H}^2$ up to isometry since $\mathrm{PGL}(2, \mathbb{R})$ acts triply transitively on $\partial \mathbb{H}^2$.

10.6 FENCHEL–NIELSEN COORDINATES

As every closed surface of negative Euler characteristic can be built from pairs of pants, we can extend Proposition 10.5 to coordinatize $\mathrm{Teich}(S_g)$ for $g \geq 2$. Using this idea, we will prove the following theorem of Fricke [67].

For $g \geq 2$, $\mathrm{Teich}(S_g)$ is homeomorphic to $\mathbb{R}^{6g-6}$.

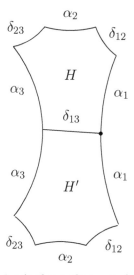

Figure 10.5 A pair of pants from a marked hexagon.

(We will give an explicit homeomorphism below; see Theorem 10.6.)

The basic idea is to decompose S_g into pairs of pants using $3g - 3$ simple closed curves. Then there are $3g - 3$ length parameters that determine the hyperbolic structure on each pair of pants, and there are $3g - 3$ twist parameters that determine how the pairs of pants are glued together. Taken together, these $6g - 6$ coordinates are the Fenchel–Nielsen coordinates [64, Section 26.9] for $\mathrm{Teich}(S_g)$. We now explain this more precisely.

10.6.1 LENGTH PARAMETERS AND TWIST PARAMETERS

In order to define the Fenchel–Nielsen coordinates we must first choose a *coordinate system of curves* on S_g. This consists of the following data:

- a pants decomposition $\{\gamma_1, \ldots \gamma_{3g-3}\}$ of oriented simple closed curves and

- a set $\{\beta_1, \ldots, \beta_n\}$ of *seams*; that is, a collection of disjoint simple closed curves in S_g so that the intersection of the union $\cup \beta_i$ with any pair of pants P determined by the $\{\gamma_j\}$ is a union of three disjoint arcs connecting the boundary components of P pairwise.

Given a pants decomposition, we can construct seams by first choosing three disjoint arcs on each pair of pants and then matching up endpoints in any fashion. See Figure 10.9 below for an example in the case $g = 2$.

Fix once and for all a coordinate system of curves on S_g consisting of an oriented pants decomposition $\{\gamma_i\}$ with seams $\{\beta_i\}$.

We define the $3g - 3$ *length parameters* of a point $X \in \mathrm{Teich}(S_g)$ to be the ordered $(3g - 3)$-tuple of positive real numbers

$$(\ell_1(X), \ldots, \ell_{3g-3}(X)),$$

where $\ell_i(X) = \ell_X(\gamma_i)$.

According to Proposition 10.5, the length parameters for a point of $\mathrm{Teich}(S_g)$ determine the isometry types of the $2g - 2$ pairs of pants cut out by the coordinate system of curves for S_g. In order to record how these pants are glued together we introduce the twist parameters $\theta_i(X)$.

Before we begin in earnest with twist parameters, let us make an observation. Suppose we have two hyperbolic pairs of pants with totally geodesic boundary, as on the left-hand side of Figure 10.6. If these pairs of pants have boundary components of the same length, then we can glue them together to obtain a compact hyperbolic surface X of genus 0 with four boundary components. It is intuitively clear that the isometry type of X depends on how much we rotate the pairs of pants before gluing. For instance, as Figure 10.6 indicates, the shortest arc connecting two boundary components of X changes as we change the gluing instructions. Thus we have a circle's worth of choices for the isometry type of X. Of course, we care about more than just the isometry type—we also care about markings. So the twist parameters we define on Teichmüller space will be real numbers, but modulo 2π, they are simply recording the angles at which we glue pairs of pants.

As a first step toward defining the twist parameters, suppose that β is an arc in a hyperbolic pair of pants P connecting boundary components γ_1 and γ_2 of P. We define the twisting number of β at γ_1 as follows. Let δ be the unique shortest arc connecting γ_1 and γ_2. Let N_1 and N_2 be regular metric neighborhoods of γ_1 and γ_2. We can modify β by isotopy (leaving the endpoints fixed) so that it agrees with δ outside of $N_1 \cup N_2$; see Figure 10.7. The *twisting number* of β at γ_1 is the signed horizontal displacement of the endpoints $\beta \cap \partial N_1$. The sign is determined by the orientation of γ_1. The twisting number of β at γ_2 is defined in the same way.

Given $X = [(X, \phi)] \in \mathrm{Teich}(S_g)$, we define the ith twist parameter $\theta_i(X)$ as follows: let β_j be one of the two seams that cross γ_i. On each side of the $\phi(\gamma_i)$ geodesic there is a pair of pants, and the $\phi(\beta_j)$ geodesic gives an arc in each of these. Let t_L and t_R be the twisting numbers of each of these arcs on the left and right sides of $\phi(\gamma_i)$, respectively. The ith *twist parameter* of X is defined to be

$$\theta_i(X) = 2\pi \frac{t_L - t_R}{\ell_X(\gamma_i)}.$$

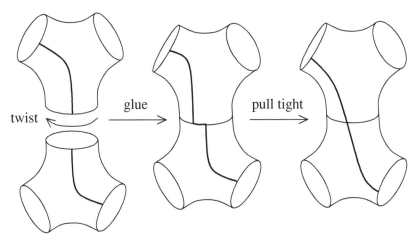

Figure 10.6 The effect of the twist parameter on geodesic arcs. If the twist parameter were instead taken to be zero, the geodesic arc at the end would be the union of the two geodesic arcs from the original pairs of pants.

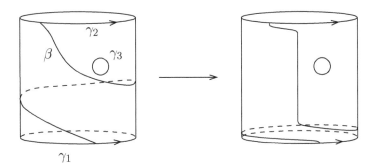

Figure 10.7 Modifying an arc on a pair of pants so that it agrees with a perpendicular arc except near its endpoints.

Since there were two choices of seams β_j, we need to check that the twist parameter is well defined. To see this, we pass to the universal cover of the neighborhood N_i of $\phi(\gamma_i)$. As in the proof of Proposition 10.5, the four geodesic arcs connecting $\phi(\gamma_i)$ to the boundary components of the adjacent pairs of pants are perpendicular to $\phi(\gamma_i)$. Also, on each side of $\phi(\gamma_i)$, the two geodesics lie on diametrically opposed points along $\phi(\gamma_i)$. If we modify the seams as in the definition of the twist parameter and pass to the universal cover of N_i, we obtain Figure 10.8. Here the geodesic arcs are dashed, and the modified seams are solid. Each lift of a seam connects two dashed arcs, and the twist parameter is the signed distance between these dashed arcs.

Combining the fact that the two perpendicular arcs lie diametrically opposite each other on $\phi(\gamma_i)$ with the fact that the seams do not cross each other, we see that the twist parameters computed from the two seams are the same.

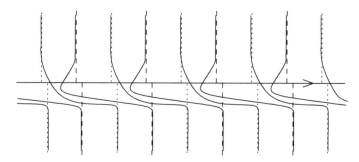

Figure 10.8 The universal cover of the annular neighborhood of a $\phi(\gamma_i)$.

10.6.2 FENCHEL–NIELSEN COORDINATES

Now that we have defined the length and twist parameters, we can give the precise statement of Fricke's theorem.

Theorem 10.6 *Let $g \geq 2$ and fix any coordinate system of curves on S_g. The map*

$$FN : \mathrm{Teich}(S_g) \to \mathbb{R}_+^{3g-3} \times \mathbb{R}^{3g-3}$$

defined by setting

$$FN(\mathcal{X}) = (\ell_1(\mathcal{X}), \ldots, \ell_{3g-3}(\mathcal{X}), \theta_1(\mathcal{X}), \ldots, \theta_{3g-3}(\mathcal{X}))$$

is a homeomorphism. In particular, $\mathrm{Teich}(S_g) \approx \mathbb{R}^{6g-6}$.

The ordered set of numbers $(\ell_1(\mathcal{X}), \ldots, \ell_{3g-3}(\mathcal{X}), \theta_1(\mathcal{X}), \ldots, \theta_{3g-3}(\mathcal{X}))$ are called the *Fenchel–Nielsen coordinates* of the point $\mathcal{X} \in \mathrm{Teich}(S_g)$.

Proof. Denote the pants decomposition of the fixed coordinate system of curves for S_g by $\{\gamma_i\}$ and the seams by $\{\beta_i\}$.

Let $(\ell_1, \ldots, \ell_{3g-3}, \theta_1, \ldots, \theta_{3g-3}) \in \mathbb{R}_+^{3g-3} \times \mathbb{R}^{3g-3}$. In order to prove that FN is a bijection, we will find a unique $\mathcal{X} \in \mathrm{Teich}(S_g)$ with these Fenchel–Nielsen coordinates with respect to the given coordinate system of curves. We construct $\mathcal{X}$ in four steps.

Step 1. Let $P_{i,j,k}$ denote the pair of pants[1] determined by γ_i, γ_j, and γ_k. Note that γ_i, γ_j, and γ_k might not be distinct. By Proposition 10.5, we can construct a hyperbolic pair of pants $X_{i,j,k}$ whose boundary components have lengths ℓ_i, ℓ_j, and ℓ_k, and there is only one way to do this up to isometry. By construction, there is a homeomorphism $P_{i,j,k} \to X_{i,j,k}$ taking each γ_i to a boundary curve of length ℓ_i. Via this homeomorphism, the boundary curves of $X_{i,j,k}$ inherit orientations from the γ_i.

Step 2. For each $X_{i,j,k}$ and each pair of its boundary components, we draw the unique geodesic arc that is perpendicular to those boundary components. For each $m \in \{i, j, k\}$, we adjust this *seam* as follows: in a small neighborhood of a boundary component corresponding to the left side of γ_m, we replace each geodesic arc with an arc that travels along that boundary component an oriented distance of $(\theta_m/2\pi)\ell_m$. The result is unique up to isotopy relative to $\partial X_{i,j,k}$.

Given a seam in $P_{i,j,k}$, that is, an intersection of some β_k with $P_{i,j,k}$, there is a unique corresponding seam in $X_{i,j,k}$, namely, the arc that connects the corresponding boundary components.

Step 3. Since the boundary curves and the seams of the $X_{i,j,k}$ are identified with the boundary curves and seams of the $P_{i,j,k}$, there is a unique way to construct a quotient

$$X = \coprod X_{i,j,k}/\sim$$

of the disjoint union of the $X_{i,j,k}$. Specifically, we identify corresponding boundary components of the $X_{i,j,k}$, and we do this in such a way that the corresponding seams match up.

Step 4. We construct a diffeomorphism from $\phi : S_g \to X$ that respects the identifications of the coordinate system of curves. The marked surface (X, ϕ) represents the desired point of $\mathrm{Teich}(S_g)$.

By construction, $[(X, \phi)]$ is a point in $\mathrm{Teich}(S_g)$ with the desired Fenchel–Nielsen coordinates. We have thus defined a map $FN' : \mathbb{R}_+^{3g-3} \times \mathbb{R}^{3g-3} \to \mathrm{Teich}(S_g)$. It is clear that FN' is an inverse of FN. That both maps are continuous is straightforward to check from the definitions. Thus FN is a homeomorphism, and we are done. □

---

[1] This is a slight abuse of notation because when $g = 2$ it is possible to have two pairs of pants determined by the same triple $\{i, j, k\}$.

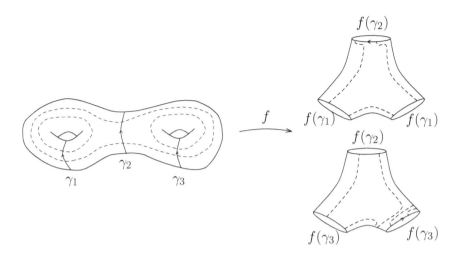

Figure 10.9 Constructing the map $\phi : S_g \to \coprod X_{i,j,k}/\sim$. Here the second twist parameter is a small positive number, and the third twist parameter is approximately -2π (the first is zero).

10.6.3 FENCHEL–NIELSEN COORDINATES FOR NONCLOSED SURFACES

Let S_g^b be a compact surface of genus g with b boundary components. Assume that $\chi(S_g^b) < 0$. As in Section 8.3, a pants decomposition for S_g^b has $3g - 3 + b$ curves (boundary curves are not included). Fenchel–Nielsen coordinates for S_g^b are given by a total of $6g - 6 + 3b$ coordinates. There are $3g - 3 + 2b$ length parameters, one for each curve of the pants decomposition and one for each boundary curve. There are $3g - 3 + b$ twist parameters, one for each curve of the pants decomposition. We thus obtain

$$\mathrm{Teich}(S_g^b) \approx \mathbb{R}^{6g-6+3b}.$$

By setting some or all of the length parameters to be zero, we can turn boundary components into punctures. So if $S_{g,n}$ is a surface of genus g with n punctures and $\chi(S_{g,n}) < 0$, then

$$\mathrm{Teich}(S_{g,n}) \approx \mathbb{R}^{6g-6+2n}.$$

Together with our determination of $\mathrm{Teich}(T^2)$, this in particular gives

$$\mathrm{Teich}(T^2) \approx \mathrm{Teich}(S_{1,1}) \approx \mathrm{Teich}(S_{0,4}) \approx \mathbb{R}^2.$$

What is more, each of these isomorphisms is natural. For example, the forgetful map $S_{1,1} \to T^2$ and the quotient map $T^2 \to S_{0,4}$ identify Fenchel–

Nielsen coordinate systems on the three surfaces.

10.6.4 FENCHEL–NIELSEN COORDINATES FOR THE TORUS

We can define Fenchel–Nielsen coordinates for T^2 using a method similar to that used for hyperbolic surfaces. Pick a cylinder decomposition (instead of a pants decomposition) of T^2, that is, an oriented simple closed curve γ. Also choose a seam, which in this case is a simple closed curve β in T^2 with $i(\beta, \gamma) = 1$.

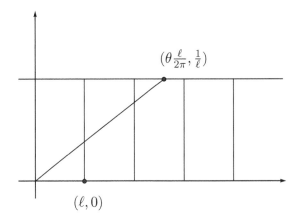

$$\left(\theta \frac{\ell}{2\pi}, \frac{1}{\ell}\right)$$

$$(\ell, 0)$$

Figure 10.10 The effect of length and twist parameters on the universal cover of the corresponding point in Teich(T^2).

The Fenchel–Nielsen coordinates for a point $\mathcal{X} = [(X, \phi)] \in \text{Teich}(T^2)$ is a pair (ℓ, θ) defined as follows. The length coordinate ℓ is the length in X of any geodesic in the homotopy class of $\phi(\gamma)$. When we cut X along any such geodesic, we obtain a flat cylinder X'. The curve $\phi(\beta)$ becomes an arc on X'. The universal cover of X' is isometric to $\mathbb{R} \times 1/\ell$. Any lift of the arc $\phi(\beta)$ to this cover is an arc, and the twist parameter θ is given by the horizontal displacement of its two endpoints. Specifically, if the displacement is d, then $\theta = (d/\ell)2\pi$.

If we identify $\text{Teich}(T^2)$ with the upper half-plane $\mathbb{H}^2$ via the bijection given in Proposition 10.1, we can write the Fenchel–Nielsen coordinates as a map $FN : \mathbb{H}^2 \to \mathbb{R}_+ \times \mathbb{R}$. Specifically, we have

$$FN(x, y) = (1/\sqrt{y}, 2\pi x).$$

It is instructive to define the inverse map $\mathbb{R}_+ \times \mathbb{R} \to \mathbb{H}^2 \approx \text{Teich}(T^2)$. Let $(\ell, \theta) \in \mathbb{R}_+ \times \mathbb{R}$. We start by constructing the unique flat, unit-area, right cylinder X' with boundary length ℓ. Call its boundary components δ_1 and δ_2. We draw a vertical arc β' on X' and then modify β' by dragging its endpoint on δ_1 an oriented distance $(\theta/2\pi)\ell$ along δ_1 (see Figure 10.10). We then obtain a torus X by identifying δ_1 and δ_2 by the unique orientation-preserving isometry that identifies the endpoints of β'. There is a homeomorphism $\phi : T^2 \to X$, unique up to isotopy, that sends γ to the image of $\delta_1 \cup \delta_2$ in X and β to the image of β' in X. Then (X, ϕ) represents the desired point in $\mathrm{Teich}(T^2)$. This point of $\mathrm{Teich}(T^2)$ corresponds to the point $(\theta/2\pi, 1/\ell^2)$ in the upper half-plane.

10.6.5 WOLPERT'S MAGIC FORMULA

The Fenchel–Nielsen coordinates $(\ell_1, \ldots, \ell_{3g-3}, \theta_1, \ldots, \theta_{3g-3})$ obviously depend in an essential way on the choice of coordinate system of curves. It follows that the same can be said for the associated 1-forms $d\ell_i$ and $d\theta_i$ on $\mathrm{Teich}(S_g)$. There are infinitely many coordinate systems to choose from, each giving a different set of coordinates (and thus different associated 1-forms) on $\mathrm{Teich}(S_g)$. Wolpert discovered the remarkable fact that the 2-form

$$\omega = \frac{1}{2} \sum_{i=1}^{3g-3} d\ell_i \wedge d\theta_i$$

on $\mathrm{Teich}(S_g)$ actually *does not* depend on the initial choice of pants decomposition inducing the coordinates $\{(\ell_i, \theta_i)\}$. Wolpert does this by proving that ω is equal to the *Weil–Petersson form* on Teichmüller space; see [216, Theorem 3.14]. Since the Weil–Petersson form is defined without any reference to a choice of pants decomposition, it follows that ω does not depend on the pants decomposition.[2]

10.7 THE $9g - 9$ THEOREM

At the beginning of the chapter we described a map

$$\ell : \mathrm{Teich}(S) \to \mathbb{R}^{\mathcal{S}},$$

where $\mathcal{S}$ is the set of isotopy classes of essential curves in the surface S. The map is given by $X \mapsto \ell_X$. It would be interesting to say that ℓ is injective,

---

[2]Actually, the twist parameters θ_i in Wolpert's formula are defined differently than our twist parameters. In our coordinates, a full twist on the ith curve corresponds to replacing θ_i with $\theta_i + 2\pi$. Under Wolpert's conventions, a full twist is $\theta_i \mapsto \theta_i + \ell_i$.

in other words, that a point $X = [(X, \phi)]$ of $\mathrm{Teich}(S)$ is completely determined by the geodesic lengths in X of the simple closed curves in S. In this section we will show something much stronger: there are finitely many simple closed curves in S whose lengths in a marked hyperbolic surface determine the corresponding point of $\mathrm{Teich}(S)$. For $S = S_g$ the next theorem states that $9g - 9$ curves suffice.

Theorem 10.7 ($9g - 9$ theorem) *There is a collection of simple closed curves $\{\delta_1, \ldots, \delta_{9g-9}\}$ in S_g so that the map from $\mathrm{Teich}(S_g)$ to $\mathbb{R}^{9g-9}$ given by*

$$ X \mapsto (\ell_X(\delta_1), \ldots, \ell_X(\delta_{9g-9})) $$

is injective.

Our proof of Theorem 10.7 generalizes to the case of any hyperbolic $S_{g,n}$, where $9g - 9$ is replaced by $3(3g - 3 + n)$.

It has been shown that in fact there are $6g - 5$ simple closed curves in S_g whose lengths determine a point in $\mathrm{Teich}(S_g)$. On the other hand, it has also been shown that no $6g - 6$ curves suffice; see [75].

The *marked length spectrum* of a hyperbolic surface X is the function $S \to \mathbb{R}$ that records the lengths of the isotopy classes of simple closed curves in X. It follows immediately from Theorem 10.7 that the marked length spectrum of X—indeed only a finite part of it—determines X up to isometry.

To begin, we give the technical statement at the heart of the proof of the $9g - 9$ theorem on the convexity of length functions. Then we state and prove the $9g - 9$ theorem, and then we prove the statement about convexity of length functions.

10.7.1 CONVEXITY OF LENGTH FUNCTIONS

How does the hyperbolic geometry of a genus $g \geq 2$ surface X change as one varies $X = [(X, \phi)]$ over $\mathrm{Teich}(S_g)$? One specific problem in this direction is to understand, for a given simple closed curve γ in S_g, the function $\mathrm{Teich}(S_g) \to \mathbb{R}_+$ defined by $X \mapsto \ell_X(\gamma)$.

Fix on S_g a pants decomposition $\{\gamma_i\}$ consisting of oriented simple closed curves. We take this pants decomposition as part of a coordinate system of curves that gives Fenchel–Nielsen coordinates on $\mathrm{Teich}(S_g)$.

Fix any point $X \in \mathrm{Teich}(S_g)$ and consider the one-parameter family $\{X_s : s \in \mathbb{R}\}$ of points in $\mathrm{Teich}(S_g)$ obtained from X by varying the twist parameter s associated to the curve $\gamma = \gamma_1$.

Proposition 10.8 *Let b be any isotopy class of simple closed curves on S_g such that $i(b, \gamma) > 0$. The function $\mathbb{R} \to \mathbb{R}_+$ given by*

$$s \mapsto \ell_{\mathcal{X}_s}(b)$$

is strictly convex.

10.7.2 PROOF OF THE $9g - 9$ THEOREM

Let $\{\gamma_1, \ldots, \gamma_{3g-3}\}$ be a pants decomposition of S_g and choose simple closed curves $\{\beta_1, \ldots, \beta_{3g-3}\}$ in S_g so that $i(\beta_i, \gamma_i) > 0$ and $i(\beta_i, \gamma_j) = 0$ for $i \neq j$. We do not require the β_i to be disjoint. Let $\alpha_i = T_{\gamma_i}(\beta_i)$.

Choose a set of Fenchel–Nielsen coordinates for $\mathrm{Teich}(S_g)$ where the coordinate system of curves consists of the pants decomposition $\{\gamma_i\}$ and any set of seams. For $\mathcal{X} \in \mathrm{Teich}(S_g)$, we will show that the set $\{\ell_{\mathcal{X}}(\alpha_i), \ell_{\mathcal{X}}(\beta_i), \ell_{\mathcal{X}}(\gamma_i)\}$ determines the Fenchel–Nielsen coordinates of $\mathcal{X}$.

The length parameters for $\mathcal{X}$ are exactly the $\ell_{\mathcal{X}}(\gamma_i)$. It therefore remains to show that the twist parameters for $\mathcal{X}$ are uniquely determined by the $\{\ell_{\mathcal{X}}(\alpha_i), \ell_{\mathcal{X}}(\beta_i), \ell_{\mathcal{X}}(\gamma_i)\}$. Let $\mathcal{X}_t$ be the point of $\mathrm{Teich}(S)$ with the same length parameters as $\mathcal{X}$ and with twist parameters $t = (t_1, \ldots, t_{3g-3})$. Up to a reparametrization of $\mathrm{Teich}(S_g)$, we can assume that $\mathcal{X} = \mathcal{X}_0$. We will show that if $t_i \neq 0$ for some i, then either $\ell_{\mathcal{X}_t}(\alpha_i) \neq \ell_{\mathcal{X}}(\alpha_i)$ or $\ell_{\mathcal{X}_t}(\beta_i) \neq \ell_{\mathcal{X}}(\beta_i)$.

Consider the functions $A(t) = \ell_{\mathcal{X}_t}(\alpha_1)$ and $B(t) = \ell_{\mathcal{X}_t}(\beta_1)$. Since $i(\alpha_1, \gamma_j) = i(\beta_1, \gamma_j) = 0$ for $j \neq 1$, both functions are simply functions of the parameter t_1, which we denote by s. By Proposition 10.8, $A(s)$ and $B(s)$ are strictly convex, hence so is their sum $(A + B)(s)$. Also, by definition, we have that $A(s + 2\pi) = B(s)$.

Assume $A(s) = A(0)$ for some $s \neq 0$. We will show that $B(s) \neq B(0)$, that is, $A(2\pi) \neq A(2\pi + s)$. For concreteness say $s > 0$. Since $A(s) = A(0)$, it follows from the strict convexity that $A(t) < A(0)$ for $t \in (0, s)$ and that $A(t)$ is strictly increasing for $t > s$. If $s < 2\pi$, then $s < 2\pi < 2\pi + s$, and it follows that $A(2\pi) < A(2\pi + s)$. If $s > 2\pi$, then $0 < 2\pi < s < 2\pi + s$, so $A(2\pi) < A(0) = A(s) < A(2\pi + s)$. Finally, if $s = 2\pi$, we certainly cannot have $A(2\pi) = A(2\pi + s)$, for then $A(t)$ would take the same value at 0, 2π, and 4π, violating strict convexity.

We have shown that if $t = (t_1, \ldots, t_{3g-3})$ and $\ell_{\mathcal{X}_t}(\alpha_1) = \ell_{\mathcal{X}}(\alpha_1)$ and $\ell_{\mathcal{X}_t}(\beta_1) = \ell_{\mathcal{X}}(\beta_1)$, then $t_1 = 0$. Since the same argument works for the other twist parameters, the theorem is proven.

10.7.3 Proof of the Convexity of Length Functions

As preparation for our proof of Proposition 10.8, we will give a way of comparing the lengths of curves in X_s versus their lengths in $X = X_0$. Recall that in our discussion of the Fenchel–Nielsen coordinates, we regarded X as the equivalence class of a marked hyperbolic surface (X, ϕ). We constructed X from a collection of hyperbolic pairs of pants $X_{i,j,k}$ whose isometry types were determined by the length parameters for X. Then we identified the $X_{i,j,k}$ along their boundary components, and the amount of rotating we did before gluing was determined by the twist parameters for X. The marking ϕ was then constructed using the seams as a guide.

Twist deformations and earthquake maps. Given the above description of X, we can construct X_s as follows. We modify the gluing of the $X_{i,j,k}$ along γ by rotating to the left by an angle $s/2\pi$. The new identification gives a new hyperbolic surface X_s. Note that X_s is isometric to $X_{s+2\pi}$.

There is then a natural way to modify the marking ϕ to obtain a marking $\phi_s : S_g \to X_s$, as we now explain. Abusing notation, let γ denote the simple closed curves in X and X_s marked by the curve γ in S_g (in X this is exactly $\phi(\gamma)$, but in X_s this curve does not yet have a name). There is a canonical isometry $\tau_0 : X - \gamma$ and $X_s - \gamma$ since both X and X_s are obtained by gluing together the same set of $X_{i,j,k}$. If we modify τ_0 by an $s/2\pi$ left-hand twist on the left side of γ, we obtain a map from $X - \gamma$ to $X_s - \gamma$ that uniquely extends to a homeomorphism $\tau_s : X \to X_s$. The marking for X_s is then $\phi_s = \tau_s \circ \phi$.

Let $\pi : \mathbb{H}^2 \to X$ be the universal covering. Just as X_s is described by cutting X along γ and regluing with a twist, the universal covering $\pi_s : \mathbb{H}_s^2 \to X_s$ can be constructed by decomposing $\mathbb{H}^2$ along the lifts of γ, sliding the pieces to the left by $(s/2\pi)\ell_X(\gamma)$, and regluing. More precisely, let $\mathbb{H}_s^2$ be the metric space obtained from $\mathbb{H}^2$ by the following inductive procedure. Choose some lift $\widetilde{\gamma}_1$ of γ in $\widetilde{X} = \mathbb{H}^2$. Decompose $\mathbb{H}^2$ into the union of the open half-space to the left of $\widetilde{\gamma}_1$ and the closed half-space to the right of $\widetilde{\gamma}_1$. We reglue the pieces after translating a distance $(s/2\pi)\ell_X(\gamma)$ to the left. Next choose some lift $\widetilde{\gamma}_2$ that is adjacent to $\widetilde{\gamma}_1$ (here adjacent means there are no other lifts between the two). We decompose the new space along the $\widetilde{\gamma}_2$ as above and reglue by the same recipe. We can perform this procedure inductively along all lifts of γ. At the end we have a new metric space $\mathbb{H}_s^2$. Each point of $\mathbb{H}_s^2$ has a neighborhood that is isometric to an open disk in $\mathbb{H}^2$. It follows that $\mathbb{H}_s^2$ is globally isometric to $\mathbb{H}^2$.

There is a built-in discontinuous map

$$E_s : \mathbb{H}^2 \to \mathbb{H}_s^2 \approx \mathbb{H}^2,$$

which is called an *earthquake map*.

Away from the preimage of γ, the covering map $\pi_s : \mathbb{H}_s^2 \to X_s$ is given by

$$\pi_s = \tau_0 \circ \pi \circ \mathrm{E}_s^{-1}$$

(it is easy to check that this composition is a local homeomorphism).

Computing lengths. We would like to compute the length of β in X_s by looking in $\mathbb{H}^2$ as opposed to $\mathbb{H}_s^2$. In other words, we want to use $\mathbb{H}^2$ as a frame of reference, independent of s. The image of β in X_s under the marking for X_s is $\phi_s(\beta) = \tau_s \circ \phi(\beta)$. The preimage of this curve in $\mathbb{H}_s^2$ under the covering map $\pi_s = \tau_0 \circ \pi \circ \mathrm{E}_s^{-1}$ is then

$$\mathrm{E}_s \circ \pi^{-1} \circ \tau_0^{-1} \circ \tau_s \circ \phi(\beta).$$

Let us unwrap this composition. The curve $\phi(\beta)$ is the image of β in X under the marking for X. The map $\tau_0^{-1} \circ \tau_s : X \to X$ is a discontinuous map that twists the left-hand side of γ by $s/2\pi$. Thus $\pi^{-1} \circ (\tau_0^{-1} \circ \tau_s) \circ \phi(\beta)$ differs from the preimage $\pi^{-1} \circ \phi(\beta)$ in $\mathbb{H}^2$ by a lift of the partial twist $\tau_0^{-1} \circ \tau_s$. So the preimage $\pi^{-1} \circ (\tau_0^{-1} \circ \tau_s) \circ \phi(\beta)$ consists of a collection of "broken paths" in $\mathbb{H}^2$ that "jump" to the left by $(s/2\pi)\ell_X(\gamma)$ every time they approach a lift of γ from the left; see Figure 10.11. The effect of $\mathrm{E}_s : \mathbb{H}^2 \to \mathbb{H}_s^2$ is to take these broken paths in $\mathbb{H}^2$ to continuous paths in $\mathbb{H}_s^2$.

Since E_s is a local isometry, we can compute the length of a continuous path in $\mathbb{H}_s^2$ by considering its image in $\mathbb{H}^2$ under E_s, computing the lengths of each piece of this broken path, and adding up. In particular, if $\widehat{\beta}$ is a path lifting of $\phi_s(\beta)$ to $\mathbb{H}_s^2$, then the length of the broken path $\mathrm{E}_s^{-1}(\widehat{\beta})$ is the same as the length of $\phi_s(\beta)$ in X_s.

We can choose $\widehat{\beta}$ to start and end on lifts of γ in $\mathbb{H}_s^2$. There is a deck transformation D_β of $\mathbb{H}_s^2$ that corresponds to the conjugacy class $\phi_s(\beta)$ and that fixes the lift of $\phi_s(\beta)$ containing $\widehat{\beta}$. The hyperbolic isometry D_β takes the start point of $\widehat{\beta}$ to the endpoint of $\widehat{\beta}$.

Now to find the value of $\ell_{X_s}(\beta)$, we modify $\widehat{\beta}$ by homotopy until length is minimized. We can perform this homotopy so that the endpoints stay in the preimage of γ throughout and so that the startpoint and endpoint differ by D_β throughout.

We know that the minimizing path is a geodesic segment that starts and ends on lifts of γ, that intersects $i(\beta, \gamma) + 1$ lifts of γ, and whose endpoints differ by D_β. Therefore, the closure of the image of the minimizing path under E_s^{-1} in $\mathbb{H}^2$ is a collection of $i(\beta, \gamma)$ geodesic segments $\delta_1, \dots, \delta_{i(\beta,\gamma)}$: $[0, 1] \to \mathbb{H}^2$ with the following properties:

1. $\delta_i(1)$ and $\delta_{i+1}(0)$ lie on a common lift of γ and differ by a displacement of $(s/2\pi)\ell_X(\gamma)$ to the left (also $\delta_i(0)$ and $\delta_{i+1}(1)$ lie on distinct lifts).

2. $\delta_1(0)$ and $\delta_{i(b,\gamma)}(1)$ differ by the map $D'_\beta = E_s^{-1} \circ D_\beta \circ E_s$.

The collection of segments $\{\delta_i\}$ (thought of as a collection of subsets of $\mathbb{H}^2$) is completely determined by the collection of points $\{\delta_i(0)\}$. Indeed, the other endpoints are determined by the two conditions above.

We have thus reduced the problem of finding the length of β in X_s to the problem of sliding the points $\{\delta_i(0)\}$ along a fixed collection of $i(b,\gamma)$ geodesics in $\mathbb{H}^2$ until the length of the corresponding piecewise geodesic path is minimized.

Say that the point $\delta_i(0)$ is restricted to the lift $\tilde\gamma_i$ of γ. We can identify $\tilde\gamma_1 \times \cdots \times \tilde\gamma_{i(b,\gamma)}$ with $\mathbb{R}^{i(b,\gamma)}$. Let

$$L : \mathbb{R}^{i(b,\gamma)} \times \mathbb{R} \to \mathbb{R}_+$$

denote the function that takes as input the points $\delta_i(0)$ and s and records the length of the corresponding piecewise geodesic path.

For a point w in some $\tilde\gamma_i \approx \mathbb{R}$, let w^s denote the point that lies an oriented distance of $(s/2\pi)\ell_X(\gamma)$ from w along $\tilde\gamma_i$. We can write L more concretely as the function

$$d(z_1, z_2^{\pm s}) + d(z_2, z_3^{\pm s}) + \cdots + d(z_{i(b,\gamma)}, D'_\beta(z_1)),$$

where each z_i lies in $\tilde\gamma_i \approx \mathbb{R}$ and the signs are determined by the orientations of the $\tilde\gamma_i$ (all distances are taken in $\mathbb{H}^2$).

We finally have

$$\ell_{X_s}(b) = \inf \left\{ L(z,s) : z \in \mathbb{R}^{i(b,\gamma)} \right\}.$$

Finishing the proof. We require the following fact, suggested by Mladen Bestvina; it is an ingredient in a new proof of the Nielsen realization theorem due to Bestvina–Bromberg–Fujiwara–Souto [19].

LEMMA 10.9 *Let $f : \mathbb{R}^m \times \mathbb{R}^n \to \mathbb{R}$ be a strictly convex function. If the function $F : \mathbb{R}^m \to \mathbb{R}$ defined by*

$$F(x) = \min\{f(x,y) : y \in \mathbb{R}^n\}$$

is well defined, that is, if the minimum always exists, then F is strictly convex.

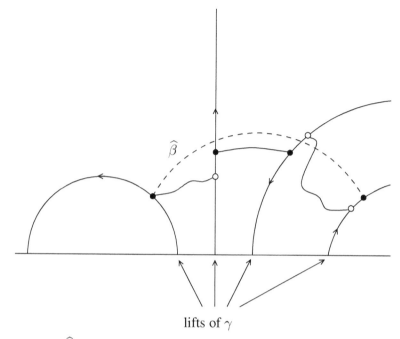

$\widehat{\beta}$

lifts of γ

Figure 10.11 If $\widehat{\beta}$ is the lift of β for the point X of Teich(S_g) as in the discussion, then the broken path is homotopic to a lift of the image of β in the new point X_s of Teich(S_g) obtained by varying the twist parameter s on one curve γ. At each lift of γ, the path jumps to the left a distance s.

Proof of Proposition 10.8. The starting point is the following basic fact from hyperbolic geometry (see [35, Chapter II, Proposition 2.2]):

> Let α and β be two disjoint geodesics in $\mathbb{H}^2$ parameterized at unit speed. The function $d : \mathbb{R}^2 \to \mathbb{R}$ given by $(s, t) \mapsto d_{\mathbb{H}^2}(\alpha(s), \beta(t))$ is strictly convex.

Given $k + 1$ disjoint oriented geodesics α_i in $\mathbb{H}^2$, each parameterized at unit speed, we consider the function $f_1 : \mathbb{R}^{2k} \to \mathbb{R}$ given by

$$f_1(x_1, y_1, \ldots, x_k, y_k) = \sum_{i=1}^{k} d(x_i, y_i),$$

where each x_i lies in α_i and each y_i lies in α_{i+1}. Since strict convexity is preserved under finite sums, we have that f_1 is strictly convex.

Next, let $f_2 : \mathbb{R}^{2k} \times \mathbb{R} \to \mathbb{R}$ be the function given by $f_2(x, s) = f_1(x)$ for any $x \in \mathbb{R}^{2k}$. The function f_2 is strictly convex in every direction except the s-direction, where it is constant.

Restrict f_2 to the hyperplane in $\mathbb{R}^{2k} \times \mathbb{R}$ described by $x_i = y_{i-1}^s$ (abusing the previous notation) for $2 \leq i \leq k$, and $y_k = \psi(x_1)^{-s}$, where ψ is some fixed isometry $\alpha_1 \to \alpha_{k+1}$. Call the new function f_3. It is straightforward to check that this hyperplane is not parallel to the s-direction, and so f_3 is strictly convex.

Finally, let $F : \mathbb{R} \to \mathbb{R}$ be the function given by $F(s) = \inf\{f_3(x, s)\}$, where x is any point on the hyperplane in $\mathbb{R}^{2k}$ where f_3 is defined. We have that $F(s)$ is strictly convex by Lemma 10.9. But, by the above discussion, for the appropriate choices of oriented geodesics α_i and isometry ψ, the function $F(s)$ is exactly the function $\ell_{x_s}(b)$, and so we are done. $\qquad\square$

Chapter Eleven

Teichmüller Geometry

Teichmüller space $\mathrm{Teich}(S)$ was defined in Chapter 10 as the space of hyperbolic structures on the surface S modulo isotopy. But $\mathrm{Teich}(S)$ parameterizes other important structures as well, for example, complex structures on S modulo isotopy and conformal classes of metrics on S up to isotopy.

We would like to have a way to compare different complex or conformal structures on S to each other. A natural way to do this is to search for a quasiconformal homeomorphism $f : S \to S$ that is homotopic to the identity map and that has the smallest possible quasiconformal dilatation with respect to the two structures. Informally, a homeomorphism with minimal dilatation is one that distorts angles least. This problem was solved by Grötzsch when S is a rectangle, and for general surfaces by Teichmüller.

After presenting the solution to this extremal problem, we will see how the least dilatation can be used to define a metric on Teichmüller space called the Teichmüller metric. Understanding the basic properties of this metric, for example, determining its geodesics, is important in a number of problems in low-dimensional topology. In particular, it will play a central role in Chapter 13, where we present Bers' proof of the Nielsen–Thurston classification of surface homeomorphisms.

The underlying objects encoding the solution to the extremal problem are holomorphic quadratic differentials and their associated measured foliations. Thus we will spend some time describing these objects.

There are many approaches to the theory of quasiconformal mappings and Teichmüller theory, each with their own advantages and disadvantages. In this chapter, we adopt an approach of Bers that is described in the lecture notes written by Abikoff [1].

11.1 QUASICONFORMAL MAPS AND AN EXTREMAL PROBLEM

In this section we define quasiconformal maps between surfaces in order to set up the extremal problem mentioned above. The natural setting for quasiconformal maps is that of complex structures on surfaces, as opposed to hyperbolic structures. Thus we begin by explaining the correspondence between these two types of structures.

11.1.1 COMPLEX STRUCTURES VERSUS HYPERBOLIC STRUCTURES

By a *Riemann surface* X we mean a 1-dimensional complex manifold. This means that X comes equipped with an atlas of charts to $\mathbb{C}$ that has biholomorphic transition maps; that is, transition maps are holomorphic with holomorphic inverses. Two Riemann surfaces X and Y are said to be *isomorphic* if there is a biholomorphic homeomorphism between them.

The uniformization theorem gives that any Riemann surface of genus $g \geq 2$ is the quotient of the unit disk Δ by a group Γ of biholomorphic automorphisms acting properly discontinuously and freely on Δ; see, for example, [198, Chapter 9]. Any group of biholomorphic automorphisms of Δ preserves the hyperbolic metric on Δ. Thus Δ/Γ has an induced hyperbolic structure, and conversely, any such hyperbolic structure gives a complex structure on X. In other words, for $g \geq 2$, there is a bijective correspondence:

$$\left\{ \begin{array}{c} \text{Isomorphism classes} \\ \text{of Riemann surfaces} \\ \text{homeomorphic to } S_g \end{array} \right\} \longleftrightarrow \left\{ \begin{array}{c} \text{Isometry classes} \\ \text{of hyperbolic surfaces} \\ \text{homeomorphic to } S_g \end{array} \right\}$$

Using isothermal coordinates, one can define a complex structure on any surface endowed with a Riemannian metric. In this way $\mathrm{Teich}(S_g)$ can also be identified with the set of conformal classes of Riemannian metrics on S_g.

11.1.2 QUASICONFORMAL MAPS

Let U and V be open subsets of $\mathbb{C}$ and let $f : U \to V$ be a homeomorphism that is smooth outside of a finite number of points. In Section 10.2, we explained how to write linear maps $\mathbb{R}^2 \to \mathbb{R}^2$ using the notation of complex analysis. We now apply this idea to describe the differential df.

Using the usual notation for maps $\mathbb{R}^2 \to \mathbb{R}^2$, we can write f as $f(x, y) = (a(x, y), b(x, y))$, where $a, b : \mathbb{R}^2 \to \mathbb{R}$. Where it is defined, the derivative df is then the real linear map

$$df = \left(\begin{array}{cc} a_x & a_y \\ b_x & b_y \end{array} \right).$$

We can also write

$$df = f_x \, dx + f_y \, dy,$$

where $f_x = (a_x, b_x)$ and $f_y = (a_y, b_y)$.

Switching to complex notation and setting $z = x + iy$, we can write

$f_x = a_x + ib_x$ and $f_y = a_y + ib_y$, and we can rewrite df as

$$df = f_z \, dz + f_{\bar{z}} \, d\bar{z},$$

where

$$f_z = \frac{1}{2}(f_x - if_y) \text{ and } f_{\bar{z}} = \frac{1}{2}(f_x + if_y).$$

Recall from Section 10.2 that the quantity $\mu_f = f_{\bar{z}}/f_z$ is called the complex dilatation of f.

The condition that $f_{\bar{z}} \equiv 0$ is equivalent to the condition that f satisfies the Cauchy–Riemann equations. Thus f is holomorphic if and only if $f_{\bar{z}} \equiv \mu_f \equiv 0$. Also, since

$$|f_z|^2 - |f_{\bar{z}}|^2 = a_x b_y - a_y b_x,$$

we see that f is orientation-preserving if and only if $|f_z| > |f_{\bar{z}}|$, which is the same as saying $|\mu_f| < 1$.

Dilatation. Suppose now that the homeomorphism $f : U \to V$ is orientation-preserving. Let p be a point of U at which f is differentiable. The *dilatation of f at p* is defined to be

$$K_f(p) = \frac{|f_z(p)| + |f_{\bar{z}}(p)|}{|f_z(p)| - |f_{\bar{z}}(p)|} = \frac{1 + |\mu_f(p)|}{1 - |\mu_f(p)|}.$$

The quantity $\log(K_f(p))/2$ is precisely the distance between $\mu_f(p)$ and 0 in the Poincaré disk model of $\mathbb{H}^2$ (this makes sense since f is orientation-preserving and so $|\mu_f| < 1$). Note in particular that $K_f(p) \geq 1$.

The quantity $K_f(p)$ can be interpreted as follows. The map df_p takes the unit circle in $TU_p \approx \mathbb{C}$ to an ellipse E in $TV_{f(p)}$, and $K_f(p)$ is the ratio of the length of the major axis of E to the length of the minor axis of E. To see this, we parameterize the unit circle in $\mathbb{C}$ as $\theta \mapsto e^{i\theta}$ for $\theta \in [0, 2\pi]$. The image of this circle under df_p is then the ellipse E and is determined by the equation $E(\theta) = f_z(p)e^{i\theta} + f_{\bar{z}}(p)e^{-i\theta}$ for $\theta \in [0, 2\pi]$. The modulus (i.e., absolute value) of a point $E(\theta)$ is

$$|E(\theta)| = \left| f_z(p)e^{i\theta} + f_{\bar{z}}(p)e^{-i\theta} \right| = |f_z(p)| \left| 1 + \mu_f(p)e^{-i2\theta} \right|.$$

Since

$$1 - |\mu_f(p)| \leq \left| 1 + \mu_f(p)e^{-i2\theta} \right| \leq 1 + |\mu_f(p)|,$$

it follows that the ratio of the maximum modulus of a point on E to the minimum modulus of a point on E is precisely $K_f(p)$.

The *dilatation* of the map f is defined to be the number

$$K_f = \sup K_f(p),$$

where the supremum is taken over all points p where f is differentiable. Thus $1 \leq K_f \leq \infty$. If $K_f < \infty$, we say that f is a *quasiconformal* or K_f-*quasiconformal* map between the domains U and V of $\mathbb{C}$. Note that biholomorphic maps are conformal with conformal inverses, hence are 1-quasiconformal. The notion of a quasiconformal homeomorphism was first considered by Grötzsch in 1928.

Quasiconformal maps. Let $f : X \to Y$ be a homeomorphism between Riemann surfaces that is smooth outside of a finite number of points. Assume further that f respects the orientations induced by the complex structures on X and Y and that f^{-1} is smooth outside of a finite number of points. Since the transition maps in any atlases for X and Y are biholomorphic (hence 1-quasiconformal) and since the local expressions for f are orientation-preserving, there is a well-defined notion of the dilatation $K_f(p)$ of f at a point $p \in X$ where f is smooth. Since f is smooth outside of a finite number of points, we can define $K_f = \sup K_f(p)$ as above. We will say that f is quasiconformal or K_f–quasiconformal if $K_f < \infty$.

A map between Riemann surfaces is *holomorphic* if, in any chart, it is given by a holomorphic map from some domain in $\mathbb{C}$ to $\mathbb{C}$. A bijective, holomorphic map between Riemann surfaces is called a *conformal map*. Conformal maps between Riemann surfaces are also *biholomorphic*; that is, they have holomorphic inverses. The last fact follows from the open mapping theorem; see Section 10.32 and Theorem 10.34 in Rudin's book [187].

Lemma 11.1 *Let $f : X \to Y$ be a homeomorphism between Riemann surfaces. Then f is a 1-quasiconformal homeomorphism if and only if it is a conformal map.*

Proof. Suppose that f is conformal. In this case, f' is defined at every point and never vanishes. Further, f takes circles in the tangent space of X to (nondegenerate) circles in the tangent space of Y [187, Theorem 14.2], and so f is 1-quasiconformal.

Now suppose that f is 1-quasiconformal. This is the same as saying that $f_{\bar{z}} \equiv 0$ wherever f is differentiable. Let $A \subset X$ be the set of points where f is not differentiable. The restriction $f|_{X-A}$ is then holomorphic. Since f is a homeomorphism, its singularities at A must be removable [187, Theorem 10.20]. Since f is continuous, it follows that f is already holomorphic. As

f is a homeomorphism, it is bijective. By our definition of a conformal map of Riemann surfaces, f is conformal.   □

The group QC(X). Let X be a Riemann surface. We would like to show that the set of quasiconformal homeomorphisms $X \to X$ forms a group $QC(X)$. We require some basic facts about the dilatations of linear maps. The first fact is that if $f : \mathbb{C} \to \mathbb{C}$ is any linear map, then the dilatations of f and f^{-1} are equal, that is, $K_f = K_{f^{-1}}$. The second fact, which we will use repeatedly, is the following.

LEMMA 11.2 *Let f and g be two linear maps $\mathbb{C} \to \mathbb{C}$. Denote the complex dilatations of f, g, and $f \circ g$ by μ_f, μ_g, and $\mu_{f \circ g}$ and denote the dilatations by K_f, K_g, and $K_{f \circ g}$. We have*

$$K_{f \circ g} \leq K_f K_g,$$

with equality if and only if either $\arg(\mu_f) = \arg(\mu_g)$ or one of μ_f and μ_g is zero.

The last statement of Lemma 11.2 can be rephrased as: $K_{f \circ g} = K_f K_g$ if and only if the directions of maximal stretch for f and g are the same or at least one of K_f or K_g is 1. Lemma 11.2 is an easy exercise in linear algebra [70, Section 1.2].

We can now deduce the following about compositions of quasiconformal homeomorphisms of X.

Proposition 11.3 *Let X be a Riemann surface and let f and g be quasiconformal homeomorphisms of X with dilatations K_f and K_g. We have:*

1. *The composition $f \circ g$ is quasiconformal and*

$$K_{f \circ g} \leq K_f K_g.$$

2. *The inverse f^{-1} is quasiconformal and*

$$K_{f^{-1}} = K_f.$$

3. *If g is conformal, then*

$$K_{f \circ g} = K_f = K_{g \circ f}.$$

In particular, the set of quasiconformal homeomorphisms $QC(X)$ forms a group.

11.1.3 TEICHMÜLLER'S EXTREMAL PROBLEM

In 1928 Grötzsch considered the following natural extremal problem, at least in the case of rectangles. Because Teichmüller later considered the case of general Riemann surfaces [203], this problem is sometimes referred to as *Teichmüller's extremal problem*.

> *Fix a homeomorphism $f : X \to Y$ of Riemann surfaces and consider the set of dilatations of quasiconformal homeomorphisms $X \to Y$ in the homotopy class of f. Is the infimum of this set realized? If so, is the minimizing map unique?*

Teichmüller's theorems (see below) give a positive solution to both questions (under the assumption of negative Euler characteristic). The minimizing map is called the *Teichmüller map*.

In Section 11.8, we will use Teichmüller's theorems to define a metric on Teichmüller space called the Teichmüller metric, as follows. Let $g \geq 2$ and let $\mathcal{X}, \mathcal{Y} \in \mathrm{Teich}(S_g)$. The points $\mathcal{X}$ and $\mathcal{Y}$ can be represented by marked Riemann surfaces X and Y. Because of the markings, there is a unique preferred homeomorphism of Riemann surfaces $X \to Y$, namely, the change of marking map, which corresponds to the identity map of S_g (for abstract Riemann surfaces without markings, there is no way to choose such a preferred map). Thus we can ask for the infimum of the dilatations of quasiconformal homeomorphisms $X \to Y$ in the preferred homotopy class. Teichmüller's theorems say that there exists a unique quasiconformal homeomorphism $h : X \to Y$ of minimal dilatation among all maps $X \to Y$ in this homotopy class. We can then define a distance function

$$d_{\mathrm{Teich}}(\mathcal{X}, \mathcal{Y}) = \frac{1}{2} \log(K_h).$$

In Section 11.8, we will prove that d_{Teich} is a metric on $\mathrm{Teich}(S_g)$.

As we will see below, the Teichmüller map is smooth outside a finite set of points in S_g but is not smooth at all points of S_g. This is precisely why we defined the notion of quasiconformality for homeomorphisms that are smooth outside a finite set of points. Quasiconformality can be defined for homeomorphisms with significantly weaker smoothness conditions than we have assumed. We chose smoothness outside a finite set of points since this is easier to work with and avoids technical difficulties, but it is still general enough for all of our applications.

11.2 MEASURED FOLIATIONS

We will see that Teichmüller maps, the maps that appear as solutions to Teichmüller's extremal problem, are homeomorphisms of a surface that stretch along one foliation of the surface and shrink along a transverse foliation. In order to make this precise, we first need to give a careful discussion of measured foliations.

11.2.1 MEASURED FOLIATIONS ON THE TORUS

Before giving the general definition of a measured foliation, we restrict our attention to the case of the torus where (as usual) the situation is much simpler. We will also explain what it means for a linear map of the torus to stretch the torus along one foliation and shrink along another.

Let ℓ be any line through the origin in $\mathbb{R}^2$. The line ℓ determines a foliation $\widetilde{\mathcal{F}}_\ell$ of $\mathbb{R}^2$ consisting of the set of all lines in $\mathbb{R}^2$ parallel to ℓ. Translations of $\mathbb{R}^2$ take lines to lines, and so any translation preserves $\widetilde{\mathcal{F}}_\ell$ in the sense that it takes leaves to leaves.

Since all of the deck transformations for the standard covering $\mathbb{R}^2 \to T^2$ are translations, the foliation $\widetilde{\mathcal{F}}_\ell$ descends to a foliation $\mathcal{F}_\ell$ of T^2. If the slope of ℓ is rational, then every leaf of $\mathcal{F}_\ell$ is a simple closed geodesic in T^2. If the slope of ℓ is irrational, then every leaf of $\mathcal{F}_\ell$ is a dense geodesic in T^2.

The foliations $\widetilde{\mathcal{F}}_\ell$ come equipped with extra structure. Let $\nu_\ell : \mathbb{R}^2 \to \mathbb{R}$ be the function that records distance from ℓ. Integration against the 1-form $d\nu_\ell$ gives a *transverse measure* on $\widetilde{\mathcal{F}}_\ell$. What this means is that any smooth arc α transverse to the leaves of $\widetilde{\mathcal{F}}_\ell$ can be assigned a length $\mu(\alpha) = \int_\alpha d\nu_\ell$. The quantity $\mu(\alpha)$ is the total variation of α in the direction perpendicular to ℓ. Thus $\mu(\alpha)$ is invariant under isotopies of α that move each point of α within the leaf of $\widetilde{\mathcal{F}}_\ell$ in which it is contained. The 1-form $d\nu_\ell$ is preserved by translations and so descends to a 1-form w_ℓ on T^2 and induces a transverse measure on the foliation $\mathcal{F}_\ell$. The structure of a foliation on T^2 together with a transverse measure is called a *transverse measured foliation* on T^2.

Note that a transverse measured foliation on T^2 is completely determined by the 1-form w_ℓ. The leaves of $\mathcal{F}_\ell$ in T^2 are simply the integral submanifolds to the distribution determined by the kernel of w_ℓ.

Consider a linear map $A \in \mathrm{SL}(2, \mathbb{Z})$ with two distinct real eigenvalues $\lambda > 1$ and $\lambda^{-1} < 1$ corresponding to eigenspaces ℓ and ℓ'. As in the proof of Theorem 2.5, A induces a homeomorphism ϕ_A of the torus T^2. The homeomorphism ϕ_A preserves the foliations $\widetilde{\mathcal{F}}_\ell$ and $\widetilde{\mathcal{F}}_{\ell'}$ and multiplies their transverse measures by λ^{-1} and λ, respectively. We think of ϕ_A as stretching by a factor of λ in the ℓ-direction and contracting by a factor of λ in the

ℓ'-direction.

On a higher-genus surface, it is not clear what it would mean for a homeomorphism to stretch in the direction of a single vector. However, we can define a foliation on a higher-genus surface, and we will see that it makes sense for a homeomorphism to stretch the surface in the direction of that foliation. Teichmüller maps will be given exactly such a description.

11.2.2 SINGULAR MEASURED FOLIATIONS

We will transfer our discussion of measured foliations on the torus to closed surfaces of genus $g \geq 2$. The Euler–Poincaré formula (see below) shows that such surfaces do not admit foliations. This can be corrected by allowing foliations with a finite number of singularities of a specific type.

Singular foliations. A *singular foliation* $\mathcal{F}$ on a closed surface S is a decomposition of S into a disjoint union of subsets of S, called the *leaves* of $\mathcal{F}$, and a finite set of points of S, called *singular points* of $\mathcal{F}$, such that the following two conditions hold.

1. For each nonsingular point $p \in S$, there is a smooth chart from a neighborhood of p to $\mathbb{R}^2$ that takes leaves to horizontal line segments. The transition maps between any two of these charts are smooth maps of the form $(x, y) \mapsto (f(x, y), g(y))$. In other words, the transition maps take horizontal lines to horizontal lines.

2. For each singular point $p \in S$, there is a smooth chart from a neighborhood of p to $\mathbb{R}^2$ that takes leaves to the level sets of a k-pronged saddle, $k \geq 3$; see Figure 11.1.

We say that a singular foliation is *orientable* if the leaves can be consistently oriented, that is, if each leaf can be oriented so that nearby leaves are similarly oriented. It is not hard to see that a foliation is *locally orientable* if and only if each of its singularities has an even number of prongs. For instance, the foliation in Figure 11.1 is not orientable in a neighborhood of the singular point. However, there do exist foliations that are locally orientable but not (globally) orientable.

The Euler–Poincaré formula. The following proposition gives a topological constraint on the total number of prongs at all singularities of a measured foliation.

Proposition 11.4 (Euler–Poincaré formula) *Let S be a surface with a singular foliation. Let P_s denote the number of prongs at a singular point s.*

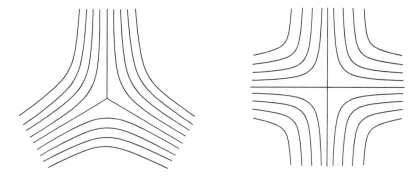

Figure 11.1 A foliation at a three-pronged singular point (left) and at a four-pronged singular
          point (right).

Then

$$2\chi(S) = \sum (2 - P_s),$$

where the sum is over all singular points of the foliation.

Since $P_s \geq 3$, Proposition 11.4 implies that a surface S with $\chi(S) > 0$ cannot carry a (singular or nonsingular) foliation. Proposition 11.4 also implies that any foliation on a surface S with $\chi(S) = 0$ must have no singular points and that any foliation on a surface S with $\chi(S) < 0$ must have at least one singular point. Because of this, we will unambiguously use the term "foliation" for foliations that have singularities as well as for those that do not.

The Euler–Poincaré formula is a straightforward consequence of the Poincaré–Hopf formula for vector fields applied to the context of line fields; see [61, Exposé 5, Section 1.6].

Measured foliations. As in the case of foliations on the torus, we would like to equip foliations on higher-genus surfaces with a transverse measure, that is, a length function defined on arcs transverse to the foliation. In order to do this precisely, we will need some preliminaries.

Let $\mathcal{F}$ be a foliation on a surface S. A smooth arc α in S is *transverse* to $\mathcal{F}$ if α misses the singular points of $\mathcal{F}$ and is transverse to each leaf of $\mathcal{F}$ at each point in its interior. Let $\alpha, \beta : I \to S$ be smooth arcs transverse to $\mathcal{F}$. A *leaf-preserving isotopy* from α to β is a map $H : I \times I \to S$ such that

- $H(I \times \{0\}) = \alpha$ and $H(I \times \{1\}) = \beta$

- $H(I \times \{t\})$ is transverse to $\mathcal{F}$ for each $t \in [0, 1]$

- $H(\{0\} \times I)$ and $H(\{1\} \times I)$ are each contained in a single leaf.

Note that the second and third conditions imply that $H(\{s\} \times I)$ is contained in a single leaf for any $s \in [0, 1]$.

A transverse measure μ on a foliation $\mathcal{F}$ is a function that assigns a positive real number to each smooth arc transverse to $\mathcal{F}$, so that μ is invariant under leaf-preserving isotopy and μ is regular (i.e., absolutely continuous) with respect to Lebesgue measure. In other words, this last condition means that each point of S has a neighborhood U and a smooth chart $U \to \mathbb{R}^2$ so that the measure μ is induced by $|dy|$ on $\mathbb{R}^2$.

A *measured foliation* $(\mathcal{F}, \mu)$ on a surface S is a foliation $\mathcal{F}$ of S equipped with a transverse measure μ.

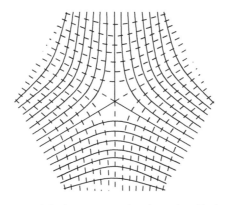

Figure 11.2 Two transverse foliations near a singular point. Each foliation has a three-pronged singularity.

We say that two measured foliations are *transverse* if their leaves are transverse away from the singularities; see Figure 11.2. Note that transverse measured foliations must have the same set of singularities.

Natural charts. There is another way of defining a measured foliation on a surface S. Let $\{p_i\}$ be a finite set of points in S. Suppose we have an atlas for $S - \{p_i\}$ where all transition maps are of the form

$$(x, y) \mapsto (f(x, y), c \pm y)$$

for some constant c depending on the transition map. Then it makes sense to pull back the horizontal foliation of $\mathbb{R}^2$ with its transverse measure $|dy|$ (the absolute variation in the y-direction). After reinserting the p_i, the result is a measured foliation on S.

Conversely, given any measured foliation, one can construct an atlas where the transition maps are given as above and where the transverse measure is given by $|dy|$. Any chart from such an atlas is called a set of *natural coordinates* for the measured foliation.

If we have an ordered pair of transverse measured foliations and there is an atlas where, away from the singular points, the first foliation is the pullback of the horizontal foliation of $\mathbb{R}^2$ with the measure $|dy|$ and the second foliation is the pullback of the vertical foliation with the measure $|dx|$, then we say that this atlas, and each of its charts, is *natural* with respect to the pair of measured foliations.

The action of Homeo(S). There is a natural action of $\mathrm{Homeo}(S)$ on the set of measured foliations of S. Namely, if $\phi \in \mathrm{Homeo}(S)$ and if $(\mathcal{F}, \mu)$ is a measured foliation of S, then the action of ϕ on $(\mathcal{F}, \mu)$ is given by

$$\phi \cdot (\mathcal{F}, \mu) = (\phi(\mathcal{F}), \phi_\star(\mu)),$$

where $\phi_\star(\mu)(\gamma)$ is defined as $\mu(\phi^{-1}(\gamma))$ for any arc γ transverse to $\phi(\mathcal{F})$. As a consequence, the mapping class group $\mathrm{Mod}(S)$ acts on the set of isotopy classes of measured foliations (the quotient of the set of measured foliations by $\mathrm{Homeo}_0(S)$).

Measured foliations as 1-forms. Any locally orientable measured foliation $(\mathcal{F}, \mu)$ can be described locally in terms of a closed 1-form as follows. In any chart where $\mathcal{F}$ is orientable there is a closed real-valued 1-form ω so that, away from the singular points of $\mathcal{F}$, the leaves of $\mathcal{F}$ are precisely the integral submanifolds of the distribution given by the kernel of ω, and μ is given by the formula

$$\mu(\gamma) = \int_\gamma |\omega|$$

for any arc γ transverse to $\mathcal{F}$. Indeed, in a neighborhood of a nonsingular point, we have seen that we can take the 1-form to be dy. A key point, though, is that $-dy$ serves the same purpose—it defines the same foliation and the same measure as dy. In the neighborhood of a singular point, the 1-form can be taken to be the derivative of a saddle function.

If a measured foliation is globally orientable, then there is a well-defined way of distinguishing between dy and $-dy$ on the entire surface. Thus the local 1-forms we described above glue together to give a globally defined closed 1-form on the surface. Conversely, the kernel of a closed 1-form on a surface defines an orientable foliation.

Punctures and boundary. The theory of measured foliations can be easily adapted to the case of surfaces with punctures and/or boundary. At a puncture, a foliation can take the form of a regular point or a k-pronged singularity with $k \geq 3$, as in the case of foliations on closed surfaces. However, at a puncture we also allow *one-pronged singularities* as in Figure 11.3.

Figure 11.3 A one-pronged singularity on a surface with a puncture.

A measured foliation on a compact surface S with nonempty boundary is defined similarly to the case when S is closed. There are four different pictures in the neighborhood of a point of ∂S depending on whether or not the point is singular and whether or not the leaves are parallel to the boundary or transverse to the boundary; see Figure 11.4.

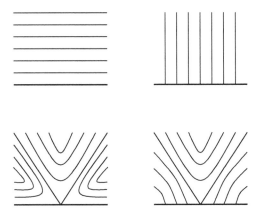

Figure 11.4 Measured foliations near the boundary of a surface.

11.2.3 FOUR CONSTRUCTIONS OF MEASURED FOLIATIONS

In this subsection we give four concrete ways of constructing measured foliations on a closed surface.

From a polygon. Given any closed surface S, we can realize S as the quotient of a polygon P in $\mathbb{R}^2$ by side identifications. We are using the Euclidean plane here and not the hyperbolic plane because we want to consider structures inherited from Euclidean geometry. We impose two additional conditions: *(i)* anytime two edges of P are identified, they are parallel, and *(ii)* the total Euclidean angle around each point of S is greater than π (the second condition needs to be checked only at the vertices of P). We do not need to assume that P is connected. One example of this is the realization of S_g as the quotient of a regular $(4g + 2)$-gon in $\mathbb{R}^2$ with opposite sides identified. Another example is given in Figure 11.6.

Any foliation of $\mathbb{R}^2$ by parallel lines restricts to give a foliation of (the interior of) P. We claim that this foliation induces a foliation of S. It is easy to see that any point of S coming from a point of P that is not a vertex of P has a regular neighborhood that satisfies the definition of a regular point of a foliation.

So what happens at a point $p \in S$ corresponding to a vertex of P? The first observation is that since identified sides of P are parallel, the total angle around p is an integer multiple of π. In particular, there is some vertex $\widetilde{p}$ of P in the preimage of p, and a vector v based at $\widetilde{p}$ that points into P (possibly along an edge) and is parallel to the foliation of P. If we sweep out an angle of π starting with v, we find a closed Euclidean half-disk in S that is foliated by lines parallel to the diameter. If we continue to sweep out angles of π, we see that a neighborhood of p looks like some number of Euclidean half-disks each foliated by lines parallel to the diameter and glued along oriented radii. By our assumption on the total angle around each point of S coming from a vertex of P, we know that there are at least two half-disks glued at p. If there are exactly two half-disks, then p is a regular point. If there are k half-disks, where $k \geq 3$, then p is a singularity with k prongs.

One measure on the induced foliation of S is the one given by the total variation of the Euclidean distance in the direction perpendicular to the foliation of P. The charts we described above are the natural charts for the nonsingular points.

Suppose that, in this construction, we orient each edge of P so that the identifications respect these orientations. If all side pairings identify sides of P that are parallel in the oriented sense (as opposed to antiparallel), then the resulting foliation of S is orientable. Indeed, either of the two orientations of the foliation on the interior of P extend to give an orientation of the entire foliation of S.

It is a fact that every measured foliation comes from this polygon construction. The idea is that the natural coordinates for a measured foliation pick out large rectangles in the surface that are foliated by horizontal lines. See Section 14.3 for further discussion.

Enlarging a simple closed curve. Let S be a closed surface of genus g. We can realize S topologically as a Euclidean $(4g + 2)$-gon with opposite sides identified. We can then straighten the sides of this polygon so that two opposite sides are horizontal and the other $4g$ sides are vertical. The result is a Euclidean rectangle R. If we identify the two horizontal edges of R we obtain an annulus A. The foliation of R by vertical lines descends to a foliation of A by curves parallel to the boundary. This foliation further descends to a foliation of S where each nonsingular leaf is a simple closed curve. All of these curves lie in the same homotopy class. There is a one-parameter family of measures obtained by scaling the rectangle horizontally.

Let α denote one of the nonsingular leaves in S. We say that the above measured foliation is obtained by *enlarging* the simple closed curve α. Note that, by change of coordinates, we can enlarge any nonseparating simple closed curve in a closed surface (it is also possible to extend the construction to separating curves).

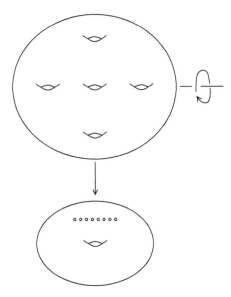

Figure 11.5 A twofold branched cover over the torus.

From a branched cover. Let $g \geq 2$ and let $p : S_g \to T^2$ be a branched covering map. For our purposes, a *branched cover* of one topological surface over another is the quotient of one orientable surface by a finite group of orientation-preserving homeomorphisms. So, for instance, orbifold coverings are branched coverings. One such example, with $2g-2$ branch points, is illustrated in Figure 11.5.

Any measured foliation $(\mathcal{F}, \mu)$ of T^2 pulls back via p to a measured foliation $(p^\star(\mathcal{F}), p^\star(\mu))$ on S_g. The singularities of $p^\star(\mathcal{F})$ are precisely the ramification points of the covering. The foliation $p^\star(\mathcal{F})$ has a singularity of order $2k$ above any branch point of order k. Since the deck transformations of the cover $S_g \to T^2$ are orientation-preserving, an orientation of the foliation on T^2 pulls back to an orientation on the foliation of S_g. Since every foliation of T^2 is orientable, every foliation of S_g obtained by this construction is orientable.

The same construction as above can be used to pull back measured foliations on any closed surface via any (branched or unbranched) cover.

From a pair of filling simple closed curves. Let α and β be two transverse simple closed curves that are in minimal position and that fill a closed surface S. Take, for instance, the example in Figure 1.7. We can think of $\alpha \cup \beta$ as a 4-valent graph in S, where the vertices are the points of $\alpha \cap \beta$. In fact, by also considering the closures of the components of $S - (\alpha \cup \beta)$ as 2-cells, we have a description of S as a 2-complex X.

We construct a dual complex X'. The complex X' is formed by taking one vertex for each 2-cell of X, one edge transverse to each edge of X, and one 2-cell for each vertex of X. Since the vertices of X are 4-valent, it follows that X' is a *square complex*, that is, each 2-cell of X' is a square. What is more, each square of X' has a segment of α running from one side to the opposite side.

We can foliate each square of X' by lines parallel to α. This gives rise to a foliation $\mathcal{F}_\alpha$ on all of S. We declare the "width" of each square to be the same fixed number, and this gives a measure on $\mathcal{F}_\alpha$. The foliation associated to β is a measured foliation $\mathcal{F}_\beta$ that is transverse to $\mathcal{F}_\alpha$.

This last construction is really just a special case of both the polygon construction and the branched cover construction. Indeed, we can think of X' as a disconnected polygon with sides identified. Also, if we think of T^2 as the unit square with sides identified, then there is a branched cover from $S \approx X' \to T^2$ that takes each square of X' to the unit square and takes the α-foliation to the foliation of the unit square by horizontal lines.

11.3 HOLOMORPHIC QUADRATIC DIFFERENTIALS

We now describe the complex-analytic counterparts to measured foliations, holomorphic quadratic differentials. Since quasiconformal maps are most easily described via complex analysis, we will be able to exploit this point of view in proving Teichmüller's theorems.

11.3.1 Quadratic Differentials and Measured Foliations

The *holomorphic cotangent bundle* to a Riemann surface X is the complex line bundle over X whose fiber above a point $p \in X$ is the space of complex linear maps $T_p(X) \to \mathbb{C}$. A holomorphic 1-form is a holomorphic section of the holomorphic cotangent bundle of X. A *holomorphic quadratic differential* on X, which is the object of interest here, is a holomorphic section of the symmetric square of the holomorphic cotangent bundle of X. For instance, the tensor square of a holomorphic 1-form on X is a holomorphic quadratic differential.

We can alternatively describe a holomorphic quadratic differential on X in terms of local coordinates, as follows. Let $\{z_\alpha : U_\alpha \to \mathbb{C}\}$ be an atlas for X. A holomorphic quadratic differential q on X is specified by a collection of expressions $\{\phi_\alpha(z_\alpha)\, dz_\alpha^2\}$ with the following properties:

1. Each $\phi_\alpha : z_\alpha(U_\alpha) \to \mathbb{C}$ is a holomorphic function with a finite set of zeros.

2. For any two coordinate charts z_α and z_β, we have

$$\phi_\beta(z_\beta) \left(\frac{dz_\beta}{dz_\alpha} \right)^2 = \phi_\alpha(z_\alpha). \qquad (11.1)$$

The second condition can be phrased as "the collection $\{\phi_\alpha(z_\alpha)\, dz_\alpha^2\}$ is invariant under change of local coordinates." To say this yet another way, if q is given in one chart by $\phi_\alpha(z)$ and in another chart by $\phi_\beta(z)$, and the change of coordinates from the first chart to the second chart is the holomorphic map ψ, then

$$\phi_\beta(z) \left(\frac{d\psi}{dz} \right)^2 = \phi_\alpha(z).$$

It follows that the order of a zero of a holomorphic quadratic differential is well defined independent of the chart.

More concretely, a holomorphic quadratic differential q on a Riemann surface X is a holomorphic map from the holomorphic tangent bundle of X to $\mathbb{C}$. To make this explicit, say that in local coordinates the holomorphic quadratic differential q is given by $\phi(z)\, dz^2$. Suppose that some tangent vector v to X is given by $\alpha \in \mathbb{C} \approx T_{z_0}(\mathbb{C})$. Then we have

$$q(v) = \phi(z_0)\alpha^2.$$

Note that, for any $v \in T_p(X)$, we have $q(v) = q(-v)$.

Measured foliations from quadratic differentials. Given a holomorphic quadratic differential q on a Riemann surface X, we obtain a foliation by taking the union of the zeros of q with the set of smooth paths in X whose tangent vectors at each point evaluate to positive real numbers under q. This foliation is called the *horizontal foliation* for q. If we instead take the paths in X whose tangent vectors evaluate to negative real numbers under q, the resulting foliation is called the *vertical foliation* for q.

Say that, within some chart, a holomorphic quadratic differential is given by the expression $\phi(z)\,dz^2$. In any given chart, the function

$$\mu(\alpha) = \int_\alpha \left| \mathrm{Im}\left(\sqrt{\phi(z)}\,dz \right) \right|$$

induces a transverse measure μ_q on the horizontal foliation for q. By taking real parts instead of imaginary parts, we obtain a transverse measure on the vertical foliation for q. Below we will define natural coordinates for a holomorphic quadratic differential, where this formula always takes a standardized form. To check that μ_q really determines a transverse measure, one can either apply (11.1) directly or appeal to natural coordinates.

Consider, for example, a holomorphic quadratic differential q that in some coordinate chart has the form $q(z) = dz^2$. Say that some tangent vector v is given by $\alpha \in \mathbb{C} \approx T_{z_0}(\mathbb{C})$ in this chart. As above, we have

$$q(v) = \alpha^2.$$

Now $\alpha^2 > 0$ precisely when α is a nonzero real number, and $\alpha^2 < 0$ precisely when α is a purely imaginary number. Therefore, in the given chart, the horizontal foliation is the union of horizontal lines and the vertical foliation is the union of vertical lines. The measures for these foliations are the ones induced by $|dy|$ and $|dx|$.

Now consider a holomorphic quadratic differential with local expression $q(z) = z^k\,dz^2$. Say $\alpha \in \mathbb{C} \approx T_{z_0}(\mathbb{C})$ is the local expression for a tangent vector v. In this case,

$$q(v) = z_0^k\,\alpha^2.$$

Some lines of the horizontal foliation are easy to spot, namely, vectors of the form $\alpha \in T_\alpha(\mathbb{C})$, where $\alpha^{k+2} = 1$. It is not hard, then, to see that the horizontal foliation has the form of a $(k+2)$-pronged singular point, as in the theory of measured foliations. The vertical foliation is the transverse foliation obtained by rotating the picture for the horizontal foliation by an angle of $\pi/(k+2)$.

Quadratic differentials versus 1-forms. Why do we consider holomorphic quadratic differentials as opposed to holomorphic 1-forms, which are differentials of the form $\phi(z)\,dz$? The reason is that, for a holomorphic quadratic differential q, we have that $q(v) > 0$ if and only if $q(-v) > 0$, and so the associated horizontal and vertical foliations are not necessarily oriented. On the other hand, the (analogously defined) horizontal and vertical foliations for a holomorphic 1-form are automatically oriented. In our study of mapping class groups and Teichmüller space, we will be forced to deal with both oriented foliations and unoriented foliations.

Natural coordinates. Let q be a holomorphic quadratic differential on a compact Riemann surface X. We will now show that every point of X has local coordinates, called *natural coordinates*, so that in these local coordinates we have $q(z) = z^k\,dz^2$ for some $k \geq 0$. Since we just showed that the horizontal and vertical foliations for $z^k\,dz^2$ satisfy the definition of a measured foliation, it will follow that the horizontal and vertical foliations for q really are transverse measured foliations, as defined above.

First consider a regular point p of q; that is, assume $q(p) \neq 0$. Let $z : U \to \mathbb{C}$ be a local coordinate with $z(p) = 0$ and write $q(z) = \phi(z)\,dz^2$ in this chart. Since q is assumed to have finitely many zeroes, we can pick the chart small enough that $\phi(z)$ does not vanish in this chart. Our goal is to show that there is a local coordinate ζ at p so that $q(\zeta) = d\zeta^2$. Such coordinates are obtained by composing z with the change of coordinates

$$\eta(z) = \int_0^z \sqrt{\phi(\omega)}\,d\omega,$$

where some branch of the square root function is chosen (this is possible since $\phi \neq 0$). Of course, when we integrate from 0 to z, we really mean to integrate along some (any) path from 0 to z.

The natural coordinates are $\zeta = \eta \circ z : U \to \mathbb{C}$. We can check that q has the desired form in the ζ-coordinates. First, by the fundamental theorem of calculus, we have

$$d\eta = \sqrt{\phi(z)}.$$

Now let $q(\zeta) = \psi(\zeta)\,d\zeta^2$ in the ζ-coordinates. By (11.1), we have

$$\psi(\zeta)(d\eta)^2 = \phi(z)$$

or

$$\psi(\zeta)\phi(z) = \phi(z),$$

and so $\psi(\zeta) \equiv 1$, as desired.

The natural coordinates ζ are unique up to translation and sign. It is therefore possible to associate measures to the horizontal and vertical foliations. Locally, the measures are given by $|dy|$ and $|dx|$, respectively.

Now suppose that $p \in X$ is a zero of q of order $k \geq 1$. By a variation of the above argument, there is a local coordinate ζ so that, in this coordinate, q is given by $q(z) = z^k \, dz^2$. As above, these coordinates are called the natural coordinates. For the details of this argument, see [200, Section 6].

We have shown that the horizontal and vertical foliations of a holomorphic quadratic differential give a pair of measured foliations that are transverse to each other. It is a deep theorem of Hubbard–Masur that, given any measured foliation $(\mathcal{F}, \mu)$, one can build a holomorphic quadratic differential whose corresponding horizontal foliation is, up to a certain equivalence, equal to $(\mathcal{F}, \mu)$ [96].

The Euler–Poincaré formula revisited. From the Euler–Poincaré formula (Proposition 11.4) and the correspondence between the order of a zero of a holomorphic quadratic differential and the number of prongs of the associated foliation, we deduce that a holomorphic quadratic differential must vanish at exactly $4g - 4$ points, where points are counted with multiplicity.

Euclidean areas and lengths. The natural coordinates for a holomorphic quadratic differential q on a Riemann surface X endow X with a singular Euclidean metric. A *singular Euclidean metric* on a surface S is a flat metric outside of a finite number of points, around each of which the metric is modeled on gluing flat rectangles together in the same way as is done to give a singular point of a measured foliation. Locally, the area form of this metric is given by

$$\frac{1}{2i}|\phi(z)| \, \overline{dz} \wedge dz = |\phi(z)| dx \wedge dy,$$

where $\phi(z) \, dz^2$ is the local expression for q.

We can also talk about the Euclidean length of a path in X with respect to q. This length form is given by

$$|\phi(z)|^{1/2}|dz| = |\phi(z)|^{1/2} \sqrt{dx^2 + dy^2}.$$

The singular Euclidean metric induced by a holomorphic quadratic differential is nonpositively curved in the sense that it is locally CAT(0). It follows that, given any arc α in X, there is a unique shortest path among all paths homotopic to α with endpoints fixed (see [35, Chapter II, proof of Corollary 4.7]).

Punctures and boundary. The modifications needed to define holomorphic quadratic differentials on a surface with punctures and/or boundary mirror the changes we made for measured foliations. In the neighborhood of a puncture, holomorphic quadratic differentials take the form $z^k \, dz^2$, where $k \geq -1$. That is, we allow simple poles at punctures.

11.3.2 QUADRATIC DIFFERENTIALS ON THE TORUS

Let X be a closed Riemann surface of genus 1. There is a lattice $\Lambda < \mathbb{C}$ so that $X \approx \mathbb{C}/\Lambda$. Let $\pi : \mathbb{C} \to X$ denote the quotient map. For a small enough open set U in X, there is an open set $\widetilde{U} \subset \mathbb{C}$ so that $\pi|_{\widetilde{U}} : \widetilde{U} \to U$ is a homeomorphism. The collection of such maps $\{\pi|_{\widetilde{U}}^{-1}\}$ is an atlas for X. All of the transition maps for this atlas are translations. For any point $x \in X$, the set of images of x under all charts is $\pi^{-1}(x)$.

Let q be a holomorphic quadratic differential on X. From (11.1) and the fact that all transition maps are translations, it follows that q can be written as a doubly periodic holomorphic function $\phi : \mathbb{C} \to \mathbb{C}$. A doubly periodic function $\mathbb{C} \to \mathbb{C}$ is bounded, and so by Liouville's theorem ϕ is constant. We therefore have that the set of holomorphic quadratic differentials on X is in bijection with $\mathbb{C}$. Under this bijection, the horizontal foliation for the differential corresponding to $z \in \mathbb{C}$ has leaves consisting of straight lines that meet the x-axis with angle $- \arg(z)/2$.

11.3.3 CONSTRUCTIONS OF QUADRATIC DIFFERENTIALS

Having explained some of the basics of holomorphic quadratic differentials, the question remains: how does one actually construct a holomorphic quadratic differential? Since we know how to derive a measured foliation from a holomorphic quadratic differential, it makes sense to generalize our two main constructions of measured foliations, namely, the construction via polygons and the construction via branched covers.

Quadratic differentials via polygons: the Swiss cross example. Just as we were able to construct measured foliations from certain polygons, we can also construct holomorphic quadratic differentials from the same polygons. Instead of explaining the construction in generality, we will explain one particular case, the Swiss cross example, in detail. This example exhibits all of the subtleties of the general case.

Consider the closed polygonal region P in $\mathbb{C}$ in Figure 11.6. Let S be the topological surface obtained by identifying sides of P by Euclidean translation, as indicated in the figure. We start by describing an atlas for S that

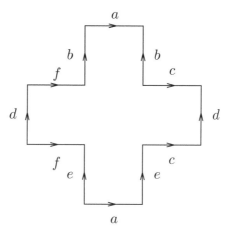

Figure 11.6 The Swiss cross.

gives S the structure of a Riemann surface. The first chart is self-evident: the subset of S corresponding to the interior of P is already identified with an open subset of $\mathbb{C}$. Now consider a point p in S corresponding to a point in the interior of an edge of P. Let U be an open neighborhood of p corresponding to a union of two half-disks in P (each half-disk is the intersection of an open disk in $\mathbb{C}$ with P). To define the chart for U, we say that one "half" of U maps to the corresponding half-disk in P, and the other half-disk maps to the image of its corresponding half-disk in P under a translation, where the translation is chosen so that the image of U under the chart is an open disk in $\mathbb{C}$.

We proceed similarly at the corners. The eight corners with angle $\pi/2$ glue together to form two disks in S, and so we can use the same method as above to glue the pieces together. Consider the other four corners. We see that in S these four vertices of P are identified to a single point, which we call s. We also see that the total Euclidean angle around the four copies of s in P is $4(3\pi/2) = 6\pi$. Thus we cannot simply glue these pieces together by translations and expect to get an open disk in $\mathbb{C}$. The solution is as follows: one by one, translate each of the corresponding vertices of P to $0 \in \mathbb{C}$, apply the map $z \mapsto z^{1/3}$, and then apply rotations so that the four corners glue together to form an open disk about $0 \in \mathbb{C}$; we take care so that the image of the kth corner lies between the rays with argument $k\pi/2$ and $(k+1)\pi/2$.

The above-defined charts indeed define a complex structure on S, that is, transition maps are biholomorphic. The only place where there could possibly be an issue is at the point s (since $z^{1/3}$ is not differentiable at 0). However, this point appears in only one chart, so there is nothing to check.

We can now use the atlas we have constructed in order to define an explicit holomorphic quadratic differential q on S chart by chart. Here the notion of holomorphic is taken with respect to the complex structure on S that we just constructed.

We define q on every chart except the last one constructed above by setting $q(z) = dz^2$. On the last chart we let $q(z) = 9z^4\, dz^2$. Note that each chart except for the last gives natural coordinates since the local expression for q in these cases is dz^2.

We can see that the point s will be a singularity of the horizontal foliation for q. The prongs at s come from eight segments of the horizontal foliation for q, two at each preimage of s. In S, two of these pairs are identified (the ones labeled c and f), and so in the end the singularity at s has six prongs (cf. Proposition 11.4). This agrees with the fact that we gave q an order 4 zero at s.

Let us now check that (11.1) holds for all transition maps. Consider a point near s that lies in the "big chart" (the first one we defined) and the "special chart" (the last one we defined). Call the big coordinate z and the special coordinate w. Then (11.1) demands that

$$(dz/dw)^2 = 9w^4.$$

But this is the same as saying that

$$(3w^2)^2 = 9w^4,$$

which is obviously true. Checking (11.1) for the other transition maps is similar. Therefore, q really is a holomorphic quadratic differential. The area and arc length forms are exactly the ones coming from the Euclidean metric in the big chart.

One way to get other holomorphic quadratic differentials on the Swiss cross Riemann surface is simply to change the expression for q on the big chart to be any $\alpha\, dz^2$ for $\alpha \in \mathbb{C}$. In this case we get other holomorphic quadratic differentials which are qualitatively different; for instance, some have closed leaves and some do not.

One can construct other examples of holomorphic quadratic differentials using the same idea as above, that is, by gluing together Euclidean polygons by Euclidean translations. For starters, the reader might like to consider the example indicated in Figure 11.7.

As with measured foliations, the polygon construction for holomorphic quadratic differentials has a converse: every holomorphic quadratic differential can be realized in this way. Indeed, the natural coordinates tell us how to cut up the Riemann surface into (finitely many) rectangles, each foliated by horizontal lines. By placing these rectangles in the Euclidean plane so that the foliations are horizontal and recording the side identifications, we obtain a polygonal description of the surface where the holomorphic quadratic differential is given by dz^2. For details, see [200, Section 11].

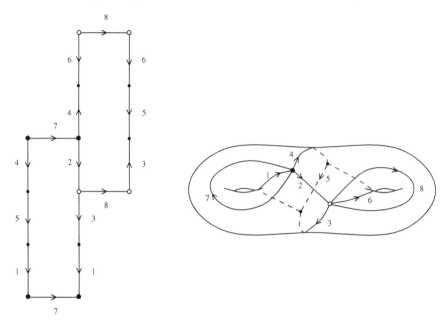

Figure 11.7   An example of a surface obtained by gluing Euclidean rectangles by Euclidean translations. As with the Swiss cross example, this surface can be given the structure of a Riemann surface with a holomorphic quadratic differential.

Quadratic differentials via branched covers. Another way to construct holomorphic quadratic differentials is via branched covers. A map $p : X \to Y$ between closed Riemann surfaces is a *branched covering map* if, for any point $x \in X$, there are local coordinates where p is given by $z \mapsto z^k$ for some $k \geq 1$. Note that, if $p : X \to Y$ is an orbifold covering in the sense of Chapter 7 or is a branched cover of topological surfaces as in Section 11.2, then we can pull back any complex structure on Y to a complex structure on X, and the map p will then be a branched covering map of Riemann surfaces. The point x is a *ramification point* of the branched covering if and only if $k > 0$, and in this case k is called the *degree of ramification*.

Let $p : X \to Y$ be a branched covering map and suppose we have a holomorphic quadratic differential q on Y. We can lift q to a holomorphic quadratic differential $\widetilde{q}$ on X as follows. If U is some open neighborhood of a point $x \in X$, and if there are charts $U \to \mathbb{C}$ and $p(U) \to \mathbb{C}$ where in these coordinates p is given by $z \mapsto \psi(z)$, then by (11.1) we have

$$\widetilde{\phi}(z) = \phi(\psi(z))(d\psi(z))^2,$$

where $\widetilde{\phi}(z)\,dz^2$ and $\phi(z)\,dz^2$ are the local expressions for $\widetilde{q}$ and q. If x is a ramification point of degree k and if q has a zero of order m at $p(x)$ (we allow $m = 0$), then we see that $\widetilde{q}$ has a zero of order $km+2(k-1)$ at x. This agrees with our discussion about branched covers and measured foliations: the foliations for q and $\widetilde{q}$ have singularities with $m+2$ prongs and $k(m+2)$ prongs, respectively.

Of course, in order to lift a holomorphic quadratic differential on a Riemann surface Y, we first need to know how to find a holomorphic quadratic differential on Y. When Y has genus 1, we already proved above that the space of holomorphic quadratic differentials on Y is in bijective correspondence with $\mathbb{C}$. We also explained how to construct branched covers of higher-genus topological surfaces over the torus (Figure 11.5), and again these give rise to branched covers of Riemann surfaces over Y.

11.3.4 THE VECTOR SPACE OF HOLOMORPHIC QUADRATIC DIFFERENTIALS

Let X be any Riemann surface. It is easy to check that the sum of two holomorphic quadratic differentials on X is a holomorphic quadratic differential on X, as is any complex multiple of a holomorphic quadratic differential. It follows that the set of all holomorphic quadratic differentials on X forms a complex vector space denoted $\mathrm{QD}(X)$.

Our first goal is to give a lower bound on the dimension of $\mathrm{QD}(X)$. Choose some finite set of points $P \subset X$. Let $K_P(X)$ denote the complex vector space of meromorphic functions $f : X \to \mathbb{C}$, where f has only simple poles, each occurring at points of P. The following theorem is a special case of Riemann's inequality (see [60, III.4.8] or [68, Section 8.3]).

THEOREM 11.5 *Let X be a closed Riemann surface of genus g and let $P \subset X$ be a finite set of points. We have*

$$\dim_{\mathbb{C}}(K_P(X)) \geq |P| + 1 - g.$$

We now obtain the desired bound on the dimension of $\mathrm{QD}(X)$.

Proposition 11.6 *Let X be a closed Riemann surface of genus g. We have*

$$\dim_{\mathbb{C}}(\mathrm{QD}(X)) \geq 3g - 3.$$

Proof. Let q_0 be an element of $\mathrm{QD}(X)$ with only simple zeros. Recall that the horizontal foliation for q_0 has three prongs at each singularity. By the Euler–Poincaré formula (Proposition 11.4), q_0 has exactly $4g - 4$ zeros.

Let P be the set of $4g - 4$ zeros of q_0. By Theorem 11.5, we have

$$\dim_{\mathbb{C}}(K_P(X)) \geq 3g - 3.$$

We claim that there is a map $\mathrm{QD}(X) \rightarrow K_P(X)$ given by $q \mapsto q/q_0$. Indeed, the ratio q/q_0 is a well-defined function on X by (11.1), and the poles of q/q_0 are precisely the zeros of q_0 at which q does not have a zero. This map is a vector space isomorphism, so we are done.   □

The inequality of Theorem 11.7 turns out to be an equality for $g \geq 2$ (in the case $g = 1$ we have already shown that $\mathrm{QD}(X) \approx \mathbb{C}$). This equality can be deduced from the Riemann–Roch theorem, a deep theorem which sharpens Riemann's inequality. On the other hand, in Section 11.4 we will define a map Ω from the open unit ball in $\mathrm{QD}(X)$ to $\mathrm{Teich}(S_g)$. It follows from the definition of Ω and Teichmüller's uniqueness theorem (Theorem 11.9 below) that Ω is injective. By Brouwer's invariance of domain theorem (see Theorem 11.15 below), we then obtain the following theorem.

THEOREM 11.7 *Let X be a closed Riemann surface of genus g. We have*

$$\dim_{\mathbb{C}}(\mathrm{QD}(X)) = 3g - 3.$$

In what follows, we will only need Proposition 11.6, and not Theorem 11.7.

A topology on $\mathrm{QD}(X)$. Let X be a Riemann surface and let $q \in \mathrm{QD}(X)$. By (11.1) the absolute value of q at a point does not depend on the local expression for q. In other words, $|q|$ is a function $X \rightarrow \mathbb{R}$. We can thus define a norm on $\mathrm{QD}(X)$ by the formula

$$\|q\| = \int_X |q|.$$

This norm induces a metric on $\mathrm{QD}(X)$ and hence a topology. With respect to this topology any n-dimensional complex subspace of $\mathrm{QD}(X)$ is homeomorphic to $\mathbb{R}^{2n}$.

A dimension count for $\mathrm{QD}(X)$. For $g \geq 2$, we can give a heuristic dimension count for $\mathrm{QD}(X)$ that is in the same spirit as our dimension counts for $\mathrm{Teich}(S_g)$ in Chapter 10.

Fix a closed Riemann surface X of genus $g \geq 2$ and let q be a holomorphic quadratic differential on X. As discussed above, it is possible to realize X by a Euclidean polygon so that $q(z) = dz^2$ in the interior chart of the polygon. We can thus count the dimension of $\mathrm{QD}(X)$ by counting the dimension of the space of polygons that give X.

Specifically, we consider connected Euclidean polygons P with the following properties:

- If we identify pairs of parallel sides of P, we obtain a closed surface S of genus g.

- Every vertex of P maps to a point on S with total Euclidean angle 3π.

As in the discussion above, P induces a complex structure on S and the quadratic differential dz^2 induces a holomorphic quadratic differential q on S. The second condition on P means that each point of S coming from a vertex of P is a simple zero of q (of the form $q(z) = z\,dz^2$) and there are no other zeros.

Examples of such polygons exist for every genus; see Figure 11.7 for one example in genus 2. The set of these polygons has codimension zero in the space of all polygons giving holomorphic quadratic differentials on S, and so we aim to count the dimension of the space of such polygons P.

Let P and q be as above. By the Euler–Poincaré formula, q has $4g - 4$ simple zeros. Since a simple zero of q accounts for a total interior angle of 3π in P and since every vertex of P corresponds to a simple zero of q, we see that the sum of the interior angles of P must be

$$3\pi(4g - 4) = (12g - 12)\pi.$$

The sum of the interior angles of a Euclidean n-gon is $(n - 2)\pi$, and so we see that P must have $12g - 10$ sides. If we think of each side of P as a vector, we get $2(12g - 10)$ dimensions worth of freedom. Since side lengths and angles must match in pairs, we are down to $12g-10$ dimensions. The last pair of sides is determined by the others, and so we lose two more dimensions, giving $12g - 12$, exactly twice what we want.

But as we change the polygon, we are also changing the complex structure on S. We are trying to compute $\mathrm{QD}(X)$ for a fixed Riemann surface X. In order to take this into account we must subtract the dimension of the space of all complex structures on S, namely, the dimension of $\mathrm{Teich}(S)$, which is $6g - 6$. We have thus given a heuristic that shows that there are $6g - 6$ dimensions worth of possible holomorphic quadratic differentials q on any fixed Riemann surface X.

Dimension count for $QD(X)$: polygons

$+2(12g - 10)$: Choose $12g - 10$ vectors for the polygon's sides.

$-(12g - 10)$: Sides must match in pairs.

-2 : The last pair of sides is determined by the others.

$-(6g - 6)$: Subtract the dimension of $\mathrm{Teich}(S)$.

$= 6g - 6$ Total dimensions

11.4 TEICHMÜLLER MAPS AND TEICHMÜLLER'S THEOREMS

We are now ready to describe the homeomorphisms that minimize the quasiconformal dilatation in a given homotopy class, thus giving a solution to Teichmüller's extremal problem. The solution to this problem was first given by Teichmüller [188, 203] and Ahlfors [3]; see also Bers [14].

11.4.1 STATEMENT OF THE THEOREMS

Let X and Y be two closed Riemann surfaces of genus g. We say that a homeomorphism $f : X \to Y$ is a *Teichmüller mapping* if there are holomorphic quadratic differentials q_X and q_Y on X and Y, respectively, and a positive real number K so that the following two conditions hold:

1. The homeomorphism f takes the zeros of q_X to the zeros of q_Y.

2. If $p \in X$ is not a zero of q_X, then with respect to a set of natural coordinates for q_X and for q_Y based at p and $f(p)$, the homeomorphism f can be written as

$$f(x + iy) = \sqrt{K}x + i\frac{1}{\sqrt{K}}y.$$

In complex notation, this can be written as

$$f(z) = \frac{1}{2}\left(\left(\frac{K + 1}{\sqrt{K}}\right)z + \left(\frac{K - 1}{\sqrt{K}}\right)\bar{z}\right).$$

Since $f_z = \frac{K+1}{2\sqrt{K}}$ and $f_{\bar{z}} = \frac{K-1}{2\sqrt{K}}$, we see that the dilatation of f is

$$K_f = \begin{cases} K & \text{if} \quad K \geq 1, \\ 1/K & \text{if} \quad K < 1. \end{cases}$$

We can concisely describe f by saying that it has *initial differential q_X*, *terminal differential q_Y*, and *horizontal stretch factor K*. (Because of the ambiguity that the horizontal stretch factors K and $1/K$ give rise to Teichmüller mappings with the same dilatation, we need to keep the distinction between horizontal stretch factor and dilatation.)

Note that the existence of a Teichmüller mapping presupposes that the initial and terminal differentials have the same Euclidean area; this is not a strong assumption, as any holomorphic quadratic differential can be scaled by a real number so as to have unit area and this rescaling does not change the corresponding horizontal or vertical foliations.

A Teichmüller mapping is not differentiable at the zeros of the initial differential, but it is smooth at all other points. This is why in our definition of quasiconformal homeomorphisms we chose to consider homeomorphisms that are smooth outside of a finite number of points instead of considering only smooth homeomorphisms.

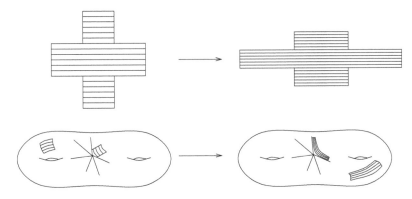

Figure 11.8 A Teichmüller mapping of the Swiss cross surface.

As a first example of a Teichmüller mapping, consider the homeomorphism between the Swiss cross Riemann surface and the stretched Swiss cross Riemann surface indicated in Figure 11.8. We can get other, more complicated Teichmüller mappings by rotating the foliation in Figure 11.8 so that it is not parallel to any sides of the polygon or even so that its slope is irrational.

Given two arbitrary complex structures on a topological surface, it is certainly not obvious that one can construct a Teichmüller mapping taking one structure to the other. However, the next theorem states that these mappings do indeed exist, and moreover they exist in every homotopy class.

THEOREM 11.8 (Teichmüller's existence theorem) *Let X and Y be closed Riemann surfaces of genus $g \geq 1$ and let $f : X \to Y$ be a homeomorphism. There exists a Teichmüller mapping $h : X \to Y$ homotopic to f.*

The main reason that Teichmüller mappings are so useful and important is that they provide a complete solution to Teichmüller's extremal problem for quasiconformal mappings.

THEOREM 11.9 (Teichmüller's uniqueness theorem) *Let $h : X \to Y$ be a Teichmüller map between two closed Riemann surfaces of genus $g \geq 1$. If $f : X \to Y$ is a quasiconformal homeomorphism homotopic to h, then*

$$K_f \geq K_h.$$

Equality holds if and only if $f \circ h^{-1}$ is conformal. In particular, if $g \geq 2$, then equality holds if and only if $f = h$.

The second statement follows from the first statement plus the fact that the only homotopically trivial conformal homeomorphism of a closed Riemann surface of genus $g \geq 2$ is the identity (cf. Proposition 7.7). For a closed Riemann surface X of genus $g = 1$, the group of conformal automorphisms of X is isomorphic to the group T^2.

We will prove both Theorems 11.8 and 11.9 later in this chapter.

Teichmüller's theorems for the torus. In Section 10.2, we showed that $\text{Teich}(T^2)$ is in natural bijective correspondence with $\text{SL}(2, \mathbb{R})/\text{SO}(2, \mathbb{R})$. Teichmüller's theorems can be interpreted in this case as saying that for each element of $\text{SL}(2, \mathbb{R})/\text{SO}(2, \mathbb{R})$ there is a distinguished representative in $\text{SL}(2, \mathbb{R})$ that is a hyperbolic matrix and whose leading eigenvalue is minimal among all hyperbolic representatives. It is instructive to find, for example, this distinguished representative for the coset given by $\left(\begin{smallmatrix} 1 & 1 \\ 0 & 1 \end{smallmatrix}\right)$; compare the proof of Theorem 11.20 below.

A minimization theorem for 1-manifolds. The analogue of Teichmüller's uniqueness theorem for 1-manifolds is nothing other than the mean value theorem. Identify S^1 with $\mathbb{R}/\mathbb{Z}$. Consider the set of all smooth homeomorphisms $S^1 \to S^1$ that fix 0. Define the dilatation of a smooth homeomorphism f by $\sup |f_x|$. By the mean value theorem, the dilatation is minimized in a given homotopy class precisely when f is linear.

11.4.2 GENERATING TEICHMÜLLER MAPS

In our definition of the Teichmüller map, we were handed two Riemann surfaces and a homeomorphism between them. It is natural to ask if, given an arbitrary closed Riemann surface X, an initial holomorphic quadratic differential q_X, and some $K > 1$, it is always possible to find a Riemann surface Y and a terminal holomorphic quadratic differential q_Y so that there is a

Teichmüller mapping $f : X \to Y$ with initial differential q_X, terminal differential q_Y, and horizontal stretch factor K. It turns out that this is always possible.

We now give the construction of the required Teichmüller mapping f for the given input is X, q_X, and K. Let X' be the complement in X of the zeroes of q_X. We will refer to the topological surfaces underlying X and X' as S and S', respectively. The surface X' is still a Riemann surface; its complex structure is given by a sufficiently large set of natural coordinates with respect to q_X, now thought of as a holomorphic quadratic differential on X'. If we compose each chart for X' with the affine map

$$f(x + iy) = \sqrt{K}x + i\frac{1}{\sqrt{K}}y,$$

we obtain a new set of charts on S', and this new set of charts defines a new complex structure on S'. Call the resulting Riemann surface Y'. In order to obtain the desired closed Riemann surface Y, we need only note that, by the removable singularity theorem (see, e.g., [48, Theorem V.1.2]), the complex structure Y' on S' extends uniquely to a complex structure Y on all of S.

There is an induced homeomorphism $f : X \to Y$ and an induced holomorphic quadratic differential q_Y on Y. By construction, f is a Teichmüller mapping with the desired properties. If we fix X and q_X but vary K in $(0, \infty)$, we obtain a one-parameter family of Riemann surfaces homeomorphic to X. Since each of these Riemann surfaces comes with an identification with X, we can think of this one-parameter family as a set of points in $\mathrm{Teich}(S)$, where S is the topological surface underlying X. The resulting subset of $\mathrm{Teich}(S)$ is called a *Teichmüller line*. The point X corresponds to $K = 1$. When we define the metric on Teichmüller space, we will see that Teichmüller lines are in fact geodesics.

Since the initial differential q_X on X specifies a unique ray in $\mathrm{Teich}(S)$, we see that we can think of q_X as giving a tangent direction and the pair (q_X, K) as giving a tangent vector to $\mathrm{Teich}(S)$ at X. Above we gave a norm on $\mathrm{QD}(X)$. The resulting map $(q_X, \|q_X\|) \to \mathrm{Teich}(S)$ can be thought of as an exponential map $T_X(\mathrm{Teich}(S)) \to \mathrm{Teich}(S)$. We remark that $\mathrm{QD}(X)$ is usually identified with the *cotangent* space of $\mathrm{Teich}(S)$ at the point X.

11.4.3 SURFACES WITH PUNCTURES AND BOUNDARY

In order to state Teichmüller's existence and uniqueness theorems (Theorems 11.8 and 11.9) in the context of punctured surfaces, we need to distinguish two types of conformal structures near a topological puncture; see Figure 11.9. The puncture on the left-hand side of the figure has a neighborhood that is conformally equivalent to the unit disk in $\mathbb{C}$ minus 0, and

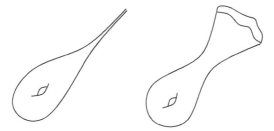

Figure 11.9 Two different conformal types in the neighborhood of a puncture: a punctured disk (left) and an annulus (right).

the other puncture has a neighborhood that is conformally equivalent to an annulus $\{z \in \mathbb{C} : r_1 \leq |z| < r_2\}$. A homeomorphism $f : X \to Y$ has a quasiconformal representative if and only if f takes punctures to punctures, and at each puncture these conformal types are preserved. In this case, both Teichmüller existence and uniqueness theorems hold, assuming that the underlying topological surface has negative Euler characteristic. One way to prove this in the case of punctures of the first type is to find a double covering of the Riemann surface in question where the complex structure can be extended over the punctures.

Since there is no quasiconformal homeomorphism homotopic to $f : X \to Y$ if the types of punctures are not preserved, one usually considers the Teichmüller space of a surface where the conformal types of the punctures are part of the given data.

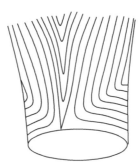

Figure 11.10 An allowable measured foliation near the boundary of a surface in Teichmüller theory.

In the case of surfaces with boundary, one must be careful about the measured foliations that are allowed. For the purposes of Teichmüller theory, we insist that

1. each component of ∂S contain at least one singularity, and

2. any leaf of not containing a singularity on ∂S but meeting a small tubular neighborhood of ∂S must be parallel to ∂S; that is, ∂S should be a union of leaves connecting singularities. See Figure 11.10.

In terms of quadratic differentials on a Riemann surface X, these conditions can be restated as: each component of ∂X must contain at least one zero and ∂X is part of the horizontal foliation away from the singularities. As a consequence, the corresponding singular Euclidean metric degenerates at the boundary.

For details on extending Teichmüller theory to the nonclosed case, see, for example, [1, Chapter II].

11.5 GRÖTZSCH'S PROBLEM

In 1928 Grötzsch proved the following precursor to the Teichmüller uniqueness theorem. It solves the quasiconformal extremal problem for rectangles. The proof of Teichmüller's uniqueness theorem is based on the solution to Grötzsch's problem.

THEOREM 11.10 (Grötzsch's problem) *Let X be the rectangle $[0,a] \times [0,1]$ in $\mathbb{R}^2$ and let Y be the rectangle $[0,Ka] \times [0,1]$ for some $K \geq 1$. If $f : X \to Y$ is any orientation-preserving homeomorphism that is smooth away from a finite number of points, that takes horizontal sides to horizontal sides, and that takes vertical sides to vertical sides, then*

$$K_f \geq K$$

with equality if and only if f is affine.

Note that Theorem 11.10 really gives a statement about general rectangles, as any rectangle is conformally equivalent to one with vertical side length 1.

Proof. Let $f : X \to Y$ be as in the statement of the theorem. See Figure 11.11. Let $K_f(x,y)$ and $j_f(x,y)$ denote the dilatation and the Jacobian of f at the point $(x,y) \in X$.

We begin with two simple inequalities. The first is

$$|f_x(x,y)|^2 \leq K_f(x,y)j_f(x,y). \tag{11.2}$$

If M and m are the supremum and infimum of

$$\{|df(v)| : v \in UT(X)\},$$

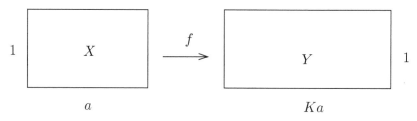

Figure 11.11 The setup for Grötzsch's problem.

then (11.2) is true because

$$K_f(x, y) = M/m \qquad \text{and} \qquad j_f(x, y) = Mm.$$

The second key inequality is

$$\int_X |f_x(x, y)|\, dA \geq K\, \text{Area}(X), \qquad (11.3)$$

which is obtained from the inequality $\int_0^a |f_x(x, y)|\, dx \geq Ka$ by integrating from 0 to 1 with respect to y.

We are now ready to show that $K_f \geq K$. Without loss of generality, assume $K \geq 1$. Then

$$
\begin{aligned}
(K\ \text{Area}(X))^2 &\leq \left(\int_X |f_x(x, y)|\, dA \right)^2 \\
&\leq \left(\int_X \sqrt{K_f(x, y)} \sqrt{j_f(x, y)}\, dA \right)^2 \\
&\leq \left(\int_X j_f(x, y)\, dA \right) \left(\int_X K_f(x, y)\, dA \right) \\
&\leq (K\ \text{Area}(X))(K_f\ \text{Area}(X)).
\end{aligned}
$$

The first three inequalities follow from (11.3), (11.2), and the Cauchy–Schwarz inequality. The fourth inequality follows from the fact that $K_f(x, y) \leq K_f$ for all $(x, y) \in X$. It follows that $K_f \geq K$.

The lower bound $K_f = K$ is achieved when f is the affine map

$$A : (x, y) \mapsto (Kx, y).$$

It remains to prove the uniqueness statement.

Let $f : X \to Y$ be an orientation-preserving homeomorphism as in the statement of the theorem and assume $K_f = K$. By replacing f with $A^{-1} \circ f$, we can assume that $K = 1$, hence $K_f = 1$. Our goal now is to show that f is the identity.

In the sequence of four inequalities above, the assumption that distinguishes the x-direction from the y-direction is the assumption that $K \geq 1$. Since we now have $K = 1$, we have attained symmetry between the horizontal and vertical directions.

We must have that all four inequalities in the above calculation are equalities. For the first of the four inequalities, namely, inequality (11.3), to be an equality, it must be that f takes horizontal line segments of X to horizontal line segments of Y. By symmetry, f must also take vertical line segments of X to vertical line segments of Y. We therefore have

$$f(x, y) = (u(x), v(y)).$$

For the second inequality above, namely, inequality (11.2), to be an equality, we must have that the direction of maximal stretch for f is the x-direction at almost every point of X. By symmetry, the direction of maximal stretch for f must also be the y-direction at almost every point. Thus $|u'(x)| = |v'(y)|$ at almost every point $(x, y) \in X$, and $j_f(x, y) = u'(x)v'(y) = u'(x)^2 = v'(y)^2$ (we are using the fact that f is orientation-preserving to say that $j_f \geq 0$) and $K_f(x, y) = 1$. Since $K_f = \sup K_f(x, y) = 1$, we must have that $K_f(x, y) \equiv 1$. For the third inequality above, the Cauchy–Schwarz inequality, to be an equality, we must have that

$$j_f(x, y)/K_f(x, y) = j_f(x, y) = u'(x)^2 = v'(y)^2$$

is constant almost everywhere. For the fourth inequality to be an equality, we must have that $j_f(x, y)$ is equal to $K = 1$ almost everywhere. Thus $u'(x) = v'(x) = 1$ almost everywhere and $f(x, y) = (x, y)$ almost everywhere. Since f is a homeomorphism, it follows that $f(x, y) = (x, y)$ everywhere, as desired. $\qquad\square$

11.6 PROOF OF TEICHMÜLLER'S UNIQUENESS THEOREM

The proof of Teichmüller's uniqueness theorem (Theorem 11.9) is almost exactly the solution to Grötzsch's problem just given. The only difficulty is to prove the analogue of the inequality (11.3) with the rectangle X replaced by a closed Riemann surface. To do this we will first need the following lemma.

In what follows ℓ_q denotes the Euclidean length of a path with respect to a holomorphic quadratic differential q.

Lemma 11.11 *Let q_Y be a holomorphic quadratic differential on a closed Riemann surface Y. Let $h : Y \to Y$ be a homeomorphism that is homotopic to the identity. Then there exists a constant $M \geq 0$ with the following property: any arc $\alpha : [0, 1] \to Y$ embedded in a leaf of the horizontal foliation for q_Y satisfies*

$$\ell_{q_Y}(h(\alpha)) \geq \ell_{q_Y}(\alpha) - M.$$

The point of Lemma 11.11 is that, while the constant M depends on the homeomorphism h, it does not depend on the arc α.

Proof. Let $H(x, t) : Y \times [0, 1] \to Y$ be any homotopy from h to the identity. For each $x \in Y$, the ℓ_{q_Y}-length of the path $H(x, t)$ is a continuous function $Y \to \mathbb{R}$. Since Y is compact, it attains a maximum $N \geq 0$.

Denote by δ_0 and δ_1 the arcs $H(\alpha(0), 1 - t)$ and $H(\alpha(1), t)$, where $0 \leq t \leq 1$. Note that $\ell_{q_Y}(\delta_0) \leq N$ and $\ell_{q_Y}(\delta_1) \leq N$. The concatenation of arcs $\delta_0 \star h(\alpha) \star \delta_1$ is homotopic to α, relative to endpoints. Since α is embedded in a horizontal leaf of q_Y, it minimizes the length of any arc in its relative homotopy class. We thus have

$$\begin{aligned}
\ell_{q_Y}(\alpha) &\leq \ell_{q_Y}(\delta_0 \star h(\alpha) \star \delta_1) \\
&= \ell_{q_Y}(\delta_0) + \ell_{q_Y}(h(\alpha)) + \ell_{q_Y}(\delta_0) \\
&\leq \ell_{q_Y}(h(\alpha)) + 2N.
\end{aligned}$$

Setting $M = 2N$ completes the proof. $\square$

We are now ready to prove the analogue for closed Riemann surfaces of inequality (11.3), which estimates the mean horizontal stretching of a map homotopic to a Teichmüller mapping. In the statement of the following proposition, f_x denotes the derivative of f in the direction of the horizontal foliation for q_X, and $\text{Area}(q)$ denotes the Euclidean area of the holomorphic quadratic differential q.

Proposition 11.12 *Let $h : X \to Y$ be a Teichmüller mapping between closed Riemann surfaces. Suppose that h has initial differential q_X, terminal differential q_Y, and horizontal stretch factor K. Let $f : X \to Y$ be any homeomorphism that is homotopic to h and that is smooth outside a finite number of points. Then*

$$\int_X |f_x| \, dA \geq \sqrt{K} \, \text{Area}(q_X).$$

We note that in the case that each of the horizontal leaves of q_X is closed, the proof of Proposition 11.12 is quite similar to the solution of Grötzsch's problem. The case when q_X has nonclosed leaves is more subtle.

Proof of Proposition 11.12. Consider the function $\delta \,:\, X \times \mathbb{R}_{\geq 0} \to \mathbb{R}_{\geq 0}$ given by

$$\delta(p, L) = \int_{-L}^{L} |f_x| \, dx.$$

If $\alpha_{p,L}$ is the horizontal arc of length $2L$ centered at p, then $\delta(p, L)$ is the integral of $|f_x|$ along $\alpha_{p,L}$. Note that δ is not defined everywhere; specifically, it is not defined at any (p, L) where p lies at a horizontal distance less than L from a zero of q_X. This is a set of measure zero in $X \times \mathbb{R}_{\geq 0}$ since there are only finitely many zeros of q_X, each meeting finitely many leaves of the horizontal foliation for q_X.

Where $\delta(p, L)$ is defined, we have

$$\delta(p, L) = \ell_{q_Y} (f(\alpha_{p,L})).$$

The Teichmüller map h takes $\alpha_{p,L}$ to an arc of q_Y-length $2L\sqrt{K}$. It follows from Lemma 11.11 that

$$\ell_{q_Y} (f(\alpha_{p,L})) \geq 2L\sqrt{K} - M,$$

where M is independent of p and of L. Thus

$$\begin{aligned}
\int_X \delta(p, L) \, dA &= \int_X \ell_{q_Y} (f(\alpha_{p,L})) \, dA \\
&\geq \int_X \left(2L\sqrt{K} - M \right) dA \\
&= \left(2L\sqrt{K} - M \right) \mathrm{Area}(q_X).
\end{aligned}$$

On the other hand, Fubini's theorem gives

$$\int_X \delta(p, L) \, dA = \int_X \left(\int_{-L}^{L} |f_x| \, dx \right) dA = 2L \int_X |f_x| \, dA.$$

Combining the above two equations gives

$$2L \int_X |f_x| \, dA \geq \left(2L\sqrt{K} - M \right) \mathrm{Area}(q_X).$$

Dividing both sides of this inequality by $2L$ and allowing L to tend to infinity gives the result. ☐

As alluded to above, the proof of Teichmüller's uniqueness theorem is now almost verbatim the solution of Grötzsch's problem. The main change is that the inequality (11.3) is replaced with Proposition 11.12. Also, the map A from Grötzsch's problem is replaced with a Teichmüller map h, and the horizontal and vertical directions must now be interpreted as the directions of the horizontal and vertical foliations determined by h. With these changes, the solution to Grötzsch's problem (the proof of Theorem 11.10) applies exactly as stated to prove Teichmüller's uniqueness theorem.

11.7 PROOF OF TEICHMÜLLER'S EXISTENCE THEOREM

The goal of this section is to prove Teichmüller's existence theorem (Theorem 11.8). The key idea here is to reinterpret the existence problem for Teichmüller maps as the problem of proving surjectivity of a natural "exponential map" $\Omega : \mathrm{QD}(X) \to \mathrm{Teich}(S_g)$. We will define such a map, prove continuity and properness, and deduce surjectivity by general topology, namely, invariance of domain. We now begin executing this strategy.

11.7.1 Proof of the Theorem

Let X be a closed Riemann surface and let $q \in \mathrm{QD}(X)$. Recall that we have a norm on $\mathrm{QD}(X)$ given by

$$\|q\| = \int_X |q|.$$

Let $\mathrm{QD}_1(X)$ denote the open unit ball in $\mathrm{QD}(X)$. For $q \in \mathrm{QD}_1(X)$, set

$$K = \frac{1 + \|q\|}{1 - \|q\|}.$$

As in Section 11.4, we can construct a Riemann surface Y and a Teichmüller mapping $h : X \to Y$ with initial differential q and horizontal stretch factor K. By identifying X with S_g, we can regard X as a point $\mathcal{X} \in \mathrm{Teich}(S_g)$. Then, regarding the Teichmüller map h as a marking $h : S_g \to Y$, we obtain a point $\mathcal{Y} = [(Y, h)]$ in $\mathrm{Teich}(S_g)$. This procedure therefore defines a function

$$\Omega : \mathrm{QD}_1(X) \to \mathrm{Teich}(S_g).$$

Teichmüller's existence theorem then amounts to the statement that Ω is surjective. Indeed, let Z be a Riemann surface and let $f : X \to Z$ be a homeomorphism. Teichmüller's existence theorem states that there is a Teichmüller map $h : X \to Z$ in the homotopy class of f. As above, by identifying X with S_g, we can regard f as a marking $S_g \to Z$. Then the pair (Z, f) represents a point $\mathcal{Z} \in \text{Teich}(S_g)$. If there exists $q \in \text{QD}_1(X)$ such that $\Omega(q) = \mathcal{Z}$, then this exactly means that there is a Teichmüller map $h : X \to Z$ in the homotopy class of f, as desired.

Proposition 11.13 *Let* $g \geq 1$. *The map* $\Omega : \text{QD}_1(X) \to \text{Teich}(S_g)$ *is continuous.*

Proposition 11.13 is far from obvious. For example, even if $q \in \text{QD}(X)$ has the property that its horizontal foliation has only closed leaves, there will be a nearby $q' \in \text{QD}(X)$ with leaves that are not closed. If we stretch along both foliations by a factor of K, there is no simple reason why the resulting points of $\text{Teich}(S_g)$ should be close to each other.

Proposition 11.13 represents the main content of our proof of Teichmüller's existence theorem. We will prove it below as a corollary of the measurable Riemann mapping theorem.

Proposition 11.14 *The map* $\Omega : \text{QD}_1(X) \to \text{Teich}(S_g)$ *is proper.*

Proof. Let $\kappa : \text{Teich}(S_g) \to \mathbb{R}$ be defined by the following formula. For $\mathcal{Y} \in \text{Teich}(S_g)$, we represent $\mathcal{Y}$ by a marked Riemann surface Y. Then we set

$$\kappa(\mathcal{Y}) = \inf\{K_h \,|\, h : X \to Y \text{ a quasiconformal homeomorphism}$$
$$\text{isotopic to the change of marking}\}.$$

We claim that the map κ is continuous. Indeed, given two nearby points $\mathcal{Y}$ and $\mathcal{Y}'$ in $\text{Teich}(S_g)$, we can represent them by nearby elements of $\text{DF}(\pi_1(S_g), \text{PSL}(2, \mathbb{R}))$. The (marked) fundamental domains for these representations can be made K-quasiconformally equivalent for any $K > 1$ by taking $\mathcal{Y}'$ sufficiently close to $\mathcal{Y}$. By the definition of $\kappa(\mathcal{Y})$, there exists a quasiconformal homeomorphism $h : X \to Y$, isotopic to the change of marking, with $K_h = \kappa(\mathcal{Y}) + \epsilon$, where ϵ is an arbitrarily small positive number. Since the composition of a $(\kappa(\mathcal{Y}) + \epsilon)$-quasiconformal map with a K-quasiconformal map is $K(\kappa(\mathcal{Y}) + \epsilon)$-quasiconformal, it follows that $\kappa(\mathcal{Y}')$ can be made arbitrarily close to $\kappa(\mathcal{Y})$ by taking $\mathcal{Y}'$ close to $\mathcal{Y}$.

Let $A \subset \text{Teich}(S_g)$ be compact. Since κ is continuous, $\kappa|A$ attains a maximum, say $M \geq 0$. We claim that $\Omega^{-1}(A)$ is contained in the closed ball of radius $(M - 1)/(M + 1)$ about the origin in $\text{QD}_1(X)$. Since $\Omega^{-1}(A)$ is closed by Proposition 11.13, this claim will imply that $\Omega^{-1}(A)$ is compact, so that Ω is proper.

We now prove the claim. Let $q \in \Omega^{-1}(A)$. By the definition of Ω, there is a Teichmüller map $h : X \to \Omega(q)$ that is isotopic to the change of marking and that has dilatation

$$K_h = \frac{1 + \|q\|}{1 - \|q\|}.$$

By Teichmüller's uniqueness theorem (Theorem 11.9), any quasiconformal homeomorphism $X \to \Omega(q)$ isotopic to the change of marking must have dilatation at least K_h. It follows then from the definition of M that

$$M \geq K_h = \frac{1 + \|q\|}{1 - \|q\|}.$$

Solving for $\|q\|$, we find that

$$\|q\| \leq \frac{M - 1}{M + 1} < 1,$$

which is what we wanted to show. $\square$

Brouwer's invariance of domain theorem [36] states that any injective continuous map $\mathbb{R}^n \to \mathbb{R}^n$ is an open map. We have the following straightforward consequence, which is the final piece needed for our proof of Teichmüller's existence theorem.

THEOREM 11.15 *Any proper injective continuous map* $\mathbb{R}^n \to \mathbb{R}^n$ *is a homeomorphism.*

With the above ingredients in place, we can prove Teichmüller's existence theorem (Theorem 11.8).

Proof of Teichmüller's existence theorem. The map

$$\Omega : \mathrm{QD}_1(X) \to \mathrm{Teich}(S_g)$$

is injective by Teichmüller's uniqueness theorem (Theorem 11.9), proper (Proposition 11.14) and continuous (Proposition 11.13). Also, $\mathrm{QD}_1(X)$ is a real vector space of dimension greater than or equal to $6g - 6$ (Theorem 11.6) and $\mathrm{Teich}(S_g) \approx \mathbb{R}^{6g-6}$ (Theorem 10.6). Since $\mathrm{QD}_1(X)$ contains a subspace homeomorphic to $\mathbb{R}^{6g-6}$, Theorem 11.15 implies that the map Ω is surjective, which is what we wanted to show. $\square$

11.7.2 Beltrami Differentials, the Measurable Riemann Mapping Theorem, and the Continuity of Ω

Fix a closed Riemann surface X of genus $g \geq 2$. To prove the continuity of $\Omega : \mathrm{QD}_1(X) \to \mathrm{Teich}(S_g)$, we will factor Ω as

$$\Omega : \mathrm{QD}_1(X) \overset{\Omega_1}{\to} L^\infty(U) \overset{\Omega_2}{\to} \mathrm{Teich}(S_g),$$

where $U \subset \mathbb{C}$ is the upper half-plane. The image of Ω_1 will consist of (equivalence classes of) complex-valued functions that come from Beltrami differentials, which we now define.

Ellipse fields. Recall that an *ellipse field* on a Riemann surface X is a choice of ellipse in each tangent space TX_p at each point $p \in X$. An ellipse field is *smooth* if, when written in local coordinates, it varies smoothly. A quasiconformal homeomorphism $f : X \to Y$ determines an ellipse field on X that is well defined and smooth almost everywhere, as follows. Given a point $p \in X$, we take the ellipse in TX_p that is the preimage of the unit circle in $TY_{f(p)}$ under the derivative of f (when the latter is defined). Since this ellipse is well defined only up to scale, we always choose the ellipse to have unit area.

In any chart we can encode an ellipse field by a complex-valued function μ called the *complex dilatation*. It is given locally by the formula

$$\mu = f_{\bar{z}}/f_z.$$

Recall from the beginning of the chapter that $|\mu| < 1$ if and only if f is orientation-preserving.

An ellipse field is smooth if and only if the corresponding μ is smooth in each chart. The dilatation $K_f(p)$ is locally given by the formula

$$K_f(p) = \frac{1 + |\mu(p)|}{1 - |\mu(p)|}.$$

It is also possible to calculate the angle, in any chart, of the direction of maximal stretch of df. It is given by $\frac{1}{2}\arg(\mu)$. This information completely determines the corresponding ellipse field; namely, the ellipse at the point p is (up to scale) the unit-area ellipse with major axis having angle $\frac{1}{2}\arg(\mu)$ and with the ratio of the lengths of the axes being $(1 + |\mu(p)|)/(1 - |\mu(p)|)$.

Beltrami differentials. To make a definition of μ that is independent of charts, we define it as a $(-1, 1)$-*form*, which simply means that μ transforms under change of coordinates by the formula

$$\mu(z) = \mu(w) \overline{\left(\frac{dw}{dz}\right)} \bigg/ \frac{dw}{dz} \,, \qquad (11.4)$$

where z and w are two overlapping sets of coordinates. Since $\overline{(dw/dz)}/(dw/dz)$ lies on the unit circle, the $(-1, 1)$-form μ gives rise to a well-defined function $|\mu| : X \to \mathbb{R}$. We say that the differential μ is a *Beltrami differential* if $|\mu|$ is essentially bounded, that is, off of a set of measure zero it is a bounded function.

We can interpret (11.4) geometrically as follows: if a transition map between charts rotates by an angle α at a point, then μ is multiplied by $e^{i2\alpha}$. Since the angle of the ellipse field is locally given by $\frac{1}{2} \arg(\mu)$, this means that the differential of a change of coordinates map from one chart to another takes the ellipse field corresponding to the first local expression for μ to the ellipse field corresponding to the second local expression for μ.

As the Riemann surface X is a quotient $X = U/\pi_1(X)$ of the upper half-plane U by conformal automorphisms, we can think of the coefficient μ of a Beltrami differential as a bounded, $\pi_1(X)$-equivariant, measurable function on U. That is, we can use the interior of some preferred fundamental domain for $\pi_1(X)$ in U as the target of a single chart, take μ as above on that chart, and extend μ to all of U using the action of $\pi_1(X)$ on U and the change-of-coordinates formula (11.4). By reflecting across the real axis, that is, by setting $\mu(\bar{z}) = \overline{\mu(z)}$, we can also think of μ as an element of $L^\infty(\mathbb{C})$. Recall that two functions represent the same point of $L^\infty(\mathbb{C})$ if they are equal almost everywhere. We give $L^\infty(\mathbb{C})$ its usual topology of almost everywhere uniform convergence.

The measurable Riemann mapping theorem. We saw above how every quasiconformal homeomorphism f of a Riemann surface X gives rise to a bounded unit-area ellipse field μ_f on X. We also saw that any such unit-area ellipse field is equivalent to giving a ($\pi_1(X)$-equivariant) element $\mu \in L^\infty(U)$. Which such $\mu \in L^\infty(\mathbb{C})$ occur? This question gives rise to the fundamental "inverse problem" for Beltrami differentials: given any $\mu \in L^\infty(\mathbb{C})$, is it possible to find a quasiconformal homeomorphism $f : \mathbb{C} \to \mathbb{C}$ so that f satisfies the *Beltrami equation*

$$\mu f_z = f_{\bar{z}}$$

almost everywhere? Very generally, the answer is yes.

THEOREM 11.16 (Measurable Riemann mapping theorem) *Let* $\mu \in L^\infty(\mathbb{C})$ *and suppose* $\|\mu\|_\infty < 1$. *There exists a unique quasiconformal* *homeomorphism* $f^\mu : \widehat{\mathbb{C}} \to \widehat{\mathbb{C}}$ *that fixes* 0, 1, *and* ∞ *and satisfies almost everywhere the Beltrami equation*

$$\mu f_z^\mu = f_{\bar{z}}^\mu.$$

Further, f^μ *is smooth wherever* μ *is, and* f^μ *varies complex analytically with respect to* μ.

By the uniqueness statement in Theorem 11.16 we see that if $\mu(\bar{z}) = \overline{\mu(z)}$, then f^μ restricts to a self-map of the upper half-plane U of $\mathbb{C}$. The uniqueness statement, together with the $\pi_1(X)$-equivariance of μ, also implies that f^μ is $\pi_1(X)$-equivariant.

There is a long history concerning the existence of solutions to the Beltrami equation. The case where μ is continuous was first proven by Lavrentiev [130], and where μ is measurable by Morrey [162]. The analytic dependence of the solution on μ is due to Ahlfors–Bers [2].

The proof of Theorem 11.16 is beyond the scope of this book. We refer the reader to [2] or [3, Chapter 5] for the proof. Assuming this theorem, we can now prove Proposition 11.13, which states that $\Omega : \mathrm{QD}_1(X) \to \mathrm{Teich}(S_g)$ is continuous.

The continuity of Ω. We now apply the measurable Riemann mapping theorem in order to prove that Ω is continuous.

Proof of Proposition 11.13. Let X be a Riemann surface of genus $g \geq 2$. We take X to be homeomorphically identified with S_g, and so X represents a point of $\mathrm{Teich}(S_g)$. After defining the maps

$$\Omega_1 : \mathrm{QD}_1(X) \to L^\infty(U) \ \text{ and } \ \Omega_2 : L^\infty(U) \to \mathrm{Teich}(S_g)$$

that we alluded to at the start of this section, we will prove that both Ω_1 and Ω_2 are continuous. We will then show that $\Omega = \Omega_2 \circ \Omega_1$.

Let $q \in \mathrm{QD}(X)$. Just as we were able to convert a Beltrami differential into a $\pi_1(X)$-equivariant function $U \to \mathbb{C}$, so are we able to convert q into a $\pi_1(X)$-equivariant function $\tilde{q} : U \to \mathbb{C}$ defined almost everywhere on U. What is more, if we fix the covering map $U \to X$ and the preferred fundamental domain in U ahead of time, then the map $\mathrm{QD}(X) \to L^\infty(U)$ given by $q \mapsto \tilde{q}$ is a well-defined function.

We can then define $\Omega_1 : \mathrm{QD}_1(X) \to L^\infty(U)$ by setting $\Omega_1(0) = 0$ and by setting

$$\Omega_1(q)(z) = \|q\| \, \overline{\tilde{q}(z)}/|\tilde{q}(z)|$$

for $q \neq 0$. Here $\|q\|$ is the norm of the vector q in the vector space $\mathrm{QD}(X)$, and $|\tilde{q}(z)|$ is the absolute value of the complex number $\tilde{q}(z)$. Note that if $\tilde{q}$ transforms by dz^2, then $\overline{\tilde{q}}/|\tilde{q}|$ transforms by $\overline{dz}/dz$, as desired. Informally, $\overline{dz^2}/(dz\overline{dz}) = \overline{dz}/dz$.

We claim that the map Ω_1 is continuous. Indeed, as a function on $\tilde{X} \approx U$, the element $\Omega_1(q) \in L^\infty(U)$ is equivariant with respect to the $\pi_1(X)$ action on U in the sense that (11.1) is satisfied. Thus, if we change $q \in \mathrm{QD}_1(X)$ by a small amount in one chart (say the chart given by the preferred fundamental domain), then by (11.1), the function $\Omega_1(q)$ changes by a small amount. It also follows from (11.1) that $\|\Omega_1(q)\|_\infty = \|q\| < 1$.

The map $\Omega_2 : L^\infty(U) \to \mathrm{Teich}(S_g)$ is given by the measurable Riemann mapping theorem. To make this precise, we begin by realizing X as a representation

$$\rho : \pi_1(X) \to \mathrm{Isom}^+(\mathbb{H}^2).$$

Let $\mu \in L^\infty(U)$ and reflect over the real axis so that $\mu \in L^\infty(\mathbb{C})$. Let $f^\mu : \mathbb{C} \to \mathbb{C}$ be the function guaranteed by the measurable Riemann mapping theorem (Theorem 11.16) and restrict it to U. If we conjugate each element in the image of ρ by f^μ, we obtain a new Riemann surface X', and f^μ induces a homeomorphism $X \to X'$ that is smooth almost everywhere. We can regard X' as a point of $\mathrm{Teich}(S_g)$. The last sentence in the statement of the measurable Riemann mapping theorem (Theorem 11.16) implies that Ω_2 is continuous.

It only remains to check that $\Omega : \mathrm{QD}_1(X) \to \mathrm{Teich}(S_g)$ is equal to the composition $\Omega_2 \circ \Omega_1$. Let $q \in \mathrm{QD}_1(X)$ and suppose that at some point $u \in U$ we have $\tilde{q}(u) = re^{i\theta}$. Then, $\Omega(q) \in \mathrm{Teich}(S_g)$ is obtained from X by stretching by a factor of $(1 + \|q\|)/(1 - \|q\|)$ in the direction $e^{-i\theta/2}$ at that point. On the other hand, $\Omega_1(q)(u)$ is equal to $\|q\|e^{-i\theta}$, and so the map $f^\mu = \Omega_2(\Omega_1(q))$ satisfies $f^\mu_{\bar{z}}/f^\mu_z = \|q\|e^{-i\theta}$ at u. That is, f^μ stretches in the direction $e^{-i\theta/2}$ by a factor $(1 + \|q\|)/(1 - \|q\|)$ at u. Thus $\Omega = \Omega_2 \circ \Omega_1$, and we are done. $\square$

Beltrami differentials versus quadratic differentials. Let X be a Riemann surface representing a point $\mathcal{X} \in \mathrm{Teich}(S_g)$. We already discussed the correspondence between $\mathrm{QD}(X)$ and the cotangent space to $\mathrm{Teich}(S_g)$ at $\mathcal{X}$. There is a natural pairing between quadratic differentials and Beltrami differentials that we can use to identify the tangent space of $\mathrm{Teich}(S_g)$. Specifically, if a holomorphic quadratic differential q is given locally by $\phi(z)\,dz^2$ and a Beltrami differential μ is given locally by $\mu(z)\,\overline{dz}/dz$, then we set

$$\langle q, \mu \rangle = \int_X \phi\mu \, |dz|^2.$$

This pairing allows us to identify the tangent space to $\mathrm{Teich}(S_g)$ at X as the space of Beltrami differentials on X modulo the subspace of infinitesimally trivial Beltrami differentials (both spaces are infinite-dimensional, but the quotient has dimension $6g - 6$). The infinitesimally trivial Beltrami differentials are the ones that are the derivatives of homeomorphisms of $S_g \approx X$ that are homotopic to the identity.

11.8 THE TEICHMÜLLER METRIC

Let $X, Y \in \mathrm{Teich}(S_g)$ and say that X and Y are represented by marked Riemann surfaces X and Y, respectively. Again, because of the markings, there is a preferred map $f : X \to Y$, the change-of-marking map. Let $h : X \to Y$ be a Teichmüller mapping in the homotopy class of f whose existence is guaranteed by Teichmüller's existence theorem (Theorem 11.8). Let $K = K_h$ be the dilatation of h. We define the *Teichmüller distance* between X and Y to be

$$d_{\mathrm{Teich}}(X, Y) = \frac{1}{2}\log(K).$$

By Teichmüller's uniqueness theorem (Theorem 11.9) the function d_{Teich} is well defined.

11.8.1 BASIC PROPERTIES

For the next proposition, recall that in Section 10.3 we defined a topology on $\mathrm{Teich}(S_g)$, the algebraic topology.

Proposition 11.17 *The Teichmüller distance d_{Teich} defines a complete metric on $\mathrm{Teich}(S_g)$ whose topology is compatible with the algebraic topology on $\mathrm{Teich}(S_g)$.*

The metric defined by d_{Teich} is called the *Teichmüller metric*.

Proof. Teichmüller's existence theorem (Theorem 11.8) implies that $d_{\mathrm{Teich}}(X, Y) = 0$ if and only if there is a Teichmüller mapping $h : X \to Y$ of dilatation 1 that is homotopic to the change of marking. By Lemma 11.1 the homeomorphism h is conformal. This is the same as saying that $X = Y$ in $\mathrm{Teich}(S_g)$. By Proposition 11.3 the inverse of a K-quasiconformal homeomorphism is a K-quasiconformal homeomorphism, and so $d_{\text{Teich}}(\mathcal{X}, \mathcal{Y}) = d_{\text{Teich}}(\mathcal{Y}, \mathcal{X})$. The triangle inequality for d_{Teich} also follows from Proposition 11.3, which states that the composition of a K-quasiconformal homeomorphism and a K'-quasiconformal homeomorphism is a KK'-quasiconformal homeomorphism. Thus d_{Teich} is a metric.

Next we show completeness of $(\text{Teich}(S_g), d_{\text{Teich}})$. Let $\mathcal{X}$ be a point of $\text{Teich}(S_g)$ represented by a marked Riemann surface X. Recall that in Section 11.7 we defined a map

$$\Omega : \text{QD}_1(X) \to \text{Teich}(S_g)$$

and showed it was a homeomorphism. Under Ω^{-1}, a point in $\text{Teich}(S_g)$ at distance $\log(K)/2$ from the basepoint $\mathcal{X}$ maps to a point of $\text{QD}_1(X)$ whose norm is $(K-1)/(K+1)$. For $K \geq 1$, we have an inequality

$$\frac{K-1}{K+1} \leq \frac{1}{2}\log(K).$$

If K is bounded from above, then $(K-1)/(K+1)$ is bounded away from 1. Thus Ω^{-1} takes closed balls about the basepoint $\mathcal{X} \in \text{Teich}(S_g)$ to compact balls about the origin in $\text{QD}_1(X)$. Since Ω^{-1} is a homeomorphism, this implies that closed balls about the basepoint in $\text{Teich}(S_g)$ are compact, and thus $(\text{Teich}(S_g), d_{\text{Teich}})$ is complete. $\square$

One may wonder about the factor of $1/2$ in the definition of d_{Teich}. The factor of $1/2$ is included so that certain 2-dimensional subspaces of $\text{Teich}(S_g)$, called *Teichmüller disks*, are isometric to the hyperbolic plane $\mathbb{H}^2$ in the unit disk model, which has curvature -4. Briefly, a Teichmüller disk is obtained as follows: start with a complex structure on a surface S coming from a polygon with parallel sides identified, for example, the Swiss cross example. If we apply an element of $\text{SL}(2, \mathbb{R})$, acting as a linear transformation of $\mathbb{R}^2$, the images of the sides of the polygon are still parallel, and so we obtain a new complex structure on S. The stabilizer of a complex structure is the orthogonal group. Since $\text{SL}(2, \mathbb{R})/\text{SO}(2, \mathbb{R}) \approx \mathbb{H}^2$, it follows that the $\text{SL}(2, \mathbb{R})$-orbit of the original marked complex structure is a copy of $\mathbb{H}^2$ in $\text{Teich}(S_g)$. It turns out that this inclusion $\mathbb{H}^2 \hookrightarrow \text{Teich}(S_g)$ is an isometric embedding.

11.8.2 TEICHMÜLLER GEODESICS

In Section 11.4, we explained how any point $\mathcal{X} = [(X, \phi)] \in \text{Teich}(S_g)$ and any holomorphic quadratic differential $q \in \text{QD}(X)$ determine an embedded copy of $\mathbb{R} \hookrightarrow \text{Teich}(S_g)$ containing $\mathcal{X}$, called a *Teichmüller line*. By the definition of d_{Teich}, this embedding is actually an isometric embedding.

Proposition 11.18 *Let $g \geq 1$. Teichmüller lines in* $\mathrm{Teich}(S_g)$ *are bi-infinite geodesics with respect to the Teichmüller metric.*

Even more is true: Teichmüller lines account for all geodesics in $(\mathrm{Teich}(S_g), d_{\mathrm{Teich}})$.

THEOREM 11.19 *Let $g \geq 1$. Every geodesic segment in the space* $(\mathrm{Teich}(S_g), d_{\mathrm{Teich}})$ *is a subsegment of some Teichmüller line. In particular, there is a unique geodesic in* $\mathrm{Teich}(S_g)$ *between any two points.*

Proof. Let $\mathcal{X}$ and $\mathcal{Z}$ be points of $\mathrm{Teich}(S_g)$ and suppose $\mathcal{Y} \in \mathrm{Teich}(S_g)$ satisfies

$$d(\mathcal{X}, \mathcal{Y}) + d(\mathcal{Y}, \mathcal{Z}) = d(\mathcal{X}, \mathcal{Z}).$$

In other words, if $K_{\mathcal{X}\mathcal{Y}}$, $K_{\mathcal{Y}\mathcal{Z}}$, and $K_{\mathcal{X}\mathcal{Z}}$ are the stretch factors of the corresponding Teichmüller maps, we have

$$\log(K_{\mathcal{X}\mathcal{Y}}K_{\mathcal{Y}\mathcal{Z}}) = \log(K_{\mathcal{X}\mathcal{Y}}) + \log(K_{\mathcal{Y}\mathcal{Z}}) = \log(K_{\mathcal{X}\mathcal{Z}}),$$

and so

$$K_{\mathcal{X}\mathcal{Y}}K_{\mathcal{Y}\mathcal{Z}} = K_{\mathcal{X}\mathcal{Z}}.$$

Say that $\mathcal{X}$, $\mathcal{Y}$, and $\mathcal{Z}$ are represented by marked Riemann surfaces (X, ϕ), (Y, ψ), and $(\mathcal{Z}, \zeta)$. Let $h_{\mathcal{X}\mathcal{Y}}$ and $h_{\mathcal{Y}\mathcal{Z}}$ denote the Teichmüller maps homotopic to the change-of-marking maps $\psi \circ \phi^{-1} : X \to Y$ and $\zeta \circ \psi^{-1} : Y \to Z$. The composition

$$X \stackrel{h_{\mathcal{X}\mathcal{Y}}}{\Rightarrow} Y \stackrel{h_{\mathcal{Y}\mathcal{Z}}}{\Rightarrow} Z$$

has dilatation at most $K_{\mathcal{X}\mathcal{Y}}K_{\mathcal{Y}\mathcal{Z}}$ (Proposition 11.3). Since $K_{\mathcal{X}\mathcal{Y}}K_{\mathcal{Y}\mathcal{Z}} = K_{\mathcal{X}\mathcal{Z}}$, the dilatation of $h_{\mathcal{Y}\mathcal{Z}} \circ h_{\mathcal{X}\mathcal{Y}}$ must in fact be equal to $K_{\mathcal{X}\mathcal{Z}}$. Teichmüller's uniqueness theorem (Theorem 11.9) then gives that $h_{\mathcal{Y}\mathcal{Z}} \circ h_{\mathcal{X}\mathcal{Y}}$ must be the Teichmüller map $X \to Z$ in the homotopy class of the change of marking $\zeta \circ \phi^{-1}$. It follows that the horizontal foliations for the terminal differential for $h_{\mathcal{X}\mathcal{Y}}$ and the initial differential for $h_{\mathcal{Y}\mathcal{Z}}$ are equal. This means that $\mathcal{Y}$ lies on the Teichmüller line passing through $\mathcal{X}$ and $\mathcal{Z}$, which proves the first statement.

The second statement of the theorem is an immediate consequence of the first statement plus Teichmüller's uniqueness theorem. $\square$

11.8.3 THE TEICHMÜLLER METRIC FOR THE TORUS

Some intuition about the Teichmüller metric and Teichmüller geodesics can be gleaned from understanding them in the special case when $g = 1$, that is, for $\mathrm{Teich}(T^2)$. The situation in this case is simple enough that it can be worked out explicitly.

Recall that we exhibited a bijection between $\mathrm{Teich}(T^2)$ and $\mathbb{H}^2$ (Proposition 10.1). We now give a significant strengthening of this fact.

Theorem 11.20 *The bijection* $\mathbb{H}^2 \to \mathrm{Teich}(T^2)$ *given in Proposition 10.1 induces an isometry*

$$(\mathbb{H}^2, d_{\mathbb{H}^2}) \to (\mathrm{Teich}(T^2), 2\, d_{\mathrm{Teich}}).$$

The factor of 2 in the theorem comes from the fact that $\mathrm{Teich}(T^2)$ is isometric to the open unit disk in $\mathbb{C}$ with the infinitesimal metric

$$\rho(z)|dz| = \frac{|dz|}{1 - |z|^2}.$$

This space is isometric to the hyperbolic plane scaled by $1/2$ and hence has curvature -4.

Proof. First, the action of $\mathrm{SL}(2, \mathbb{R})$ on equivalence classes of marked lattices in $\mathbb{R}^2$ is the same as the action of $\mathrm{SL}(2, \mathbb{R})$ on $\mathrm{Teich}(T^2) \approx \mathbb{H}^2$ via Möbius transformations. Indeed, $\mathrm{SL}(2, \mathbb{R})$ is generated by matrices of the form

$$\begin{pmatrix} 1 & t \\ 0 & 1 \end{pmatrix} \qquad \begin{pmatrix} 0 & 1 \\ -1 & 0 \end{pmatrix},$$

where $t \in \mathbb{R}$. Therefore, this is a slight generalization of the fact that $\mathrm{Mod}(T^2)$ acts on $\mathrm{Teich}(T^2)$ by Möbius transformations (see the proof of Proposition 12.1).

Let $\mathcal{X}, \mathcal{Y} \in \mathrm{Teich}(T^2)$ and let x and y be the corresponding points in $\mathbb{H}^2$. Say that $d_{\mathbb{H}^2}(x, y) = \delta$. Let A be an element of $\mathrm{SL}(2, \mathbb{R})$ that corresponds to the hyperbolic element of $\mathrm{Isom}^+(\mathbb{H}^2)$ with axis passing through x and y and with translation distance δ. The matrix A is unique up to sign. We can write A as

$$A = C \begin{pmatrix} e^{\delta/2} & 0 \\ 0 & e^{-\delta/2} \end{pmatrix} C^{-1}$$

for some $C \in \mathrm{SL}(2, \mathbb{R})$. As above, we can also think of A as acting by a linear transformation on $\mathbb{R}^2$. What is more, if $\mathcal{X}$ and $\mathcal{Y}$ are represented by marked flat tori X and Y, then A can be regarded as a map $X \to Y$. Indeed, if we represent $\mathcal{X}$ and $\mathcal{Y}$ by marked lattices in $\mathbb{R}^2$ the action of A on $\mathbb{R}^2$ takes any $\mathcal{X}$-lattice to some $\mathcal{Y}$-lattice.

Since A has two real eigenvalues, there are two 1-dimensional eigenspaces on X along which A expands and contracts. Thus A is a Teichmüller map from X to Y. The holomorphic quadratic differential corresponding to this Teichmüller map is the one whose horizontal foliation lies in the direction of the eigenspace for A corresponding to the leading eigenvalue.

As an isometry of $\mathbb{H}^2$, the matrix A is a hyperbolic isometry that translates along its axis a distance δ in the hyperbolic metric $d_{\mathbb{H}}^2$. On the other hand, the dilatation of the action of A as a map $X \to Y$ is e^δ, so that

$$d_{\text{Teich}}(\mathcal{X}, \mathcal{Y}) = \frac{1}{2}\log(e^\delta) = \delta/2.$$

It follows that the Teichmüller metric on $\mathbb{H}^2 \approx \text{Teich}(T^2)$ is the hyperbolic metric scaled by a factor of $1/2$. $\square$

Another way to show that the Teichmüller metric on $\text{Teich}(T^2)$ is equivalent to the hyperbolic metric on $\mathbb{H}^2$ of curvature -4 is to use the Poincaré disk model. We already explained in Section 11.1 that the complex dilatation μ of an orientation-preserving linear map f is a point of the open unit disk. Further, in the metric ρ of curvature -4 on the unit disk, the dilatation K_f of f satisfies

$$d_\rho(0, \mu) = \frac{1}{2}\log(K_f).$$

From this fact one can deduce Theorem 11.20.

Chapter Twelve

Moduli Space

The moduli space of Riemann surfaces is one of the fundamental objects of mathematics. It is ubiquitous, appearing as a basic object in fields from low-dimensional topology to algebraic geometry to mathematical physics. The moduli space $\mathcal{M}(S)$ parameterizes, among other things: isometry classes of hyperbolic structures on S, conformal classes of Riemannian metrics on S, biholomorphism classes of complex structures on S, and isomorphism classes of smooth algebraic curves homeomorphic to S.

We will access $\mathcal{M}(S)$ as the quotient of $\mathrm{Teich}(S)$ by an action of $\mathrm{Mod}(S)$. A key result of this chapter is the theorem (due to Fricke) that $\mathrm{Mod}(S)$ acts properly discontinuously on $\mathrm{Teich}(S)$, with a finite-index subgroup of $\mathrm{Mod}(S)$ acting freely. As such, $\mathcal{M}(S)$ is finitely covered by a smooth aspherical manifold.

In this chapter we will also prove some of the basic topological properties of moduli space. While moduli space is not compact, Mumford's compactness criterion describes precisely what it means to go to infinity in $\mathcal{M}(S)$. We will also see that moduli space has only one end, that is, it is connected at infinity. Fricke's theorem and Mumford's compactness criterion are crucial ingredients in our proof of the Nielsen–Thurston classification of elements of $\mathrm{Mod}(S)$ (see Chapter 13).

One reason for the importance of moduli space is that it plays a fundamental role in the classification of surface bundles. In this chapter we will also explain the connection between the cohomology of moduli space and characteristic classes of surface bundles.

12.1 MODULI SPACE AS THE QUOTIENT OF TEICHMÜLLER SPACE

Recall that a point $\mathcal{X} \in \mathrm{Teich}(S)$ is the equivalence class of a pair (X, ϕ) where X is a hyperbolic surface and $\phi : S \to X$ is a diffeomorphism. An element $f \in \mathrm{Mod}(S)$ acts on $\mathrm{Teich}(S)$ as follows: choose a representative $\psi \in \mathrm{Diff}^+(S)$ of f and set

$$f \cdot \mathcal{X} = [(X, \phi \circ \psi^{-1})].$$

This formula is encoded in the following diagram:

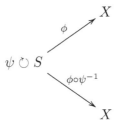

Note that the element $[(X, \psi \circ \phi^{-1})]$ is well defined since homotopic markings determine equivalent points of $\mathrm{Teich}(S)$. Also, we use ψ^{-1} instead of ψ so that we have a well-defined group action.

It follows easily from the definition of the Teichmüller metric that the action of $\mathrm{Mod}^{\pm}(S)$ on $\mathrm{Teich}(S)$ is an isometric action. In particular, the action is by diffeomorphisms.

We see from the definition of the $\mathrm{Mod}(S)$ action on $\mathrm{Teich}(S)$ that the orbit of a point $\mathcal{X} = [(X, \phi)]$ is the set of points $[(X, \psi)]$ where the marking ψ ranges over all homotopy classes of diffeomorphisms $S \to X$.

Another way to see the action of $\mathrm{Mod}(S)$ on $\mathrm{Teich}(S)$ is by recalling that $\mathrm{Teich}(S)$ can also be thought of as the quotient $\mathrm{HypMet}(S)/\mathrm{Diff}_0(S)$. Now $\mathrm{Diff}^+(S)$ acts on $\mathrm{HypMet}(S)$ by pullback. The action of $\mathrm{Diff}^+(S)$ on $\mathrm{HypMet}(S)$ induces an action of $\mathrm{Diff}^+(S)$ on $\mathrm{Teich}(S)$. This action factors through an action of $\mathrm{Mod}(S) = \mathrm{Diff}^+(S)/\mathrm{Diff}_0(S)$ on $\mathrm{Teich}(S)$. A quick trace through the definitions gives that this action agrees with the definition in terms of markings given above.

The *moduli space* of hyperbolic surfaces homeomorphic to S is defined to be the quotient space

$$\mathcal{M}(S) = \mathrm{Teich}(S)/\mathrm{Mod}(S).$$

Thinking of $\mathrm{Teich}(S)$ as the space of marked hyperbolic surfaces homeomorphic to S, the group $\mathrm{Mod}(S)$ acts on $\mathrm{Teich}(S)$ simply by changing the markings. In this way we can think of $\mathcal{M}(S)$ as the space of (unmarked) hyperbolic surfaces up to oriented isometry. Thinking of $\mathrm{Teich}(S)$ as $\mathrm{Teich}(S) = \mathrm{HypMet}(S)/\mathrm{Diff}_0(S)$, we see that

$$\mathcal{M}(S) = \mathrm{HypMet}(S)/\mathrm{Diff}^+(S).$$

The extended action. The action of $\mathrm{Mod}(S)$ on $\mathrm{Teich}(S)$ extends to an action of $\mathrm{Mod}^{\pm}(S)$ on $\mathrm{Teich}(S)$. We emphasize, though, that $\mathcal{M}(S)$ is the quotient of $\mathrm{Teich}(S)$ by the (unextended) mapping class group, so $\mathcal{M}(S)$

is a twofold orbifold cover of $\mathrm{Teich}(S)/\mathrm{Mod}^{\pm}(S)$. The latter quotient can be identified with the space of isometry classes of hyperbolic metrics on S, where isometries are not required to be orientation-preserving.

The kernel of the action. Let S be a surface with $\chi(S) < 0$. We claim that the kernel of the $\mathrm{Mod}^{\pm}(S)$ action on $\mathrm{Teich}(S)$ is precisely the subgroup of $\mathrm{Mod}^{\pm}(S)$ consisting of elements that fix the isotopy class of each essential simple closed curve in S. So, when $S = S_{g,n}$, the kernel of the action is the cyclic group of order 2 generated by a hyperelliptic involution if $S \in \{S_{1,1}, S_{1,2}, S_{2,0}\}$, it is the Klein four group generated by hyperelliptic involutions if $S = S_{0,4}$, and it is trivial otherwise (cf. the discussion after Theorem 3.10). Thus, in all cases other than $S = S_{0,4}$, the kernel of the $\mathrm{Mod}^{\pm}(S)$ action on $\mathrm{Teich}(S)$ is $Z(\mathrm{Mod}(S))$.

Suppose $h \in \mathrm{Mod}^{\pm}(S)$ acts as the identity on $\mathrm{Teich}(S)$. If $\psi \in \mathrm{Diff}(S)$ is a representative of h, then ψ is isotopic to an isometry of S with respect to every hyperbolic metric on S. It follows that h fixes the length of every isotopy class of simple closed curves in S. Since for any pair of distinct isotopy classes of essential simple closed curves in S one can find a metric on S where the lengths of the corresponding geodesics are not equal to each other, it follows that h in fact fixes the isotopy class of each simple closed curve in S. Conversely, it follows from Theorem 10.7 that if $h \in \mathrm{Mod}^{\pm}(S)$ fixes the isotopy class of every simple closed curve in S, then h acts trivially on $\mathrm{Teich}(S)$.

The geometric point of view. As mentioned above, the action of $\mathrm{Mod}^{\pm}(S)$ on $\mathrm{Teich}(S)$ is an isometric action. That is, we have a map

$$\Upsilon : \mathrm{Mod}^{\pm}(S) \to \mathrm{Isom}(\mathrm{Teich}(S)).$$

It is a theorem of Royden that, except in the cases $S = S_{1,1}$ and $S = S_{0,4}$, the map Υ is actually surjective [186]. We described above the kernel of Υ. Since Υ is injective when S has genus at least 3, it follows that

$$\mathrm{Mod}^{\pm}(S_g) \approx \mathrm{Isom}(\mathrm{Teich}(S_g)) \text{ for } g \geq 3.$$

For $g = 2$, we have

$$\mathrm{Mod}^{\pm}(S_2)/Z(\mathrm{Mod}(S_2)) \approx \mathrm{Isom}(\mathrm{Teich}(S_2)).$$

Finally, since $\mathrm{Teich}(T^2)$ is isometric to $\mathbb{H}^2$ (up to scale), it follows that $\mathrm{Mod}^{\pm}(T^2)/\ker \Upsilon \approx \mathrm{PGL}(2, \mathbb{Z})$ has infinite index in $\mathrm{Isom}(\mathrm{Teich}(T^2)) \approx \mathrm{PGL}(2, \mathbb{R})$.

Stabilizers of points. Assume $\chi(S) < 0$. Let $\mathfrak{X} \in \mathrm{Teich}(S)$ and say that $\mathfrak{X}$ is represented by the marked surface (X, ϕ). We would like to determine the stabilizer in $\mathrm{Mod}(S)$ of $\mathfrak{X}$. Let $h \in \mathrm{Mod}(S)$ and say that h is represented by a diffeomorphism ψ. We have $h \cdot \mathfrak{X} = \mathfrak{X}$ if and only if the marked surfaces (X, ϕ) and $(X, \phi \circ \psi^{-1})$ are equivalent, which is the case if and only if $\phi \circ \psi \circ \phi^{-1} : X \to X$ is freely isotopic to an isometry τ_h of X. Note that τ_h is well defined since no two distinct isometries of a hyperbolic surface are isotopic. Also, τ_h is orientation-preserving since ψ is. The correspondence $h \leftrightarrow \tau_h$ is an isomorphism between the stabilizer of $\mathfrak{X}$ in $\mathrm{Mod}(S)$ and $\mathrm{Isom}^+(X)$. In particular, by Proposition 7.7, the stabilizer of $\mathfrak{X}$ in $\mathrm{Mod}(S)$ is finite.

For $g \in \{1, 2\}$, the hyperelliptic involution fixes the isotopy class of every simple closed curve in S_g. Thus, as above, the hyperelliptic involution stabilizes every point of $\mathrm{Teich}(S_g)$. For $g \geq 3$, one can show that the stabilizer of a generic point of $\mathrm{Teich}(S_g)$ is trivial.

The algebraic point of view. Recall that the Dehn–Nielsen–Baer theorem states that the natural map $\sigma : \mathrm{Mod}^{\pm}(S_g) \to \mathrm{Out}(\pi_1(S_g))$ given by the (outer) action of $\mathrm{Mod}^{\pm}(S_g)$ on $\pi_1(S_g)$ is an isomorphism. There is a natural action

$$\mathrm{Out}(\pi_1(S_g)) \circlearrowleft \mathrm{DF}(\pi_1(S_g), \mathrm{PSL}(2, \mathbb{R})) / \mathrm{PGL}(2, \mathbb{R})$$

defined in the following way. Given any $\Phi \in \mathrm{Aut}(\pi_1(S_g))$ and any $\rho \in \mathrm{DF}(\pi_1(S_g), \mathrm{PSL}(2, \mathbb{R}))$, we define

$$[\Phi] \cdot [\rho] = [\rho \circ \Phi^{-1}].$$

It is also easy to check that this action corresponds to the action of $\mathrm{Mod}^{\pm}(S_g)$ on $\mathrm{Teich}(S_g)$ defined above. To be more precise, let $\eta : \mathrm{Teich}(S_g) \to \mathrm{DF}(\pi_1(S_g), \mathrm{PSL}(2, \mathbb{R})) / \mathrm{PGL}(2, \mathbb{R})$ be the homeomorphism defined in Proposition 10.2. Then for each $f \in \mathrm{Mod}^{\pm}(S_g)$ and each $\mathfrak{X} \in \mathrm{Teich}(S_g)$, we have

$$\eta(f \cdot \mathfrak{X}) = \sigma(f) \cdot \eta(\mathfrak{X}).$$

12.2 MODULI SPACE OF THE TORUS

The moduli space $\mathcal{M}(T^2)$ of flat, unit-area metrics on the torus T^2 is a particularly important example of a moduli space. It is known as the *modular surface*. It is an object of central importance in mathematics, one reason being that it is the moduli space of elliptic curves. For us, it is useful as an explicitly computable example of a moduli space.

We saw in Section 10.1 that $\mathrm{Teich}(T^2)$ can be identified with the hyperbolic plane $\mathbb{H}^2$. We will now see that the action of $\mathrm{Mod}(T^2) \approx \mathrm{SL}(2, \mathbb{Z})$ on $\mathrm{Teich}(T^2) \approx \mathbb{H}^2$ is simply the following action of $\mathrm{SL}(2, \mathbb{Z})$ on $\mathbb{H}^2$ by Möbius transformations:

$$\begin{pmatrix} a & b \\ c & d \end{pmatrix} \mapsto f(z) = \frac{az - b}{-cz + d}.$$

A straightforward calculation shows that this indeed defines a group action.

Proposition 12.1 *Let $\sigma : \mathrm{Mod}(T^2) \to \mathrm{SL}(2, \mathbb{Z})$ be the isomorphism of Theorem 2.5 and let $\eta : \mathrm{Teich}(T^2) \to \mathbb{H}^2$ be the identification from Proposition 10.1. For any $\mathcal{X} \in \mathrm{Teich}(T^2)$ and any $f \in \mathrm{Mod}(T^2)$, we have*

$$\eta(f \cdot \mathcal{X}) = \sigma(f) \cdot \eta(\mathcal{X}).$$

In other words, Proposition 12.1 states that η *semiconjugates* the action of $f \in \mathrm{Mod}(T^2)$ on $\mathrm{Teich}(T^2)$ to the action of $\sigma(f) \in \mathrm{SL}(2, \mathbb{Z})$ on $\mathbb{H}^2$.

Proof. It is enough to check the statement of the proposition on a set of generators of $\mathrm{Mod}(T^2)$, say

$$M = \begin{pmatrix} 1 & 1 \\ 0 & 1 \end{pmatrix} \quad \text{and} \quad N = \begin{pmatrix} 0 & -1 \\ 1 & 0 \end{pmatrix}.$$

Let α and β be based loops in T^2 representing generators for $\pi_1(T^2)$ with $\hat{i}(\alpha, \beta) = 1$ (this makes sense if we identify α and β with their images in $H_1(T^2; \mathbb{Z})$). The isomorphism of Theorem 2.5 identifies M with the mapping class T_α^{-1}, thinking of α as an unoriented simple closed curve; it also identifies N with the order 4 mapping class $(T_\alpha T_\beta T_\alpha)^{-1}$, which can be described by cutting T^2 along α and β, rotating the square by $\pi/2$, and regluing.

Given a point $[(X, \phi)] \in \mathrm{Teich}(T^2)$, we can represent it, as in Proposition 10.1, by a unique marked lattice in $\mathbb{C} \approx \mathbb{R}^2$ with basis vector 1 corresponding to the oriented curve α and basis vector $\tau \in \mathbb{C}$ in the upper half-plane corresponding to β. We know that

$$T_\alpha^{-1} \cdot [(X, \phi)] = [(X, \phi \circ T_\alpha)],$$

where we appropriately regard T_α as either a mapping class or a homeomorphism. The formula $T_{\phi(\alpha)} = \phi \circ T_\alpha \circ \phi^{-1}$ (Fact 3.7) gives that

$$(\phi \circ T_\alpha)(\beta) \sim T_{\phi(\alpha)}(\phi(\beta)),$$
$$(\phi \circ T_\alpha)(\alpha) \sim \phi(\alpha),$$

where $\sim$ denotes the isotopy relation. In other words, the effect of T_α^{-1} on the marked lattice is to keep 1 fixed and to send τ to $\tau - 1$. But this means that T_α^{-1} acts on $\mathbb{H}^2$ by the Möbius transformation $z \mapsto z - 1$, which is what we wanted to show.

By similar reasoning, the mapping class associated to N acts on the marked lattice $(1, \tau)$ by sending it to the marked lattice $(-\tau, 1)$. To get the induced action on $\mathbb{H}^2$ we need to put the latter into standard form (rotate/flip so the first complex number is 1). If we write $\tau = re^{i\theta}$, then the resulting lattice corresponds to

$$\frac{1}{r}e^{i(\pi-\theta)} = -\frac{1}{r}e^{-i\theta}.$$

But this is nothing other than $-\frac{1}{\tau}$, which is what we needed to show. □

We thus have

$$\mathcal{M}(T^2) = \text{Teich}(T^2)/\text{Mod}(T^2) \approx \mathbb{H}^2/\text{SL}(2, \mathbb{Z}),$$

where the action is given by Proposition 12.1. The kernel of the $\text{SL}(2, \mathbb{Z})$ action on $\mathbb{H}^2$ is $\{\pm I\} = Z(\text{SL}(2, \mathbb{Z}))$, and so $\mathcal{M}(T^2)$ can also be written as $\mathbb{H}^2/\text{PSL}(2, \mathbb{Z})$.

A fundamental domain. A well-known fundamental domain for the $\text{SL}(2, \mathbb{Z})$ action on $\mathbb{H}^2$ is shown in Figure 12.1. One way to see this is as follows. The action of $\text{SL}(2, \mathbb{Z})$ on $\mathbb{H}^2$ is cellular with respect to the ideal triangulation of $\mathbb{H}^2$ with vertices at $\mathbb{Q} \cup \{\infty\}$ (cf. Figure 4.3). Further, this action is transitive on triangles, and the stabilizer of each triangle is the full group of rotations. Thus a fundamental domain is given by one third of one triangle, or, what is shown in Figure 12.1, one sixth of one triangle plus one sixth of an adjacent triangle.

Stabilizers. As in the higher-genus case, the stabilizer in $\text{Mod}(T^2)$ of a point $[(X, \phi)] \in \text{Teich}(T^2)$ corresponds precisely to the set of isotopy classes of isometries of X. This in turn can be identified with a finite subgroup of $\text{SL}(2, \mathbb{Z})$. Recall from Section 7.1.1 that, up to powers, there are only two conjugacy classes of finite-order elements of $\text{SL}(2, \mathbb{Z})$. The first is that of the matrix N, which fixes the point i and rotates by an angle of π, thus identifying the two halves of the circular boundary of the fundamental domain. This fixed point corresponds to the isometry of the square torus obtained by rotating the square by an angle $\pi/2$. The second conjugacy class

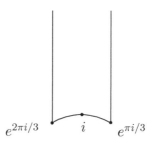

$e^{2\pi i/3}$ i $e^{\pi i/3}$

Figure 12.1 The fundamental domain for the modular surface.

is that of the matrix

$$\begin{pmatrix} -1 & 1 \\ -1 & 0 \end{pmatrix},$$

whose class in $\mathrm{PSL}(2,\mathbb{Z})$ has order 3 and whose unique fixed point in $\mathbb{H}^2$ is the point $e^{i\pi/3}$. This fixed point corresponds to the order 3 symmetry of the hexagonal torus (the relationship between the point $e^{i\pi/3}$ and the hexagonal torus is explained in Figure 12.3).

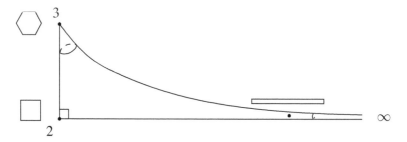

Figure 12.2 A schematic of the modular surface with sample points labeled.

The modular curve. We can also see how $\mathrm{SL}(2,\mathbb{Z})$ identifies the sides of its fundamental domain: the left side is identified with the right side by the translation $z \mapsto z+1$ corresponding to M, and the two halves of the bottom side are identified by a rotation of angle π about i, corresponding to N. Therefore, topologically, $\mathcal{M}(T^2)$ is a punctured sphere. Taking into account the fixed points, we see that $\mathcal{M}(T^2)$ has the structure of an orbifold with signature $(0; 2, 3, \infty)$, where ∞ signifies the puncture. That is, we can think of $\mathcal{M}(T^2)$ as a punctured sphere with cone points of order 2 and 3; see Figure 12.2.

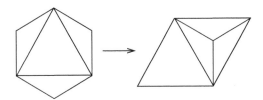

Figure 12.3 The cut-and-paste operation from a regular hexagon to the parallelogram spanned by 1 and $e^{i\pi/3}$.

The above discussion in particular gives the important fact that the action of $\mathrm{Mod}(T^2)$ on $\mathrm{Teich}(T^2)$ is properly discontinuous. In particular, the modular curve is an orbifold. One of the main results in this chapter is the analogous result for higher-genus surfaces.

By the theory of orbifolds, the orbifold fundamental group $\pi_1^{orb}(\mathcal{M}(T^2))$ of $\mathcal{M}(T^2)$ is isomorphic to the group of covering transformations, namely $\mathrm{PSL}(2,\mathbb{Z})$. By the Van Kampen theorem for orbifolds, $\pi_1^{orb}(\mathcal{M}(T^2))$ is generated by two loops, one around each cone point [116, Section 6.1]. Further, these generators have order 2 and 3, respectively, and there are no other relations. We have thus recovered the classical isomorphism

$$\mathrm{PSL}(2,\mathbb{Z}) \approx \mathbb{Z}/2\mathbb{Z} * \mathbb{Z}/3\mathbb{Z}.$$

As is true in higher genus, $\mathcal{M}(T^2)$ parameterizes *oriented* isometry classes of marked tori/lattices. Note that $i + \epsilon$ and $i - \epsilon$ (for $\epsilon \in \mathbb{R}$) correspond to isometric tori that are not oriented-isometric. A fundamental domain for the quotient of $\mathrm{Teich}(T^2)$ by $\mathrm{Mod}^{\pm}(T^2)$ would be, say, the left half of the fundamental domain for $\mathrm{Mod}(T^2)$.

12.3 PROPER DISCONTINUITY

Recall that the action of a group G on a topological space X by homeomorphisms is *properly discontinuous* if, for any compact set $B \subset X$, the set

$$\{g \in G : g \cdot B \cap B \neq \emptyset\}$$

is finite. The main goal of this section is to prove the following.

THEOREM 12.2 (Fricke) *Let $g \geq 1$. The action of $\mathrm{Mod}(S_g)$ on $\mathrm{Teich}(S_g)$ is properly discontinuous.*

Theorem 12.2 (and its proof) extends to case of surfaces $S_{g,n}$ with $\chi(S_{g,n}) \leq 0$. We comment on the required modifications at the end of the proof.

Before proving Theorem 12.2 we use it to deduce some basic properties of $\mathcal{M}(S)$.

First, whenever a group acts by isometries on a metric space, the quotient has an induced pseudometric. The distance between any two orbits is defined to be the infimum of the distance between any pair of representatives. When the action is properly discontinuous, one has the additional property that two orbits have distance 0 if and only if they are equal; in other words, the induced pseudometric is a metric. We thus have:

> *For $g \geq 1$, the Teichmüller metric on $\mathrm{Teich}(S_g)$ induces a metric on $\mathcal{M}(S_g)$.*

We call this metric the *Teichmüller metric* on $\mathcal{M}(S_g)$.

Second, when a group acts properly discontinuously by homeomorphisms on a manifold, the quotient is called an orbifold. If the original manifold is aspherical (i.e., has contractible universal cover), then the orbifold is *aspherical*. If the action is also free, then the quotient is again a manifold. As we said at the beginning of the chapter, $\mathrm{Mod}(S_g)$ has a finite-index subgroup that acts freely on $\mathrm{Teich}(S_g)$. Thus, as another consequence of Theorem 12.2, we have the following.

THEOREM 12.3 *For $g \geq 1$, the space $\mathcal{M}(S_g)$ is an aspherical orbifold and is finitely covered by an aspherical manifold.*

It is in fact known that $\mathcal{M}(S)$ is a complex orbifold, finitely covered by a complex manifold, and that it is in fact a quasiprojective variety, finitely covered by a smooth variety [52].

12.3.1 THE RAW LENGTH SPECTRUM OF A HYPERBOLIC SURFACE

The proof of Theorem 12.2 relies on a few lemmas regarding the lengths of curves in a hyperbolic surface. The first lemma we will need concerns the *raw length spectrum* of a hyperbolic surface X, which is defined to be the set of positive real numbers

$$\mathrm{rls}(X) = \{\ell_X(c)\} \subset \mathbb{R}_+,$$

where c ranges over all isotopy classes of essential (or peripheral) simple closed curves in X. In other words, $\mathrm{rls}(X)$ is the set of lengths of simple closed geodesics in X.

Lemma 12.4 (Discreteness of the length spectrum) *Let X be any closed hyperbolic surface. The set $\mathrm{rls}(X)$ is a closed discrete subset of $\mathbb{R}$. Further, for each $L \in \mathbb{R}$, the set*

$$\{c : c \text{ an isotopy class of simple closed curves in } X \text{ with } \ell_X(c) \leq L\}$$

is finite.

Proof. The hyperbolic surface X is the quotient of $\mathbb{H}^2$ by a free, properly discontinuous isometric action of $\pi_1(X)$. Let $K \subset \mathbb{H}^2$ be a fundamental domain for this action. Since X is closed, K is compact. Since K is a fundamental domain, every closed geodesic γ in X has a lift $\tilde{\gamma}$ that intersects K. There is then a unique (up to sign) $\gamma_0 \in \pi_1(X)$ that acts on $\tilde{\gamma}$ with translation length $\ell_X(\gamma)$. As a closed loop, γ_0 is freely homotopic to γ.

Let $R > 0$ be given. Let γ in X be any closed geodesic of length at most R. As in the previous paragraph, choose any lift $\tilde{\gamma}$ that intersects K and let $\langle \gamma_0 \rangle$ be the corresponding cyclic subgroup of $\pi_1(X)$. Any point $p \in \tilde{\gamma} \cap K$ is moved by the hyperbolic translation γ_0 a distance $\ell_X(\gamma)$ in $\mathbb{H}^2$. Let K_R denote the closed R-neighborhood of the compact set K. Then K_R is a compact subset of $\mathbb{H}^2$ with the property that $(\gamma_0 \cdot K_R) \cap K_R \neq \emptyset$. Since the action of $\pi_1(X)$ on $\mathbb{H}^2$ is properly discontinuous, there are only finitely many such γ_0, hence only finitely many such γ. This proves the second statement. The first statement follows. $\qquad\square$

12.3.2 WOLPERT'S LEMMA

The next lemma, due to Wolpert [215], gives the basic fact that any K-quasiconformal map distorts hyperbolic lengths of closed curves by a factor of at most K.

Lemma 12.5 (Wolpert's lemma) *Let X_1 and X_2 be hyperbolic surfaces and let $\phi : X_1 \to X_2$ be a K-quasiconformal homeomorphism. For any isotopy class c of simple closed curves in X_1, the following inequalities hold:*

$$\frac{\ell_{X_1}(c)}{K} \leq \ell_{X_2}(\phi(c)) \leq K\ell_{X_1}(c).$$

Lemma 12.5 has the following immediate consequence: for any $\mathfrak{X}_1, \mathfrak{X}_2 \in \mathrm{Teich}(S)$ with $d_{\mathrm{Teich}}(\mathfrak{X}_1, \mathfrak{X}_2) \leq \log(K)/2$, and any isotopy class c of simple closed curves in S, we have $\ell_{\mathfrak{X}_1}(c)/K \leq \ell_{\mathfrak{X}_2}(c) \leq K\ell_{\mathfrak{X}_1}(c)$.

Proof. Let $\gamma_1, \gamma_2 \in \mathrm{Isom}^+(\mathbb{H}^2)$ be isometries of $\widetilde{X}_1 \approx \widetilde{X}_2 \approx \mathbb{H}^2$ corresponding to c and $\phi(c)$, respectively. Consider the annuli A_1 and A_2 obtained by taking the quotient of $\mathbb{H}^2$ by $\langle \gamma_1 \rangle \approx \mathbb{Z}$ and $\langle \gamma_2 \rangle \approx \mathbb{Z}$, respectively. Since the map $\pi_1(X_i) \to \mathrm{Isom}^+(\mathbb{H}^2)$ is well defined only up to conjugacy in $\mathrm{PGL}(2, \mathbb{R})$, we can take γ_1 to be the map $z \mapsto e^{\ell_{X_1}(c)} z$ and γ_2 to be $z \mapsto e^{\ell_{X_2}(\phi(c))} z$, where we think of $\mathbb{H}^2$ in the upper-half plane model.

We can put the annuli A_1 and A_2 in a standard form; that is, for each i, we can find the unique (open) Euclidean annulus A_{m_i} of circumference 1 and height m_i so that A_i is conformally equivalent to A_{m_i}. We call the number m_i the *modulus* of the annulus A_i. To find the standard form of A_1, note that we can choose a branch of the natural logarithm that takes the upper half-plane $\mathbb{H}^2$ to the infinite strip of points in $\mathbb{C}$ with imaginary part in $(0, \pi)$. Under this identification, the group $\langle \gamma_1 \rangle$ corresponds to the infinite cyclic group of translations generated by $z \mapsto z + \ell_{X_1}(c)$. Since the natural logarithm is a conformal map, A_1 is conformally equivalent to the annulus obtained by identifying vertical sides of a rectangle whose width is $\ell_{X_1}(c)$ and whose height is π; thus the modulus m_1 is equal to $\pi / \ell_{X_1}(c)$. Similarly, $m_2 = \pi / \ell_{X_2}(\phi(c))$.

The map ϕ lifts to a K-quasiconformal mapping

$$\widetilde{\phi} : A_1 \to A_2.$$

Note that since $\langle \gamma_i \rangle < \pi_1(X_i)$, this is formally weaker than saying that ϕ is a K-quasiconformal map from $X_1 = \mathbb{H}^2/\pi_1(X_1)$ to $X_2 = \mathbb{H}^2/\pi_1(X_2)$. The solution to Grötzsch's problem can be modified (slightly) to prove that $\widetilde{\phi}$ changes the modulus by at most a multiplicative factor of K (in the solution to Grötzsch's problem, replace the x-direction in the rectangle with the S^1-direction in the annulus). In other words, we have

$$\frac{1}{K} m_2 \leq m_1 \leq K m_2.$$

The lemma follows.   □

12.3.3 THE PROOF OF PROPER DISCONTINUITY

We are ready to demonstrate the proper discontinuity of the action of $\mathrm{Mod}(S_g)$ on $\mathrm{Teich}(S_g)$.

Proof of Theorem 12.2. Let B be a compact set in $\mathrm{Teich}(S_g)$. We need to show that the set of $f \in \mathrm{Mod}(S_g)$ such that $(f \cdot B) \cap B \neq \emptyset$ is finite. Let $\mathcal{X}$ be some arbitrary point in B and let D denote the diameter of B.

Let c_1 and c_2 be isotopy classes of essential simple closed curves in S_g that fill S_g (Proposition 3.5). Let $L = \max\{\ell_{\mathcal{X}}(c_1), \ell_{\mathcal{X}}(c_2)\}$.

Suppose $f \in \mathrm{Mod}(S_g)$ satisfies $(f \cdot B) \cap B \neq \emptyset$. It follows that $d_{\mathrm{Teich}}(\mathcal{X}, f \cdot \mathcal{X}) \leq 2D$. By Wolpert's lemma (Lemma 12.5), $\ell_{f \cdot \mathcal{X}}(c_i) \leq KL$ for $i = 1, 2$, where $K = e^{4D}$. But since $\ell_{f \cdot \mathcal{X}}(c_i) = \ell_{\mathcal{X}}(f^{-1}(c_i))$, we have that $\ell_{\mathcal{X}}(f^{-1}(c_i)) \leq KL$.

By Lemma 12.4, there are finitely many isotopy classes of simple closed curves b in S_g so that $\ell_{\mathcal{X}}(b) \leq KL$. Thus there are only finitely many possibilities for $f^{-1}(c_1)$ and $f^{-1}(c_2)$. But by the Alexander method (Proposition 2.8), there are finitely many choices for f^{-1} once the isotopy classes $f^{-1}(c_i)$ are determined. Thus there are finitely many possibilities for f that satisfy $(f \cdot B) \cap B \neq \emptyset$, and we are done. □

Punctures and boundary components. Theorem 12.2 extends with little difficulty to the case of surfaces with punctures and/or boundary. Lemma 12.4 needs a slight modification in the noncompact case. The key is that one can choose sufficiently small disjoint open horoball neighborhoods of the cusps so that every essential geodesic is disjoint from these neighborhoods. Since the complement of these horoball neighborhoods is a compact surface, the proof then proceeds as in the closed case.

12.4 MUMFORD'S COMPACTNESS CRITERION

We can see from our explicit description of $\mathcal{M}(T^2)$ that it is not compact. For instance, the ray $ti \in \mathbb{H}^2 \approx \mathrm{Teich}(T^2)$, where $t \geq 1$, projects to a ray X_t in $\mathcal{M}(T^2)$ that leaves every compact set. Even more, the distance between X_0 and X_t tends to infinity as t tends to infinity, and so $\mathcal{M}(T^2)$ has infinite diameter.

In the above example, we can think of X_t as the set of flat tori obtained from the square torus by pinching one of the simple closed curves to ever smaller lengths. We will use a similar idea to show that $\mathcal{M}(S_g)$ has infinite diameter with respect to the Teichmüller metric. This in particular will demonstrate that $\mathcal{M}(S_g)$ is not compact.

First we introduce a useful function on $\mathcal{M}(S_g)$. For $X \in \mathcal{M}(S_g)$, the injectivity radius of X at a point x is the largest r for which the r-ball in X centered at x is isometrically embedded. Then the *injectivity radius* of X is the infimum of these injectivity radii over all points of X.

A related function is $\ell(X)$, the length of the shortest essential closed geodesic in X. It is not hard to see that the number $\ell(X)$ is twice the injectivity radius of X and that any geodesic realizing $\ell(X)$ is necessarily simple. It follows from Lemma 12.4 that $\ell(X)$ is strictly positive.

Fix some $X \in \mathcal{M}(S_g)$ and let $\mathcal{X} \in \mathrm{Teich}(S_g)$ be some lift. As above,

$\ell(X)$ is positive and is realized by a simple closed curve γ in S_g in the sense that $\ell_X(\gamma) = \ell(X)$. We can use γ as part of a coordinate system of curves for Fenchel–Nielsen coordinates on $\mathrm{Teich}(S_g)$. Then, for $t \geq 1$, we can construct $\mathcal{X}_t \in \mathrm{Teich}(S_g)$ with the property that $\ell_{\mathcal{X}_t}(\gamma) = \ell(X)/t$. Let X_t denote the image of $\mathcal{X}_t$ in $\mathcal{M}(S_g)$. We have $\ell(X_t) \leq \ell(X)/t$. It then follows from Wolpert's lemma (Lemma 12.5) that the distance between X and X_t in $\mathcal{M}(S_g)$ tends to infinity as t tends to infinity. In particular, we note that

> *The diameter of $\mathcal{M}(S_g)$ with respect to the Teichmüller metric is infinite.*

We have just shown one way to construct a sequence of points in $\mathcal{M}(S_g)$ leaving every compact set: starting from any given hyperbolic surface, choose some simple closed curve and pinch it to have smaller and smaller length. Our goal in this section is to prove Mumford's compactness criterion, which says that this is essentially the only way a sequence of points in $\mathcal{M}(S_g)$ can leave every compact set.

The *ϵ-thick part* of $\mathcal{M}(S_g)$ is the set

$$\mathcal{M}_\epsilon(S_g) = \{X \in \mathcal{M}(S_g) : \ell(X) \geq \epsilon\}.$$

Since the length spectrum of each closed hyperbolic surface is discrete (Lemma 12.4), it follows that $\{\mathcal{M}_\epsilon(S_g) : \epsilon > 0\}$ is an *exhaustion* of $\mathcal{M}(S_g)$:

$$\mathcal{M}(S_g) = \bigcup_\epsilon \mathcal{M}_\epsilon(S_g).$$

The following theorem is due to Mumford; see [166].

THEOREM 12.6 (Mumford's compactness criterion) *Let $g \geq 1$. For each $\epsilon > 0$, the space $\mathcal{M}_\epsilon(S_g)$ is compact.*

In other words, Theorem 12.6 states that, in order for a sequence $\{X_n\} \subset \mathcal{M}(S)$ to leave every compact set in $\mathcal{M}(S)$, the injectivity radii of X_n must tend to 0 as $n \to \infty$.

12.4.1 MAHLER'S COMPACTNESS CRITERION

The case $g = 1$ for Theorem 12.6 is a special case of a classical theorem of Mahler about lattices in $\mathbb{R}^n$ [139]. A *lattice* in $\mathbb{R}^n$ is the $\mathbb{Z}$-span of a basis for $\mathbb{R}^n$. We say that a lattice is *marked* if it comes equipped with a basis (as a $\mathbb{Z}$-module). Recall that we discussed marked lattices in $\mathbb{R}^2$ in Section 10.2.

The *injectivity radius* of a lattice $\Lambda \subset \mathbb{R}^n$ is half the length of the shortest nonzero vector in Λ. The injectivity radius of Λ can also be viewed as half the length of the shortest essential closed curve in the flat n-dimensional torus $\mathbb{R}^n/\Lambda$. The *volume* of Λ is the Riemannian volume of $\mathbb{R}^n/\Lambda$.

The group $\mathrm{SL}(n, \mathbb{R})$ acts transitively on the space of marked unit volume lattices in $\mathbb{R}^n$. The *moduli space of unit volume lattices in $\mathbb{R}^n$* is the quotient

$$\mathcal{L}_n = \mathrm{SL}(n, \mathbb{R})/\mathrm{SL}(n, \mathbb{Z})$$

endowed with the quotient topology from the Lie group $\mathrm{SL}(n, \mathbb{R})$. The moduli space of isometry classes of unit volume, flat, n-dimensional tori can be identified as the quotient $\mathrm{SO}(n)\backslash\mathcal{L}_n$.

As above, we can define $\mathcal{L}_n(\epsilon)$ to be the subspace of $\mathcal{L}_n$ consisting of lattices with injectivity radius bounded below by ϵ.

THEOREM 12.7 (Mahler's compactness criterion) *Let $n \geq 1$. For any $\epsilon > 0$, the space $\mathcal{L}_n(\epsilon)$ is compact.*

We now give the proof of Mahler's compactness criterion for $n = 2$, which is exactly Mumford's compactness criterion for the torus. The proof contains all the ideas needed for the general case $n \geq 2$ but is much simpler notationally.

Proof. Suppose $\Lambda \approx \mathbb{Z}^2$ is any lattice in $\mathbb{R}^2$ with injectivity radius bounded below by ϵ. Let v be the shortest nonzero vector in Λ and let w be the shortest vector among those with smallest nonzero distance to the (real) subspace spanned by v. By our choice of v and w, there are no points of Λ in the interior of the parallelogram spanned by v and w, and so v and w generate Λ.

We will show that the norms of v and w are bounded above by a function of ϵ (independent of Λ) and that w_2, the projection of w to $v^\perp$, is bounded from below by a function of ϵ. The first property will ensure that any infinite sequence of lattices has a convergent subsequence, and the second property will ensure that the limiting lattice is nondegenerate.

Let w_1 be the projection of w to the real subspace spanned by v. The set of vectors in Λ with smallest nonzero distance to the real span of v is $\{w + kv : k \in \mathbb{Z}\}$, and so $|w_1| \leq |v|/2$.

We have

$$|v| \leq |w| \leq |w_1| + |w_2| \leq \frac{1}{2}|v| + |w_2|,$$

and so $|w_2| \geq |v|/2 \geq \epsilon$. Since $|v||w_2| = 1$, we have $|v| = 1/|w_2| \leq 1/\epsilon$. Also $|w_2| = 1/|v| \leq 1/2\epsilon$ and $|w_1| \leq |v|/2 \leq 1/2\epsilon$. This completes the proof.   □

12.4.2 BERS' CONSTANT

In order to prove Mumford's compactness criterion we will need the following theorem of Bers [16].

THEOREM 12.8 (Bers' constant) *Let S be a compact surface with $\chi(S) <$ 0. There is a constant $L = L(S)$ such that for any hyperbolic surface X (with totally geodesic boundary) homeomorphic to S, there is a pants decomposition $\{\gamma_i\}$ of X with $\ell_X(\gamma_i) \leq L$ for each i.*

Bers' constant is the smallest L that satisfies the conclusion of the theorem. Buser has shown that Bers' constant is at most $21(g - 1)$ for a closed surface of genus g; he suggests that the actual bound should be on the order of $\sqrt{g}$) [42, Section 5.2.5]. Our proof of Theorem 12.8 gives a bound that grows faster than exponentially in g, but this suffices for our purposes.

Proof. Suppose that S has genus g and b boundary components. Recall that a pants decomposition for S has $3g - 3 + b$ simple closed curves. We will prove the following statement by induction on k for $0 \leq k \leq 3g - 3 + b$: there is a constant $L_k = L_k(S)$ so that for every hyperbolic surface X with totally geodesic boundary that is homeomorphic to S, there is a set of k distinct, disjoint, essential closed geodesics each of length at most L_k. This inductive statement is true for $k = 0$ since we may take $L_k = 0$.

Now assume the inductive hypothesis for some fixed $k \geq 0$. Let X be a hyperbolic surface with totally geodesic boundary that is homeomorphic to S. Choose a collection of k closed geodesics in X as in the inductive hypothesis and cut X along these curves. Let Y be any component of the cut surface that is not homeomorphic to a pair of pants and let y be a point on Y that is furthest from ∂Y. We must find an essential closed geodesic in Y whose length is bounded above by a function of S.

Let $D(y, \rho)$ be the disk of radius ρ in Y centered at y. More precisely, $D(y, \rho)$ is the image under the exponential map of the ball of radius ρ in the tangent space $T_y(Y)$. For small ρ, this is an embedded disk isometric to a disk of radius ρ in $\mathbb{H}^2$. Therefore, its area is given by the formula

$$\int_0^{2\pi} \int_0^\rho \sinh(r) \, dr \, d\theta = 2\pi(\cosh(\rho) - 1).$$

The key point for us is that the area of $D(y, \rho)$ is a proper function of ρ.

Let ρ_y denote the supremum over ρ so that $D(y, \rho)$ is an embedded disk in Y disjoint from ∂Y. Since the area of Y is less than or equal to the area of X (which is $-2\pi\chi(S)$), we have that ρ_y is finite and is bounded above by a function of S. The disk $D(y, \rho_y)$ either is not embedded in Y or intersects ∂Y.

In the first case, there are two radii of $\partial D(y, \rho_y)$ that meet at both end-points. The union of these two arcs is a closed geodesic of length $2\rho_y$, which, as discussed, is bounded above by a function of S. The geodesic is necessarily essential by the uniqueness of geodesics in a hyperbolic surface.

In the second case, we note that $D(y, \rho_y)$ must intersect ∂Y in at least two points, for otherwise, we could find a point in Y that is further from ∂Y. Thus we have two arcs from y to ∂Y, which we think of as an arc γ between components δ_1 and δ_2 of ∂Y (possibly $\delta_1 = \delta_2$). Let N be a regular metric neighborhood of $\gamma \cup \delta_1 \cup \delta_2$. By making N arbitrarily small, the length of the simple closed curve $\alpha = \partial N$ is arbitrarily close to $2\ell_Y(\gamma) + \ell_Y(\delta_1) + \ell_Y(\delta_2)$, and so the geodesic in the class of α has length strictly less than this. Since Y is not a pair of pants (and is not an annulus), α is essential in Y, hence in X, and we are done. $\square$

12.4.3 THE PROOF OF MUMFORD'S COMPACTNESS CRITERION

We can now prove Mumford's compactness criterion, that the ϵ-thick part of moduli space is compact.

Proof of Theorem 12.6. As noted above, the case $g = 1$ is a restatement of Mahler's compactness criterion (Theorem 12.7) for $n = 2$. So we assume that $g \geq 2$.

Since $\mathcal{M}(S_g)$ inherits the Teichmüller metric from $\mathrm{Teich}(S_g)$, it suffices to show that the space $\mathcal{M}_\epsilon(S_g)$ is sequentially compact. Let $\epsilon > 0$. Let $\{X_i\}$ be a sequence in $\mathcal{M}_\epsilon(S_g)$ and let $\mathcal{X}_i \in \mathrm{Teich}(S_g)$ be a lift of X_i for each i. To prove that some subsequence of $\{X_i\}$ converges in $\mathcal{M}_\epsilon(S_g)$, we will show that, for a fixed choice of Fenchel–Nielsen coordinates, the $\mathcal{X}_i$ can be chosen to lie in a compact rectangular region of the Euclidean space $\mathbb{R}_+^{3g-3} \times \mathbb{R}^{3g-3}$.

By Theorem 12.8, for each $\mathcal{X}_i$ there is a pants decomposition $\mathcal{P}_i$ of S_g with $\ell_{\mathcal{X}_i}(\gamma) \in [\epsilon, L]$ for each $\gamma \in \mathcal{P}_i$ (L is Bers' constant). Since there are only finitely many topological types of pants decompositions of S_g, we can choose a subsequence, also denoted $\{\mathcal{X}_i\}$, and a sequence $f_i \in \mathrm{Mod}(S_g)$ so that $f_i(\mathcal{P}_i) = \mathcal{P}_1$.

Now, in Fenchel–Nielsen coordinates adapted to $\mathcal{P}_1$ (with arbitrary seams chosen for twist coordinates), the $\mathcal{Y}_i = f_i \cdot \mathcal{X}_i$ have length parameters in $[\epsilon, L]$.

Since Dehn twists about the curves of $\mathcal{P}_1$ change the twist parameters by 2π, there is for each i a product h_i of Dehn twists about the curves of $\mathcal{P}_1$ so that the twist parameters of $h_i \cdot \mathcal{Y}_i$ lie in the interval $[0, 2\pi]$. This finishes the proof. $\square$

The proof of Mumford's compactness criterion generalizes easily to compact surfaces with finitely many boundary components and finitely many points removed.

12.4.4 ISOSPECTRAL SURFACES

Since the marked length spectrum of a hyperbolic surface X determines X up to isometry (Theorem 10.7), one might wonder if the raw length spectrum $\mathrm{rls}(X)$ also determines X up to isometry. Vignéras proved, however, that this is not the case: for each $g \geq 2$ there exist $X \neq Y \in \mathcal{M}(S_g)$ with $\mathrm{rls}(X) = \mathrm{rls}(Y)$. Such surfaces are said to be *isospectral*. Sunada later proved that for $g \geq 3$ the set of such X having a distinct isospectral Y is a positive-dimensional subset of $\mathcal{M}(S_g)$ [201]. We would like to mention the deep theorem of Huber that two closed hyperbolic surfaces are isospectral if and only if their Laplacians have the same spectrum [98, 99, 100].

While surfaces isospectral to a given closed hyperbolic surface X can exist, it is a theorem of McKean [149, 150] that there are only finitely many such surfaces. The proof we give is due to Wolpert [215].

THEOREM 12.9 *Let* $g \geq 2$. *For any* $X \in \mathcal{M}(S_g)$, *the set*

$$\{Y \in \mathcal{M}(S_g) : \mathrm{rls}(X) = \mathrm{rls}(Y)\}$$

is finite.

Theorem 12.9 is also true for $g = 1$, but we will not need this fact.

Proof. Let $X \in \mathcal{M}(S_g)$ and let $\mathcal{X}$ be a lift of X to $\mathrm{Teich}(S_g)$. We first prove that, for any compact set $B \subset \mathrm{Teich}(S_g)$, the set

$$\{\mathcal{Y} \in B : \mathrm{rls}(\mathcal{X}) = \mathrm{rls}(\mathcal{Y})\}$$

is finite.

Let $\{\gamma_1, \ldots, \gamma_{9g-9}\}$ be the finite set of simple closed curves in S_g whose lengths determine any hyperbolic structure on S_g; such a set of curves is guaranteed by Theorem 10.7. Let $L = \max\{\ell_{\mathcal{X}}(\gamma_i)\}$ and let $K = e^{2R}$, where B is contained in the ball of radius R around $\mathcal{X}$.

Wolpert's lemma (Lemma 12.5) gives that for any $\mathcal{Y} \in B$ the bound $\ell_{\mathcal{Y}}(\gamma_i) \leq KL$ holds for each i. Now if $\mathrm{rls}(\mathcal{Y}) = \mathrm{rls}(\mathcal{X})$, then $\ell_{\mathcal{Y}}(\gamma_i) \in \mathrm{rls}(\mathcal{X})$. But by discreteness of the raw length spectrum (Lemma 12.4), there are only finitely many points in $\mathrm{rls}(\mathcal{X}) \cap (0, KL]$. Thus for any $\mathcal{Y} \in B$ with $\mathrm{rls}(\mathcal{Y}) = \mathrm{rls}(\mathcal{X})$, there are only finitely many choices for the values $\ell_{\mathcal{Y}}(\gamma_i)$, and hence finitely many such $\mathcal{Y}$, as desired.

By the discreteness of $\mathrm{rls}(X)$, we can choose some $\epsilon > 0$ so that $X \in \mathcal{M}_\epsilon(S_g)$. Any $Y \in \mathcal{M}(S_g)$ with $\mathrm{rls}(Y) = \mathrm{rls}(X)$ must also lie in $\mathcal{M}_\epsilon(S)$. Now $\mathcal{M}_\epsilon(S)$ is compact by Mumford's compactness criterion (Theorem 12.6), and so there is a compact set B in $\mathrm{Teich}(S_g)$ that projects onto $\mathcal{M}_\epsilon(S)$. But since there are finitely many $\mathcal{Y} \in B$ with $\mathrm{rls}(\mathcal{Y}) = \mathrm{rls}(\mathcal{X})$, there are finitely many $Y \in \mathcal{M}_\epsilon(S_g)$ with $\mathrm{rls}(Y) = \mathrm{rls}(X)$, and we are done. $\square$

We record the following consequence of the proof of Theorem 12.9: for any $\mathcal{X} \in \mathrm{Teich}(S)$, the set

$$\{\mathcal{Y} \in \mathrm{Teich}(S) : \mathrm{rls}(\mathcal{X}) = \mathrm{rls}(\mathcal{Y})\}$$

is discrete.

12.5 THE TOPOLOGY AT INFINITY OF MODULI SPACE

A basic measure of the noncompactness of a space is its number of ends, defined below. One can consider this as computing "π_0 at infinity." There is also a version of "π_1 at infinity" of a space. In this section we define and compute these basic invariants for $\mathcal{M}(S)$.

12.5.1 THE MAIN TECHNICAL RESULT

The various connectedness properties for $\mathcal{M}(S)$ at infinity will all be deduced from the following.

Proposition 12.10 *Let $g \geq 2$. Let $\mathcal{X}, \mathcal{Y} \in \mathrm{Teich}(S_g)$ and suppose that their images $X, Y \in \mathcal{M}(S_g)$ lie in $\mathcal{M}(S_g) - \mathcal{M}_\epsilon(S_g)$. Then there is a path from $\mathcal{X}$ to $\mathcal{Y}$ in $\mathrm{Teich}(S_g)$ whose projection to $\mathcal{M}(S_g)$ lies in $\mathcal{M}(S_g) - \mathcal{M}_\epsilon(S_g)$.*

In other words, Proposition 12.10 tells us that, given any two points in $\mathrm{Teich}(S_g)$ each of which has some short essential closed curve, these points are connected by a path in $\mathrm{Teich}(S_g)$ every point of which has some short essential closed curve. Of course, the specific closed curve which is short will change depending on where we are on the connecting path.

Proof. By the assumptions on $\mathcal{X}$ and $\mathcal{Y}$, there are nontrivial simple closed curves α and β in S_g with $\ell_{\mathcal{X}}(\alpha) < \epsilon$ and $\ell_{\mathcal{Y}}(\beta) < \epsilon$. By Theorem 4.3, there is a sequence of essential simple closed curves $\alpha = \gamma_1, \ldots, \gamma_n = \beta$ such that $\gamma_i \cap \gamma_{i+1} = \emptyset$ for all i.

Take γ_1 and γ_2 to be part of a Fenchel–Nielsen coordinate system of curves. By decreasing only the length parameter of γ_2 in this coordinate system, while keeping the other parameters fixed, we obtain a connected path in $\text{Teich}(S_g)$, starting at $\mathcal{X}$ and ending at some point $\mathcal{X}_2$ with the property that $\ell_{\mathcal{X}_2}(\gamma_2) < \epsilon$ and $\ell_{\mathcal{Z}}(\gamma_1) < \epsilon$ for all points $\mathcal{Z}$ on the path.

Repeating this procedure from γ_2 to γ_3, and so on, we obtain a path in $\text{Teich}(S_g)$ from $\mathcal{X}$ to some $\mathcal{Y}'$ where each point on the path projects to $\mathcal{M}(S_g) - \mathcal{M}_\epsilon(S_g)$ and in particular where the length of $\gamma_n = \beta$ in $\mathcal{Y}'$ is less than ϵ. We can then vary the last set of Fenchel–Nielsen coordinates to obtain a path from $\mathcal{Y}'$ to $\mathcal{Y}$ where the length of β remains less than ϵ. The concatenation of these paths satisfies the conclusion of the proposition. $\square$

12.5.2 THE END OF MODULI SPACE

The theory of ends of spaces is a way to encode the number of "noncompact directions" of a space. We will need only the notion of one end. A connected, locally compact topological space X has *one end* if, for every compact set $B \subset X$, the space $X \setminus B$ has only one component whose closure is noncompact. For example, compact spaces do not have one end; neither does the real line, as the complement of a closed interval has two unbounded components.

Suppose that X is a connected, locally compact metric space and that X_i is an exhaustion of X by compact sets with $X \setminus X_i$ path-connected. Then X has one end. This holds, for example, for $X = \mathbb{R}^d$ with $d \geq 2$, where one can choose X_i to be the ball of radius i about any fixed point.

Proposition 12.10 allows us to deduce the following.

Corollary 12.11 *Let $g \geq 1$. The moduli space $\mathcal{M}(S_g)$ has one end.*

Proof. In the case $g = 1$, the fact that moduli space has one end follows directly from the explicit description of $\mathcal{M}(T^2)$ given in Section 12.2.

Let $g \geq 2$. Then $\mathcal{M}(S_g) - \mathcal{M}_\epsilon(S_g)$ is connected for any $\epsilon > 0$ by Proposition 12.10. Since the $\mathcal{M}_\epsilon(S_g)$ form an exhaustion of $\mathcal{M}(S_g)$, we conclude that $\mathcal{M}(S_g)$ has one end. $\square$

The key fact used in the proof of Proposition 12.10 is that the complex of curves $\mathcal{C}(S_g)$ is connected. Proposition 12.10 and Corollary 12.11 both hold for $S_{g,n}$ with $3g - 3 + n \geq 2$ since $\mathcal{C}(S_{g,n})$ is connected in these cases.

12.5.3 LOOPS IN MODULI SPACE

Taking the fundamental group of the topological space underlying $\mathcal{M}(S)$ misses its salient features. Indeed, we have the following fact:

$\mathcal{M}(S_g)$ *is simply connected for all $g \geq 1$.*

For $g = 1$, this follows from the fact that $\mathcal{M}(T^2)$ is the $(0; 2, 3, \infty)$ hyperbolic orbifold, so that the underlying topological space is a once-punctured sphere, which is homeomorphic to $\mathbb{R}^2$. The fact that $\mathcal{M}(S_g)$ is simply connected is due to Maclachlan [135] and follows from the following three facts:

- $\text{Mod}(S_g)$ is generated by finite-order elements (Theorem 7.16).

- The action of each finite-order element on $\text{Teich}(S_g)$ has a fixed point (see Section 12.1).

- The cover $\text{Teich}(S_g) \to \mathcal{M}(S_g)$ enjoys the path-lifting property (the path-lifting property holds any time we take the quotient of a simply connected space by a properly discontinuous action [4, 39]).

To get simple connectivity of $\mathcal{M}(S_g)$ from these three facts, take any loop in $\mathcal{M}(S_g)$ based at the image in $\mathcal{M}(S_g)$ of a fixed point of one of the generators of $\text{Mod}(S_g)$. The lift of this loop is a closed loop in $\text{Teich}(S_g)$, and any null homotopy in $\text{Teich}(S_g)$ descends to a null homotopy in $\mathcal{M}(S_g)$.

The more useful notion to consider is the orbifold fundamental group of $\mathcal{M}(S)$. The *orbifold fundamental group* of the quotient X/Γ of a simply connected space X by a group Γ acting properly discontinuously (but not necessarily freely) is defined to be

$$\pi_1^{orb}(X/\Gamma) \approx \Gamma.$$

Since $\mathcal{M}(S) = \text{Teich}(S)/\text{Mod}(S)$, the group $\text{Mod}(S)$ acts properly discontinuously on $\text{Teich}(S)$, and $\text{Teich}(S)$ is simply connected, we have

$$\pi_1^{orb}(\mathcal{M}(S)) \approx \text{Mod}(S).$$

Two loops α, β in X/Γ are homotopic in the orbifold sense if they have path liftings $\widetilde{\alpha}, \widetilde{\beta} : [0, 1] \to X$ such that $\widetilde{\alpha}(0) = \widetilde{\beta}(0)$ and $\widetilde{\alpha}(1) = \widetilde{\beta}(1)$. For example, a loop around the cone point of order 2 in $\mathcal{M}(T^2)$ is trivial in the topological category but nontrivial in the orbifold category; it has order 2 in $\pi_1^{orb}(\mathcal{M}(T^2))$.

With the above comments in hand, we now consider loops in $\mathcal{M}(S)$ in the orbifold sense. The orbifold $\mathcal{M}(T^2)$ has a unique homotopy class of loops that can be freely homotoped (in the orbifold sense) outside every compact subset of $\mathcal{M}(T^2)$; namely, the free homotopy class represented by the conjugacy class of the element $\left(\begin{smallmatrix} 1 & 1 \\ 0 & 1 \end{smallmatrix}\right)$ in $\text{SL}(2, \mathbb{Z}) \approx \text{Mod}(T^2)$. This contrasts greatly with the behavior of $\mathcal{M}(S_g)$ when $g \geq 2$.

Corollary 12.12 *Let $g \geq 2$. Any loop in $\mathcal{M}(S_g)$ can be freely homotoped (even in the orbifold sense) outside every compact set in $\mathcal{M}(S_g)$.*

Proof. It suffices to consider loops that are essential and compact sets that are of the form $\mathcal{M}_\epsilon(S_g)$. Let any $\epsilon > 0$ be given. Let α be any essential loop in $\mathcal{M}(S_g)$, and X any point in $\mathcal{M}(S_g) - \mathcal{M}_\epsilon(S_g)$. Since $\mathcal{M}(S_g)$ is path-connected, α can be freely homotoped to a loop β based at X. The loops α and β are homotopic in the orbifold sense.

As above, β can be lifted to a path $\widetilde{\beta}$ in $\mathrm{Teich}(S_g)$. Proposition 12.10 gives a path γ between the endpoints of $\widetilde{\beta}$ with projection $\overline{\gamma}$ in $\mathcal{M}(S_g) - \mathcal{M}_\epsilon(S_g)$. The loop $\overline{\gamma}$ is homotopic to β, hence α, in the orbifold sense. $\square$

Another way to state Corollary 12.12 is that, for $g \geq 2$, the (orbifold) fundamental group of $\mathcal{M}(S_g)$ "relative to infinity" is trivial. More formally, the inclusion map $\mathcal{M}(S_g) - \mathcal{M}_\epsilon(S_g) \hookrightarrow \mathcal{M}(S_g)$ induces an isomorphism of orbifold fundamental groups.

We also remark that both Corollaries 12.11 and 12.12 are true with $\mathcal{M}(S_g)$ replaced by the manifold $\mathrm{Teich}(S_g)/\Gamma$, where Γ is a finite-index torsion-free subgroup of $\mathrm{Mod}(S_g)$.

12.6 MODULI SPACE AS A CLASSIFYING SPACE

In this section we explain the close relationship between $\mathcal{M}(S_g)$ and the classifying space of $\mathrm{Mod}(S_g)$. This connection relates the cohomology of these two objects.

Classifying spaces and covers of $\mathcal{M}(S_g)$. In Section 5.6, we proved that the space $\mathrm{BHomeo}^+(S_g)$ that classifies S_g-bundles is homotopy-equivalent to $\mathrm{BMod}(S_g)$ for $g \geq 2$. In particular, the elements of the cohomology groups $H^*(\mathrm{Mod}(S_g); \mathbb{Z})$ give (integral) characteristic classes for orientable S_g-bundles. In this section we explain the close relationship of these results to the topology of $\mathcal{M}(S_g)$.

$\mathrm{Mod}(S_g)$ acts properly discontinuously on the contractible space $\mathrm{Teich}(S_g)$, with quotient $\mathcal{M}(S_g)$. However, $\mathcal{M}(S_g)$ is not a $\mathrm{BMod}(S_g)$-space. The problem is that the action of $\mathrm{Mod}(S_g)$ on $\mathrm{Teich}(S_g)$ is not free. This is unfortunate because we have a decent geometric and topological picture of $\mathcal{M}(S_g)$: it is finitely covered by a $(6g - 6)$-dimensional manifold, it has the homotopy type of a finite cell complex (see the comment on page 127), and it occurs naturally elsewhere in mathematics. In contrast, any $\mathrm{BMod}(S_g)$-space cannot have the homotopy type of a finite cell complex since $\mathrm{Mod}(S_g)$ has torsion (combine Proposition VIII.2.2 and Corollary VIII.2.5 in [38]). However, we now use a standard method in topology to show that $\mathcal{M}(S_g)$ is rationally a classifying space for $\mathrm{Mod}(S_g)$.

Let Γ be any finite-index, torsion-free normal subgroup of $\mathrm{Mod}(S_g)$. For example, in Theorem 6.9 we proved that for any $g \geq 1$ and $m \geq 3$ the group $\mathrm{Mod}(S_g)[m]$ of elements of $\mathrm{Mod}(S_g)$ acting trivially on $H_1(S_g; \mathbb{Z}/m\mathbb{Z})$ is such a subgroup. Since the point stabilizers of the action of $\mathrm{Mod}(S_g)$ on $\mathrm{Teich}(S_g)$ are finite, it follows that any such Γ acts freely and properly discontinuously on $\mathrm{Teich}(S_g)$. In particular, $\mathrm{Teich}(S_g)/\Gamma$ is a $K(\Gamma, 1)$-space. It follows from the discussion in Section 5.6 that the characteristic classes for oriented S_g-bundles with monodromy lying in Γ are precisely the elements of $H^*(\mathrm{Teich}(S_g)/\Gamma; \mathbb{Z})$.

We now want to convert this back into information about $\mathcal{M}(S_g)$. We use the Borel construction (cf. the proof of Proposition 5.6) as follows. The diagonal action of the group $\mathrm{Mod}(S_g)$ on the contractible space $\mathrm{EMod}(S_g) \times \mathrm{Teich}(S_g)$ is free and properly discontinuous, and so the quotient space is a $K(\mathrm{Mod}(S_g), 1)$-space; we denote it by $\mathrm{BMod}(S_g)$ (even though $\mathrm{BMod}(S_g)$ usually denotes the quotient $\mathrm{EMod}(S_g)/\mathrm{Mod}(S_g)$). But the projection map $\mathrm{EMod}(S_g) \times \mathrm{Teich}(S_g) \to \mathrm{Teich}(S_g)$ is $\mathrm{Mod}(S_g)$-equivariant and so induces a continuous map

$$h : \mathrm{BMod}(S_g) \to \mathcal{M}(S_g).$$

If $\mathfrak{X} \in \mathrm{Teich}(S_g)$ maps to $X \in \mathcal{M}(S_g)$, then $h^{-1}(X)$ is a classifying space for the stabilizer of $\mathfrak{X}$ in $\mathrm{Mod}(S_g)$.

Using the same construction with $\mathrm{Mod}(S_g)$ replaced by the finite-index subgroup Γ, we obtain a continuous map

$$\widetilde{h} : \mathrm{B}\Gamma \to \mathrm{Teich}(S_g)/\Gamma.$$

The map $\widetilde{h}$ is a homotopy equivalence by Whitehead's theorem since $\mathrm{B}\Gamma$ and $\mathrm{Teich}(S_g)/\Gamma$ are classifying spaces and $\widetilde{h}_\star : \pi_1(\mathrm{B}\Gamma) \to \pi_1(\mathrm{Teich}(S_g)/\Gamma)$ is an isomorphism.

Rational cohomology. Let $G = \mathrm{Mod}(S_g)/\Gamma$. The finite group G acts by covering space automorphisms on $\mathrm{B}\Gamma$ and on $\mathrm{Teich}(S_g)/\Gamma$. By construction, the map $\widetilde{h}$ is G-equivariant. We thus have the following commutative diagram:

$$
\begin{array}{ccc}
\mathrm{B}\Gamma & \xrightarrow{\ \widetilde{h}\ } & \mathrm{Teich}(S_g)/\Gamma \\
\downarrow & & \downarrow \\
\mathrm{BMod}(S_g) & \xrightarrow{\ h\ } & \mathcal{M}(S_g)
\end{array}
$$

Since $\widetilde{h}$ is a G-equivariant homotopy equivalence, it induces a G-

equivariant isomorphism

$$\widetilde{h}^* : H^*(\mathrm{Teich}(S_g)/\Gamma; \mathbb{Q}) \to H^*(\mathrm{B}\Gamma; \mathbb{Q}).$$

For a vector space V equipped with a G-action, denote by V^G the G-*invariants* of the action, that is,

$$V^G = \{v \in V : gv = v \text{ for all } g \in G\}.$$

Since $\widetilde{h}^*$ is G-equivariant, it restricts to an isomorphism of the corresponding invariants.

Now the covering map $\mathrm{B}\Gamma \to \mathrm{BMod}(S_g)$ induces an isomorphism

$$H^*(\mathrm{BMod}(S_g); \mathbb{Q}) \approx H^*(\mathrm{B}\Gamma; \mathbb{Q})^G,$$

and the covering map $\mathrm{Teich}(S_g)/\Gamma \to \mathcal{M}(S_g)$ induces an isomorphism

$$H^*(\mathcal{M}(S_g); \mathbb{Q}) \approx H^*(\mathrm{Teich}(S_g)/\Gamma; \mathbb{Q})^G.$$

These isomorphisms come from the basic transfer argument in cohomology [91, Proposition 3G.1]. We have thus proven the following theorem relating the rational cohomology of $\mathrm{BHomeo}^+(S_g) \simeq \mathrm{BMod}(S_g)$ to the rational cohomology of $\mathcal{M}(S_g)$.

THEOREM 12.13 *Let $g \geq 2$ and let $h : \mathrm{BMod}(S_g) \to \mathcal{M}(S_g)$ be the map constructed above. Then the induced homomorphism*

$$h^* : H^*(\mathcal{M}(S_g); \mathbb{Q}) \to H^*(\mathrm{BMod}(S_g); \mathbb{Q})$$

is an isomorphism.

Thus the rational characteristic classes of surface bundles are precisely elements of $H^*(\mathcal{M}(S); \mathbb{Q})$. We emphasize that rational coefficients are crucial here; they are used in the transfer argument.

PART 3

The Classification and Pseudo-Anosov Theory

Chapter Thirteen

The Nielsen–Thurston Classification

In this chapter we explain and prove one of the central theorems in the study of mapping class groups: the Nielsen–Thurston classification of elements of $\mathrm{Mod}(S)$. This theorem is the analogue of the Jordan canonical form for matrices. It states that every $f \in \mathrm{Mod}(S)$ is one of three special types: periodic, reducible, or pseudo-Anosov. The knowledge of individual mapping classes is essential to our understanding of the algebraic structure of $\mathrm{Mod}(S)$. As we will soon explain, it is also essential for our understanding of the geometry and topology of many 3-dimensional manifolds.

We begin this chapter with a classification of elements of $\mathrm{Mod}(T^2)$. We then describe higher-genus analogues for each of the three types of elements of $\mathrm{Mod}(T^2)$, after which we are able to state the Nielsen–Thurston classification theorem in various forms, as well as a connection to 3-manifold theory. The rest of the chapter is devoted to Bers' proof of the Nielsen–Thurston classification. Bers' proof is an analogue of the geometric classification of elements of $\mathrm{Isom}^+(\mathbb{H}^2)$ by their translation lengths. Our treatment of the proof is self-contained and presents a combined application of the material from Chapters 10–12. The collar lemma is highlighted as a new ingredient, as it is also a fundamental result in the hyperbolic geometry of surfaces.

13.1 THE CLASSIFICATION FOR THE TORUS

Recall that in Chapter 1 we classified the nontrivial elements of $\mathrm{Isom}^+(\mathbb{H}^2)$ into three types: elliptic, parabolic, and hyperbolic. We also know that $\mathrm{Mod}(T^2) \approx \mathrm{SL}(2, \mathbb{Z})$ (Theorem 2.5) and that $\mathrm{PSL}(2, \mathbb{R}) \approx \mathrm{Isom}^+(\mathbb{H}^2)$. We thus obtain a classification of elements of $\mathrm{Mod}(T^2)$ by considering the type of the corresponding element of $\mathrm{Isom}^+(\mathbb{H}^2)$.

What we would really like, though, is a classification of elements of $\mathrm{Mod}(T^2)$ that is intrinsic to the torus. As we now show, the three types of hyperbolic isometries correspond to three qualitatively different types of homeomorphisms of T^2.

The trichotomy. Let $f \in \mathrm{Mod}(T^2)$, let A be the corresponding element of $\mathrm{SL}(2, \mathbb{Z})$, and let τ be the corresponding element of $\mathrm{Isom}^+(\mathbb{H}^2)$. We consider the three cases for τ in turn. Recall that the standard isomorphism $\mathrm{PSL}(2, \mathbb{R}) \rightarrow \mathrm{Isom}^+(\mathbb{H}^2)$ sends the equivalence class of the matrix

$$\begin{pmatrix} a & b \\ c & d \end{pmatrix}$$

to the Möbius transformation

$$z \mapsto \frac{az + b}{cz + d}$$

acting on the upper half-plane.

If τ is elliptic, this means that τ fixes a point of $\mathbb{H}^2$ and is thus a rotation. By the proper discontinuity of the action of $\mathrm{SL}(2, \mathbb{Z})$ on $\mathbb{H}^2$, we see that τ must be a finite-order rotation. Thus A, hence f, has finite order. We say that f is *periodic*.

If τ is parabolic, then τ fixes a unique point in $\partial \mathbb{H}^2$. This is the same as saying that A has a unique real eigenvector. It follows that A has exactly one real eigenvalue and that this eigenvalue has multiplicity 2. Since the product of the eigenvalues of A is equal to the determinant, which is 1, the eigenvalue for A is ± 1. This means that, up to sign, A fixes a vector in $\mathbb{R}^2$. Since A is an integer matrix, it follows that A fixes a rational vector in $\mathbb{R}^2$ up to sign. From this it follows that f fixes the corresponding isotopy class of (unoriented) simple closed curves in T^2. In this case we say that f is *reducible*.

If τ is hyperbolic, then τ fixes two points in $\partial \mathbb{H}^2$. This is equivalent to the statement that A has two linearly independent real eigenvectors or that A has two distinct real eigenvalues. Since the determinant of A is 1, it follows that its two eigenvalues are inverses, say λ and $1/\lambda$, where $\lambda > 1$. Therefore, A has two eigenspaces in $\mathbb{R}^2$, one of which is stretched by a factor of λ and one of which is contracted by a factor of λ. This data gives a bundle of information about $f \in \mathrm{Mod}(T^2)$; we call this information an *Anosov package*. Specifically, there is on T^2 a pair of foliations $\mathcal{F}^s$ and $\mathcal{F}^u$, called the *stable* and *unstable* foliations for f, that satisfy the following properties.

1. Each leaf of $\mathcal{F}^s$ and of $\mathcal{F}^u$ is the image of an injective map $\mathbb{R} \rightarrow T^2$.

2. The foliations $\mathcal{F}^s$ and $\mathcal{F}^u$ are transverse at all points.

3. There is a natural transverse measure μ_s (respectively μ_u) assigning a measure to each arc transverse to $\mathcal{F}^s$ (respectively μ_u) obtained by realizing the foliations by straight lines in some flat metric on T^2, and declaring the measure of a transverse arc to be the total variation in the direction perpendicular to the foliation.

4. There is an affine representative $\phi \in \text{Homeo}^+(T^2)$ of f with

$$\phi(\mathcal{F}^u, \mu_u) = (\mathcal{F}^u, \lambda\mu_u) \quad \text{and} \quad \phi(\mathcal{F}^s, \mu_s) = (\mathcal{F}^s, \lambda^{-1}\mu_s),$$

where $\lambda > 1$ is the leading eigenvalue of A.

In this case we say that f is *Anosov*. The discussion so far can be summarized by the following.

THEOREM 13.1 *Each nontrivial element $f \in \text{Mod}(T^2)$ is of exactly one of the following types: periodic, reducible, Anosov.*

We can be even more specific in the first two cases. A nontrivial finite-order element of $\text{Mod}(T^2)$ has order 2, 3, 4, or 6. Also, a nontrivial reducible element of $\text{Mod}(T^2)$ is either a power of a Dehn twist or the product of a power of a Dehn twist with the hyperelliptic involution.

The linear algebra approach. Using just the isomorphism $\text{Mod}(T^2) \approx \text{SL}(2, \mathbb{Z})$, and without appealing to hyperbolic geometry, we can give a more algebraic approach to the classification for $\text{Mod}(T^2)$. Let $A \in \text{SL}(2, \mathbb{Z})$ and let $f \in \text{Mod}(T^2)$ denote the corresponding mapping class. The characteristic polynomial for A is $x^2 - \text{trace}(A)x + 1$. It follows that the eigenvalues of A are inverses of each other—call them λ and λ^{-1}. There are then three cases to consider:

1. $|\text{trace}(A)| \in \{0, 1\}$

2. $|\text{trace}(A)| = 2$

3. $|\text{trace}(A)| > 2$.

The three cases are equivalent to the cases: λ and λ^{-1} are complex, $\lambda = \lambda^{-1} = \pm 1$, and λ and λ^{-1} are distinct reals. In the first case it follows from the Cayley–Hamilton theorem that A, hence f, has finite order. In the second case A has a rational eigenvector, and from this it follows that f is reducible. In the third case we see that A has two real eigenvalues, and so f is Anosov.

We summarize the results of this section in the following table.

Mapping class	$\mathbb{H}^2$ isometry	|Trace|	Sample matrix
Periodic	Elliptic	$0, 1$	$\left(\begin{smallmatrix} 0 & 1 \\ -1 & 1 \end{smallmatrix}\right)$
Reducible	Parabolic	2	$\left(\begin{smallmatrix} 1 & n \\ 0 & 1 \end{smallmatrix}\right)$
Anosov	Hyperbolic	$3, 4, \ldots$	$\left(\begin{smallmatrix} 2 & 1 \\ 1 & 1 \end{smallmatrix}\right)$

13.2 THE THREE TYPES OF MAPPING CLASSES

We now describe three kinds of elements of the mapping class group of a surface. Each is an analogue of one of the three types of elements of $\mathrm{Mod}(T^2)$ in the statement of Theorem 13.1. The Nielsen–Thurston classification theorem (Section 13.3) states that every mapping class falls into (at least) one of these three categories.

13.2.1 PERIODIC MAPPING CLASSES

We have already studied *periodic*, or finite-order, elements of the mapping class group. A basic example is shown in Figure 13.1; see also Figures 2.1–2.3. See Chapter 7 for a general discussion of finite-order mapping classes.

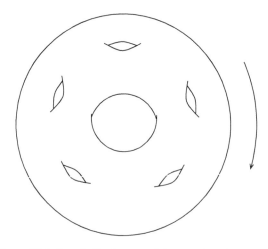

Figure 13.1 A periodic element of the mapping class group.

Theorem 7.1 states that every periodic mapping class has a representative diffeomorphism that has finite order. Note that a priori we know only that there is a representative with a power *isotopic to* the identity. We are now prepared to give a proof of this theorem. Indeed, we will show that each periodic element of $\mathrm{Mod}(S)$ can be realized as an isometry of S with respect to some hyperbolic metric. The idea is to show that a periodic element of $\mathrm{Mod}(S)$, thought of as an isometry of $\mathrm{Teich}(S)$, has a fixed point in $\mathrm{Teich}(S)$. Nielsen gave a direct proof of Theorem 7.1 in 1942 [171]. The proof we present here was first suggested by Fenchel [62, 63] and Macbeath [134].

The following proof is one case where the language of markings muddles the argument. Thus we will regard points $\mathcal{X}, \mathcal{Y} \in \mathrm{Teich}(S)$ as complex structures X and Y on S and Teichmüller maps $X \to Y$ as dilatation-minimizing maps in the homotopy class of the identity map $S \to S$.

Proof of Theorem 7.1. Recall from Section 12.1 that an element $f \in \mathrm{Mod}(S)$ fixes a point of $\mathrm{Teich}(S)$ if and only if it has a representative $\phi \in \mathrm{Homeo}^+(S)$ that is an isometry of S with respect to some hyperbolic metric; equivalently, f has a representative that is conformal with respect to some complex structure. So to prove the theorem it suffices to show that each periodic element $f \in \mathrm{Mod}(S)$ fixes some point of $\mathrm{Teich}(S)$.

Let $n \in \mathbb{N}$ denote the order of $f \in \mathrm{Mod}(S)$. Since $\mathrm{Teich}(S)$ is contractible (Theorem 10.6), the finite cyclic group $\langle f \rangle$ cannot act freely on $\mathrm{Teich}(S)$, for otherwise the quotient would be a finite-dimensional $K(\mathbb{Z}/n\mathbb{Z}, 1)$. Thus $f^k \cdot \mathcal{X} = \mathcal{X}$ for some $1 \le k < n$ and some $\mathcal{X} \in \mathrm{Teich}(S)$.

In the case that n is prime (or even if $\gcd(n, k) = 1$), we have that f is a power of f^k. Thus f fixes $\mathcal{X} \in \mathrm{Teich}(S)$, and we are done.

Now assume that $n = p_1 p_2 \cdots p_s$, where each p_i is prime and the p_i's are not necessarily distinct. We induct on the number of (not necessarily distinct) prime factors of n. The mapping class $f' = f^{p_s}$ has order $p_1 p_2 \cdots p_{s-1}$. By induction, f' fixes a point of $\mathrm{Teich}(S)$ and hence is realized by a conformal automorphism ϕ of a complex structure X on S. Since ϕ acts on S by conformal automorphisms, the quotient $X/\langle \phi \rangle$ is a Riemann surface with distinguished points, namely, the images of the fixed points of ϕ. Let X' denote the Riemann surface which is the complement of the distinguished points and let S' denote the underlying topological surface. Denote the fixed set of f' in $\mathrm{Teich}(S)$ by $\mathrm{Fix}(f')$.

Claim: There is a well-defined map $\theta : \mathrm{Fix}(f') \to \mathrm{Teich}(S')$.

Proof of claim: Let $\mathcal{Y} \in \mathrm{Fix}(f')$. Represent $\mathcal{Y}$ by a complex structure Y on S and let $h : X \to Y$ be the Teichmüller map in the homotopy class of the identity map $S \to S$. Denote its dilatation by K_h. Since ϕ is conformal, the dilatation of $\phi \circ h \circ \phi^{-1}$ is equal to K_h. Also, since h is homotopic to the identity, $\phi \circ h \circ \phi^{-1}$ is homotopic to the identity. It then follows from Teichmüller's uniqueness theorem that h and $\phi \circ h \circ \phi^{-1}$ are equal as elements of $\mathrm{Homeo}^+(S)$, which is to say that ϕ commutes with h. It follows that h descends to a Teichmüller map from the complex structure X' on S' to a unique complex structure Y' on S'. We define $\theta(\mathcal{Y})$ to be $\mathcal{Y}' = [Y']$.

Given the claim, it is not hard to see that θ is in fact a homeomorphism. As $\mathrm{Teich}(S')$ is contractible, it follows that $\mathrm{Fix}(f')$ is contractible. Since f commutes with f', it follows that $\langle f \rangle$ acts on $\mathrm{Fix}(f')$. As $\langle f' \rangle$ is contained in the kernel of this action, the action of $\langle f \rangle$ factors through an action of $\langle f \rangle / \langle f' \rangle \approx \mathbb{Z}/p_s \mathbb{Z}$ on $\mathrm{Fix}(f')$. As above, the latter action must have a fixed point, and so since p_s is prime, we are again able to deduce that f has a fixed point in $\mathrm{Fix}(f') \subseteq \mathrm{Teich}(S)$.   □

Our argument for Theorem 7.1 can be easily adapted to prove an even stronger result: every finite solvable subgroup of $\mathrm{Mod}(S)$ is realized as a subgroup of the isometry group of S for some hyperbolic metric on S. Of course, as we noted in Theorem 7.2, it is now known (and much harder to prove) that any finite subgroup of $\mathrm{Mod}(S)$ can be realized as a group of conformal automorphisms of some conformal structure on S.

13.2.2 REDUCIBLE MAPPING CLASSES

We say that an element f of $\mathrm{Mod}(S)$ is *reducible* if there is a nonempty set $\{c_1, \ldots, c_n\}$ of isotopy classes of essential simple closed curves in S so that $i(c_i, c_j) = 0$ for all i and j and so that $\{f(c_i)\} = \{c_i\}$. The collection is called a *reduction system* for f. In this case, we can further understand f via the following procedure:

1. Choose representatives $\{\gamma_i\}$ of the $\{c_i\}$ with $\gamma_i \cap \gamma_j = \emptyset$ for $i \neq j$.

2. Choose a representative ϕ of f with $\{\phi(\gamma_i)\} = \{\gamma_i\}$.

3. Consider the homeomorphism of the noncompact, possibly disconnected surface $S - \cup \gamma_i$ induced by ϕ.

Note that the second step is an application of the Alexander method plus Proposition 1.11. As each connected component of $S - \cup \gamma_i$ is simpler than S itself, as measured for example by Euler characteristic, we can hope to understand f by induction on the complexity of S. In particular, we can decompose f into irreducible pieces (cf. Corollary 13.3 below). Of course, in order to do this, even when S is closed, one must extend the theory to nonclosed S.

Examples. A typical reducible mapping class is obtained as follows. Suppose S is a closed genus 2 surface. Let γ be a separating simple closed curve in S and let S' and S'' be the two embedded subsurfaces of S bounded by γ. Choose ϕ' and ϕ'' to be homeomorphisms of S' and S'' that fix γ pointwise. Even better, choose ϕ' and ϕ'' so that neither fixes the isotopy class of any essential simple closed curve in S' or S'', respectively; since S' and S'' are tori, it suffices to choose ϕ' and ϕ'' so that the induced actions on $H_1(S'; \mathbb{Z})$ and $H_1(S''; \mathbb{Z})$ are without fixed vectors. Let σ be a homeomorphism of S that switches the two sides of γ. The homeomorphism

$$\sigma \circ T_\gamma \circ \phi' \circ \phi''$$

represents a reducible element of $\mathrm{Mod}(S)$. The isotopy class of γ is the unique reduction system in this case.

A simple example of a reducible mapping class is a Dehn twist T_a: any collection of distinct isotopy class of curves $\{c_i\}$ satisfying $i(c_i, c_j) = i(c_i, a) = 0$ is a reduction system.

Another example is the mapping class given in Figure 13.1 (which curves, or collections of curves, are fixed?). Thus we see that there is an overlap between the set of periodic and reducible elements of $\mathrm{Mod}(S)$.

Canonical reduction systems. In each of the last two examples, there are many choices for the reduction system, as there are many collections of curves fixed by either mapping class. A reduction system for $f \in \mathrm{Mod}(S)$ is called *maximal* if it is maximal with respect to inclusion of reduction systems for f. We can then consider the intersection of all maximal reduction systems for f. This intersection is clearly canonical, in that no choices are involved in its construction. We call it the *canonical reduction system* for f.

As a first example, we show that the canonical reduction system for a periodic $f \in \mathrm{Mod}(S)$ is empty. For simplicity, assume $\chi(S) < 0$. By Theorem 7.1, f is represented by a finite-order homeomorphism ϕ. Let X denote the quotient orbifold $S/\langle \phi \rangle$. Suppose that c is an isotopy class of simple closed curves in S that is part of some reduction system for f. Then c has a representative γ that descends to an essential simple closed curve $\overline{\gamma}$ in X. But by the classification of surfaces, once X has one essential simple closed curve $\overline{\gamma}$, it has another one $\overline{\delta}$ with $i(\overline{\gamma}, \overline{\delta}) > 0$. The isotopy class of the preimage δ of $\overline{\delta}$ is a reduction system for f, and $i(\delta, \gamma) > 0$. It follows that γ and δ do not belong to a common maximal reduction system. In particular, c is not an element of the canonical reduction system for f. Thus the canonical reduction system for f is empty.

We can also show that the canonical reduction system for T_a is a. It follows immediately from Proposition 3.2 that, for any isotopy class of simple closed curves b with $i(a, b) > 0$, we have $T_a^k(b) \neq b$. It follows that b cannot belong to a reduction system for a. In other words, any reduction system for a consists of isotopy classes of curves that are disjoint from a. As $T_a(a) = a$, it follows that any maximal reduction system for T_a contains a, and so a is in the canonical reduction system. Now, let b be any other element of some reduction system. As above, we have $i(a, b) = 0$. But we can find another isotopy class c such that $i(c, a) = 0$ and $i(b, c) > 0$. Since $i(a, c) = 0$, it follows that $T_a(c) = c$, and so c is part of some reduction system for T_a. On the other hand, since $i(b, c) > 0$, the isotopy classes b and c cannot belong to the same reduction system. Therefore, any maximal reduction system for T_a that contains c does not contain b, and so b is not an element of the canonical reduction system.

Canonical reduction systems were introduced by Birman, Lubotzky, and McCarthy [28] and by Handel and Thurston [81]. In the language of Birman–Lubotzky–McCarthy, the isotopy class c of an essential simple closed curve in S is in the canonical reduction system for f exactly when it satisfies the following two criteria: *(i)* c is part of some reduction system for f, and *(ii)* $f^k(b) \neq b$ whenever $i(b, c) \neq 0$ and $k \neq 0$. The advantage of their definition is that it gives qualitative information about the isotopy classes in an essential reduction system. It is possible to show that their definition is equivalent to ours.

Periodic versus reducible. Dehn twists are examples of mapping classes that are reducible but not periodic. The example in Figure 13.1 is reducible and periodic. One element of $\mathrm{Mod}(S_g)$ that is periodic but not reducible is the example that realizes the upper bound of Theorem 7.5, that is, the periodic element of maximal order in $\mathrm{Mod}(S_g)$. Recall that this mapping class is realized by representing S_g as a $(4g + 2)$-gon and rotating the polygon by one click, that is, by $2\pi/(4g + 2)$. The quotient surface is a sphere with three cone points: one corresponding to the center of the polygon, one corresponding to the vertices of the polygon (all of which are identified in the quotient), and one of which corresponds to the midpoints of the edges of the polygon (again, all of these are identified in the quotient). In the complement of the cone points, there are no essential curves on the sphere. It follows that the mapping class is not reducible.

The question remains: what can we say about mapping classes that are neither periodic nor reducible?

13.2.3 PSEUDO-ANOSOV MAPPING CLASSES

An element $f \in \mathrm{Mod}(S)$ is called *pseudo-Anosov* if there is a pair of transverse measured foliations $(\mathcal{F}^u, \mu_u)$ and $(\mathcal{F}^s, \mu_s)$ on S, a number $\lambda > 1$, and a representative homeomorphism ϕ so that

$$\phi \cdot (\mathcal{F}^u, \mu_u) = (\mathcal{F}^u, \lambda\mu_u) \quad \text{and} \quad \phi \cdot (\mathcal{F}^s, \mu_s) = (\mathcal{F}^s, \lambda^{-1}\mu_s).$$

The measured foliations $(\mathcal{F}^u, \mu_u)$ and $(\mathcal{F}^s, \mu_s)$ are called the *unstable foliation* and the *stable foliation*, respectively, and the number λ is called the *stretch factor*[1] of ϕ (or of f). The map ϕ is a *pseudo-Anosov homeomor-*

---

[1] In the literature the number λ is often called the *dilatation* of the mapping class f. However, this terminology is not consistent with our usage of the word "dilatation." As we shall see, pseudo-Anosov homeomorphisms with stretch factor λ correspond to Teichmüller maps with dilatation λ^2.

phism.

Of course, the representative ϕ of f is not unique; we can change the stable and unstable foliations by an isotopy and then conjugate ϕ by the homeomorphism at the end of the isotopy. It is a theorem, however, that this is the only nonuniqueness: any two homotopic pseudo-Anosov homeomorphisms are conjugate by a homeomorphism that is isotopic to the identity [61, Exposé 12, Theorem III].

The map ϕ is a diffeomorphism away from the singularities of the stable and unstable foliations. Since both the stable and unstable foliations span the tangent space at the singularities, ϕ is not smooth at the singularities. One should compare the definition of a pseudo-Anosov homeomorphism with the definition of a Teichmüller map.

Are the leaves of the stable foliation stretched by ϕ or are they shrunk by ϕ? To check, let α be an arc of the stable foliation. We want to compare $\mu_u(\phi(\alpha))$ with $\mu_u(\alpha)$. By definition of the action of $\mathrm{Homeo}^+(S)$ on the set of measured foliations on S (see Section 11.2), the former is equal to $\phi^{-1} \cdot \mu_u(\alpha) = \lambda^{-1}\mu_u(\alpha)$, where $\lambda > 1$. Thus the correct statement is that ϕ shrinks the leaves of the stable foliation and stretches the leaves of the unstable foliation. One way to remember this is that, if we take a point p on a stable leaf that emanates from a singularity x of $\mathcal{F}_s$, then $\phi^n(p)$ approaches x as n goes to infinity; we think of this as a stability condition.

It turns out that the above structure has strong implications for the dynamical, topological, and geometric structure of pseudo-Anosov homeomorphisms. Indeed, the study of pseudo-Anosov homeomorphisms admits a rich theory, some of which we present in Chapter 14.

Punctures and boundary. The definition of a pseudo-Anosov homeomorphism carries over for surfaces with punctures and/or boundary. For surfaces with boundary, however, we must restrict to the class of measured foliations described at the end of Section 11.4. As with Teichmüller maps, the definition of a pseudo-Anosov homeomorphism is not so natural for surfaces with boundary, and so if we prefer, we can define a pseudo-Anosov homeomorphism for a surface with boundary as a homeomorphism that restricts to a pseudo-Anosov homeomorphism on the punctured surface obtained by removing the boundary. Note, for example, that, given any pseudo-Anosov homeomorphism on any surface, we can remove any finite orbit to obtain a pseudo-Anosov homeomorphism on a punctured surface.

13.3 STATEMENT OF THE NIELSEN–THURSTON CLASSIFICATION

As discussed above, the following theorem is one of the central results in the study of mapping class groups. The theorem is due to Thurston, but it has a somewhat involved history, which we discuss below. The theorem gives a classification of elements of $\mathrm{Mod}(S)$, where by classification we mean that each element of $f \in \mathrm{Mod}(S)$ is shown to have a representative in $\mathrm{Homeo}^+(S)$ that is one of three very specific forms from which one can read off a great deal of information.

Theorem 13.2 (Nielsen–Thurston classification)   *Let $g, n \geq 0$. Each mapping class $f \in \mathrm{Mod}(S_{g,n})$ is periodic, reducible, or pseudo-Anosov. Further, pseudo-Anosov mapping classes are neither periodic nor reducible.*

The main content of Theorem 13.2 is that every irreducible, infinite-order mapping class has a representative that is pseudo-Anosov and so automatically has a great deal of structure. We will further explore this structure and many of its implications in Chapter 14.

A canonical form.   One of the useful aspects of Theorem 13.2 comes from the fact that when f is reducible, one can cut along a reduction system of simple closed curves to obtain a homeomorphism of a (possible disconnected, possibly with boundary) surface, and one can again apply Theorem 13.2 to each of these components. Repeating this process, one finally obtains that any $f \in \mathrm{Mod}(S)$ has a representative that breaks up into finite-order pieces and pseudo-Anosov pieces. In fact, it is possible to do this in such a way that the only curves we cut along are the curves of the canonical reduction system for f [28].

Corollary 13.3 (Canonical form for a mapping class)   *Let $g, n \geq 0$ and let $S = S_{g,n}$. Let $f \in \mathrm{Mod}(S)$ and let $\{c_1, \dots, c_m\}$ be its canonical reduction system. Choose representatives of the c_i with pairwise disjoint closed neighborhoods $R_1, \dots, R_m$. Let $R_{m+1}, \dots, R_{m+p}$ denote the closures of the connected components of $S - \cup_{i=1}^m R_i$. Let $\eta_i : \mathrm{Mod}(R_i) \to \mathrm{Mod}(S)$ denote the homomorphism induced by the inclusion $R_i \to S$ (see Theorem 3.18). Then there is a representative ϕ of f that permutes the R_i, so that some power of ϕ leaves each R_i invariant. What is more, there exists a $k \geq 0$ so that $\phi^k(R_i) = R_i$ for all i and*

$$f^k = \prod_{i=1}^{m+p} \eta_i(f_i),$$

where $f_i \in \mathrm{Mod}(R_i)$ is a power of a Dehn twist for $1 \leq i \leq m$ and $f_i \in$
$\mathrm{Mod}(R_i)$ *is either pseudo-Anosov or the identity for $m + 1 \leq i \leq m + p$.*

The decomposition of f given in Corollary 13.3 is analogous to the Jordan canonical form of a matrix. The $\phi^k|_{R_i}$ are the analogues of Jordan blocks. A example schematic of the normal form is shown in Figure 13.2.

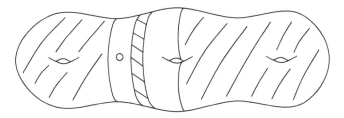

Figure 13.2 A schematic of the normal form of f^k in Corollary 13.3. Each subsurface is fixed. A shaded region indicates a pseudo-Anosov component or a Dehn twist component. An unshaded region indicates an identity component.

Ivanov showed that if f lies in the finite-index subgroup $\mathrm{Mod}(S_g)[m]$ with $m \geq 3$, then the integer k in Corollary 13.3 can always be taken to be 1 [106, Corollary 1.8].

We can sharpen Corollary 13.3 to say that an arbitrary element of $\mathrm{Mod}(S)$ has a normal form without having to take powers. The cost is that we need to deal with the mapping class group of a disconnected surface. Let $\phi^k|_{R_i}$ be a "Jordan block" for $f \in \mathrm{Mod}(S)$ as in Corollary 13.3 and say that $\phi^k|_{R_i}$ is pseudo-Anosov with stretch factor λ and stable foliation $(\mathcal{F}, \mu)$. Say that $k_0 \leq k$ is the smallest positive integer so that ϕ^{k_0} preserves R_i. The $\langle \phi \rangle$-orbit of R_i is $R_i, \phi(R_i), \ldots, \phi^{k_0-1}(R_i)$. We can push forward the foliation $\mathcal{F}$ to each $\phi^j(R_i)$ in order to obtain a foliation on the disconnected subsurface $R_i \cup \phi(R_i) \cup \cdots \cup \phi^{k_0-1}(R_i)$. Then, for $0 \leq j < k_0$, we can define the measure on $\phi_\star^j(\mathcal{F})$ to be the pushforward of μ multiplied by λ^{j/k_0}. We can then regard ϕ as a pseudo-Anosov homeomorphism of the disconnected surface $R_i \cup \phi(R_i) \cup \cdots \cup \phi^{k_0-1}(R_i)$.

We can now deduce the sharp normal form for mapping classes. The starting point is the statement of Corollary 13.3 as given. Instead of raising f to the power k, though, we analyze f itself. The mapping class f acts on the set $\{R_1, \ldots, R_{m+p}\}$. We consider each orbit as a single surface which is in general disconnected. Denote these surfaces by $R'_1, \ldots, R'_q$. By the previous paragraph, f acts as either a periodic or a pseudo-Anosov mapping class on each R_i' that is not a disjoint union of annuli. In other words, we have

$$f = \prod_{i=1}^{q} \eta_i(f_i'),$$

where each $f_i' \in \mathrm{Mod}(R_i')$ is either periodic, pseudo-Anosov, or (in the case that R_i' is a disjoint union of annuli) a root of a multitwist.

Let S' denote the surface obtained from S by deleting a representative of the canonical reduction system for f. If we instead consider the induced action of f on S', then we lose the information of $f_1, \ldots f_m$ (Proposition 3.19). So f induces an element of $\mathrm{Mod}(S')$ with only pseudo-Anosov and periodic blocks.

Historical remarks. Nielsen wrote a series of papers on the classification of surface homeomorphisms in the 1920s–1940s [168, 169, 170, 172]. His approach to classifying elements of $\mathrm{Mod}(S)$ was to consider their induced action on $\partial \mathbb{H}^2$. Because Nielsen's work is lengthy (spanning over 400 pages) and lacks sufficient organizing perspective, this work was largely ignored by topologists for many years.

In 1974 Thurston developed the theory of measured foliations on surfaces and used this to prove Theorem 13.2 as stated above. A posteriori, it became clear that all of the required tools for the classification had already been discovered by Nielsen. A paper by Miller explains how to understand the pseudo-Anosov case (the most important case) of the classification from the Nielsen point of view [157].

Thurston did not publish the details of his proof of Theorem 13.2, although he did distribute an announcement of his results, which appeared years later in print [207]. This announcement is remarkable for both its brevity and its richness. The first complete published proof of the classification is due to Bers in 1978 [15], who proved the theorem from the point of view of Teichmüller theory; see Section 13.6 below. Around the same time, the *Séminaire sur les difféomorphismes des surfaces d'après Thurston*, held at L'Université de Paris-sud à Orsay and led by Albert Fathi, François Laudenbach, and Valentin Poénaru, worked out the full details of Thurston's proof. The result is a 284-page monograph known as FLP [61]. The relationship between the works of Thurston, Bers, and Nielsen is explained in a paper by Gilman [72]. Other points of view on Nielsen–Thurston theory are contained in the writings of Handel and Thurston [81], Casson and Bleiler [44], and Bonahon [31]. The key objects in these works are geodesic laminations, which are implicit in Nielsen's work.

13.4 THURSTON'S GEOMETRIC CLASSIFICATION OF MAPPING TORI

One of Thurston's original motivations for studying homeomorphisms of surfaces was to understand the possible geometric structures on surface bundles over the circle. Recall that an S_g-*bundle over* S^1 is a fiber bundle with fiber S_g and base S^1. Such spaces provide a rich collection of closed 3-dimensional manifolds.

Since S^1 minus one point is contractible, any bundle over $S^1 - \{\text{point}\}$ is trivial. It follows that every S_g-bundle over S^1 is homeomorphic (even isomorphic as an S_g-bundle) to some mapping torus of some $\phi \in \mathrm{Homeo}(S_g)$. The *mapping torus* for $f \in \mathrm{Mod}(S_g)$ is defined as

$$M_f = \frac{S_g \times [0,1]}{(x,0) \sim (\phi(x),1)},$$

where $\phi \in \mathrm{Homeo}^+(S_g)$ is a representative for f. The obvious projection $M_f \to S^1$ with fiber S_g gives M_f the structure of an S_g-bundle over S^1. The element $f \in \mathrm{Mod}(S_g)$ is called the *monodromy* of this bundle. Note that we have restricted to the case of orientation-preserving ϕ, so that M_f is a closed, orientable 3-manifold.

It is not difficult to prove that the homeomorphism type of M_f does not depend on the choice of representative for f; one can use an isotopy between elements of $\mathrm{Homeo}^+(S_g)$ to construct the desired homeomorphism. Similarly, if f and h are conjugate in $\mathrm{Mod}(S_g)$, then M_f is homeomorphic to M_h. However, the converse is not true: there exist many examples of nonconjugate elements $f, h \in \mathrm{Mod}(S_g)$ for which M_f is homeomorphic to M_h. In fact, there are examples of mapping tori M_f where the set of genera of the fibers in different fiberings over S^1 is unbounded. The following theorem of Thurston says that the Nielsen–Thurston type of the monodromy $f \in \mathrm{Mod}(S_g)$ alone determines the geometry that the manifold M_f admits.

Theorem 13.4 *Let* $g \geq 2$, *let* $f \in \mathrm{Mod}(S_g)$, *and let* M_f *denote the mapping torus for* f.

1. *f is periodic $\Longleftrightarrow$ M_f admits a metric locally isometric to $\mathbb{H}^2 \times \mathbb{R}$.*

2. *f is reducible $\Longleftrightarrow$ M_f contains an incompressible (i.e., π_1-injective) torus.*

3. *f is pseudo-Anosov $\Longleftrightarrow$ M_f admits a hyperbolic metric.*

The forward implications of statements 1 and 2 are easy. Indeed, if f is periodic, then M_f is finitely covered by $S_g \times S^1$, which has universal cover

$\mathbb{H}^2 \times \mathbb{R}$. If f is reducible, say (for simplicity) the representative ϕ fixes a curve α, then M_f contains an incompressible torus, namely, $\alpha \times S^1$. These facts, together with the Nielsen–Thurston classification (Theorem 13.2), imply the reverse implication of statement 3, since no hyperbolic manifold has a finite cover locally isometric to $\mathbb{H}^2 \times \mathbb{R}$ and no hyperbolic manifold contains an incompressible torus.

The reverse implications in statements 1 and 2 are not difficult to prove. The forward implication in statement 3 is a deep theorem of Thurston; see [173, 205].

The torus case. Every orientable torus bundle over S^1 is homeomorphic to a mapping torus M_f for some $f \in \mathrm{Mod}(T^2)$. In this case, the classification of geometric structures on mapping tori is as follows (for descriptions of Nil and Sol geometries, see [208]).

Theorem 13.5 *Let M_f denote the mapping torus for $f \in \mathrm{Mod}(T^2)$.*

1. *ϕ is periodic $\Longleftrightarrow$ M_f is locally isometric to Euclidean 3-space.*

2. *ϕ is reducible $\Longleftrightarrow$ M_f is locally isometric to Nil geometry.*

3. *ϕ is Anosov $\Longleftrightarrow$ M_f is locally isometric to Sol geometry.*

For a further discussion of these topics, we refer the reader to the paper by Scott [190].

13.5 THE COLLAR LEMMA

The following useful lemma in hyperbolic geometry implies that if a closed geodesic α in a hyperbolic surface is very short, then every closed geodesic β with $i(\alpha, \beta) > 0$ must be long. This lemma will be an essential ingredient in our proof of the Nielsen–Thurston classification.

Lemma 13.6 (Collar lemma) *Let γ be a simple closed geodesic on a hyperbolic surface X. Then $N_\gamma = \{x \in X : d(x, \gamma) \leq w\}$ is an embedded annulus, where w is given by*

$$w = \sinh^{-1}\left(\frac{1}{\sinh(\frac{1}{2}\ell(\gamma))}\right).$$

Proof. We assume that X is compact; the general case is similar. Choose a pants decomposition for X where each of the curves in the decomposition is a geodesic and where γ is one of the curves. Let P be a pair of pants that has γ as one of its boundary components. As in Proposition 10.5, we cut P into two isometric right-angled hexagons H and H'. Label the alternating sides of H that correspond to the boundary curves of P by c_1, c_2, and c_3. For each i, let N_i be the metric neighborhood of c_i of width $\sinh^{-1}(1/\sinh(\ell(c_i)))$ in H. If we show that the N_i are disjoint, then there cannot possibly be any identifications for the N_i in P (or in X). We therefore obtain the desired annulus by considering the above for pants on each side of γ and taking two metric neighborhoods in each of these pants, one for each hexagon.

For the following argument, refer to Figure 13.3. Let α be the shortest geodesic from c_3 to the opposite side of H. The arc α cuts H into two right-angled pentagons. Let P_1 be the pentagon that contains c_1. The *right-angled pentagon formula* says that if a and b are adjacent sides of a right-angled hyperbolic pentagon and c is the side opposite their common vertex, then

$$\sinh(a)\sinh(b) = \cosh(c)$$

(see equation (V.1) in [64, Section 8.1]). Applied to P_1, the right-angled pentagon formula gives

$$\sinh(d)\sinh(\ell(c_1)) = \cosh(c),$$

where d is the distance between c_1 and α and c is the length of the intersection of c_3 with P_1. Since $\cosh(c) > 1$, we have that

$$d > \sinh^{-1}\left(\frac{1}{\sinh(\ell(c_1))}\right).$$

Thus N_1 is strictly contained in P_1. Similarly, N_2 is also disjoint from a, and so $N_1 \cap N_2 = \emptyset$. By symmetry, N_1 and N_2 are both disjoint from N_3 and the lemma follows.   □

Note that our proof of the collar lemma really shows something stronger than the statement of the lemma: we found annuli of the given size that are not only embedded but are also disjoint from each other.

The collar construction for Riemann surfaces first appears in the work of Linda Keen [120]. The sharp version given in Lemma 13.6 is due to Matelski [142].

For our proof of the Nielsen–Thurston classification, we will not need the precise statement of Lemma 13.6. Rather, we will need only the following much weaker statement sometimes attributed to Margulis.

Corollary 13.7 *Let S be a surface with $\chi(S) < 0$. There is a constant $\delta = \delta(S)$ such that if X is any (complete, finite-area) hyperbolic surface*

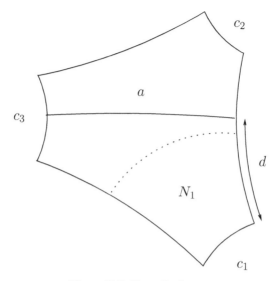

Figure 13.3 The collar lemma.

homeomorphic to S, then any two distinct closed geodesics of length less than δ are disjoint.

13.6 PROOF OF THE CLASSIFICATION THEOREM

In order to prove the Nielsen–Thurston classification theorem (Theorem 13.2), it makes sense to try to mimic the proof we gave for the torus case (Theorem 13.1). Since $\mathrm{Mod}(S)$ can be identified with a subgroup of $\mathrm{Isom}^+(\mathrm{Teich}(S))$, we can hope to classify elements of $\mathrm{Mod}(S)$ by using the geometry of $\mathrm{Teich}(S)$ to classify elements of $\mathrm{Isom}^+(\mathrm{Teich}(S))$.

In Chapter 1 we classified elements of $\mathrm{Isom}^+(\mathrm{Teich}(T^2)) \approx \mathrm{Isom}^+(\mathbb{H}^2)$ as follows. We used the fact that any isometry $\phi \in \mathrm{Isom}^+(\mathbb{H}^2)$ induces a homeomorphism $\overline{\phi}$ of the closed disk $\mathbb{H}^2 \cup \partial \mathbb{H}^2$. We then applied Brouwer's fixed point theorem to $\overline{\phi}$ to obtain a fixed point in $\mathbb{H}^2 \cup \partial \mathbb{H}^2$. Analyzing the number and location of the fixed points of $\overline{\phi}$ gave the desired trichotomy. In order to pursue this idea for higher-genus S, we would need a compactification of $\mathrm{Teich}(S)$ that is homeomorphic to a closed ball. Thurston's original proof of the Nielsen–Thurston classification theorem was in fact to construct such a compactification. The points of his compactification (besides the points of $\mathrm{Teich}(S)$) are projective classes of measured foliations on S; see [61] for full details and Chapter 15 of this book for a brief discussion.

Translation length. There is a different way of classifying elements of $\mathrm{Isom}(\mathbb{H}^2)$. For any metric space X and any $\phi \in \mathrm{Isom}(X)$, we define the *translation length* $\tau(\phi)$ by

$$\tau(\phi) = \inf_{x \in X} \{d(x, \phi(x))\}.$$

The isometry ϕ then falls into exactly one of the following three categories:

1. *Elliptic*: $\tau(\phi) = 0$ and is realized.

2. *Parabolic*: $\tau(\phi)$ is not realized.

3. *Hyperbolic*: $\tau(\phi) > 0$ and is realized.

When $X = \mathbb{H}^2$, this classification is compatible with the classification via the number of fixed points in $\mathbb{H}^2 \cup \partial \mathbb{H}^2$ discussed above.

The Bers proof of the Nielsen–Thurston classification theorem proceeds by analyzing elements $f \in \mathrm{Mod}(S)$ via their isometric action on $\mathrm{Teich}(S)$.

13.6.1 SETUP

Let $S = S_{g,n}$ and let $f \in \mathrm{Mod}(S)$. We recall that f acts on $\mathrm{Teich}(S)$ as an isometry of the Teichmüller metric. We will derive the Nielsen–Thurston classification via the trichotomy of isometries of the metric space $\mathrm{Teich}(S)$. Specifically, we will show:

1. f is elliptic in $\mathrm{Isom}(\mathrm{Teich}(S)) \implies f$ is periodic in $\mathrm{Mod}(S)$.

2. f is parabolic in $\mathrm{Isom}(\mathrm{Teich}(S)) \implies f$ is reducible in $\mathrm{Mod}(S)$.

3. f is hyperbolic in $\mathrm{Isom}(\mathrm{Teich}(S)) \implies f$ is pseudo-Anosov in $\mathrm{Mod}(S)$.

At the end, we will show that pseudo-Anosov mapping classes can be neither periodic nor reducible.

We remark that, in contrast to the case of parabolic isometries of $\mathbb{H}^2$, there exist reducible mapping classes $f \in \mathrm{Mod}(S)$ for which $\tau(f)$ is positive and not realized. See the notes at the end of the proof for a discussion of this case.

13.6.2 THE ELLIPTIC CASE

If f is an elliptic element of $\mathrm{Isom}(\mathrm{Teich}(S))$, then by definition f fixes a point of $\mathrm{Teich}(S)$. We have already observed at the start of Chapter 12 that if f fixes a point of $\mathrm{Teich}(S)$, then f is periodic.

13.6.3 THE PARABOLIC CASE

Assume that f is parabolic as an element of $\mathrm{Isom}(\mathrm{Teich}(S))$. We need to find an f-invariant collection of isotopy classes of pairwise disjoint simple closed curves in S. Let $(\mathfrak{X}_i)$ be a sequence in $\mathrm{Teich}(S)$ with the property that $d(\mathfrak{X}_i, f \cdot \mathfrak{X}_i) \to \tau(f)$. We will produce the required reducing system as a set of short simple closed curves on $\mathfrak{X}_i$ for i large.

Step 1. The projection of $(\mathfrak{X}_i)$ to $\mathcal{M}(S)$ leaves every compact set in $\mathcal{M}(S)$.

Suppose to the contrary that these projections lie in a fixed compact set in $\mathcal{M}(S)$. Then for some choice of $h_i \in \mathrm{Mod}(S)$, the sequence $(h_i \cdot \mathfrak{X}_i)$ stays in a fixed compact region of $\mathrm{Teich}(S)$. Denote $h_i \cdot \mathfrak{X}_i$ by $\mathcal{Y}_i$. By compactness, there is a subsequence of $(\mathcal{Y}_i)$ that converges to a point $\mathcal{Y} \in \mathrm{Teich}(S)$. Since $\mathrm{Mod}(S)$ acts on $\mathrm{Teich}(S)$ by isometries, we have

$$d(\mathcal{Y}_i, h_i f h_i^{-1} \cdot \mathcal{Y}_i) = d(h_i^{-1} \cdot \mathcal{Y}_i, f \cdot (h_i^{-1} \cdot \mathcal{Y}_i)) = d(\mathfrak{X}_i, f \cdot \mathfrak{X}_i),$$

and so

$$\lim_{i \to \infty} d(\mathcal{Y}_i, h_i f h_i^{-1} \cdot \mathcal{Y}_i) = \lim_{i \to \infty} d(\mathfrak{X}_i, f \cdot \mathfrak{X}_i) = \tau(f).$$

We claim that

$$d(\mathcal{Y}, h_k f h_k^{-1} \cdot \mathcal{Y}) = \tau(f)$$

for some k.

Let i be fixed. Applying the triangle inequality to the four points $\mathcal{Y}$, $\mathcal{Y}_i$, $h_i f h_i^{-1} \cdot \mathcal{Y}_i$, and $h_i f h_i^{-1} \cdot \mathcal{Y}$ we obtain

$$d(\mathcal{Y}, h_i f h_i^{-1} \cdot \mathcal{Y}) \le d(\mathcal{Y}, \mathcal{Y}_i) + d(\mathcal{Y}_i, h_i f h_i^{-1} \cdot \mathcal{Y}_i)$$
$$+ d(h_i f h_i^{-1} \cdot \mathcal{Y}_i, h_i f h_i^{-1} \cdot \mathcal{Y}).$$

Now let $i \to \infty$. Since $\mathcal{Y}_i \to \mathcal{Y}$, the first and last terms on the right-hand side tend to zero, and the middle term tends to $\tau(f)$. Therefore,

$$\lim_{i \to \infty} d(\mathcal{Y}, h_i f h_i^{-1} \cdot \mathcal{Y}) = \tau(f).$$

By proper discontinuity of the $\mathrm{Mod}(S)$ action on $\mathrm{Teich}(S)$ (Theorem 12.2), we have that the sequence $h_i f h_i^{-1}$ is eventually constant; that is, there is some N so that $h_i f h_i^{-1} = h_N f h_N^{-1}$ when $i \ge N$. Thus

$$d(\mathcal{Y}, h_N f h_N^{-1} \cdot \mathcal{Y}) = \tau(f),$$

which proves the claim.

It follows that

$$d(h_N^{-1} \cdot \mathcal{Y}, f \cdot (h_N^{-1} \cdot \mathcal{Y})) = \tau(f),$$

which contradicts the assumption that $\tau(f)$ is not realized. Thus it must be the case that $(\mathcal{X}_i)$ leaves every compact set of $\mathcal{M}(S)$.

Step 2. Finding a reduction system for f.

For $\mathcal{X} \in \mathrm{Teich}(S)$, let $\ell(\mathcal{X})$ denote the length of the shortest essential simple closed curve in $\mathcal{X}$. By Mumford's compactness criterion (Theorem 12.6),

$$\lim_{i \to \infty} \ell(\mathcal{X}_i) = 0.$$

By Wolpert's lemma (Lemma 12.5), there exists $K > 1$, depending only on $\tau(f)$, such that for any $\mathcal{X}, \mathcal{Y} \in \mathrm{Teich}(S)$, if $d(\mathcal{X}, \mathcal{Y}) \leq \tau(f) + 1$, then $\ell_{\mathcal{X}}(c) \leq K\ell_{\mathcal{Y}}(c)$ for any isotopy class of simple closed curves c in S. Choose M large enough so that

- $d(\mathcal{X}_M, f \cdot \mathcal{X}_M) < \tau(f) + 1$, and

- $\ell(\mathcal{X}_M) < (1/K)^{3g-3+n}\delta$,

where δ is the constant from the corollary of the collar lemma (Corollary 13.7).

Let c_0 be an isotopy class of simple closed curves in S with $\ell_{\mathcal{X}_M}(c_0) = \ell(\mathcal{X}_M)$. For each $1 \leq i \leq 3g - 3 + n$, let

$$c_i = f^{-1}c_{i-1} = f^{-i}c_0.$$

Then

$$\ell_{\mathcal{X}_M}(f^{-i}c_0) = \ell_{f^i\mathcal{X}_M}(c_0) \leq K^i\ell_{\mathcal{X}_M}(c_0) < \delta.$$

By the definition of the constant δ from Corollary 13.7, the simple closed curves $\{c_0, \ldots, c_{3g-3+n}\}$ must be mutually disjoint. But there are at most $3g - 3 + n$ isotopy classes of pairwise disjoint essential simple closed curves in S (cf. Section 8.3). Thus it must be that two of the c_i are in the same homotopy class. It follows that $f^k(c_0) = c_0$ for some $k > 0$, so that f permutes the collection of isotopy classes $\{c_0, c_1, \ldots c_{k-1}\}$. Thus f is reducible.

13.6.4 THE HYPERBOLIC CASE

Assume that f is hyperbolic as an element of $\mathrm{Isom}(\mathrm{Teich}(S))$. Let $\mathcal{X}$ be a point of $\mathrm{Teich}(S)$ that satisfies $d(\mathcal{X}, f \cdot \mathcal{X}) = \tau(f) > 0$. By Theorem 11.19 there is a unique bi-infinite geodesic γ passing through $\mathcal{X}$ and $f \cdot \mathcal{X}$, and moreover γ is a Teichmüller line.

Our goal is to show that f is pseudo-Anosov. We will do this by first proving that f leaves γ invariant and acts on it by translation by $\tau(f)$. Intuitively, this means that f should have a representative that is a Teichmüller map and hence "looks like" a pseudo-Anosov homeomorphism. We now make this precise.

Step 1. f leaves γ invariant.

Let $\mathcal{Y} \in \gamma$ be any point in the interior of the geodesic segment from $\mathcal{X}$ to $f \cdot \mathcal{X}$. By the triangle inequality and the fact that f acts by isometries on $\mathrm{Teich}(S)$, we have

$$\begin{aligned} d(\mathcal{Y}, f \cdot \mathcal{Y}) &\le d(\mathcal{Y}, f \cdot \mathcal{X}) + d(f \cdot \mathcal{X}, f \cdot \mathcal{Y}) \\ &= d(\mathcal{Y}, f \cdot \mathcal{X}) + d(\mathcal{X}, \mathcal{Y}) \\ &= d(\mathcal{X}, f \cdot \mathcal{X}) \\ &= \tau(f). \end{aligned}$$

Refer to Figure 13.4.

The minimality of $\tau(f)$ implies that $d(\mathcal{Y}, f \cdot \mathcal{Y}) = \tau(f)$. In particular, this means that the first inequality above is an equality; that is,

$$d(\mathcal{Y}, f \cdot \mathcal{Y}) = d(\mathcal{Y}, f \cdot \mathcal{X}) + d(f \cdot \mathcal{X}, f \cdot \mathcal{Y}).$$

Thus $f \cdot \mathcal{X}$ must lie on the unique Teichmüller line γ' passing through $\mathcal{Y}$ and $f \cdot \mathcal{Y}$. As γ and γ' agree on the nontrivial geodesic segment from $\mathcal{Y}$ to $f \cdot \mathcal{X}$, and since all geodesics in $\mathrm{Teich}(S)$ are Teichmüller lines (Theorem 11.19), it follows that γ' and γ are the same Teichmüller line, and in particular $f \cdot \mathcal{Y}$ lies on γ. Since we have shown that $\mathcal{Y}$ realizes $\tau(f)$, we can apply the same argument to the geodesic segment from $\mathcal{Y}$ to $f \cdot \mathcal{Y}$. Thus the image of this segment is contained in γ, and in particular $f^2 \cdot \mathcal{X}$ lies on γ. Repeating the argument inductively, we see that $f \cdot \gamma = \gamma$ and in particular that $f^i \cdot \mathcal{X}$ lies on γ for all $i \in \mathbb{Z}$.

Fix a marked Riemann surface (X, ψ) representing $\mathcal{X}$ and let $\phi : X \to X$ be the Teichmüller mapping in the homotopy class $\psi \circ f \circ \psi^{-1}$ (this is the map whose dilatation determines $d_{\mathrm{Teich}}(\mathcal{X}, f \cdot \mathcal{X})$). Say that the horizontal stretch factor of ϕ is equal to the dilatation K_ϕ (as opposed to $1/K_\phi$).

Step 2. The map $\phi^2 : X \to X$ is a Teichmüller mapping with horizontal stretch factor K_ϕ^2.

The quasiconformal homeomorphism ϕ^2 lies in the homotopy class $\psi \circ f^2 \circ \psi^{-1}$. Since $K_{\phi_1 \circ \phi_2} \le K_{\phi_1} K_{\phi_2}$ for any two quasiconformal home-

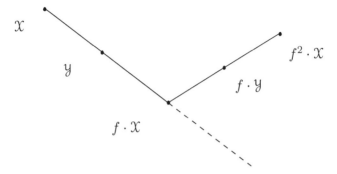

Figure 13.4 When f is hyperbolic and $\tau(f)$ is realized at X, then X, $f \cdot X$, and $f^2 \cdot X$ must be colinear.

omorphisms ϕ_1 and ϕ_2 (Proposition 11.3), we have $K_{\phi^2} \le K_\phi^2$. Therefore,

$$d_{\mathrm{Teich}}(X, f^2 \cdot X) \le \frac{1}{2} \log K_{\phi^2}$$
$$\le \frac{1}{2} \log K_\phi^2$$
$$= 2 \left(\frac{1}{2} \log K_\phi \right)$$
$$= 2\, d_{\mathrm{Teich}}(X, f \cdot X).$$

It follows from step 1 that $d_{\mathrm{Teich}}(X, f^2 \cdot X) = 2\, d_{\mathrm{Teich}}(X, f \cdot X)$. Thus the first inequality above is an equality. By Teichmüller's uniqueness theorem (Theorem 11.9), the map ϕ^2 is the unique Teichmüller map in the homotopy class of $\psi \circ f^2 \circ \psi^{-1}$.

Step 3. The initial and terminal quadratic differentials for ϕ on X are equal.

Let $q, q' \in \mathrm{QD}(X)$ denote the initial and terminal quadratic differentials for ϕ. For $p \in X$ such that $q(p) \neq 0$, the image under ϕ of the unit circle in $T_p(X)$ is an ellipse E in $T_{\phi(p)}(X)$. The direction in $T_p(X)$ of maximal stretch for $d\phi$ is the direction of the horizontal foliation for ϕ. The major axis of E has length $\sqrt{K_\phi}$ and lies in the direction of the horizontal foliation for q'.

Since ϕ^2 is a Teichmüller mapping with horizontal stretch factor K_ϕ^2, it follows that the direction in $T_{\phi(p)}$ of maximal stretch for ϕ at $\phi(p)$ must be the direction of the major axis for E (Lemma 11.2), that is, the direction of the horizontal foliation for q'. But again the direction of maximal stretch for ϕ at any point of X is the direction of the horizontal foliation for q. We have thus shown that the horizontal foliations for q and q' coincide at every point $p \in X$ for which $q(p) \neq 0$. Thus the horizontal foliations for q and q' coincide on all of X.

Away from the zeros of q there are natural coordinates for q where q is given by dz^2. In these coordinates the horizontal foliation for q is represented by horizontal lines. In these same coordinates, the horizontal foliation for q' must then also be given by horizontal lines, and so it must be that q' is given by $C\,dz^2$ for some C. It follows that the continuous function $q'/q : X \to \mathbb{C}$ is equal to the constant function C. Since the initial and terminal quadratic differentials for a Teichmüller mapping have the same Euclidean area, it must be that $C = 1$. Thus $q = q'$.

Let $(\mathcal{F}, \mu)$ be the transverse measured foliation on X given by the horizontal foliation for q. In natural coordinates around a nonsingular point, the foliation $\mathcal{F}$ is the horizontal foliation and the measure μ is $|dy|$.

Step 4. We have $\phi \cdot (\mathcal{F}, \mu) = (\mathcal{F}, \sqrt{K_\phi}\,\mu)$.

Any Teichmüller mapping takes the horizontal foliation for its initial differential to the horizontal foliation for its terminal differential. We showed that these differentials, hence the associated horizontal foliations, are equal. Thus $\phi(\mathcal{F}) = \mathcal{F}$. It remains to show that if α is an arc in X transverse to $\mathcal{F}$, then $\mu(\phi^{-1}(\alpha)) = \sqrt{K_\phi}\,\mu(\alpha)$. But this immediately follows from the fact that ϕ is a Teichmüller mapping with horizontal measured foliation $(\mathcal{F}, \mu)$ and horizontal stretch factor K_ϕ.

Step 5. The mapping class f is pseudo-Anosov with stretch factor $\sqrt{K_\phi}$.

We have shown in step 4 that $\phi \cdot (\mathcal{F}, \mu) = (\mathcal{F}, \sqrt{K_\phi}\,\mu)$, where $(\mathcal{F}, \mu)$ is the measured foliation on X coming from the horizontal foliation for q. However, there is a symmetry between the horizontal and vertical foliations for q: we can also describe the Teichmüller map ϕ as having initial differential $-q$ and horizontal stretch factor $1/K_\phi$. Thus, by symmetry, ϕ fixes the horizontal foliation $\mathcal{F}'$ for $-q$, which is the same as the vertical foliation for q, and multiplies the measure μ' by $1/\sqrt{K_\phi}$. The measured foliations $(\mathcal{F}, \mu)$ and $(\mathcal{F}', \mu')$ are unstable and stable foliations for ϕ. Thus ϕ is a pseudo-Anosov homeomorphism of X, and $\psi^{-1} \circ \phi \circ \psi$ is a pseudo-Anosov homeomorphism of S, with stable and unstable foliations $(\psi^{-1}(\mathcal{F}), \psi^\star(\mu))$ and $(\psi^{-1}(\mathcal{F}'), \psi^\star(\mu'))$ and with stretch factor $\lambda = \sqrt{K_\phi}$.

13.6.5 EXCLUSIVITY

The only thing left to prove is the exclusivity statement of the theorem, namely, that pseudo-Anosov mapping classes are neither periodic nor reducible. As part of Theorem 14.23 below, we will prove that, if $f \in \mathrm{Mod}(S)$

is pseudo-Anosov, and α is a simple closed curve in S, and we endow S with the singular Euclidean metric induced by the stable and unstable foliations, then the length of the geodesic isotopic to $f^n(\alpha)$ tends to infinity as n tends to infinity. On the other hand, this is false for periodic and reducible mapping classes because, in either case, there is at least one isotopy class of simple closed curves that is fixed by a power of f. We emphasize that the proof of Theorem 14.23 relies only on the definition of a pseudo-Anosov mapping class. This completes the proof of Theorem 13.2.

13.6.6 NOTES ON THE PROOF OF THEOREM 13.2

We can distinguish among the parabolic isometries of $\mathrm{Teich}(S)$ those with $\tau = 0$ and those with $\tau > 0$. The latter correspond to mapping classes that have a pseudo-Anosov component in the sense of Corollary 13.3.

A geodesic in Teichmüller space fixed by a pseudo-Anosov mapping class is called an *axis* for that mapping class. It turns out that the axis for a pseudo-Anosov mapping class is unique, but this does not follow from our proof of Theorem 13.2. For a proof see [140, Theorem 9.2].

Bers' approach to proving the exclusivity statement of Theorem 13.2 is to show that a reducible mapping class gives a parabolic isometry of $\mathrm{Teich}(S)$—this is the converse of the parabolic case of the proof. The idea is that continually shrinking the reducing curves continually reduces the corresponding stretch factors.

Chapter Fourteen

Pseudo-Anosov Theory

The power of the Nielsen–Thurston classification is that it gives a simple criterion for an element $f \in \mathrm{Mod}(S)$ to be pseudo-Anosov: f is neither finite-order nor reducible. This fact, however, is only as useful as the depth of our knowledge of pseudo-Anosov homeomorphisms. The purpose of this chapter is to study pseudo-Anosov homeomorphisms: their construction, their algebraic properties, and their dynamical properties.

Anosov maps of the torus. An Anosov homeomorphism of the torus T^2 is a linear representative of an Anosov mapping class. As discussed in Section 13.1, an Anosov homeomorphism $\phi : T^2 \to T^2$ has an associated Anosov package. The geometric picture of the action of ϕ on T^2 is quite explicit.

The map ϕ, considered as an element of $\mathrm{SL}(2, \mathbb{Z})$, has two distinct real eigenvalues $\lambda > 1$ and λ^{-1}. These eigenvalues are *quadratic integers*; that is, they are roots of degree 2 integer polynomials. The diffeomorphism ϕ preserves two foliations $\mathcal{F}^u$ and $\mathcal{F}^s$ on T^2; these are the projections to T^2 of the foliations of $\mathbb{R}^2$ by lines parallel to the λ and λ^{-1} eigenspaces of the matrix ϕ. The map ϕ stretches each leaf of $\mathcal{F}^u$ by a factor of λ and contracts each leaf of $\mathcal{F}^s$ by a factor of λ. The eigenspaces are lines with irrational slope, from which it follows that each leaf of $\mathcal{F}^u$ and of $\mathcal{F}^s$ is dense in T^2.

It is an easy exercise to check that ϕ-periodic points are dense. From basic linear algebra one can see that, for a generic vector $v \in \mathbb{R}^2$, the vector $\phi^n(v)$ converges to a vector $\lambda^n v_u$, where v_u is a unit vector pointing in the direction of $\mathcal{F}^u$. More precisely, the directions converge:

$$\frac{\phi^n(v)}{|\phi^n(v)|} \to \pm v_u,$$

and the magnitudes converge:

$$\sqrt[n]{|\phi^n(v)|} \to \lambda.$$

One main goal of this chapter is to describe a *pseudo-Anosov package*, which extends the above picture of Anosov homeomorphisms on T^2 to pseudo-Anosov homeomorphisms on S_g, where $g \geq 2$.

14.1 FIVE CONSTRUCTIONS

The first basic question to address is: do pseudo-Anosov mapping classes actually exist? The answer is of course yes, although it is a nontrivial matter to give explicit examples. In this section we explain five different constructions of pseudo-Anosov mapping classes.

14.1.1 BRANCHED COVERS

One way to construct a pseudo-Anosov homeomorphism is to lift an Anosov homeomorphism of the torus via a branched covering map. Recall from Chapter 7 that an orbifold cover $p : S \to S'$ is a map obtained by a finite group action on S. As in Section 11.2, we also call such a cover a *branched cover*. A branched cover $p : S \to S'$ is a true covering map over the complement of some finite collection of points B in S'. Elements of B are called *branch points*. Around each point of $p^{-1}(B)$ the map p is given in local coordinates by the map $z \mapsto z^k$, with $k \in \{2, 3, \dots\}$. Recall from Section 11.2 that, for $g \geq 2$, there is a twofold branched cover $S_g \to T^2$ with $2g - 2$ branch points.

Fix a branched covering map $p : S \to T^2$. Let ϕ be an Anosov homeomorphism of the torus, for example, the linear map of T^2 associated to any $A \in \mathrm{SL}(2, \mathbb{Z})$ with $|\operatorname{tr}(A)| > 2$. Since each rational point of T^2 is a periodic point of ϕ, we can change ϕ by isotopy and pass to a power of ϕ so that ϕ fixes pointwise the set B of branch points of p. Passing to a further power of ϕ if necessary, we can assume that ϕ lifts to a homeomorphism ψ of S (to see this, consider the action of ϕ on the finite set of index 2 subgroups of $\pi_1(T^2 - B)$).

As ψ has the same local properties as ϕ, we see that ψ is a pseudo-Anosov homeomorphism of S; indeed, the stable and unstable foliations for ψ are the preimages under p of those for ϕ. Above each branch point in T^2, the foliations for ψ each have a singularity with an even number of prongs.

By considering branched (or unbranched) coverings over higher-genus surfaces, this construction can be used to convert pseudo-Anosov mapping classes on any surface to pseudo-Anosov mapping classes on higher-genus surfaces.

14.1.2 DEHN TWIST CONSTRUCTIONS

We now present an elementary construction of pseudo-Anosov mapping classes due to Thurston [207], and a related one due to Penner. Let $S = S_{g,n}$. We say that a collection of isotopy classes of simple closed curves in S *fills* S if any simple closed curve in S has positive geometric intersection with some isotopy class in the collection (see, e.g., Figure 1.7).

If $A = \{\alpha_1, \ldots, \alpha_n\}$ is a multicurve in a surface S (that is, a set of pairwise disjoint simple closed curves), we denote the product $\prod_{i=1}^{n} T_{\alpha_i}$ by T_A. Such a mapping class is often called a *multitwist*.

Theorem 14.1 (Thurston's construction) *Suppose A and B are multicurves in S so that $A \cup B$ fills S. There is a real number $\mu = \mu(A, B)$ and a representation $\rho : \langle T_A, T_B \rangle \to \mathrm{PSL}(2, \mathbb{R})$ given by*

$$
T_A \mapsto \begin{pmatrix} 1 & -\mu^{1/2} \\ 0 & 1 \end{pmatrix} \qquad T_B \mapsto \begin{pmatrix} 1 & 0 \\ \mu^{1/2} & 1 \end{pmatrix}.
$$

The representation ρ has the following properties:

1. *An element $f \in \langle T_A, T_B \rangle$ is periodic, reducible, or pseudo-Anosov according to whether $\rho(f)$ is elliptic, parabolic, or hyperbolic.*

2. *When $\rho(f)$ is parabolic, f is a multitwist.*

3. *When $\rho(f)$ is hyperbolic, the stretch factor of the pseudo-Anosov mapping class f is equal to the larger of the two eigenvalues of $\rho(f)$.*

In the special case where A and B are single curves, say $A = \{\alpha\}$ and $B = \{\beta\}$, the real number μ in Theorem 14.1 is equal to $i(\alpha, \beta)^2$, and so the representation ρ becomes

$$
T_\alpha \mapsto \begin{pmatrix} 1 & -i(a, b) \\ 0 & 1 \end{pmatrix} \qquad T_\beta \mapsto \begin{pmatrix} 1 & 0 \\ i(a, b) & 1 \end{pmatrix}.
$$

It is nontrivial to construct a pair of filling curves for any given surface; see Proposition 3.5 for such a construction. On the other hand, it is quite easy to find a pair of multicurves that fill a given surface.

Recall from Section 3.5 that, for $n \geq 2$, the matrices $\left(\begin{smallmatrix} 1 & n \\ 0 & 1 \end{smallmatrix} \right)$ and $\left(\begin{smallmatrix} 1 & 0 \\ n & 1 \end{smallmatrix} \right)$ generate a free group of rank 2 inside $\mathrm{SL}(2, \mathbb{Z})$. From this, Theorem 14.1, and the Hopfian property for free groups [138, Theorem 2.13], it follows that, when $\mu \geq 4$, the group $\langle T_A, T_B \rangle$ is isomorphic to a free group of rank 2 and that ρ is injective. Also, one can deduce that $\langle T_A, T_B \rangle$ contains no periodic elements, and any reducible element it contains is conjugate to a power of T_A, T_B or (when $\mu = 4$) to $T_A T_B$. Finally, one can show that the image of ρ is a discrete subgroup of $\mathrm{PSL}(2, \mathbb{R})$.

Perron–Frobenius matrices. The proof of Theorem 14.1 relies on the basic theory of Perron–Frobenius matrices, which we now explain.

We say that a matrix is *positive* (respectively *nonnegative*) if each of its entries is positive (respectively nonnegative). A nonnegative matrix is *primitive* if it has a power that is a positive matrix.

The following is a fundamental theorem in the study of primitive integer matrices. See, e.g., [69, Section XIII.2].

THEOREM 14.2 (Perron–Frobenius theorem) *Let A be an $n \times n$ matrix with integer entries. If A is primitive, then A has a unique nonnegative unit eigenvector v. The vector v is positive and has a positive eigenvalue that is larger in absolute value than all other eigenvalues.*

The eigenvector of A in the statement of Theorem 14.2 is called the *Perron–Frobenius eigenvector* of A. The eigenvalue in the theorem is called the *Perron–Frobenius eigenvalue* for A.

We will now prove Theorem 14.1 as an application of Theorem 14.2.

Proof of Theorem 14.1. The idea for proving the theorem is to find a singular Euclidean structure (cf. Sections 11.2 and 11.3) on S with respect to which $\langle T_A, T_B \rangle$ acts by affine transformations. Here an affine map is one that, in local charts away from the singularities, is of the form $Mx + b$, where M is a linear map and b is a vector.

The singular Euclidean structure we construct will have the added feature that it comes equipped with an orthonormal frame field well defined up to sign. Thus, given any affine map on S, its derivative can be described by a 2×2 matrix well defined up to sign. The representation ρ will assign to each affine map in $\langle T_A, T_B \rangle$ its differential: $\rho(h) = Dh$.

Assume that the components of A and B are in minimal position. Since A and B fill S, each complementary component of $A \cup B$ is a disk, each with at most one marked point. As in Section 11.2, the union $A \cup B$ gives a cell decomposition of S, and the dual cell complex C is another cell decomposition of S (if a 2-cell of the first cell decomposition has a marked point in its interior, then that marked point is taken to be a vertex of the dual decomposition C). Each 2-cell of C is a square, corresponding to a point of intersection of an element of A with an element of B.

In order to go from the cell decomposition C of S to a singular Euclidean structure on S, we simply need to assign a length to each 1-cell of C. We can do this by assigning a "width" to each curve of A and to each curve of B. Then the length of a 1-cell of C is declared to be the width of the unique curve of $A \cup B$ that intersects it.

Consider the case where A and B are single curves. If we take the length and width of each square of C to be 1, then T_A and T_B act affinely on the resulting Euclidean structure. In the general case it is a more delicate matter to find a singular Euclidean structure on which T_A and T_B act affinely. We now explain how to do this.

Say that $A = \{\alpha_1, \ldots, \alpha_m\}$ and $B = \{\beta_1, \ldots, \beta_n\}$. Let N be the matrix with (j, k) entry

$$N_{j,k} = i(\alpha_j, \beta_k).$$

Given N, let G be the abstract bipartite graph with m red vertices and n blue vertices, and $N_{j,k}$ edges between the jth red vertex and the kth blue vertex. Then the (j, k) entry of the dth power $(NN^t)^d$ is equal to the number of paths in G of length $2d$ between the jth and kth red vertices in G.

We claim that NN^t is primitive. Indeed, this is equivalent to the statement that the graph G is connected. If G were not connected, that would mean that $A \cup B$ is not connected, and so the pair $\{A, B\}$ does not fill S.

We can thus apply the Perron–Frobenius theorem (Theorem 14.2). Denote the Perron–Frobenius eigenvalue and eigenvector for NN^t by μ and V, respectively. So

$$NN^t V = \mu V.$$

Interchanging the roles of A and B, the Perron–Frobenius eigenvalue for $N^t N$ is still μ:

$$N^t N V' = \mu V'.$$

Here V' is chosen to be $\mu^{-1/2} N^t V$. In this case we have the formula $V = \mu^{-1/2} N V'$.

The singular Euclidean structure on which T_A and T_B act affinely can now be given: assign a width of V_i to α_i and a width of V_j' to β_j. The union of rectangles of C intersecting α_i is an annulus of width V_i and circumference $\mu^{1/2} V_i$. Similarly, the annulus along β_j has width V_j' and circumference $\mu^{1/2} V_j'$.

After orienting A and B, the singular Euclidean structure has an obvious choice of orthogonal frame field well defined up to multiplication by ± 1. Specifically, we choose a positively oriented basis so that the first vector is parallel to A and the second vector is parallel to B. To see that the ambiguity of ± 1 is really an issue, consider a situation where two curves of A and B intersect twice with opposite sign and follow the frame field along the corresponding loop.

In this singular Euclidean structure, the multitwists T_A and T_B can be chosen to be affine. These affine maps fix the 1-cells of C parallel to A and B, respectively. The actions of T_A and T_B on equivalence classes of frame fields are then given exactly by the classes of the matrices given in the statement of Theorem 14.1. This action can be verified by checking on the generators.

What we have just described is indeed a well-defined map from $\langle T_A, T_B \rangle$ to the group of affine automorphisms of S. The reason for this is that if an affine map of S is isotopic to the identity, then it is the identity.

We now finish the proof of the theorem. Let $f \in \langle T_A, T_B \rangle$. The classification of elements of $\mathrm{SL}(2, \mathbb{R})$ induces a classification for $\mathrm{SL}(2, \mathbb{R})$, and so $\rho(f)$ is elliptic, parabolic, or hyperbolic.

If $\rho(f)$ is elliptic, then f has a power that fixes the orthonormal frame field of S (up to sign) at every point. Also, by construction, f fixes each singular point of the metric. Thus f has a power that acts as the identity in the neighborhood of some singular point. Since f is affine, it follows that f is periodic.

If $\rho(f)$ is parabolic, then it has a 1-dimensional eigenspace, and the eigenvalue for this eigenspace must be 1 (in $\mathrm{PSL}(2, \mathbb{R})$, eigenvalues are well-defined only up to sign). The eigenspace induces a singular foliation on S. Up to replacing f with a power, we may assume that (the affine representative of) f fixes each singularity of the foliation and preserves each leaf emanating from each singularity. Let L be one such leaf. Since the eigenvalue is 1, it follows that f fixes L pointwise. If the leaf L had an accumulation point, then it would follow that f fixes a neighborhood of this accumulation point, and so (a power of) f would be the identity. Thus we may assume that the collection of all leaves starting from singular points is a collection of closed curves in S. As these closed curves are geodesics in the singular Euclidean metric, they are all simple and homotopically nontrivial. Since f fixes this collection, it follows that f is reducible. What is more, if we cut S along the reducing curves, we obtain a foliation that does not have any singularities. By the Euler–Poincaré formula (Proposition 11.4), the cut surface must be a collection of annuli. In particular, f is a multitwist about the reducing curves.

Finally, if $\rho(f)$ is hyperbolic, then the eigenspaces of $\rho(f)$ define two transverse measured foliations, f multiplies the measure of one foliation by the larger eigenvalue of $\rho(f)$, and f multiplies the measure of the other foliation by the smaller eigenvalue of $\rho(f)$ (the foliations have singularities at the singular points of the Euclidean structure). Thus f is pseudo-Anosov, and its stretch factor is given by the larger eigenvalue of $\rho(f)$. $\qquad \square$

Pseudo-Anosov mapping classes in the Torelli group. Nielsen conjectured that there are no pseudo-Anosov elements of the Torelli group $\mathcal{I}(S_g)$ [168]. That is, he conjectured there are no homeomorphisms ϕ of a surface S with the property that for every simple closed curve α and every $n \neq 0$, the curves $\phi^n(\alpha)$ and α are homologous but not homotopic.

One application of Thurston's construction (Theorem 14.1), which accompanied the announcement of his proof of the classification [207], is that the construction makes it easy to see that Nielsen's conjecture is false: one just takes each curve in the construction to be separating.

Corollary 14.3 *Let* $g \geq 2$. *The Torelli subgroup of* $\mathrm{Mod}(S_g)$ *contains pseudo-Anosov elements.*

The example of a product of twists of separating curves actually lies in the infinite-index subgroup $\mathcal{K}(S_g)$ of $\mathcal{I}(S_g)$ (recall from Section 6.5 that $\mathcal{K}(S_g)$ is defined to be the subgroup of $\mathrm{Mod}(S_g)$ generated by twists about separating curves). It is also possible to use the Thurston construction to find a pseudo-Anosov element of $\mathcal{I}(S_g) - \mathcal{K}(S_g)$. This is tricky because, in order to use a bounding pair map, the two curves of the bounding pair must belong to different multicurves of the construction.

Consider the mapping class $T_{a_2}^{-1} T_{a_1}^{-1} T_{b_2} T_{b_1}$, where the curves are as shown in Figure 14.1. This mapping class is pseudo-Anosov by Theorem 14.1, it is in $\mathcal{I}(S_g)$ since it is the product of a bounding pair map with a pair of Dehn twists about separating curves, and it is not in $\mathcal{K}(S_g)$ since it is not in the kernel of the Johnson homomorphism (see Section 6.5).

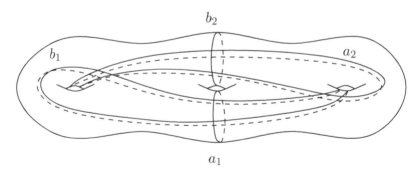

Figure 14.1 A multicurve that yields a pseudo-Anosov element of $\mathcal{I}(S_3) - \mathcal{K}(S_3)$ via the Thurston construction.

Penner's construction. Penner gives the following very general construction of pseudo-Anosov mapping classes.

Theorem 14.4 (Penner's construction) *Let* $A = \{\alpha_1, \ldots, \alpha_n\}$ *and* $B = \{\beta_1, \ldots, \beta_m\}$ *be multicurves in a surface* S *that together fill* S. *Any product* *of positive powers of the T_{α_i} and negative powers of the T_{β_i}, where each α_i and each β_i appear at least once, is pseudo-Anosov.*

In the statement of Theorem 14.4, the twists can appear in any order, for example, $T_{\alpha_1} T_{\beta_1}^{-3} T_{\alpha_2}^2$.

Penner has conjectured that every pseudo-Anosov element of the mapping class group has a power that is given by this construction [177, p. 195]. This is a difficult conjecture to disprove. For instance, one can use the Thurston construction to find pseudo-Anosov mapping classes that are not a priori given by Theorem 14.4. However, how can one tell if there is or is not another way to write the same element as a product of Dehn twists so that, in that form, it is given by the Penner construction?

The idea of Penner's proof of Theorem 14.4 is that one can explicitly find the train track (see Chapter 15) associated to the square of any such element. The train track is obtained by "smoothing out" the subset $A \cup B$ of S; see [177].

14.1.3 HOMOLOGICAL CRITERION

Let $S = S_g$ or $S = S_{g,1}$, with $g \geq 2$. We now explain how to detect pseudo-Anosov mapping classes via the symplectic representation of $\mathrm{Mod}(S)$ given in Section 6.5. The original version of this criterion is due to Casson–Bleiler [44].

We say that a polynomial is *symplectically irreducible* over $\mathbb{Z}$ if it cannot be written as a product of two polynomials, each of which is the characteristic polynomial of a matrix in $\mathrm{Sp}(2g, \mathbb{Z})$. In particular, irreducible polynomials are symplectically irreducible.

As noted in Section 6.1, the roots of the characteristic polynomial of a symplectic matrix come in pairs λ, λ^{-1}. Since the coefficients of a polynomial are symmetric functions of its roots and since the roots are paired, an easy argument gives that the characteristic polynomial $f(x) = x^n + a_{n-1}x^{n-1} + \cdots + a_1 x + 1$ of any integral symplectic matrix is monic and *palindromic*, which means that $a_k = a_{n-k}$ for each k. Thus it is much easier to be symplectically irreducible than to be irreducible.

Theorem 14.5 *Let $f \in \mathrm{Mod}(S)$ and let $\Psi(f)$ be its image in $\mathrm{Sp}(2g, \mathbb{Z})$ under the standard symplectic representation. Let $P_f(x)$ denote the characteristic polynomial of the matrix $\Psi(f)$. Suppose that each of the following conditions holds:*

1. *$P_f(x)$ is symplectically irreducible over $\mathbb{Z}$.*

2. *$P_f(x)$ is not a cyclotomic polynomial.*

3. $P_f(x)$ is not a polynomial in x^k for any $k > 1$.

Then $f \in \mathrm{Mod}(S)$ is pseudo-Anosov.

Note that if f satisfies the criteria of Theorem 14.5, then every element of the coset $f\mathcal{I}(S)$ satisfies the criteria, hence is pseudo-Anosov. Of course, in consideration of Corollary 14.3, there is no hope for any kind of converse to Theorem 14.5.

Proof. We show that if f is not pseudo-Anosov, then $P_f(x)$ fails to satisfy one of the given conditions. By the Nielsen–Thurston classification (Theorem 13.2), if f is not pseudo-Anosov, then either f is periodic or f is reducible (or both). We deal with each in turn.

If f is periodic of order n, then $\Psi(f)^n$ is the identity matrix, and so each root of $P_f(x)$ is an nth root of unity. Let ζ be one such root. The cyclotomic polynomial associated to ζ divides $P_f(x)$. Since cyclotomic polynomials are monic, palindromic, and of even degree, they are characteristic polynomials of symplectic matrices. It follows that $P_f(x)$ is either symplectically reducible over $\mathbb{Z}$ or is a cyclotomic polynomial, and so we are done in this case.

If f is reducible, then we have two (again, overlapping) subcases: either some power of f fixes the isotopy class of a nonseparating reducing curve in S, or f permutes a collection of nontrivial isotopy classes of disjoint separating curves.

For the first subcase, say $f^n(c) = c$ for some isotopy class c of (oriented) nonseparating simple closed curves in S. Since c represents a nontrivial element of $H_1(S; \mathbb{Z})$, it follows that $\Psi(f)^n$ has an eigenvalue of 1, and so $\Psi(f)$ has an eigenvalue that is an nth root of unity. As in the periodic case, this implies that $P_f(x)$ is either symplectically reducible or cyclotomic, and this completes the proof in this subcase.

For the second subcase, suppose that f permutes the isotopy classes of some collection of disjoint, essential, separating simple closed curves. Let γ be a simple closed curve in this collection and assume that the other curves in this collection all lie on one side of γ, that is, γ is an innermost curve in the collection. Let R be a closed subsurface of S that has γ as its boundary and that does not contain any other curves in the collection. It follows that the subsurfaces $\{f^i(R)\}$ are mutually disjoint (there may be only one of them); see Figure 14.2. Suppose that f^n is the smallest positive power of f that fixes the isotopy class of γ and let T be the complement of the $\cup f^i(R)$. By the Mayer–Vietoris sequence for homology, we have

$$H_1(S; \mathbb{Z}) = V_0 \oplus \cdots \oplus V_{n-1} \oplus V_n,$$

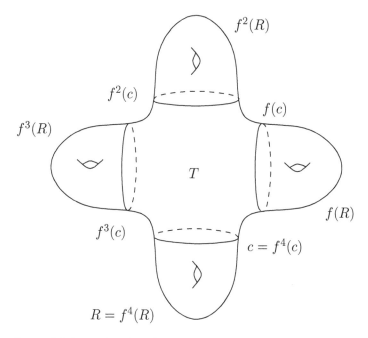

Figure 14.2 A mapping class f that fixes a union of disjoint separating curves.

where V_i is the image of $H_1(f^i(R); \mathbb{Z})$ in $H_1(S; \mathbb{Z})$ under the map induced by inclusion $f^i(R) \to S$ for $0 \le i \le n-1$, and V_n is similarly the image of $H_1(T; \mathbb{Z})$. This decomposition gives rise to a choice of basis for $H_1(S; \mathbb{Z})$; namely, the first set of basis elements is an arbitrary basis for V_0, the next set of basis elements is the image of the first basis under f_*, and so on for the first $n-1$ factors. Finally, we add an arbitrary basis for V_n. Under such a basis, $\Psi(f)$ is of the form

$$\begin{pmatrix} 0 & 0 & 0 & B & 0 \\ I & 0 & 0 & 0 & 0 \\ 0 & I & 0 & 0 & 0 \\ 0 & 0 & I & 0 & 0 \\ 0 & 0 & 0 & 0 & C \end{pmatrix},$$

where C is the induced action of f on $H_1(T; \mathbb{Z})$. In this case

$$P_f(x) = \det(\Psi(f) - xI) = \det(B - x^n I)\det(C - xI).$$

If T has genus 0, then C is a 0×0 matrix (meaning it is not really there), and $P_f(x)$ is a polynomial in x^n with $n > 1$. If T has positive genus, then C is an $m \times m$ matrix with $m \geq 1$, and so $P_f(x)$ is symplectically reducible over $\mathbb{Z}$. This completes the proof. $\qquad\square$

Explicit examples satisfying the homological criterion. It is not always easy to tell when a specific product of Dehn twists is pseudo-Anosov. The homological criterion of Theorem 14.5 can sometimes be useful for doing this. For example, we now give an infinite list of products of Dehn twists, and we use Theorem 14.5 to verify that each element in the list is pseudo-Anosov.

Let $([a_1], [b_1], [a_2], [b_2])$ be the usual homology basis for S_2 (cf. Figure 6.1). We consider the product

$$f_k = T_{a_1} T_{b_1} T_{a_1+a_2} T_{b_2} T_{a_2}^{1-k},$$

where, for example, $T_{a_1+a_2}$ denotes the Dehn twist about any simple closed curve in the homology class $[a_1] + [a_2]$. Note that f_k does not fall under either Thurston's construction or Penner's construction of pseudo-Anosov mapping classes.

We compute that $\Psi(f_k)$ is equal to

$$\begin{pmatrix} 1 & 1 & 0 & 0 \\ 0 & 1 & 0 & 0 \\ 0 & 0 & 1 & 0 \\ 0 & 0 & 0 & 1 \end{pmatrix} \begin{pmatrix} 1 & 0 & 0 & 0 \\ -1 & 1 & 0 & 0 \\ 0 & 0 & 1 & 0 \\ 0 & 0 & 0 & 1 \end{pmatrix} \begin{pmatrix} 1 & 1 & 0 & 1 \\ 0 & 1 & 0 & 0 \\ 0 & 1 & 1 & 1 \\ 0 & 0 & 0 & 1 \end{pmatrix} \begin{pmatrix} 1 & 0 & 0 & 0 \\ 0 & 1 & 0 & 0 \\ 0 & 0 & 1 & 0 \\ 0 & 0 & -1 & 1 \end{pmatrix} \begin{pmatrix} 1 & 0 & 0 & 0 \\ 0 & 1 & 0 & 0 \\ 0 & 0 & 1 & 1-k \\ 0 & 0 & 0 & 1 \end{pmatrix} = \begin{pmatrix} 0 & 1 & 0 & 0 \\ -1 & 0 & 1 & -k \\ 0 & 1 & 0 & 1 \\ 0 & 0 & -1 & k \end{pmatrix}.$$

Thus the characteristic polynomial for $\Psi(f_k)$ is

$$P_k(t) = t^4 - kt^3 + t^2 - kt + 1.$$

We now check that $P_k(t)$ satisfies the hypotheses of Theorem 14.5 for $|k| > 1$. First of all, it is obvious that $P_k(t)$ is not a polynomial in t^m for any $m > 1$. If $P_k(t)$ is a nontrivial product of characteristic polynomials of symplectic matrices, then it factors into two integral quadratic polynomials, each of the form

$$P_{k,i}(t) = t^2 - (\lambda_i + \lambda_i^{-1})t + 1 = (t - \lambda_i)(t - \lambda_i^{-1}),$$

where $\lambda_i + \lambda_i^{-1}$ is an integer. To check that this is not the case, we consider the polynomial $Q_k(x)$ obtained from $P_k(t)$ by dividing by t^2 and substituting $x + x^{-1}$ for t:

$$Q_k(x) = x^2 - kx - 1.$$

The polynomials $P_{k,i}(t)$ have integral coefficients if and only if the roots of $Q_k(x)$ are integers. It is easy to check that this is not the case when $k \neq 0$. It remains to check that $P_k(t)$ is not a cyclotomic polynomial. Since the degree of the nth cyclotomic polynomial is Euler's totient $\phi(n)$ and since $\phi(mn) = \phi(m)\phi(n)\gcd(m,n)/\phi(\gcd(m,n))$, we see that the only degree 4 cyclotomic polynomials are the fifth, eighth, tenth, and twelfth: $t^4 + t^3 + t^2 + t + 1$, $t^4 + 1$, $t^4 - t^3 + t^2 - t + 1$, and $t^4 - t^2 + 1$. Thus, if $|k| \neq 1$, then $P_k(t)$ is not cyclotomic.

14.1.4 KRA'S CONSTRUCTION

Let S be a compact surface of negative Euler characteristic, perhaps with finitely many punctures. Let $\mathrm{Mod}(S,p)$ denote the mapping class group of S with one marked point p. Recall from Section 4.2 the Birman exact sequence

$$1 \to \pi_1(S,p) \overset{Push}{\to} \mathrm{Mod}(S,p) \to \mathrm{Mod}(S) \to 1.$$

We say that an element γ of $\pi_1(S,p)$ *fills* S if every closed curve in S that represents γ intersects every essential simple closed curve in S.

Theorem 14.6 (Kra's construction) *Let* $S = S_{g,n}$ *and assume that* $\chi(S) < 0$. *Let* $\gamma \in \pi_1(S,p)$. *The mapping class* $Push(\gamma) \in \mathrm{Mod}(S,p)$ *is pseudo-Anosov if and only if* γ *fills* S.

Since each $Push(\gamma) \in \mathrm{Mod}(S,p)$ acts trivially on $H_1(S;\mathbb{Z})$ when $S = S_g$, Theorem 14.6 gives examples of pseudo-Anosov elements of the Torelli group $\mathcal{I}(S,p)$ and hence also provides counterexamples to the conjecture of Nielsen mentioned earlier in this section.

Kra's original proof is Teichmüller-theoretic: he shows directly that if γ fills, then the translation distance of the action of $Push(\gamma)$ on Teichmüller space is realized [127]. We now give an elementary proof of Theorem 14.6. This proof is apparently new, but it was inspired by an algebraic proof due to Kent–Leininger–Schleimer [121].

Proof. One direction of the theorem is obvious: if γ does not fill, then we can find an isotopy class of simple closed curves that is fixed by $Push(\gamma)$.

Now assume that γ fills S. We will show that $Push(\gamma)$ is not reducible. Note that whenever γ fills S, then γ^n fills S for every $n \neq 0$. It then follows by the same argument that $Push(\gamma)^n = Push(\gamma^n)$ is not reducible for any $n \neq 0$. By the Nielsen–Thurston classification, if an element of $\mathrm{Mod}(S,p)$ has no nontrivial power that is reducible, then it is pseudo-Anosov. We will thus be able to conclude that $Push(\gamma)$ is pseudo-Anosov.

Now let δ be any simple closed curve in $S - p$ and let $\tilde{\delta}$ denote the full preimage of δ in the universal cover $\widetilde{(S,p)}$, by which we mean $\mathbb{H}^2$ with an infinite collection of marked points (the lifts of p). Any representative of the mapping class $\mathcal{P}ush(\gamma)$ can be lifted to $\widetilde{(S,p)}$. It is possible to choose the representative homeomorphism so that its lift to $\mathbb{H}^2$ pushes each of the marked points in $\mathbb{H}^2$ along a path lifting of γ. We denote the resulting relative homeomorphism of $\widetilde{(S,p)}$ by $\widetilde{\mathcal{P}ush}(\gamma)$.

Since γ intersects δ essentially, it follows that for any component $\widetilde{\delta}_i$ of $\widetilde{\delta}$ there is a path lifting of γ that connects marked points on different sides of $\widetilde{\delta}_i$. Therefore, there are marked points in $\widetilde{(S,p)}$ that lie between $\widetilde{\delta}_i$ and $\widetilde{\mathcal{P}ush}(\gamma)(\widetilde{\delta}_i)$.

Any isotopy between $\mathcal{P}ush(\gamma)(\delta)$ and δ relative to p would lift to an equivariant relative isotopy of $\widetilde{\mathcal{P}ush}(\gamma)(\widetilde{\delta}_i)$ to $\widetilde{\delta}_i$ since $\widetilde{\mathcal{P}ush}(\gamma)(\widetilde{\delta}_i)$ is the only lift of $\mathcal{P}ush(\gamma)(\delta)$ with the same endpoints at infinity as $\widetilde{\delta}_i$. However, because there are marked points between $\widetilde{\delta}_i$ and $\widetilde{\mathcal{P}ush}(\gamma)(\widetilde{\delta}_i)$, no such isotopy exists. This proves the theorem. $\qquad\square$

14.1.5 A CONSTRUCTION FOR BRAID GROUPS

The following gives a construction of pseudo-Anosov homeomorphisms for $\mathrm{Mod}(S_{0,n})$, where $S_{0,n}$ denotes the sphere with n punctures.

Theorem 14.7 *Let n be a prime number. If f is an infinite-order element of* $\mathrm{Mod}(S_{0,n})$ *that permutes the punctures cyclically, then f is pseudo-Anosov.*

Proof. Suppose that f is reducible. Since n is prime, the partition of the punctures induced by the reducing curves must have sets of different size. Note that f does not preserve this partition because it may permute the reducing curves. However, there is another nontrivial partition of the punctures where we group together punctures that lie in subsets of the same size in the first partition. The mapping class f preserves this partition and hence does not permute the punctures cyclically, contradicting the assumption. Since $f \in \mathrm{Mod}(S_{0,n})$ is also assumed to have infinite order, the Nielsen–Thurston classification implies that f is pseudo-Anosov. $\qquad\square$

In Section 7.1.1, we completely classified the finite-order elements of $\mathrm{Mod}(S_{0,n})$. Since such elements are easy to avoid, it is not hard to write down explicit elements of $\mathrm{Mod}(S_{0,n})$ that satisfy the criteria of Theorem 14.7. The construction in the theorem can be easily modified to work for braid groups: for n prime, if an element of $B_n \approx \mathrm{Mod}(D_n)$ permutes the n marked points of D_n cyclically and is not a root of a central element, then it is pseudo-Anosov.

14.2 PSEUDO-ANOSOV STRETCH FACTORS

To each pseudo-Anosov $f \in \mathrm{Mod}(S_g)$ we have attached a real number $\lambda > 1$, namely, the stretch factor of f. The set of real numbers that occur as the stretch factor of some pseudo-Anosov is quite restricted. It is still not known precisely which λ can occur; indeed, this and related problems are currently an active area of research. The purpose of this section is to prove a few fundamental facts concerning stretch factors of pseudo-Anosov homeomorphisms.

14.2.1 PSEUDO-ANOSOV STRETCH FACTORS ARE ALGEBRAIC INTEGERS

The following theorem appears in Thurston's announcement of his proof of the Nielsen–Thurston classification [207].

Theorem 14.8 *Let* $g \geq 2$. *If* λ *is the stretch factor of a pseudo-Anosov* $f \in \mathrm{Mod}(S_g)$, *then* λ *is an algebraic integer whose degree is bounded above by* $6g - 6$.

In his paper Thurston states that the examples of Theorem 14.1 show that the bound of Theorem 14.8 is sharp [207]. Franks and Rykken showed that the stretch factor of a pseudo-Anosov mapping class is a quadratic integer if and only if it is obtained by lifting through an n-fold branched cover over T^2, as explained in Section 14.1.1 [66].

Theorem 14.8 can be generalized to punctured surfaces $S_{g,n}$. We leave the computation of the maximal degree in this case as an exercise.

In order to prove Theorem 14.8 we will need the definition of the orientation cover for a foliation.

Orientation covers for foliations. Let $(\mathcal{F}, \mu)$ be a measured foliation of a surface S and let P be the set of singularities of $\mathcal{F}$. Pick a basepoint $z \in S - P$. We can use $\mathcal{F}$ to define a homomorphism

$$\tau : \pi_1(S - P, z) \to \mathbb{Z}/2\mathbb{Z}$$

as follows. Pick one of the two unit vectors $v_z \in TS_z$ that is tangent to the leaf of $\mathcal{F}$ containing z. Given $[\gamma] \in \pi_1(S - P, z)$, pick a representative loop $\gamma : [0,1] \to S - P$ with $\gamma(0) = \gamma(1) = z$. Since $\mathcal{F}$ is defined via local charts, it makes sense to continue choosing unit vectors $v_{\gamma(t)}$ to obtain a continuous vector field along γ, with each $v_{\gamma(t)}$ being tangent to the leaf of $\mathcal{F}$ through $\gamma(t)$. Now $v_{\gamma(1)}$ is tangent to the leaf of $\mathcal{F}$ through z, so it is equal to either v_z or $-v_z$. In the first case define $\tau([\gamma]) = 0$; in the latter case define $\tau([\gamma]) = 1$. The map τ is a well-defined homomorphism because any homotopy of loops in $S - P$ gives a continuous deformation of the corresponding vector fields $v_{\gamma(t)}$.

The homomorphism τ is called the *orientation homomorphism* associated to $\mathcal{F}$. This terminology comes from the fact that τ is precisely the obstruction to orienting the leaves of $\mathcal{F}$ in a consistent way. A measured foliation is orientable in the sense of Section 11.2 if and only if its orientation homomorphism is trivial.

Recall that a measured foliation is locally orientable if and only if each singularity has an even number of prongs. In agreement with this, if γ bounds a disk in S containing one singularity, then $\tau([\gamma]) = 0$ if and only if that singularity has an even number of prongs.

If $\mathcal{F}$ is not orientable, then by extending over the singularities, the orientation homomorphism gives rise to a connected twofold branched cover

$$p : \widetilde{S} \to S$$

called the *orientation cover* of S for $\mathcal{F}$. What is more, there is an induced measured foliation $(\widetilde{\mathcal{F}}, \widetilde{\mu})$ on $\widetilde{S}$ that is orientable, and so that p maps leaves of $\widetilde{\mathcal{F}}$ to leaves of $\mathcal{F}$ and $p_*\widetilde{\mu} = \mu$. The branch points of the cover are exactly the preimages under p of the singularities of $\mathcal{F}$ with an odd number of prongs.

An alternate construction. It is possible to construct the orientation cover $\widetilde{S}$ for a foliation $\mathcal{F}$ in a way that is similar to the standard construction of the orientation double cover of a nonorientable manifold. Thus one lets $\widetilde{S}$ be the set of pairs (z, v) where $z \in S - P$ and v is tangent to the leaf of $\mathcal{F}$ through z. Then one must also define the cover over P.

We will now use orientation covers to prove that stretch factors are algebraic integers.

Proof of Theorem 14.8. Let ϕ be a pseudo-Anosov representative of f and let $(\mathcal{F}^u, \mu)$ be its unstable foliation. We first prove the theorem in the special case that $\mathcal{F}^u$ is an orientable foliation. As explained in Section 11.2, this means that there is a closed 1-form ω on S_g so that

$$\phi^*\omega = \lambda\omega,$$

where ϕ^* denotes the pullback of a differential form.

We claim that the cohomology class $[\omega] \in H^1(S_g; \mathbb{R})$ is nonzero. Indeed, suppose that $\omega = dF$, where $F : S_g \to \mathbb{R}$ is a smooth nonzero function. The formula $\phi^*\omega = \lambda\omega$ implies that $\phi^*F = \lambda F + C$, where C is a constant. However, since S_g is compact, this is impossible (for instance, the difference between the maximum and minimum values of F is invariant under pullback). The claim is thus proven.

Since $\phi^*\omega = \lambda\omega$, we in particular have

$$\phi^*([\omega]) = \lambda[\omega],$$

where $\phi^* : H^1(S_g; \mathbb{R}) \to H^1(S_g; \mathbb{R})$ now denotes the induced action of ϕ on cohomology. That is, λ is a nontrivial eigenvalue for ϕ^*. But ϕ^* preserves the integer lattice $H^1(S_g; \mathbb{Z})$, so the matrix for ϕ^* in the standard basis has integer entries. The characteristic polynomial for this matrix is thus a $2g \times 2g$ matrix with integer entries, and it has λ as an eigenvalue. Thus λ is an algebraic integer of degree at most $2g$. Note that the assumption $g \geq 2$ implies that $2g < 6g - 6$.

We now prove the theorem in the case where $\mathcal{F}^u$ is nonorientable. The idea is to pass to the orientation double cover of $\mathcal{F}^u$ and then to quote the argument above.

Let $\widetilde{S}$ be the orientation cover for the unstable foliation $\mathcal{F}^u$ for ϕ. The induced foliation on $\widetilde{S}$ is orientable, and so it is given by a 1-form ω, which represents a nontrivial element of $H^1(\widetilde{S}; \mathbb{R})$. Since ϕ preserves $\mathcal{F}^u$, it follows that ϕ lifts to a map $\widetilde{\phi}$ of $\widetilde{S}$. Since $\widetilde{\phi}^*\omega = \pm\lambda\omega$, it follows that $\pm\lambda$ is an eigenvalue of the map induced by $\widetilde{\phi}$ on $H^1(\widetilde{S}; \mathbb{R})$. Thus λ is an algebraic integer. We now prove the claimed bound on the degree of this algebraic integer.

Say the singularities of $\mathcal{F}^u$ are $s_1, \cdots, s_k$. By the Euler–Poincaré formula (Section 11.2) each singularity contributes at least $-1/2$ to $\chi(S)$, so that $k \leq -2\chi(S)$. By the Riemann–Hurwitz formula (Section 7.2), we have

$$\chi(\widetilde{S}) \geq 2\chi(S) - k \geq 8 - 8g.$$

and so the dimension of $H^1(\widetilde{S}; \mathbb{R})$ is at most $8g - 6$. Let τ be the deck transformation for $\widetilde{S}$ over S. Since τ has order 2, its action on $H^1(\widetilde{S}; \mathbb{R})$ has eigenvalues of 1 and -1, with eigenspaces V_+ and V_-, respectively.

Now V_+ is isomorphic to $H^1(S; \mathbb{R}) \approx \mathbb{R}^{2g}$, the isomorphism being given by τ. Thus the dimension of V_- is at most $6g - 6$. As $\mathcal{F}^u$ is not orientable, the 1-form ω does not descend to S, and so it is an element of V_-. Since $\widetilde{\phi}$ commutes with τ, we have that V_- is an integral subspace invariant under $\widetilde{\phi}^*$. Since λ is a root of the characteristic polynomial for the action of $\widetilde{\phi}^*$ on V_-, the theorem follows.   □

Perron numbers. Theorem 14.8 can be strengthened further: each pseudo-Anosov stretch factor is a special kind of algebraic integer called a Perron number. A *Perron number* is an algebraic integer that is real, that is greater than 1, and that is larger than the absolute value of each of its Galois conjugates. The reason this is true is that every pseudo-Anosov stretch factor is the Perron–Frobenius eigenvalue of a Perron–Frobenius matrix (the matrix is the transition matrix for a Markov partition; see below), and all Perron–Frobenius eigenvalues are Perron numbers.

It has been conjectured that a real number $\lambda > 1$ is a pseudo-Anosov stretch factor if and only if λ is an algebraic unit and all conjugates of λ except $1/\lambda$ have absolute value lying in $(1/\lambda, \lambda)$ [152].

A consequence of the fact that every pseudo-Anosov stretch factor is a Perron number is that pseudo-Anosov stretch factors are completely determined by their minimal polynomials. This was pointed out to us by Joan Birman.

14.2.2 THE SPECTRUM OF PSEUDO-ANOSOV STRETCH FACTORS

If a pseudo-Anosov $f \in \mathrm{Mod}(S_g)$ has stretch factor λ, then any $h \in \mathrm{Mod}(S)$ conjugate to f is pseudo-Anosov with stretch factor λ. Thus we can associate to any conjugacy class of pseudo-Anosov mapping classes in $\mathrm{Mod}(S)$ a stretch factor λ.

A conjugacy class in $\mathrm{Mod}(S)$ corresponds to a free homotopy class of loops in moduli space $\mathcal{M}(S)$. Here we should recall that we need to consider homotopies in the orbifold sense or to lift to a finite manifold cover of $\mathcal{M}(S)$.

Any pseudo-Anosov $f \in \mathrm{Mod}(S)$ acts on $\mathrm{Teich}(S)$ by translating along an axis by a Teichmüller distance of $\frac{1}{2}\log(\lambda^2) = \log(\lambda)$. Since the axis for f is a Teichmüller line, the free homotopy class in $\mathcal{M}(S)$ corresponding to the conjugacy class of f in $\mathrm{Mod}(S)$ contains a closed geodesic. It turns out that the axis for f is unique, so that this closed geodesic is unique.

Conversely, considering the isometry types (elliptic, parabolic, hyperbolic) as in the proof of the Nielsen–Thurston classification, we see that the only homotopy classes of loops in $\mathcal{M}(S)$ that contain a geodesic are those corresponding to conjugacy classes of pseudo-Anosov mapping classes.

Thus the set

$$\mathrm{Spec}(\mathcal{M}(S)) = \{\log(\lambda) : \lambda \text{ is the stretch factor of some}$$
$$\text{pseudo-Anosov } f \in \mathrm{Mod}(S)\}$$

can be thought of as the Teichmüller length spectrum of moduli space $\mathcal{M}(S)$. Of course, knowing $\mathrm{Spec}(\mathcal{M}(S))$ is equivalent to knowing the set of possible stretch factors λ themselves. In analogy with the case of hyperbolic surfaces, we have the following theorem of Arnoux–Yoccoz [5] and Ivanov

[104].

THEOREM 14.9 *Let $g, n \geq 0$. For any $D \geq 1$, there exists only finitely many conjugacy classes of pseudo-Anosov elements of $\mathrm{Mod}(S_{g,n})$ with stretch factor at most D. In particular, $\mathrm{Spec}(\mathcal{M}(S_{g,n}))$ is a closed, discrete subset of $\mathbb{R}$.*

Proof. Let $S = S_{g,n}$. Let $D \geq 1$ be given. Choose $\epsilon < \delta / D^{3g-3+n}$, where δ has the property that any two distinct geodesics of length less than δ in any hyperbolic surface homeomorphic to S are disjoint; such a δ exists by Corollary 13.7.

We claim that if a pseudo-Anosov $f \in \mathrm{Mod}(S)$ has stretch factor at most D, then f has an axis in $\mathrm{Teich}(S)$ whose projection to $\mathcal{M}(S)$ lies entirely in the ϵ–thick part $\mathcal{M}_\epsilon(S)$ (cf. Section 12.4). If this were not true, then the projection of an axis $A(f)$ for f has some point X lying in $\mathcal{M}(S) - \mathcal{M}_\epsilon(S)$. In other words, there is a marked hyperbolic surface $X \in A(f)$ whose shortest simple closed curve α has length $\ell_X(\alpha) < \epsilon$. Consider the simple closed curves $\alpha, f(\alpha), f^2(\alpha), \ldots, f^{3g-3+n}(\alpha)$. By Wolpert's lemma (Lemma 12.5), each of these curves has length at most

$$\ell_X(f^i(\alpha)) \leq D^{3g-3+n} \cdot \ell_X(\alpha) < \delta \text{ for } i = 1, \ldots, 3g - 3 + n.$$

Since there are at most $3g - 3 + n$ distinct disjoint isotopy classes of simple closed curves on S, it must be that two of the curves on the list are isotopic, which implies that $f^j(\alpha)$ is isotopic to α for some $j > 0$, contradicting the fact that f is pseudo-Anosov. This proves the claim.

Let K be a compact subset of $\mathrm{Teich}(S)$ with the property that K surjects onto $\mathcal{M}_\epsilon(S)$. By the previous paragraph, each conjugacy class in $\mathrm{Mod}(S)$ with pseudo-Anosov stretch factor less than or equal to D has a representative with an axis that intersects K. Denote this list of representatives by $\{f_i\}$. Let K' be the set of points in $\mathrm{Teich}(S)$ with distance at most $\log(D)/2$ from K. Since the Teichmüller metric is proper, K' is again compact. Each f_i clearly satisfies $(f_i \cdot K') \cap K' \neq \emptyset$. Since the $\mathrm{Mod}(S)$ action on $\mathrm{Teich}(S)$ is properly discontinuous (Theorem 12.2), it follows that $\{f_i\}$ is finite, which is what we wanted to show. □

It follows from Theorem 14.9 that if we fix g, there is a smallest number λ_g that appears as the stretch factor of any pseudo-Anosov element of $\mathrm{Mod}(S_g)$. Penner proved the following beautiful theorem about λ_g [175] (see also [151]). In the statement, we write $f(x) \asymp g(x)$ for real-valued functions f and g with $f(x)/g(x) \in [1/C, C]$ for some $C > 1$.

Theorem 14.10 *The function* $\lambda_g : \mathbb{N} \to \mathbb{R}$ *satisfies*

$$\log \lambda_g \asymp 1/g.$$

It follows from Theorem 14.10 that $\bigcup_{g=1}^{\infty} \mathrm{Spec}(\mathcal{M}(S_g))$ has elements arbitrarily close to 0. Further, since a multiple of a loop in $\mathcal{M}(S_g)$ is another loop in $\mathcal{M}(S_g)$, we see that $\cup\mathrm{Spec}(\mathcal{M}(S_g))$ is dense in $(0, \infty)$.

14.3 PROPERTIES OF THE STABLE AND UNSTABLE FOLIATIONS

In this section we will explore special properties of those measured foliations that are the stable (or unstable) foliations of some pseudo-Anosov homeomorphism. Much of our treatment follows that of FLP [61].

The statements in this section are given for compact surfaces. Recall, though, that a pseudo-Anosov homeomorphism of a surface with boundary gives a pseudo-Anosov homeomorphism of the surface obtained by collapsing each boundary component to a marked point. Thus all of the results in this section hold for marked surfaces $S_{g,n}$ as well.

14.3.1 FIRST PROPERTIES

Since a pseudo-Anosov mapping class contracts the leaves of its stable foliation (with respect to the measure of the unstable foliation), we immediately obtain the following. In the statement, a *peripheral leaf* of a foliation is one contained in the boundary of the surface.

Lemma 14.11 *Let $\mathcal{F}$ be the stable or unstable foliation of a pseudo-Anosov homeomorphism of a compact surface S. Let L be any leaf of $\mathcal{F}$ that is not peripheral. Then L is not closed. Also, L does not connect two singularities of $\mathcal{F}$, two boundary components of S, or a singularity of $\mathcal{F}$ to a boundary component of S.*

Let $(\mathcal{F}, \mu)$ be a measured foliation on a surface S. Any simple closed curve in S that is not completely contained in a finite union of leaves of $\mathcal{F}$ must have nonzero μ-measure. Lemma 14.11 therefore implies the following.

Corollary 14.12 *Let $(\mathcal{F}, \mu)$ be the stable or unstable measured foliation of a pseudo-Anosov homeomorphism of a compact surface S. Then $\mu(\alpha) > 0$ for every essential simple closed curve α in S.*

In Section 11.2, we explained that any foliation that has a singularity with an odd number of prongs is not orientable. This leaves open the question of whether the stable foliation for a pseudo-Anosov homeomorphism can fail to be orientable. In the proof of Theorem 14.8 we saw that if a pseudo-Anosov $f \in \mathrm{Mod}(S_g)$ has an orientable stable foliation, then its action on $H^1(S_g; \mathbb{R})$ has an eigenvalue $\lambda > 1$. We thus have the following fact.

Corollary 14.13 *The stable and unstable foliations for any pseudo-Anosov element of the Torelli group $\mathcal{I}(S_g)$ are not orientable.*

Combined with Corollary 14.3, Corollary 14.13 in particular shows that there do exist pseudo-Anosov mapping classes with nonorientable stable foliations. The converse to Corollary 14.13 is not true: one can use the Thurston construction to find counterexamples.

14.3.2 POINCARÉ RECURRENCE FOR FOLIATIONS

A measured foliation $\mathcal{F}$ on a surface S can be viewed as a dynamical system. Fix a point $p \in S$ and follow the leaf L of $\mathcal{F}$ passing through p in one direction. What does the resulting path look like? Does it always return to points near p? What does the distribution of return times look like? If $\mathcal{F}$ is the stable foliation of a pseudo-Anosov homeomorphism f, how do these answers relate to dynamical properties of f? One can phrase these and more subtle questions in the language of dynamical systems. In this book we will concern ourselves with only a few fundamental properties. We begin with a version of Poincaré recurrence. This gives an answer to the first two questions above.

Theorem 14.14 (Poincaré recurrence for foliations) *Let $(\mathcal{F}, \mu)$ be a measured foliation on a compact surface S. Let L be an infinite half-leaf. Then any arc α transverse to $\mathcal{F}$ and intersecting L at least once must intersect L infinitely many times.*

We emphasize that the measure μ in the hypothesis of Theorem 14.14 is necessary. One can build a (singular) foliation $\mathcal{F}$ on S with nonclosed leaves that can limit to a closed leaf by spiraling about it, in particular, never returning near their starting points. It is not hard to prove that such spiraling behavior cannot occur if in addition $\mathcal{F}$ is equipped with a transverse measure μ: one simply uses the spiraling leaves and the isotopy invariance of μ to build an arc with infinite measure.

We can replace the hypothesis of Theorem 14.14 with the hypothesis that $(\mathcal{F}, \mu)$ is the stable (or unstable) foliation for a pseudo-Anosov homeomorphism. Indeed, by Lemma 14.11, every nonsingular interior point of such a foliation is the starting point for an infinite half-leaf.

Good atlases. Our proof of Theorem 14.14 uses the notion of a good atlas, which we now define.

We call a subset $P \subset S$ a *polygon* with respect to a foliation $\mathcal{F}$ if each of the following holds:

1. P is a closed, simply connected region in S.

2. P contains at most one singularity of $\mathcal{F}$.

3. ∂P is the union of finitely many arcs that alternate between being subarcs of leaves $\mathcal{F}$ and being arcs transverse to $\mathcal{F}$.

We call the transverse arcs in ∂P the *faces* of the polygon and call the other arcs of ∂P the *sides*. A polygon P is *standard* if it contains at most one singularity of $\mathcal{F}$ in its interior and if there is one face of P for each prong of the singularity (two faces total if there are no singularities); see Figure 14.3.

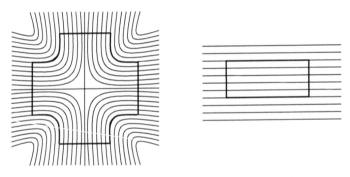

Figure 14.3 Standard polygons: charts for a good atlas.

Following [61], a *good atlas* for a foliation $\mathcal{F}$ of S consists of two collections of standard polygons, $\{U_i\}$ and $\{V_i\}$, with the following properties.

1. S is the union of the interiors of the U_i.

2. For each i, the polygon U_i is contained in V_i, and the faces of U_i are contained in the faces of V_i.

3. For each i, the measure of any transverse arc that connects a side of U_i to a side of V_i is at least ϵ_0, where ϵ_0 is some fixed number.

4. Each singular point belongs to exactly one U_i.

5. Whenever $i \neq j$, the intersection $U_i \cap U_j$ is either empty or a rectangle; see Figure 14.4.

We leave it as an exercise to show that any measured foliation on a compact surface has a good atlas.

Figure 14.4 The intersections of two charts of a good atlas.

The proof. We now give the proof of Poincaré recurrence for foliations.

Proof of Theorem 14.14. It is enough to show that $\alpha \cap L$ can never be a single endpoint of α since we can apply this statement to any subinterval of α. For the sake of contradiction, assume that $\alpha \cap L$ is a single endpoint of α. Cut S along α. The result is a surface S' with one new boundary component, which we call α'. The surface S' is equipped with an induced measured foliation that has two singular points on the boundary, corresponding to the endpoints of the arc α; see Figure 14.5. Note that the boundary component α' is transverse to $\mathcal{F}'$.

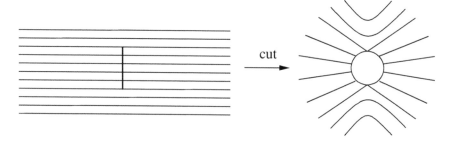

Figure 14.5 Cutting along a transverse arc.

Denote by L' the leaf of $\mathcal{F}'$ corresponding to L. The leaf L' starts from one of these singular points of $\mathcal{F}'$ on the boundary component α', say s. We see that L' does not return to s, for that would mean that L is closed and hence not infinite. Our assumptions on L on $\alpha \cap L$ now translate to the statement that L' does not return to $\partial S'$.

Choose a good atlas for $\mathcal{F}'$ with constant ϵ_0. Let β be an arc of the boundary component α' that has s as one of its endpoints and whose μ'-length is $\epsilon < \epsilon_0$. Further assume that any leaf of $\mathcal{F}'$ starting from β avoids the singularities of $\mathcal{F}'$; this is possible because each leaf starting from a singularity can hit $\partial S'$ at most once.

In order to obtain a contradiction, we will prove that we can push β along the foliation $\mathcal{F}'$ for infinite time; that is, we can create a strip of $(\mathcal{F}', \mu')$ of width ϵ and of infinite height. To make this formal we define a map

$$P : \beta \times \mathbb{R}_{\geq 0} \to S'$$

by the rule that $P(x \times \mathbb{R}_{\geq 0})$ is the entire leaf of $\mathcal{F}'$ starting from x.

Let L_s be the leaf of $\mathcal{F}'$ emanating from the singularity s on $\partial S'$. Given any particular point of L_s, there is a chart U_i of the good atlas containing that point. By the properties of a good atlas we obtain an entire strip of width ϵ in the interior of the corresponding V_i. It follows that P is an immersion. Since P never hits a singularity (by the assumption on β) and since $P^{-1}(\beta) = \beta \times \{0\}$, we see that in fact P is injective.

Since $\mathcal{F}'$ has no closed leaves and since there are finitely many V_i, the embedded, infinite strip $P(\beta \times \mathbb{R}_{\geq 0})$ passes through some V_i infinitely many times, which is a contradiction. $\square$

Poincaré recurrence shows us that the imposition of a transverse measure on a foliation greatly constrains what the foliation can look like. It precludes, for example, the possibility that a leaf spirals toward a singularity or spirals toward a curve, or that the foliation has a *Reeb component* (a closed annulus foliated by its two boundary curves and infinitely many leaves homeomorphic to $\mathbb{R}$). The existence of the good atlas is indeed a very strong condition.

14.3.3 Each Leaf of the (Un)stable Foliation is Dense

Our next goal is to prove the basic fact that each nonperipheral leaf of the (un)stable foliation for a pseudo-Anosov homeomorphism of a surface S is dense in S. In order to prove this we first give a combinatorial method for dealing with foliations.

A combinatorial description of the (un)stable foliation. In Chapter 11, we made the claim that every measured foliation of a surface could be obtained by decomposing the surface into polygons and foliating each polygon by horizontal lines. We now explain how to do this in the case where $\mathcal{F}$ is the stable or unstable foliation for a pseudo-Anosov homeomorphism. In this construction all of the polygons will be rectangles.

Let τ be a small arc transverse to $\mathcal{F}$. If τ contains any singularities of $\mathcal{F}$, then we assume that it contains exactly one, at an endpoint. Subdivide τ by placing finitely many extra vertices at the points found by the following two procedures.

1. From each endpoint of τ and for each of the two directions along $\mathcal{F}$, follow $\mathcal{F}$ to the point of first return on τ.

2. From each singularity of $\mathcal{F}$, follow each half-leaf of $\mathcal{F}$ to the first point of intersection with τ.

Denote the closed segments of the resulting subdivision of τ by τ_i.

By Poincaré recurrence for foliations (Theorem 14.14), any leaf starting at a point of τ eventually returns to τ. By the assumptions on τ, we can push each τ_i along $\mathcal{F}$ until it "hits" some τ_j. The result of this process is a union of rectangles in S, each foliated horizontally by subarcs of leaves of $\mathcal{F}$. What is more, the union of these rectangles covers S, for otherwise, the boundary of the union of rectangles would be a cycle of leaves of $\mathcal{F}$, which does not exist by Lemma 14.11. Thus we have obtained the desired rectangle decomposition.

An example of such a rectangle decomposition on the torus is given on the left side (and also the right side) of Figure 15.16.

Leaves are dense. The following consequence of Poincaré recurrence for foliations will be used to show that every pseudo-Anosov homeomorphism has a dense orbit (Theorem 14.17 below).

Corollary 14.15 *Let S be a compact surface and let $\mathcal{F}$ be the stable or unstable foliation for a pseudo-Anosov homeomorphism of S. Then any non-peripheral leaf of $\mathcal{F}$ is dense in S.*

Since the stable and unstable foliations for an Anosov map of T^2 have irrational slope, Corollary 14.15 is well known in the case of Anosov maps of the torus.

Proof of Corollary 14.15. Let τ be an arbitrary arc in S that is transverse to $\mathcal{F}$. It suffices to show that L intersects τ. Using Poincaré recurrence, we can construct a rectangle decomposition of S from τ as in the previous subsection. Since L is half-infinite (Lemma 14.11) and since L is contained in the horizontal foliations of the rectangles, it follows that L must hit one face of at least one rectangle. But this face is contained in τ by construction, so we are done.   □

The following theorem is much stronger than Corollary 14.15. A measured foliation is *uniquely ergodic* if it admits only one transverse measure up to scale.

Theorem 14.16 *Let $(\mathcal{F}^s, \mu_s)$ and $(\mathcal{F}^u, \mu_u)$ be the stable and unstable foliations for a pseudo-Anosov mapping class on a compact surface. Then $\mathcal{F}^s$ and $\mathcal{F}^u$ are uniquely ergodic.*

The proof of Theorem 14.16 relies on the theory of Markov partitions. It can be found in [61, Exposé 12, Théorème I].

14.4 THE ORBITS OF A PSEUDO-ANOSOV HOMEOMORPHISM

A basic feature of any dynamical system is its set of orbits. For example, does a transformation have dense orbits? How many periodic points does it have? Does the system exhibit extremal properties? In this section we give answers to these questions for pseudo-Anosov homeomorphisms acting on surfaces.

Existence of a dense orbit. The following theorem can be viewed as a first indication that f has a kind of mixing behavior.

Theorem 14.17 *Let* $f \in \mathrm{Homeo}^+(S)$ *be a pseudo-Anosov homeomorphism of a compact surface* S. *Then* f *has a dense orbit in* S.

We remind the reader that not every orbit of f is dense; for example, f fixes the set of singularities of its stable foliation. Indeed, we show in Theorem 14.19 below that the set of periodic points of f is dense in S. Both proofs are taken from [61, Exposé 9].

Proof. We first show that if U is a nonempty open set that is invariant under f, then U is dense in S. By taking a power of f, we may assume without loss of generality that f fixes the singular points of the stable and unstable foliations $\mathcal{F}^s$ and $\mathcal{F}^u$ of f. Let L be a nonperipheral leaf of $\mathcal{F}^s$ containing a singularity s. By Corollary 14.15, L is dense, and so U contains a point x of L. We may choose a segment $J \subset U$ of a leaf of $\mathcal{F}^u$ so that $x \in J$. Since $x \in L$, it follows that

$$\lim_{n \to \infty} f^n(x) = s.$$

Further, for each n, the segment $f^n(J)$ is a subset of a leaf of $\mathcal{F}^u$ and is also contained in U. Since f is stretching along $\mathcal{F}^u$, it follows that, as $n \to \infty$, the $f^n(J)$ approach the union of the singular leaves of $\mathcal{F}^u$ bounding the "sector" containing $\{f^n(x)\}$. At least one of these singular leaves is nonperipheral. As each nonperipheral leaf of $\mathcal{F}^u$ is dense in S (Corollary 14.15), it follows that U is dense in S.

Now let $\{U_i\}$ be a countable basis for S. Each set

$$V_i = \bigcup_{n \in \mathbb{Z}} f^n(U_i)$$

is a nonempty open set that is invariant under f and hence is dense in S. By the Baire category theorem, the set $\bigcap_i V_i$ is dense, in particular, nonempty. Let x be any point in this intersection. Then for each i there is an n_i so that $x \in f^{n_i}(U_i)$ or $f^{-n_i}(x) \in U_i$. Since each basis element U_i contains a point in the orbit of x, it follows that the orbit of x is dense in S. $\qquad\square$

Density of periodic points. We will need the following standard tool from the theory of dynamical systems; see, e.g., [196, p. 7].

Theorem 14.18 (Poincaré recurrence) *Let $\mathcal{M}$ be a finite measure space and let $T : \mathcal{M} \to \mathcal{M}$ be measure-preserving. For every $A \subseteq \mathcal{M}$ with positive measure and for almost every $x \in A$, there is an infinite increasing sequence of integers $\{n_i\}$ so that $T^{n_i}(x) \in A$ for every i.*

Theorem 14.18 is a more general principle than Poincaré recurrence for foliations (Theorem 14.14), and its proof is much more simple. The two theorems are related in that they both address the question: given a dynamical system, under what conditions can we expect a point to return close to its starting point?

The following theorem gives another property that pseudo-Anosov homeomorphisms share with chaotic dynamical systems.

Theorem 14.19 *For any pseudo-Anosov homeomorphism ϕ of a compact surface S, the periodic points of ϕ are dense in S.*

For a standard linear Anosov homeomorphism $M \in \mathrm{SL}(2, \mathbb{Z})$ of $T^2 = \mathbb{R}^2/\mathbb{Z}^2$, the set of periodic points of M is precisely the image of $\mathbb{Q}^2 \subset \mathbb{R}^2$ under the projection $\mathbb{R}^2 \to T^2$; in particular, the set of periodic points of M is dense in T^2.

Proof. Let $\mathcal{F}^s$ and $\mathcal{F}^u$ denote the stable and unstable foliation for ϕ. Choose a good atlas with respect to these foliations, as explained on page 410. Let U be a standard square in the interior of S with respect to the chosen good atlas. By assumption, U does not contain a singularity of $\mathcal{F}^u$ (or, of course, of $\mathcal{F}^s$). It suffices to show that U contains a periodic point for ϕ.

Let V be a standard square contained strictly inside U. Consider the area measure which is locally the product of the transverse measures associated to $\mathcal{F}^s$ and $\mathcal{F}^u$. As ϕ leaves this measure invariant, we can apply the principle of Poincaré recurrence (Theorem 14.18). This gives that for any N there is an $n > N$ so that $\phi^n(V) \cap V \neq \emptyset$.

Choose a point $x_1 \in V$ so that $\phi^n(x_1) \in V$. Let J be a nontrivial closed subarc in $\mathcal{F}^s$ crossing U at x_1. Since ϕ contracts stable arcs by a fixed

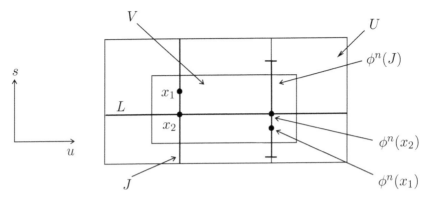

Figure 14.6 The diagram for the proof of Theorem 14.19.

multiplicative factor, we can (retroactively) choose N large enough so that $\phi^n(J) \subset U$. Pushing points along the leaves of $\mathcal{F}^u$ gives a map from $\phi^n(J)$ to J. Composing this map with ϕ^n gives a map $J \to J$. By the Brouwer fixed point theorem, this map has a fixed point $x_2 \in J$.

Let L be a nontrivial closed subarc of $\mathcal{F}^u$ that crosses U through x_2. Increasing N if necessary (again, retroactively), we can assume that $\phi^n(L) \supset L$. To see that ϕ^n has a fixed point in U, it suffices to apply the following fact to ϕ^n.

> Let I be an interval of $\mathbb{R}$. If $f : I \to \mathbb{R}$ is a continuous function with $f(I) \supseteq I$, then f has a fixed point.

Thus ϕ has a periodic point of order n in U. This completes the proof. □

A minimality property. Nielsen's original approach to the classification of surface homeomorphisms involved an extensive analysis of the action of a homeomorphism on the closed disk $\mathbb{H}^2 \cup \partial\mathbb{H}^2$. In his announcement [207], Thurston states that this type of analysis can be used to derive the following theorem. It states that while the periodic points of a pseudo-Anosov homeomorphism ϕ are prevalent enough to be dense in S, their number is as small as possible, in the sense stated in the following theorem. Recall that the *period* of a periodic point x for $\phi \in \mathrm{Homeo}(S)$ is the smallest $n \geq 1$ for which $\phi^n(x) = x$.

Theorem 14.20 *Let $g \geq 2$ and let ϕ be a pseudo-Anosov homeomorphism of $S = S_g$. For each $n > 0$, the homeomorphism ϕ has the minimum number of periodic points of period n among all homeomorphisms in its homotopy class.*

Our proof of Theorem 14.20 follows Handel [80].

Proof. Suppose that $\psi \in \mathrm{Homeo}^+(S)$ is homotopic to ϕ. Then there is a canonical bijection

$$\{\text{lifts } \widetilde{\phi} \text{ of } \phi \text{ to } \widetilde{S}\} \leftrightarrow \{\text{lifts } \widetilde{\psi} \text{ of } \psi \text{ to } \widetilde{S}\}$$

where $\widetilde{\phi}$ is identified with $\widetilde{\psi}$ if and only if they induce the same homeomorphism of $\partial\mathbb{H}^2$. We say that $\widetilde{\phi}$ and $\widetilde{\psi}$ *agree* on $\partial\mathbb{H}^2$.

We first prove that $|\mathrm{Fix}(\phi)| \leq |\mathrm{Fix}(\psi)|$. For $x \in \mathrm{Fix}(\phi)$ and $y \in \mathrm{Fix}(\psi)$, define x to be *Nielsen-equivalent* to y if there exist lifts $\widetilde{\phi}, \widetilde{\psi}$ that agree on $\partial\mathbb{H}^2$ and lifts $\widetilde{x}, \widetilde{y}$ such that $\widetilde{x} \in \mathrm{Fix}(\widetilde{\phi})$ and $\widetilde{y} \in \mathrm{Fix}(\widetilde{\psi})$. We define Nielsen equivalence between fixed points of ϕ in a similar way, so that the Nielsen equivalence classes for ϕ are precisely the projections to S of sets of the form $\mathrm{Fix}(\widetilde{\phi})$ for different lifts $\widetilde{\phi}$ of ϕ.

We first claim that every fixed point of ϕ is Nielsen-equivalent to some fixed point of ψ. To see this, choose a fixed point of ϕ and let $\widetilde{\phi}$ be some lift of ϕ. Some fixed point of $\widetilde{\phi}$ projects to the given fixed point of ϕ. Extend $\widetilde{\phi}$ to a homeomorphism of the closed disk $\mathbb{H}^2 \cup \partial\mathbb{H}^2$ (cf. Theorem 8.7). Thinking of $\mathbb{H}^2 \cup \partial\mathbb{H}^2$ as embedded in S^2 as a hemisphere, extend $\widetilde{\phi}$ to $\Phi \in \mathrm{Homeo}^+(S^2)$ by reflecting across the equator. Let $L(\Phi)$ denote the Lefschetz number of Φ, that is, the sum of the indices of the fixed points of Φ. Then (as in Section 6.3)

$$L(\Phi) = 2L(\widetilde{\phi}) + L_\infty(\widetilde{\phi}),$$

where $L_\infty(\widetilde{\phi})$ is the sum of the indices of the fixed points of $\widetilde{\phi}$ on the equator. Let $\widetilde{\psi}$ be the unique lift that agrees with with $\widetilde{\phi}$ on $\partial\mathbb{H}^2$ and let $\Psi \in \mathrm{Homeo}^+(S^2)$ be the corresponding doubled homeomorphism as constructed above. Both L and L_∞ are homotopy-invariants. Further, our choice of lifts implies that $\widetilde{\phi}$ and $\widetilde{\psi}$ induce the same homeomorphism on $\partial\mathbb{H}^2$. Thus

$$2L(\widetilde{\phi}) + L_\infty(\widetilde{\phi}) = 2L(\widetilde{\psi}) + L_\infty(\widetilde{\psi}) = 2L(\widetilde{\psi}) + L_\infty(\widetilde{\phi})$$

(actually, by the formula in Section 6.3, all terms are equal to 2, but we do not need this). Since ϕ is pseudo-Anosov, each of its fixed points must have nonzero index. It follows that $L(\widetilde{\psi}) < 0$, and in particular $\widetilde{\psi}$ has a fixed point. This proves the claim.

We now claim that each Nielsen-equivalence class of fixed points for ϕ has at most one element. Indeed, if some Nielsen class of fixed points for ϕ had at least two elements, we would have some lift $\widetilde{\phi}$ fixing two points. But the singular Euclidean metric on S coming from $\mathcal{F}^s$ and $\mathcal{F}^u$ lifts to a singular Euclidean metric on $\mathbb{H}^2$. In this metric there is a unique geodesic between any two points, and $\widetilde{\phi}$ acts affinely on this metric. Thus if $\widetilde{\phi}$ fixes two points, it would have to fix pointwise the unique geodesic between these two points. This contradicts the fact that $\widetilde{\phi}$ acts by expansion by $\lambda \neq 1$, proving the claim.

Since every fixed point of ϕ is Nielsen-equivalent to some fixed point of ψ and since each Nielsen class of fixed points for ϕ has at most one element, it follows immediately that $|\operatorname{Fix}(\phi)| \leq |\operatorname{Fix}(\psi)|$.

Let $k \geq 1$. Since ϕ^k and ψ^k are homotopic and since ϕ^k is a pseudo-Anosov homeomorphism, what we just proved gives that $|\operatorname{Fix}(\phi^k)| \leq |\operatorname{Fix}(\psi^k)|$. Unfortunately, this does not prove the theorem since some points of $\operatorname{Fix}(\psi^k)$ might have period strictly less than k.

So suppose that x is a periodic point of ϕ with period k. We have proved that ψ^k has at least one fixed point y that is Nielsen-equivalent to x. We need to show that y has period k as a periodic point of ψ. So suppose that $\psi^j(y) = y$ with $j|k$. Let $\widetilde{\phi^k}$ and $\widetilde{\psi^k}$ be equivariantly homotopic lifts of ϕ^k and ψ^k fixing lifts $\widetilde{x}$ and $\widetilde{y}$ of x and y. Such a lift exists since ψ is homotopic to ϕ. If $\widetilde{\psi^j}$ is a lift of ψ^j fixing $\widetilde{y}$, then $(\widetilde{\psi^j})^{k/j}$ is a lift of ψ^k fixing $\widetilde{y}$. By the uniqueness of lifts fixing a given point, it follows that $(\widetilde{\psi^j})^{k/j} = \widetilde{\psi^k}$.

Since ϕ is homotopic to ψ, there is a lift $\widetilde{\phi^j}$ of ϕ^j that is equivariantly homotopic to $\widetilde{\psi^j}$. Taking powers gives that $(\widetilde{\phi^j})^{k/j}$ is equivariantly homotopic to $(\widetilde{\psi^j})^{k/j} = \widetilde{\psi^k}$. But $\widetilde{\phi^k}$ is also equivariantly homotopic to $\widetilde{\psi^k}$, and so it must be that $(\widetilde{\phi^j})^{k/j} = \widetilde{\phi^k}$. In particular, $\widetilde{\phi^j}$ commutes with $\widetilde{\phi^k}$. It now follows that $\widetilde{\phi^k}$ fixes each point of the $\widetilde{\phi^j}$-orbit of $\widetilde{x}$ since

$$\widetilde{\phi^k}\left(\left(\widetilde{\phi^j}\right)^i(\widetilde{x})\right) = \left(\widetilde{\phi^j}\right)^i\left(\widetilde{\phi^k}(\widetilde{x})\right) = \left(\widetilde{\phi^j}\right)^i(\widetilde{x}).$$

As above, a lift of a pseudo-Anosov homeomorphism can fix only one point of $\mathbb{H}^2$. Thus the $\widetilde{\phi^j}$-orbit of $\widetilde{x}$ is a single point. Since x has period k, it must be that $j = k$. $\square$

Ergodicity. Pseudo-Anosov homeomorphisms satisfy another strong mixing property. Let $\mathcal{M}$ be a measure space and $T : \mathcal{M} \to \mathcal{M}$ a measure-preserving transformation. We say that T is *ergodic* if the only measurable sets in $\mathcal{M}$ that are invariant under T have either full measure or zero measure.

A pseudo-Anosov homeomorphism of S_g is itself a measure-preserving transformation of S_g, as it preserves the area measure induced by its stable and unstable foliations. Thus it makes sense to ask if pseudo-Anosov homeomorphisms are ergodic.

Theorem 14.21 *Let $g \geq 2$ and let $\phi : S_g \to S_g$ be a pseudo-Anosov homeomorphism. Then ϕ is ergodic with respect to the area measure induced by its stable and unstable foliations.*

Pseudo-Anosov homeomorphisms satisfy an even stronger property than ergodicity—they are Bernoulli processes. For a proof of this stronger property, see FLP [61, Exposé 10, Section VI].

14.5 LENGTHS AND INTERSECTION NUMBERS UNDER ITERATION

Let $A \in \mathrm{SL}(2, \mathbb{Z})$ have two distinct real eigenvalues $\lambda > 1$ and λ^{-1} and let $v \in \mathbb{R}^2$ be any vector that does not lie in the eigenspace for λ^{-1}. As discussed above, we have

$$\lim_{n \to \infty} \sqrt[n]{|A^n(v)|} = \lambda.$$

We say that $|A^n(v)|$ "grows like" λ^n. It follows that the length of any simple closed curve in T^2 grows like λ^n under iteration of an Anosov homeomorphism. Here we have used the fact that the eigenspaces for a hyperbolic element of $\mathrm{SL}(2, \mathbb{Z})$ have irrational slope, whereas the lines in $\mathbb{R}^2$ that are lifts of simple closed curves in T^2 have rational slope.

We also know that if v_u denotes one of the two unit vectors in the eigenspace for λ, then

$$\lim_{n \to \infty} \frac{A^n(v)}{|A^n(v)|} = \pm v_u.$$

On the torus, this means that if ϕ is an Anosov map of T^2 and α is a geodesic simple closed curve in T^2, then the slopes of the simple closed curves $\phi^n(\alpha)$ approach the slope of the unstable foliation for ϕ.

Our goal in this section is to prove that analogous results hold for pseudo-Anosov homeomorphisms of higher-genus surfaces. In the general case, eigenvalues will be replaced by stretch factors, and eigendirections by measured foliations.

For both theorems we need a definition. Let a be an isotopy class of simple closed curves in S and let $(\mathcal{F}, \mu)$ be a measured foliation on S. We define

$$I((\mathcal{F}, \mu), a) = \inf\{\mu(\alpha) : \alpha \text{ is in the homotopy class } a\}$$

where the closed curves α in the infimum are not assumed to be simple.

LEMMA 14.22 *Let ϕ be a pseudo-Anosov homeomorphism of a compact surface S. If $(\mathcal{F}, \mu)$ is the stable foliation for ϕ and a is any isotopy class of essential simple closed curves in S, then $I((\mathcal{F}, \mu), a) > 0$.*

Note that Lemma 14.22 does not follow immediately from Corollary 14.12 since the definition of $I((\mathcal{F}, \mu), a)$ involves an infimum. However, Lemma 14.22 can be deduced from part (2) of Proposition II.6 in Exposé 5 of FLP [61].

Lengths under iteration. We are ready to give the first theorem, about lengths of curves. If ρ is a Riemannian metric on a compact surface S and a is a homotopy class of simple closed curves in S, then we denote by $\ell_\rho(a)$ the length of a shortest representative in the homotopy class.

Theorem 14.23 *Let $g \geq 2$. Suppose that $f \in \mathrm{Mod}(S_g)$ is pseudo-Anosov with stretch factor λ. Let ρ be any Riemannian metric on S_g. If a is any isotopy class of simple closed curves in S_g, then*

$$\lim_{n \to \infty} \sqrt[n]{\ell_\rho(f^n(a))} = \lambda.$$

Proof. Let μ^s and μ^u be the measures associated to the stable and unstable foliations for f and let

$$d\mu = \sqrt{(d\mu^s)^2 + (d\mu^u)^2}$$

be the corresponding singular Euclidean metric.

We first prove the analogue of the theorem for the metric μ. Let ℓ_μ denote the length function with respect to μ. For an isotopy class b, the number $\ell_\mu(b)$ is defined in the same way as for a Riemannian metric.

Let α be a representative simple closed curve for the isotopy class a and let ϕ be a pseudo-Anosov homeomorphism representing f. From the definitions we have

$$\ell_\mu(f^n(a)) \leq \int_{\phi^n(\alpha)} d\mu^s + \int_{\phi^n(\alpha)} d\mu^u = \lambda^n \int_\alpha d\mu^s + \lambda^{-n} \int_\alpha d\mu^u$$

and

$$\ell_\mu(f^n(a)) \geq I((\mathcal{F}^s, \mu^s), \phi^n(a)) = \lambda^n I((\mathcal{F}^s, \mu^s), a).$$

Lemma 14.22 gives that $I((\mathcal{F}^s, \mu^s), a) > 0$. We thus have

$$\lim_{n \to \infty} \sqrt[n]{\ell_\mu(\phi^n(\alpha))} = \lambda.$$

To complete the proof of the theorem, we need to relate the metric μ to the given metric ρ. First, any two norms on a vector space are comparable, by which we mean there is a constant M so that the length of a vector with respect to the two different norms differs by a multiplicative factor of at most M. It follows that if we prove the theorem for any one Riemannian metric ρ_0, then we have proven it for the given Riemannian metric ρ.

Unfortunately, the metric for which we have proven the theorem, namely, μ, is not Riemannian. However, we would like to apply the same kind of reasoning to say that ρ is comparable to μ. More precisely, the theorem will follow once we prove that there exist constants m and M so that

$$m \leq \frac{\ell_\rho(c)}{\ell_\mu(c)} \leq M$$

for every isotopy class of simple closed curves c in S_g.

Let $\{s_i\}$ be the set of singularities of the foliation $\mathcal{F}^s$. Choose a radius r small enough so that the closed balls $B(s_i, r)$ are embedded, pairwise disjoint, and small enough so that the geodesic (in either the ρ- or the μ-metric) between two points on $\partial B(s_i, r)$ lies entirely in $B(s_i, r)$.

Since norms on a vector space are comparable in the above sense, there are constants m' and M' so that if β is any rectifiable curve, then in the complement C of the union of the $B(s_i, r/2)$, we have

$$m' \leq \frac{\ell_\rho(\beta \cap C)}{\ell_\mu(\beta \cap C)} \leq M'. \tag{14.1}$$

Note that in (14.1) we are using ℓ_ρ and ℓ_μ to denote the length of actual paths, as opposed to infima.

Now we must estimate lengths of paths lying near the singularities. We claim that there exist constants m'' and M'' so that if x and y are any distinct points in the same $\partial B(s_i, r)$, then

$$m'' \leq \frac{d_\rho(x, y)}{d_\mu(x, y)} \leq M''. \tag{14.2}$$

If x and y are sufficiently close, then the inequalities (14.1) apply. For (x, y) outside an open neighborhood of the diagonal of $\partial B(s_i, r) \times \partial B(s_i, r)$, the fraction in the middle of (14.2) is a well-defined, continuous, positive function on a compact set. The claim follows.

Set $m = \min\{m', m''\}$ and $M = \max\{M', M''\}$.

Let γ be a ρ-geodesic representative for c and let γ' be the curve obtained from γ by replacing each segment of the intersection $\cup B(s_i, r) \cap \gamma$ with the corresponding μ-geodesic segment. Combining the two inequalities (14.1) and (14.2) gives

$$\frac{\ell_\rho(\gamma)}{\ell_\mu(\gamma')} \geq m.$$

Thus

$$\ell_\rho(c) = \ell_\rho(\gamma) \geq m\ell_\mu(\gamma') \geq m\ell_\mu(c).$$

For the other direction, choose γ to be a μ-geodesic for c and let γ' be the curve obtained by substituting ρ-geodesics inside the $B(s_i, r)$. We then have

$$\ell_\rho(c) \leq \ell_\rho(\gamma') \leq M\ell_\mu(\gamma) = M\ell_\mu(c).$$

This completes the proof of the theorem. $\square$

Intersection numbers under iteration. We now explain what it means for a pseudo-Anosov homeomorphism to pull a simple closed curve in the direction of its unstable foliation. We will come back to this idea in Section 15.1.

Theorem 14.24 *Let $g \geq 2$. Let $f \in \mathrm{Mod}(S_g)$ be a pseudo-Anosov mapping class. Denote the stable and unstable foliations of f by $(\mathcal{F}^s, \mu_s)$ and $(\mathcal{F}^u, \mu_u)$. Normalize so that the area of S_g with respect to the area form induced by μ_u and μ_s is equal to 1. Then, for any two isotopy classes of curves a and b in S_g, we have*

$$\lim_{n \to \infty} \frac{i(f^n(a), b)}{\lambda^n} = I((\mathcal{F}^s, \mu_s), a)\, I((\mathcal{F}^u, \mu_u), b).$$

In particular,

$$\lim_{n \to \infty} \sqrt[n]{i(f^n(a), b)} = \lambda.$$

For a proof of Theorem 14.24, see [61, Exposé 12, Section IV].

Let $\mathcal{MF}$ denote the set of equivalence classes of measured foliations on a surface S where the equivalence relation is generated isotopy and Whitehead moves (see Figure 15.11).

Let S denote the set of isotopy classes of simple closed curves in the surface S. Since Whitehead moves and isotopy do not affect the function I, we can think of I as giving a map

$$I : \mathcal{MF} \to \mathbb{R}^S.$$

The geometric intersection number i also gives a map

$$\mathcal{S} \to \mathbb{R}^{\mathcal{S}},$$

where a maps to $i(a, \cdot)$.

With this setup, the notion of convergence to a measured foliation hinted at above can be formalized to give the following immediate corollary of Theorem 14.24. In the statement, $P(\mathbb{R}^{\mathcal{S}})$ is the space of projective classes in $\mathbb{R}^{\mathcal{S}}$, where two functions $\mathcal{S} \to \mathbb{R}$ are defined to be projectively equivalent if they differ by a constant multiple. We use brackets to denote the projective class of an element of $\mathbb{R}^{\mathcal{S}}$.

Corollary 14.25 *Let $g \geq 2$. Let $f \in \mathrm{Mod}(S_g)$ be a pseudo-Anosov mapping class. Denote the unstable foliation for f by $(\mathcal{F}^u, \mu_u)$. Then for any isotopy class a of simple closed curves in S_g, we have*

$$\lim_{n \to \infty} [f^n(a)] = [(\mathcal{F}^u, \mu_u)]$$

in $P(\mathbb{R}^{\mathcal{S}})$.

Chapter Fifteen

Thurston's Proof

In this chapter we give some indication of how Thurston originally discovered the Nielsen–Thurston classification theorem. We begin with a concrete, accessible example that illustrates much of the general theory. We then provide a sketch of how that general theory works. Our goal is not to give a formal treatment as per the rest of the text. Rather, we hope to convey to the reader part of the beautiful circle of ideas surrounding the Nielsen–Thurston classification, including Teichmüller's theorems, Markov partitions, train tracks, foliations, laminations, and more.

15.1 A FUNDAMENTAL EXAMPLE

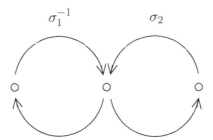

Figure 15.1 The mapping class f is $\sigma_1^{-1}\sigma_2$.

We start by giving an in-depth analysis of a fundamental and beautiful example that has gained a certain amount of fame in the world of low-dimensional topology and dynamical systems.

Let $S_{0,4}$ denote the sphere with four punctures. Taking one of the punctures to lie at infinity, we can regard $S_{0,4}$ as the thrice-punctured plane. The surface $S_{0,4}$ is the simplest surface that admits a pseudo-Anosov homeomorphism. There is a particularly simple pseudo-Anosov mapping class $f \in \mathrm{Mod}(S_{0,4})$ given by

$$f = \sigma_1^{-1}\sigma_2,$$

where σ_2 and σ_1^{-1} are the half-twists indicated in Figure 15.1. One sense in which f is simple is that its conjugacy class uniquely realizes the smallest stretch factor of any pseudo-Anosov mapping class in $\mathrm{Mod}(S_{0,4})$.

15.1.1 ITERATION, SIMPLE CLOSED CURVES, AND TRAIN TRACKS

Thurston's first main idea is to understand homeomorphisms by iterating them on simple closed curves. We now do this for f, keeping track of what happens to a chosen isotopy class c of simple closed curves in $S_{0,4}$ as f is iterated. Such an isotopy class c is shown in Figure 15.2 along with its images $f(c)$ and $f^2(c)$.

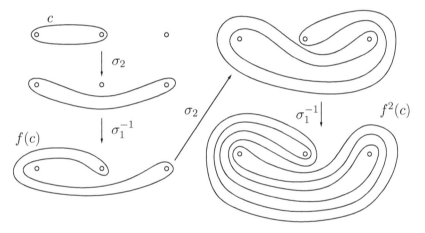

Figure 15.2 The first two iterates of c under f.

At this point the figures become harder to draw, as the number of strands is increasing quite rapidly as we iterate f. The number of horizontal strands in the figure for $f^2(c)$ is 10 (count them!). The number of horizontal strands for $f^5(c)$ is 188, for $f^{10}(c)$ is 21,892, and for $f^{100}(c)$ is

$$9079473888330615906394593939394821238467652.$$

One concrete way to measure the complexity of the curves $f^n(c)$ is to draw horizontal rays from the left- and right-hand punctures that travel outward, and to count the number of times these rays intersect $f^n(c)$. When we do this, we see that $f^n(c)$ intersects the left- and right-hand rays F_{2n+1} and F_{2n} times, respectively, where $F_0 = 0$, $F_1 = 1$, and $F_i = F_{i-1} + F_{i-2}$ is the ith Fibonacci number. In particular, the number of strands in a picture of $f^n(c)$ grows exponentially. What is worse, this is just one isotopy class, but we would like to understand how f acts on all isotopy classes of simple closed curves in $S_{0,4}$.

Thurston discovered a simple but powerful combinatorial device that completely solves the problem. The first observation is that the isotopy class $f^2(c)$, as shown in Figure 15.2, can be represented by the data in Figure 15.3 as follows. Replace n parallel strands of $f^2(c)$ by a single strand labeled n; we can think of this as pinching down (or homotoping) parallel strands into one strand and labeling this strand with an integer that records how many strands are pinched together. Sometimes this process is called *zipping* strands together. If this is to be done in a continuous manner so as not to cross the punctures, then at four points it will be necessary for the pinched-together strands to split into two strands, as shown in Figure 15.3. We emphasize that this process is not canonical; for instance, by drawing the figure differently, different parts of the curve might look parallel.

After performing this process of homotoping together parallel strands, we obtain a finite graph τ embedded in $S_{0,4}$ with the following properties:

1. Each edge of τ is the smooth image of an interval.

2. At each vertex of τ there is a well-defined tangent line; the data of a vertex with its tangent line is called a *switch*. The (half-)edges meeting each switch are divided into two sets, one on each side of the switch.

3. Each edge of τ is labeled with a nonnegative integer called a *weight*. We denote the set of weights by ν, and we sometimes refer to ν as a *measure*.

4. The weights satisfy the *switch condition* at each switch: the sums of the weights on each side of the switch are equal to each other.

The pair (τ, ν) is called a *measured train track* for $f^2(c)$. The graph τ itself is called simply a *train track*. We also say that τ carries the isotopy class $f^2(c)$. In general, we say that a multicurve b is carried by the train track τ if τ can be obtained by performing the above zipping procedure on b.

If a measured train track τ carries an isotopy class of a multicurve b with weights ν, then we can reconstruct b from the pair (τ, ν) as follows. Replace each edge of τ of weight n with n parallel line segments. After doing this we see that the switch condition ensures that there is a well-defined way to glue the endpoints of all segments coming from the edges incident to that particular switch. The result is a finite union of disjoint simple closed curves, which one can check lies in the isotopy class b. We encourage the reader to perform this process for the curve $f^2(c)$ carried by the train track (τ, ν) given in Figure 15.3.

We also note that four of the weights on the edges of the measured train track (τ, ν) for $f^2(c)$ are redundant. Indeed, given the edges with weights 6

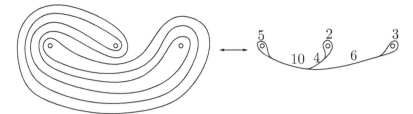

Figure 15.3 Converting $f^2(c)$ into a train track.

and 4, the weights of all the other edges are completely determined by the switch conditions.

15.1.2 THE LINEAR ALGEBRA OF TRAIN TRACKS

The power of the above setup is that it is easy to keep track of the isotopy class of any simple closed curve carried by τ under any number of iterations of f. To see how to do this, consider the train track τ endowed with an arbitrary measure ν. Such a measure is given by two weights x and y, as shown in Figure 15.4. In these (x, y)-coordinates the curves c, $f(c)$, and $f^2(c)$ are given by $(0, 2)$, $(2, 2)$, and $(6, 4)$, respectively.

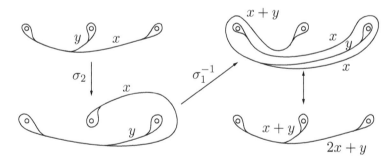

Figure 15.4 Applying the map f directly to the train track.

We now apply f to the measured train track (τ, ν) directly. On the image $f(\tau, \nu)$, we then perform the pinching (or zipping) process as described above: homotope parallel edges of $f(\tau, \nu)$ together, keeping track of the sum of the weights of the edges that are homotoped to each single edge. This process is illustrated in Figure 15.4. The result will clearly be another train track. In this case something very special happens: the resulting measured train track has the same underlying train track τ but with different weights! See Figure 15.4.

We describe this situation by saying that $f(\tau, \nu)$ is carried by τ. The weights transform as follows: the edge of (τ, ν) with weight x has weight $2x + y$ in $f(\tau, \nu)$, while the edge with weight y has weight $x + y$ in $f(\tau, \nu)$. As noted above, the weights on all other edges are determined by the weights on these two edges. Thus the homeomorphism f acts on the original measured train track (τ, ν) by changing the edge weights in a linear way. This action can therefore be completely described by the *train track matrix* for f:

$$
M = \begin{pmatrix} 2 & 1 \\ 1 & 1 \end{pmatrix}.
$$

This gives us a quick and simple way to encode the action of iterates of f not only on c but also on any isotopy class of simple closed curves that is carried by τ, as follows. If b is an isotopy class that corresponds to the train track τ with weights $(x, y) = (x_0, y_0)$, then $f^n(b)$ is the isotopy class of simple closed curves corresponding to τ with weights (x_n, y_n) given by

$$
\begin{pmatrix} x_n \\ y_n \end{pmatrix} = \begin{pmatrix} 2 & 1 \\ 1 & 1 \end{pmatrix}^n \begin{pmatrix} x_0 \\ y_0 \end{pmatrix}.
$$

The weights (x_n, y_n) then determine a measured train track (τ, ν_n) via the switch conditions on τ. From (τ, ν_n) we can directly build the simple closed curve $f^n(b)$, as above. The image $f^n(c)$ is the special case obtained by plugging in $(x_0, y_0) = (0, 2)$.

The train track matrix M has eigenvalues

$$
\lambda = \frac{3 + \sqrt{5}}{2} \qquad \text{and} \qquad \lambda^{-1} = \frac{3 - \sqrt{5}}{2}
$$

with eigenvectors

$$
v_\lambda = \begin{pmatrix} \frac{1+\sqrt{5}}{2} \\ 1 \end{pmatrix} \qquad \text{and} \qquad v_{\lambda^{-1}} = \begin{pmatrix} \frac{1-\sqrt{5}}{2} \\ 1 \end{pmatrix}.
$$

Since $\frac{1-\sqrt{5}}{2}$ is negative, the eigenvector $v_{\lambda^{-1}}$ does not correspond to a measured train track. The eigenvalue $\lambda > 1$ with its eigenvector v_λ is the geometrically meaningful eigenvector for us. It tells us, for example, that the norm of the vector (x_n, y_n) grows like λ^n as n tends to infinity.

Since any essential simple closed curve in $S_{0,4}$ intersects τ, we can see, for example, that the geometric intersection number of any isotopy class of simple closed curves b with $f^n(c)$ grows like $\lambda^n i(b, c)$, as promised by Theorem 14.24. Indeed, our discussion here suggests a proof of that theorem. Note that it is the case that the eigenvalue $\lambda = \frac{3+\sqrt{5}}{2}$ is the stretch factor of f; in fact, f is the image of the $\left(\begin{smallmatrix} 2 & 1 \\ 1 & 1 \end{smallmatrix}\right)$ map of T^2 under the hyperelliptic involution, as in Section 9.4.

15.1.3 FOUR TRAIN TRACKS SUFFICE

We now have a detailed picture of how f acts on every isotopy class of simple closed curves carried by τ. However, not every isotopy class of simple closed curves in $S_{0,4}$ is carried by τ, even varying (x, y) arbitrarily. Consider, for example, the isotopy class of a convex simple closed curve surrounding the second and third punctures in the plane. How can we analyze the action of f on such curves?

The answer is simple: every simple closed curve in $S_{0,4}$ is clearly carried by some train track. What is more, we claim that there exist four train tracks $\tau_1, \tau_2, \tau_3, \tau_4$ in $S_{0,4}$ with the property that every simple closed curve in $S_{0,4}$ is carried by one of the τ_i. We now prove this claim.

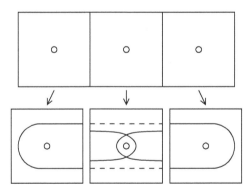

Figure 15.5 Any simple closed curve in $S_{0,4}$ can be broken up into canonical pieces.

Up to isotopy, any simple closed curve γ in $S_{0,4}$ can be drawn inside the union of the three squares shown at the top of Figure 15.5. Up to further isotopy, we can assume that γ does not form any bigons with the vertical edges of the three squares. At this point, a connected component of the intersection of γ with one of the squares is one of the six types of arcs shown at the bottom of Figure 15.5. Since γ is essential, it cannot use both types of dashed arcs, for otherwise γ would be isotopic to the nonessential curve that surrounds all three punctures. Since the other two types of arcs in the middle square intersect, γ can use at most one of those.

We therefore see that there are four types of simple closed curves in $S_{0,4}$, depending on which of each of the two pairs of arcs they use in the middle square. This information is exactly the same as saying that any simple closed curve in $S_{0,4}$ is carried by one of the train tracks shown in Figure 15.6.

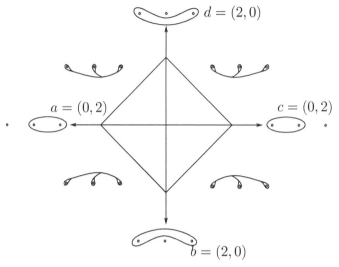

Figure 15.6 $\mathcal{PMF}$ for $S_{0,4}$.

The four train tracks $\tau_1, \tau_2, \tau_3, \tau_4$ in Figure 15.6 give four coordinate charts on the set of isotopy classes of simple closed curves in $S_{0,4}$. Each coordinate patch corresponding to a train track τ_i is given by the weights (x, y) of two chosen edges of τ_i. If we allow the coordinates x and y to be arbitrary nonnegative real numbers, then we obtain for each τ_i a closed quadrant in $\mathbb{R}^2$. Arbitrary points in this quadrant are measured train tracks. Notice that in some cases we can put weights on two different train tracks and (after deleting train track edges with weight zero) obtain equivalent measured train tracks. By identifying these measured train tracks, we obtain an identification of the four quadrants along their edges. The resulting space is homeomorphic to $\mathbb{R}^2$; see Figure 15.6. The integral points in this $\mathbb{R}^2$ correspond to isotopy classes of multicurves in $S_{0,4}$.

15.1.4 THE ACTION ON $\mathcal{PMF}$

The action of f on the isotopy classes of simple closed curves in $S_{0,4}$ induces an action on the integer points of $\mathbb{R}^2$. This action extends to a homeomorphism of $\mathbb{R}^2$; indeed, just as every simple closed curve in $S_{0,4}$ is carried by one of the τ_i, every measured train track is equivalent to another measured train track that is carried by one of the τ_i (see below). Since this homeomorphism commutes with the multiplicative action of $\mathbb{R}_+$ on $\mathbb{R}^2$, it induces a homeomorphism of the space of rays in $\mathbb{R}^2$, endowed with the appropriate topology. This space of rays is homeomorphic to a circle, which we denote by $\mathcal{PMF}$ (the notation will be explained in the next section); see Figure 15.6. Note that the rational points of $\mathcal{PMF}$ represented by pairs of integers (p, q) correspond to isotopy classes of multicurves in $S_{0,4}$.

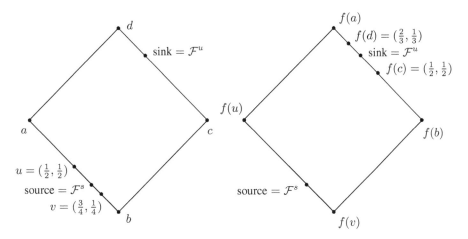

Figure 15.7 The action of f on $\mathcal{PMF}$ for $S_{0,4}$.

Figure 15.7 gives a partial depiction of the action of f on $\mathcal{PMF}$ using the coordinates and notation established in Figure 15.6. It turns out that f acts on $\mathcal{PMF}$ with source-sink dynamics, with the two fixed points of f in $\mathcal{PMF}$ corresponding to the stable and unstable foliations for the pseudo-Anosov f on $S_{0,4}$. Let us explain how this works for the fixed point that is a sink, corresponding to the unstable foliation $\mathcal{F}^u$ for f.

We saw in the last section that f fixes the train track τ corresponding to the upper-right quadrant of $\mathcal{PMF}$. We also saw that f acts on the weights of τ by the matrix M and that the unique (up to scale) positive eigenvector for this action is $\left(\frac{1+\sqrt{5}}{2}, 1\right)$, with eigenvalue $\frac{3+\sqrt{5}}{2}$. What this means for the action of f on $\mathcal{PMF}$ is that f leaves the upper-right quadrant of $\mathbb{R}^2$ invariant and acts on it via M. The eigenvector $\left(\frac{1+\sqrt{5}}{2}, 1\right)$ gives a fixed point for the action of f on $\mathcal{PMF}$. The fixed point is represented by the measured train track (τ, ν_u) with measure ν_u given by the weights $(x, y) = \left(\frac{1+\sqrt{5}}{2}, 1\right)$ and

$$f(\tau, \nu_u) = (\tau, \lambda\nu_u).$$

What is more, the fixed point represented by (τ, ν_u) corresponds to a projective class of measured foliations on $S_{0,4}$, invariant under the action of the pseudo-Anosov homeomorphism in the homotopy class of f. To construct the foliation, one first uses the weights $(x, y) = \left(\frac{1+\sqrt{5}}{2}, 1\right)$ to find the weights on all other branches of τ. For each edge of weight $r > 0$, build a rectangle of width 1 and height r with measured foliations given by the 1-forms dx and dy. The switch conditions imply that these rectangles can be glued together in a consistent way to give a foliation of $S_{0,4}$ minus four once-marked disks; see Figure 15.8. Collapsing each disk to its marked point gives a measured foliation on $S_{0,4}$. Since $f(\tau, \nu_u) = (\tau, \lambda\nu_u)$, this measured foliation is indeed the unstable foliation $\mathcal{F}^u$ for the pseudo-Anosov homeomorphism representing f.

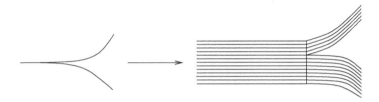

Figure 15.8 The first step in converting a train track into a foliation.

The fixed point of f in $\mathcal{PMF}$ corresponding to the source can be described in a similar way. It is the projective train track given by $\left(\frac{1+\sqrt{5}}{2}, 1\right)$ in the coordinates of the bottom-left quadrant in Figure 15.6. If we allow the (x, y)-coordinates in this quadrant to vary in such a way that $1 < x/y < 2$, then we obtain an open interval that is sent by f into the bottom-left quadrant. On this coordinate patch one can check that f acts by the transition matrix

$$\begin{pmatrix} 1 & -1 \\ -1 & 2 \end{pmatrix}.$$

One of the two eigenvectors for this matrix is positive; its corresponding eigenvalue is $\lambda^{-1} = \frac{3-\sqrt{5}}{2}$. The (x, y)-coordinates of this positive eigenvector determine a measured train track invariant by f. This in turn determines a measured foliation whose measure is multiplied by λ^{-1} under f. This foliation is nothing but the stable foliation $\mathcal{F}^s$ for f.

The reason we call the fixed point $\mathcal{F}^u \in \mathcal{PMF}$ a source for f and $\mathcal{F}^s \in \mathcal{PMF}$ a sink for f is that, for any point $z \in \mathcal{PMF} - \mathcal{F}^s$ and any neighborhood U of $\mathcal{F}^u$, we have that $f^n(z) \in U$ for large enough n. As a result, every point of $\mathcal{PMF} - \mathcal{F}^s$ is repelled from the source $\mathcal{F}^s$. We remark that, while a sink is considered to be a stable point in a dynamical system, the reason that the sink is the unstable foliation is due to the nature of the action of f on the surface; see Section 13.2.

While $\mathcal{PMF}$ has a seemingly natural structure as a simplicial complex with four edges, f does not act simplicially on $\mathcal{PMF}$. Indeed, the source-sink dynamics of f on $\mathcal{PMF}$ precludes this. Consider, for example, the action of f on the edge in $\mathcal{PMF}$ corresponding to the lower-left quadrant in $\mathbb{R}^2$. In the notation of Figure 15.7, we have that $f(u) = a$ and that f fixes the source point $\mathcal{F}^s$. Thus, using the projective coordinates of this quadrant, the point $f(1, 1 + \epsilon)$ for ϵ small enough lies in the upper-left quadrant and $f(1 + \epsilon, 1)$ remains in the lower-left quadrant.

It may seem counterintuitive that different weights on the same train track can lead to combinatorially different tracks after applying f, but the sequences of pictures in Figures 15.9 and 15.10 explain this phenomenon.

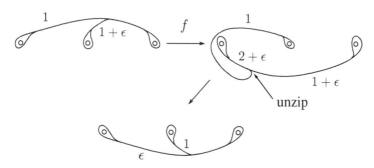

Figure 15.9 Finding the image of the point $(1, 1 + \epsilon)$ from the bottom-left quadrant.

In the calculations of $f(1, 1 + \epsilon)$ and $f(1 + \epsilon, 1)$, we are forced to use an "unzipping" procedure. If we read the arrows backward we see that this is just the opposite of the zipping procedure used earlier. The key point as to why we get different combinatorial train tracks in Figures 15.9 and 15.10 is that, when we unzip, we are forced to "peel off" the track of smaller weight. There is not enough track to peel off the one of larger weight, and so we get different unzipping sequences and hence different combinatorial types of tracks at the end.

The above description gives us a fairly thorough understanding of f and its action on all isotopy classes of simple closed curves in $S_{0,4}$, as well as on all measured foliations on $S_{0,4}$. Continuing this line of reasoning, one can prove that the source-sink dynamics do actually hold. In particular, one can check that for any isotopy class c of simple closed curves, the curves

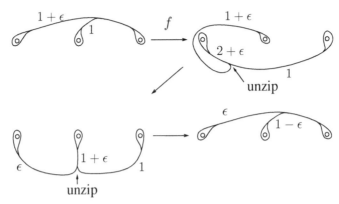

Figure 15.10 Finding the image of the point $(1 + \epsilon, 1)$ from the bottom-left quadrant. We warn the reader that it requires a clever isotopy to realize the second arrow.

$f^n(c)$ converge projectively to the projective class of the unstable measured foliation $\mathcal{F}^u$.

15.2 A SKETCH OF THE GENERAL THEORY

There are several big ideas to take away from the example in Section 15.1. The first idea is that one can understand what a homeomorphism "does" to a surface by looking at what it does to a single simple closed curve under iteration. This is analogous to the fact that one can approximate the eigenvector for a Perron–Frobenius matrix M by iterating M on almost any vector v; indeed, $M^n(v)$ converges exponentially quickly to the Perron–Frobenius eigenspace of M.

Thurston's remarkable discovery is that this analogy can be made into a reality. For any pseudo-Anosov $f \in \mathrm{Mod}(S)$, one can find an invariant train track τ for f. One can then compute the associated train track matrix M. The matrix M is Perron–Frobenius and so has a unique largest eigenvalue $\lambda > 1$ with positive eigenvector v. The eigenvalue λ is precisely the stretch factor of f. The eigenvector v specifies a measure on τ from which the unstable foliation for f can be built directly, just as is explained in Section 15.1. The invariant train track τ for f is thus a combinatorial tool that converts an a priori nonlinear problem where, for example, iteration is difficult to understand, into a linear problem about which we have essentially complete knowledge.

What is more, the analysis carried out in Section 15.1 can be used to give a proof of the Nielsen–Thurston classification. Thurston's original proof of the Nielsen–Thurston classification was actually phrased in terms of measured foliations, not train tracks. Train tracks are a technological innovation of Thurston's that appeared after his original proof. Since train tracks are combinatorial objects, they are easier to work with in practice than the more abstract measured foliations. On the other hand, some aspects of the general theory are more easily dealt with in the context of measured foliations. As such, we will present Thurston's original approach, which uses measured foliations.

We already explained how to convert a measured train track into a measured foliation, and it is not too hard to see that any measured foliation can be pinched down to a measured train track. Thus in some sense the two theories are equivalent. Indeed, it is possible to present the entire proof of the Nielsen–Thurston classification in the language of measured train tracks; see the book by Penner and Harer for the details [176].

15.2.1 Thurston's Original Proof

In this subsection we present an outline of Thurston's original proof of the Nielsen–Thurston classification (Theorem 13.2). We start with a broad overview and then proceed to explain more of the details. The full details of this approach are given in the book FLP [61].

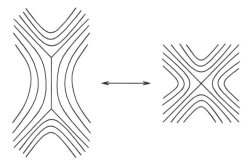

Figure 15.11 A Whitehead move.

Let $g \geq 2$ and let $S = S_g$. One space associated to the surface S is the Teichmüller space $\mathrm{Teich}(S)$. Another important space associated to S is the *measured foliation space* $\mathcal{MF}(S)$, which is the space of equivalence classes of measured foliations, where the equivalence is generated by isotopy and by Whitehead moves; see Figure 15.11.

Let $\mathcal{S}$ denote the set of isotopy classes of essential simple closed curves in S. One key idea in the approach we are describing is that by taking lengths/measures of curves, both $\mathrm{Teich}(S)$ and $\mathcal{MF}(S)$ map disjointly and injectively into $\mathbb{R}_{\geq 0}^{\mathcal{S}} - 0$, the space of nonzero functions $\mathcal{S} \to \mathbb{R}_{\geq 0}$.

There is a natural action of $\mathbb{R}_+$ on $\mathbb{R}_{\geq 0}^{\mathcal{S}} - 0$. Taking the quotient of $\mathbb{R}_{\geq 0}^{\mathcal{S}} - 0$ by this action gives a projective space $P(\mathbb{R}^{\mathcal{S}})$. The image of $\mathcal{MF}(S)$ in $P(\mathbb{R}^{\mathcal{S}})$ is denoted by $\mathcal{PMF}(S)$. It is homeomorphic to a sphere of dimension $6g - 7$. We will also see that the projectivization map restricted to $\mathrm{Teich}(S)$ is a homeomorphism onto its image; we will also denote this image in $P(\mathbb{R}^{\mathcal{S}})$ by $\mathrm{Teich}(S)$.

What is more, the subspaces $\mathrm{Teich}(S)$ and $\mathcal{PMF}(S)$ of $P(\mathbb{R}^{\mathcal{S}})$ are disjoint, and their union $\mathrm{Teich}(S) \cup \mathcal{PMF}(S)$ has the topology of a closed ball of dimension $6g - 6$. Each element of $\mathrm{Mod}(S)$ acts continuously on this ball, and so the Brouwer fixed point theorem implies the existence of a fixed point. The Nielsen–Thurston classification is then obtained by analyzing the various possibilities for this fixed point.

We now explain more of the details of this idea. As much as possible, we give references to the appropriate points in FLP.

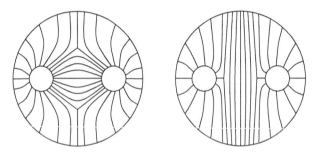

Figure 15.12 Two types of good measured foliations on a pair of pants. On the left-hand side the measures of the three boundary components satisfy the triangle inequality. On the right-hand side the measure of the outer curve is greater than the sum of the other two measures.

Step 1 (Measured foliations on a pair of pants). Just as we showed $\mathrm{Teich}(P) \approx \mathbb{R}^3$, we now explain that there is a certain subset of $\mathcal{MF}(P)$ that is homeomorphic to $\mathbb{R}_{\geq 0}^3 - 0$. Let $\mathcal{MF}_0(P)$ denote the subset of $\mathcal{MF}(P)$ represented by foliations $\mathcal{F}$ where no boundary component of P is a nonsingular leaf. We claim that

$$\mathcal{MF}_0(P) \approx \mathbb{R}_{\geq 0}^3 - 0$$

(see [61, Exposé 6, Théorème II.4]). Indeed, given any nonzero $(s, t, u) \in \mathbb{R}_{\geq 0}^3$, we can find a unique element of $\mathcal{MF}_0(P)$ where the measures of the three boundary components of P are s, t, and u. Up to Whitehead equivalence and isotopy, there are two different pictures, corresponding to whether or not the triple (s, t, u) satisfies the triangle inequality. See Figure 15.12. In both pictures we see that the foliation is obtained by gluing together three horizontally foliated rectangles. To obtain pictures of all elements of $\mathcal{MF}_0(P)$ from these two, one must allow for permutations of the boundary components of P and also allow the transverse measures of one or more rectangles to degenerate to zero.

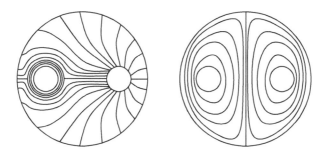

Figure 15.13 Two measured foliations on a pair of pants. The foliation on the left-hand side is obtained from the foliation on the left-hand side of Figure 15.12 by allowing two of the rectangles there to degenerate, and then adjoining a smoothly foliated annulus to one of the boundary components of P. The foliation on the right-hand side is obtained by enlarging two boundary components of P.

In order to describe all of $\mathcal{MF}(P)$, we need to consider foliations where one or more boundary components of P are nonsingular leaves. One way to obtain such a foliation is to start with a foliation in $\mathcal{MF}_0(P)$ and adjoin an annulus foliated by parallel circles to any of the boundary components of P that are closed singular leaves (left-hand side of Figure 15.13). Or we can start with the empty foliation of P and then enlarge one or more boundary components in the sense of Section 11.2 (right-hand side of Figure 15.13). It turns out that these two constructions account for all of $\mathcal{MF}(P)$ [61, Exposé 6, Proposition II.5].

Step 2 (Global coordinates for $\mathcal{MF}(S)$). The Fenchel–Nielsen coordinates on $\mathrm{Teich}(S)$ involve two sets of parameters: the length parameters determine the isometry type of each pair of pants; the twist parameters determine how the pants are glued together. We employ an analogous approach here. In step 1 we described parameters for $\mathcal{MF}(P)$, and so now we need to give twist parameters to encode how foliated pairs of pants can be glued together.

Fix a pants decomposition $\{\gamma_i\}$ of S. For each γ_i, choose an annulus A_i in S with γ_i as one of its boundary components. Cut each annulus A_i into two triangles using arcs δ_i and ϵ_i; see Figure 15.14. Each triangle is bounded by γ_i, δ_i, and ϵ_i.

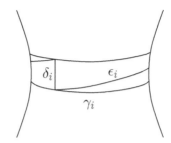

Figure 15.14 The arcs γ_i, δ_i, and ϵ_i used to define coordinates on $\mathcal{MF}(S)$.

Let $[(\mathcal{F}, \mu)] \in \mathcal{MF}(S)$. We can put $(\mathcal{F}, \mu)$ into normal form with respect to the γ_i, δ_i, and ϵ_i [61, Exposé 6, Section IV]. The idea of a normal form is that on each pair of pants the normal form should restrict to one of the elements of $\mathcal{MF}(P)$ that we already understand.

Once we have the normal form, we can define a map

$$\Theta : \mathcal{MF}(S) \rightarrow \mathbb{R}_{\geq 0}^{9g-9} - 0$$

that records the measures $(\ell_i, \theta_i, \theta_i')$ of each γ_i, δ_i, and ϵ_i with respect to the normal form of any given $[(\mathcal{F}, \mu)] \in \mathcal{MF}(S)$. The ℓ_i are thought of as length parameters and the θ_i and θ_i' as twist parameters. The map Θ is a homeomorphism onto its image. We describe the image of Θ in step 3.

One subtlety is that, in order to obtain the normal form of a foliation, we may have to modify (unglue) our foliation so that it does not cover the whole surface anymore. As a simple example, if $(\mathcal{F}, \mu)$ is a measured foliation obtained by enlarging one of the γ_i, then the normal form of $(\mathcal{F}, \mu)$ will be supported on an annular neighborhood of γ_i and will consist of nonsingular leaves parallel to γ_i. As such, all of the length parameters of this $(\mathcal{F}, \mu)$ are zero, and all of the twist parameters except the ith are zero.

Step 3 ($\mathcal{PMF}(S)$ is a sphere). We have just parameterized $\mathcal{MF}(S)$ with $9g - 9$ numbers. So how do we end up with $\mathcal{PMF}(S) \approx S^{6g-7}$? Well, since, for each i, the arcs γ_i, δ_i, and ϵ_i bound a (null homotopic) triangle T_i in S, the parameters $(\ell_i, \theta_i, \theta_i')$ satisfy a degenerate triangle inequality. That is, one of the three numbers is the sum of the other two. This is because any leaf of a foliation entering T_i along one edge must exit T_i along some other edge.

The points $(s_1, t_1, u_1, \ldots, s_{3g-3}, t_{3g-3}, u_{3g-3})$ in $\mathbb{R}_{\geq 0}^{9g-9}$ where each triple (s_i, t_i, u_i) satisfies a degenerate triangle inequality form a cone B that is homeomorphic to $\mathbb{R}^{6g-6}$. Note, for instance, that the set of points in $\mathbb{R}^3_{\geq 0}$ that satisfy a degenerate triangle inequality is a cone homeomorphic to $\mathbb{R}^2$.

We claim that the image of Θ is the entire punctured cone $B - 0$. Indeed, given a point of $B - 0$ we can directly construct a measured foliation with the specified length and twist parameters uniquely up to equivalence. For example, suppose we are given a point of $B - 0$ where each of the length coordinates is nonzero. In this case we can foliate each pair of pants with the corresponding element of $\mathcal{MF}_0(S)$. The twist parameters then tell us how to glue the foliations along the curves of the pants decomposition. For details, see [61, Exposé 6, Section V].

We thus have $\mathcal{MF}(S) \approx B - 0$, and so $\mathcal{PMF}(S) \approx S^{6g-7}$.

Step 4 ($\mathrm{Teich}(S) \cup \mathcal{PMF}(S)$ is a closed ball). We now explain how both $\mathrm{Teich}(S)$ and $\mathcal{PMF}(S)$ naturally embed in $P(\mathbb{R}^{\mathcal{S}})$. Define a map

$$\ell : \mathrm{Teich}(S) \to P(\mathbb{R}^{\mathcal{S}})$$

as the composition of the map $\mathrm{Teich}(S) \to (\mathbb{R}^{\mathcal{S}} - 0)$ given by $\mathcal{X} \mapsto \ell_{\mathcal{X}}(\cdot)$ with the projectivization map. The $9g - 9$ theorem (Theorem 10.7) implies that the map $\mathcal{X} \mapsto \ell_{\mathcal{X}}(\cdot)$ is injective. No two points in $\mathrm{Teich}(S)$ can have length functions that differ by a multiplicative factor [61, Exposé 7, Proposition 6]. Thus the image of $\mathrm{Teich}(S)$ in $\mathbb{R}^{\mathcal{S}} - 0$ intersects each $\mathbb{R}_+$-orbit in a single point. This proves that ℓ is injective.

Recall the notion of the geometric intersection number $I((\mathcal{F}, \mu), c)$ of a measured foliation with an isotopy class of simple closed curves (see Section 14.5). This gives a well-defined map $\mathcal{MF}(S) \to \mathbb{R}^{\mathcal{S}}$ via $[(\mathcal{F}, \mu)] \mapsto I((\mathcal{F}, \mu), \cdot)$. This map is injective by the above description of $\mathcal{MF}(S)$ as the punctured cone $B - 0$, so that the induced map

$$I : \mathcal{PMF}(S) \to P(\mathbb{R}^{\mathcal{S}})$$

is injective.

By tracing through the definitions one can check that the injective maps ℓ and I are in fact continuous and are homeomorphisms onto their images. We will henceforth identify $\mathrm{Teich}(S)$ and $\mathcal{PMF}(S)$ with their images in $P(\mathbb{R}^{\mathcal{S}})$.

We claim that $\mathrm{Teich}(S)$ and $\mathcal{PMF}(S)$ are disjoint in $P(\mathbb{R}^{\mathcal{S}})$. By discreteness of the raw length spectrum of a hyperbolic surface (Lemma 12.4), every point in $\mathrm{Teich}(S)$ has a shortest simple closed curve. On the other hand, given any measured foliation on S, one can use Poincaré recurrence for foliations (Theorem 14.14) to construct simple closed curves that have arbitrarily small measure with respect to that foliation [61, Exposé 8, Proposition I.1]. In other words, the length spectrum for a measured foliation, is not bounded away from zero. This is enough to distinguish points of $\mathrm{Teich}(S)$ from points of $\mathcal{PMF}(S)$ in $P(\mathbb{R}^S)$.

We now claim that the union $\mathrm{Teich}(S) \cup \mathcal{PMF}(S)$ in $P(\mathbb{R}^S)$ can be naturally topologized so that it is homeomorphic to a closed ball of dimension $6g - 6$. Let $\pi : \mathbb{R}^S_{\geq 0} \to P(\mathbb{R}^S_{\geq 0})$ denote the projectivization map. The open sets of the union $\mathrm{Teich}(S) \cup \mathcal{PMF}(S)$ as a subspace of $P(\mathbb{R}^S)$ are the open sets of $\mathrm{Teich}(S)$ together with sets of the form

$$(\mathrm{Teich}(S) \cap \pi^{-1}(U)) \cup (\mathcal{PMF}(S) \cap U),$$

where U is an open set of $P(\mathbb{R}^S_{\geq 0})$; see [61, Exposé 8, Théorème III.3]. The key to showing that $\mathrm{Teich}(S) \cup \mathcal{PMF}(S)$ is a closed ball is to show that it is a manifold with boundary and then to apply the generalized Schönflies theorem. The closed ball $\mathrm{Teich}(S) \cup \mathcal{PMF}(S)$ is called the *Thurston compactification* of $\mathrm{Teich}(S)$.

As a demonstration of the topology on the Thurston compactification of $\mathrm{Teich}(S)$, consider a sequence of points $\mathcal{X}_n$ in $\mathrm{Teich}(S)$ where $\ell_{\mathcal{X}_n}(\alpha) \to 0$ for some simple closed curve α in S. Then $\mathcal{X}_n$ limits to a point of $\mathcal{PMF}(S)$ corresponding to a (projective class of) measured foliation of S containing α as a closed leaf.

Step 5 (Applying Brouwer). In Section 12.1, we explained the properly discontinuous action of $\mathrm{Mod}(S)$ on $\mathrm{Teich}(S)$. The group $\mathrm{Mod}(S)$ also acts by homeomorphisms on $\mathcal{MF}(S)$: the action of $\mathrm{Homeo}^+(S)$ on $\mathcal{MF}(S)$ given by

$$\phi \cdot [(\mathcal{F}, \mu)] = [(\phi(\mathcal{F}), \phi_*(\mu))]$$

factors through an action of $\mathrm{Mod}(S)$. In contrast to its action on $\mathrm{Teich}(S)$, the action of $\mathrm{Mod}(S)$ on $\mathcal{MF}(S)$ is not properly discontinuous. This action does commute with the action of $\mathbb{R}_+$ on $\mathcal{MF}(S)$ by scaling measures, and so it induces an action of $\mathrm{Mod}(S)$ on $\mathcal{PMF}(S)$ by homeomorphisms. This action is also far from properly discontinuous; indeed, it has dense orbits.

One can check that for each $f \in \mathrm{Mod}(S)$ the homeomorphisms just discussed are compatible in the sense that the map of the closed ball $\mathrm{Teich}(S) \cup \mathcal{PMF}(S)$ to itself induced by f is a homeomorphism, giving us an action of $\mathrm{Mod}(S)$ on $\mathrm{Teich}(S) \cup \mathcal{PMF}(S)$ by homeomorphisms. One should think of this action in analogy with a discrete group of isometries acting by homeomorphisms on the visual compactification of n-dimensional hyperbolic space: the action on the interior is properly discontinuous, while the action on the boundary has dense orbits.

We now have that any $f \in \mathrm{Mod}(S)$ induces a self-homeomorphism (which we also call f) of the closed ball $\mathrm{Teich}(S) \cup \mathcal{PMF}(S)$ of dimension $6g - 6$. We can thus apply the Brouwer fixed point theorem to conclude that f has a fixed point in $\mathrm{Teich}(S) \cup \mathcal{PMF}(S)$. This means either that f fixes a point of $\mathrm{Teich}(S)$ or there exists $[(\mathcal{F}, \mu)] \in \mathcal{MF}(S)$ so that $f \cdot [(\mathcal{F}, \mu)] = [(\mathcal{F}, \lambda\mu)]$ for some $\lambda \in \mathbb{R}_+$ (here brackets denote the equivalence generated by isotopy and by Whitehead moves, but not by the $\mathbb{R}_+$-action).

Step 6 (Analyzing the fixed point). We call a measured foliation *arational* if it does not contain any closed leaves. We have the following cases for the fixed point of f:

1. $f \cdot \mathcal{X} = \mathcal{X}$, where $\mathcal{X} \in \mathrm{Teich}(S)$

2. $f \cdot [(\mathcal{F}, \mu)] = [(\mathcal{F}, \lambda\mu)]$, where $(\mathcal{F}, \mu)$ is not arational

3. $f \cdot [(\mathcal{F}, \mu)] = [(\mathcal{F}, \lambda\mu)]$, where $(\mathcal{F}, \mu)$ is arational and $\lambda = 1$

4. $f \cdot [(\mathcal{F}, \mu)] = [(\mathcal{F}, \lambda\mu)]$, where $(\mathcal{F}, \mu)$ is arational and $\lambda > 1$.

We have already seen in Chapter 13 that in case 1 the mapping class f is periodic. In case 3 we deduce that f permutes the finite collection of rectangles in some rectangle decomposition of S induced by $\mathcal{F}$, and so f must again be periodic. In case 2 the finite number of homotopy classes of closed leaves of $\mathcal{F}$ must be permuted by f, and so f is reducible. In case 4 one can build a Markov partition (see below) for f and use it to find a unique measured foliation that is transverse to $\mathcal{F}$ and that is also projectively invariant by f. This proves that f is pseudo-Anosov. This last step is the most technically involved part of Thurston's proof. See Exposé 9 of FLP for the details.

The Thurston compactification for the torus. The Thurston compactification of $\mathrm{Teich}(T^2)$ can be described quite explicitly as follows. Let α, β, and γ denote the $(1, 0)$, $(0, 1)$, and $(1, 1)$ curves in T^2. Recording the lengths of α, β, and γ gives an injective map $\mathrm{Teich}(T^2) \rightarrow \mathbb{R}_{\geq 0}^3$. Each point of $P(\mathbb{R}_{\geq 0}^3)$ can be represented by a unique point in the plane $x + y + z = 1$. For any flat metric on T^2, there are representatives of α, β, and γ that form a Euclidean triangle, and so their lengths satisfy the triangle inequality. Thus the image of $\mathrm{Teich}(T^2)$ in the plane $x + y + z = 1$ is the open triangular region T consisting of positive points that satisfy the triangle inequality.

Taking the measures of α, β, and γ gives an injective map $\mathcal{MF}(T^2) \to \mathbb{R}^3_{\geq 0}$. As in step 3, the image of this map is the set of points in $\mathbb{R}^3_{\geq 0}$ that satisfy a degenerate triangle inequality. This set is precisely the cone on ∂T, punctured at the origin. By projectivizing, we can see concretely how $\mathcal{PMF}(T^2) \cup \text{Teich}(T^2)$ is homeomorphic to a closed disk; see Figure 15.15 and [61, Exposé 1]. We remark that, as we move toward the boundary of T, the corresponding points of $\text{Teich}(T^2)$ (before projectivization) move further and further from the origin.

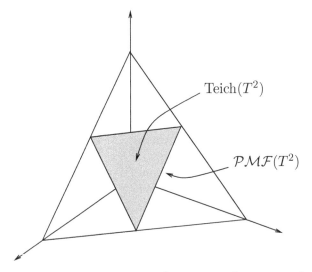

Figure 15.15 The closed ball $\text{Teich}(T^2) \cup \mathcal{PMF}(T^2)$ sitting in $\mathbb{R}^3_{\geq 0}$.

15.3 MARKOV PARTITIONS

It has already been mentioned that part of Thurston's analysis is the construction of a Markov partition for a pseudo-Anosov mapping class. As we explained above, the example in Section 15.1 is the image of

$$A = \begin{pmatrix} 2 & 1 \\ 1 & 1 \end{pmatrix} \in \text{SL}(2, \mathbb{Z}) \approx \text{Mod}(T^2)$$

under the homomorphism $\text{Mod}(T^2) \to \text{Mod}(S_{0,4})$ induced by the twofold branched cover $T^2 \to S_{0,4}$. In this section we illustrate the idea by constructing the Markov partition for this simple example.

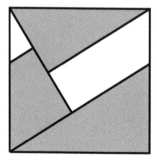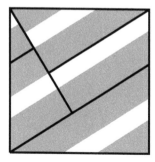

Figure 15.16 An Anosov map of the torus.

Again, the unstable and stable foliations $\mathcal{F}^u$ and $\mathcal{F}^s$ for A are the projections to T^2 of the foliations of $\mathbb{R}^2$ by lines parallel to the (irrational) eigenspaces for the eigenvalues $\lambda > 1$ and λ^{-1} of A. Choose a small subarc τ of a leaf of $\mathcal{F}^s$. Similar to the construction in Section 14.3, we can use τ, $\mathcal{F}^u$, and $\mathcal{F}^s$ to construct a rectangle decomposition of T^2 adapted to $\mathcal{F}^u$ and $\mathcal{F}^s$. The situation here is slightly simpler since $\mathcal{F}^u$ is orientable and nonsingular.

First we subdivide τ along all *backward* images of endpoints of τ. Then each segment of the subdivision gives a rectangle in T^2 obtained by pushing that segment forward along $\mathcal{F}^u$. For one such example of a rectangle decomposition, see the left-hand side of Figure 15.16. In that figure we can locate the arc τ by taking the union of the vertical sides of the rectangles.

The linear map $A \in \mathrm{Homeo}^+(T^2)$ takes the picture on the left-hand side of Figure 15.16 to the picture on the right-hand side. We see that a lot of the structure is preserved. In particular, A takes rectangles to unions of subrectangles, A preserves the horizontal and vertical directions, and A takes sides of rectangles lying in $\mathcal{F}^s$ to other such sides.

Decomposing T^2 into its constituent rectangles gives a picture as at the top of Figure 15.17 (note the identifications).

Figure 15.17 taken all at once gives another view of the linear homeomorphism A. Here we can see how the combinatorial structure for $\mathcal{F}^u$ given by the rectangles translates into a purely combinatorial description of A: it stretches the gray rectangle twice over itself and once over the white rectangle, and it stretches the white rectangle once over each. Turning this

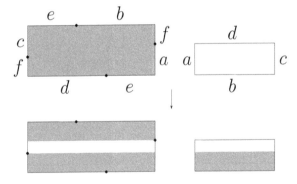

Figure 15.17 An Anosov map demystified(?).

information into a matrix in the obvious way gives a *transition matrix*:

$$\begin{pmatrix} 2 & 1 \\ 1 & 1 \end{pmatrix}.$$

Much (in this case, all) of the original information about the mapping class $[A] \in \mathrm{Mod}(T^2)$ is contained in the transition matrix. In general, we should not expect the transition matrix to bear any resemblance to the original Anosov map. In fact, we could have chosen a rectangle decomposition with more rectangles and gotten a larger matrix.

We remark that the two rectangles of the above Markov partition are similar. This is related to the fact that in this case the transition matrix is equal to its transpose; in general, the lengths and widths of the rectangles come from the transition matrix and its transpose, respectively.

As discussed above, one ingredient in Thurston's approach to the Nielsen–Thurston classification is that such a scheme as the one above is always possible. That is, given a pseudo-Anosov $f \in \mathrm{Mod}(S_g)$, there is on S_g a particular rectangle decomposition, called a *Markov partition*, so that f has a combinatorial description as above. Instead of cutting along a single stable arc, however, one typically needs to cut along several. The construction of Markov partitions is much more technically involved for pseudo-Anosov homeomorphisms than for linear Anosov homeomorphisms. But once a Markov partition is constructed, it is a powerful tool.

For example, from the transition matrix of a Markov partition for f, one can determine various properties of f. In particular, the stretch factor of f is the largest real eigenvalue of the transition matrix. This approach can be used quite easily to give proofs of Theorems 14.8 and 14.9 and one direction

of Theorem 14.10. The point is to show that any pseudo-Anosov mapping
class of a fixed surface has a transition matrix whose size (number of rows)
is uniformly bounded from above and use the fact that the set of eigenvalues
of integral $n \times n$ matrices is discrete. The other direction of Theorem 14.10
is by explicit construction.

In addition to the above, Theorems 14.16 and 14.24 and Corollary 14.25
can all be proved using the theory of Markov partitions. The idea for the
latter two is that, as we iterate a pseudo-Anosov mapping class, any curve
gets closer and closer to the horizontal foliation in each rectangle, and the
number of horizontal components in each rectangle grows like λ^n, where
here we are thinking of λ as the largest eigenvalue of the transition matrix.
We refer the reader to [61] for the details.

The theory of Markov partitions is closely related to the theory of train
tracks. We encourage the reader to find the train track hidden in Figure 15.17.

15.4 OTHER POINTS OF VIEW

There are other approaches to proving the Nielsen–Thurston classification.
One aspect we have not touched on is the theory of measured laminations.
A *geodesic lamination* in a hyperbolic surface S is a nonempty closed subset of S that is a disjoint union of simple (possibly infinite) geodesics in S.
It is also possible to endow a geodesic lamination with a transverse measure. The dictionary relating train tracks to foliations can be extended to
relate both of these objects to geodesic laminations. For instance, to obtain
a geodesic lamination from a foliation, we simply replace each leaf with its
corresponding geodesic (as determined by the endpoints of the lift of a leaf
to $\mathbb{H}^2$).

One can prove the Nielsen–Thurston classification using geodesic laminations. The Hausdorff metric gives a metric on the set of geodesic laminations in S_g. Since S_g is compact, the resulting topological space $\mathcal{L}(S_g)$ is
compact. As with the other approaches, for a given $f \in \mathrm{Mod}(S)$ we consider the sequence $f^n(c)$, where c is an isotopy class of simple closed curves
in S. Each $f^n(c)$ has a unique geodesic representative in S_g, which can be
regarded as a point in $\mathcal{L}(S)$. Since $\mathcal{L}(S)$ is compact, the sequence $(f^n(c))$
has a convergent subsequence. The limit L is the *unstable lamination* for
f. If L has closed leaves then f is reducible or finite order. Otherwise, f is
pseudo-Anosov. This point of view is discussed in the book by Casson and
Bleiler [44] and in the book by Calegari [43]. While this method quickly
finds an invariant lamination, one now needs to do some work in order to
find an invariant foliation. It is the finer structure of a projectively invari-

ant measured foliation that allowed us to prove most of the properties of pseudo-Anosov homeomorphisms.

Nielsen's original point of view was to look directly at the action of a mapping class on $\partial \mathbb{H}^2$. A streamlined version of his approach can be found in the paper by Handel and Thurston [81]. Bestvina and Handel introduced a combinatorial algorithm for finding a train track for a mapping class, and in particular for determining the Nielsen–Thurston type of a mapping class [17]. Also, the Markov partition for a pseudo-Anosov homeomorphism yields a great amount of dynamical information [61, Exposé 10]. Ivanov proved a number of structural theorems about $\mathrm{Mod}(S)$ using the dynamics of the action on the Thurston boundary [106].

The Nielsen–Thurston classification is the starting point for a number of active research directions. This is analogous to the fact that the Jordan canonical form for a matrix is one basic fact used in the vast study of Lie groups and other groups of matrices. Some places to find open problems in these and many related directions are Kirby's problem list [123], Bestvina's problem list [18], Mosher's problem list [163], and the book *Problems on Mapping Class Groups and Related Topics* [58].

Bibliography

[1] William Abikoff. *The real analytic theory of Teichmüller space*, volume 820 of *Lecture Notes in Mathematics*. Springer, Berlin, 1980.

[2] Lars Ahlfors and Lipman Bers. Riemann's mapping theorem for variable metrics. *Ann. of Math. (2)*, 72:385–404, 1960.

[3] Lars V. Ahlfors. On quasiconformal mappings. *J. Analyse Math.*, 3:1–58; correction, 207–208, 1954.

[4] M. A. Armstrong. The fundamental group of the orbit space of a discontinuous group. *Proc. Cambridge Philos. Soc.*, 64:299–301, 1968.

[5] Pierre Arnoux and Jean-Christophe Yoccoz. Construction de difféomorphismes pseudo-Anosov. *C. R. Acad. Sci. Paris Sér. I Math.*, 292(1):75–78, 1981.

[6] E. Artin. Theory of braids. *Ann. of Math. (2)*, 48:101–126, 1947.

[7] M. F. Atiyah. The signature of fibre-bundles. In *Global Analysis (Papers in Honor of K. Kodaira)*, pages 73–84. University of Tokyo Press, Tokyo, 1969.

[8] R. Baer. Kurventypen auf Flächen. *J. Reine Angew. Math.*, 156:231–246, 1927.

[9] R. Baer. Isotopie von Kurven auf orientierbaren, geschlossenen Flächen und ihr Zusammenhang mit der topologischen Deformation der Flächen. *J. Reine Angew. Math.*, 159:101–111, 1928.

[10] Walter L. Baily, Jr. On the moduli of Jacobian varieties. *Ann. of Math. (2)*, 71:303–314, 1960.

[11] Hyman Bass and Alexander Lubotzky. Automorphisms of groups and of schemes of finite type. *Israel J. Math.*, 44(1):1–22, 1983.

[12] Hyman Bass and Alexander Lubotzky. Linear-central filtrations on groups. In *The mathematical legacy of Wilhelm Magnus: groups,*

geometry and special functions (Brooklyn, NY, 1992), volume 169 of *Contemp. Math.*, pages 45–98. American Mathematical Society, Providence, RI, 1994.

[13] Alan F. Beardon. *The geometry of discrete groups*, volume 91 of *Graduate Texts in Mathematics*. Springer-Verlag, New York, 1995. Corrected reprint of the 1983 original.

[14] Lipman Bers. Quasiconformal mappings and Teichmüller's theorem. In *Analytic functions*, pages 89–119. Princeton University Press, Princeton, NJ, 1960.

[15] Lipman Bers. An extremal problem for quasiconformal mappings and a theorem by Thurston. *Acta Math.*, 141(1–2):73–98, 1978.

[16] Lipman Bers. An inequality for Riemann surfaces. In *Differential geometry and complex analysis*, pages 87–93. Springer, Berlin, 1985.

[17] M. Bestvina and M. Handel. Train-tracks for surface homeomorphisms. *Topology*, 34(1):109–140, 1995.

[18] Mladen Bestvina. Questions in geometric group theory. Available at http://www.math.utah.edu/~bestvina.

[19] Mladen Bestvina, Kenneth Bromberg, Koji Fujiwara, and Juan Souto. Shearing coordinates and convexity of length functions on Teichmueller space. arXiv:0902.0829, 2009.

[20] Mladen Bestvina, Kai-Uwe Bux, and Dan Margalit. Dimension of the Torelli group for $\mathrm{Out}(F_n)$. *Invent. Math.*, 170(1):1–32, 2007.

[21] Joan S. Birman. Abelian quotients of the mapping class group of a 2-manifold. *Bull. Amer. Math. Soc.*, 76:147–150, 1970.

[22] Joan S. Birman. Errata: Abelian quotients of the mapping class group of a 2-manifold. *Bull. Amer. Math. Soc.*, 77:479, 1971.

[23] Joan S. Birman. On Siegel's modular group. *Math. Ann.*, 191:59–68, 1971.

[24] Joan S. Birman. *Braids, links, and mapping class groups*. Annals of Mathematics Studies, No. 82. Princeton University Press, Princeton, NJ, 1974.

[25] Joan S. Birman. Mapping class groups of surfaces. In *Braids (Santa Cruz, CA, 1986)*, volume 78 of *Contemporary Mathematics*, pages 13–43. American Mathematical Society, Providence, RI, 1988.

[26] Joan S. Birman and Hugh M. Hilden. On the mapping class groups of closed surfaces as covering spaces. In *Advances in the theory of Riemann surfaces (Stony Brook, N.Y., 1969)*, Annals of Mathematics Studies, No. 66, pages 81–115. Princeton University Press, Princeton, NJ, 1971.

[27] Joan S. Birman and Hugh M. Hilden. On isotopies of homeomorphisms of Riemann surfaces. *Ann. of Math. (2)*, 97:424–439, 1973.

[28] Joan S. Birman, Alex Lubotzky, and John McCarthy. Abelian and solvable subgroups of the mapping class groups. *Duke Math. J.*, 50(4):1107–1120, 1983.

[29] Joan S. Birman and Bronislaw Wajnryb. Presentations of the mapping class group. Errata: "3-fold branched coverings and the mapping class group of a surface" [*Geometry and topology (College Park, MD, 1983/1984)*, 24–46, Lecture Notes in Mathematics, 1167, Springer, Berlin, 1985] and "A simple presentation of the mapping class group of an orientable surface" [Israel J. Math. **45** 2–3, 157–174, 1983]. *Israel J. Math.*, 88(1–3):425–427, 1994.

[30] Søren Kjærgaard Boldsen. Different versions of mapping class groups of surfaces. arXiv:0908.2221, 2009.

[31] Francis Bonahon. The geometry of Teichmüller space via geodesic currents. *Invent. Math.*, 92(1):139–162, 1988.

[32] Tara E. Brendle and Benson Farb. Every mapping class group is generated by 6 involutions. *J. Algebra*, 278(1):187–198, 2004.

[33] Tara E. Brendle and Dan Margalit. Commensurations of the Johnson kernel. *Geom. Topol.*, 8:1361–1384, 2004 (electronic).

[34] Tara E. Brendle and Dan Margalit. Addendum to: "Commensurations of the Johnson kernel" (Geom. Topol. **8**: (2004); 1361–1384, MR2119299). *Geom. Topol.*, 12(1):97–101, 2008.

[35] Martin R. Bridson and André Haefliger. *Metric spaces of nonpositive curvature*, volume 319 of *Grundlehren der Mathematischen Wissenschaften (Fundamental Principles of Mathematical Sciences)*. Springer-Verlag, Berlin, 1999.

[36] L. E. J. Brouwer. Beweis der invarianz des n-dimensionalen gebiets. *Math. Ann.*, 71(3):305–313, 1911.

[37] Kenneth S. Brown. Presentations for groups acting on simply-connected complexes. *J. Pure Appl. Algebra*, 32(1):1–10, 1984.

[38] Kenneth S. Brown. *Cohomology of groups*, volume 87 of *Graduate Texts in Mathematics*. Springer-Verlag, New York, 1994. Corrected reprint of the 1982 original.

[39] Ronald Brown and Philip J. Higgins. The fundamental groupoid of the quotient of a Hausdorff space by a discontinuous action of a discrete group is the orbit groupoid of the induced action. University of Wales, Bangor, Maths Preprint 02.25.

[40] Heinrich Burkhardt. Grundzüge einer allgemeinen Systematik der hyperelliptischen Functionen I. Ordnung. *Mathematische Annalen*, 35:198–296, 1890.

[41] W. Burnside. *Theory of groups of finite order, 2nd ed.* Dover Publications, New York, 1955.

[42] Peter Buser. *Geometry and spectra of compact Riemann surfaces*, volume 106 of *Progress in Mathematics*. Birkhäuser, Boston, MA, 1992.

[43] Danny Calegari. *Foliations and the geometry of 3-manifolds*. Oxford Mathematical Monographs. Oxford University Press, Oxford, 2007.

[44] Andrew J. Casson and Steven A. Bleiler. *Automorphisms of surfaces after Nielsen and Thurston*, volume 9 of *London Mathematical Society Student Texts*. Cambridge University Press, Cambridge, 1988.

[45] Jean Cerf. Topologie de certains espaces de plongements. *Bull. Soc. Math. France*, 89:227–380, 1961.

[46] D. R. J. Chillingworth. Winding numbers on surfaces. II. *Math. Ann.*, 199:131–153, 1972.

[47] Thomas Church and Benson Farb. Parametrized Abel-Jacobi maps, a question of Johnson, and a homological stability conjecture for the Torelli group. arXiv:1001.1114, 2010.

[48] John B. Conway. *Functions of one complex variable, 2nd ed*, volume 11 of *Graduate Texts in Mathematics*. Springer-Verlag, New York, 1978.

[49] M. Dehn. Lecture notes from the Breslau Mathematics Colloquium. *Archives of the University of Texas at Austin*, 1922.

[50] M. Dehn. Die Gruppe der Abbildungsklassen. *Acta Math.*, 69(1):135–206, 1938. Das arithmetische Feld auf Flächen.

[51] Max Dehn. *Papers on group theory and topology.* Springer-Verlag, New York, 1987. Translated from the German and with introductions.

[52] P. Deligne and D. Mumford. The irreducibility of the space of curves of given genus. *Inst. Hautes Études Sci. Publ. Math.*, (36):75–109, 1969.

[53] Clifford J. Earle and James Eells. A fibre bundle description of Teichmüller theory. *J. Differential Geometry*, 3:19–43, 1969.

[54] James Eells, Jr. and J. H. Sampson. Harmonic mappings of Riemannian manifolds. *Amer. J. Math.*, 86:109–160, 1964.

[55] V.A. Efremovič. On the proximity geometry of Riemannian manifolds. *Uspekhi Math Nauk*, 8:189, 1953.

[56] D. B. A. Epstein. Curves on 2-manifolds and isotopies. *Acta Math.*, 115:83–107, 1966.

[57] Edward Fadell and Lee Neuwirth. Configuration spaces. *Math. Scand.*, 10:111–118, 1962.

[58] Benson Farb, editor. *Problems on mapping class groups and related topics*, volume 74 of *Proceedings of Symposia in Pure Mathematics.* American Mathematical Society, Providence, RI, 2006.

[59] Benson Farb and Nikolai V. Ivanov. The Torelli geometry and its applications: research announcement. *Math. Res. Lett.*, 12(2-3):293–301, 2005.

[60] H. M. Farkas and 2nd ed Kra, I. *Riemann surfaces*, volume 71 of *Graduate Texts in Mathematics.* Springer-Verlag, New York, 1992.

[61] A. Fathi, F. Laudenbach, and V. Poénaru, editors. *Travaux de Thurston sur les surfaces*, volume 66 of *Astérisque.* Société Mathématique de France, Paris, 1979. Séminaire Orsay, with an English summary.

[62] W. Fenchel. Estensioni di gruppi discontinui e trasformazioni periodiche delle superficie. *Atti Accad. Naz Lincei. Rend. Cl. Sci. Fis. Mat. Nat. (8)*, 5:326–329, 1948.

[63] W. Fenchel. Remarks on finite groups of mapping classes. *Mat. Tidsskr. B.*, 1950:90–95, 1950.

[64] Werner Fenchel and Jakob Nielsen. *Discontinuous groups of isometries in the hyperbolic plane*, volume 29 of *de Gruyter Studies in Mathematics*. Walter de Gruyter, Berlin, 2003. Edited and with a preface by Asmus L. Schmidt, and a biography of the authors by Bent Fuglede.

[65] Ralph H. Fox. Free differential calculus. I. Derivation in the free group ring. *Ann. of Math. (2)*, 57:547–560, 1953.

[66] J. Franks and E. Rykken. Pseudo-Anosov homeomorphisms with quadratic expansion. *Proc. Amer. Math. Soc.*, 127(7):2183–2192, 1999.

[67] Robert Fricke and Felix Klein. *Vorlesungen über die Theorie der automorphen Funktionen. Band 1: Die gruppentheoretischen Grundlagen. Band II: Die funktionentheoretischen Ausführungen und die Andwendungen*, volume 4 of *Bibliotheca Mathematica Teubneriana, Bände 3*. Johnson Reprint, New York, 1965.

[68] William Fulton. *Algebraic curves*. Advanced Book Classics. Addison-Wesley, Redwood City, CA, 1989. An introduction to algebraic geometry, with notes written with the collaboration of Richard Weiss. Reprint of 1969 original.

[69] F. R. Gantmacher. *The theory of matrices. Vols. 1 and 2*. Translated by K. A. Hirsch. Chelsea Publishing, New York, 1959.

[70] Frederick P. Gardiner and Nikola Lakic. *Quasiconformal Teichmüller theory*, volume 76 of *Mathematical Surveys and Monographs*. American Mathematical Society, Providence, RI, 2000.

[71] Sylvain Gervais. Presentation and central extensions of mapping class groups. *Trans. Amer. Math. Soc.*, 348(8):3097–3132, 1996.

[72] Jane Gilman. On the Nielsen type and the classification for the mapping class group. *Adv. Math.*, 40(1):68–96, 1981.

[73] André Gramain. Le type d'homotopie du groupe des difféomorphismes d'une surface compacte. *Ann. Sci. École Norm. Sup. (4)*, 6:53–66, 1973.

[74] Edna K. Grossman. On the residual finiteness of certain mapping class groups. *J. London Math. Soc. (2)*, 9:160–164, 1974/75.

[75] Ursula Hamenstädt. Length functions and parameterizations of Teichmüller space for surfaces with cusps. *Ann. Acad. Sci. Fenn. Math.*, 28(1):75–88, 2003.

[76] Hessam Hamidi-Tehrani. Groups generated by positive multi-twists and the fake lantern problem. *Algebr. Geom. Topol.*, 2:1155–1178, 2002 (electronic).

[77] Mary-Elizabeth Hamstrom. Some global properties of the space of homeomorphisms on a disc with holes. *Duke Math. J.*, 29:657–662, 1962.

[78] Mary-Elizabeth Hamstrom. The space of homeomorphisms on a torus. *Illinois J. Math.*, 9:59–65, 1965.

[79] Mary-Elizabeth Hamstrom. Homotopy groups of the space of homeomorphisms on a 2-manifold. *Illinois J. Math.*, 10:563–573, 1966.

[80] Michael Handel. Global shadowing of pseudo-Anosov homeomorphisms. *Ergodic Theory Dynam. Systems*, 5(3):373–377, 1985.

[81] Michael Handel and William P. Thurston. New proofs of some results of Nielsen. *Adv. in Math.*, 56(2):173–191, 1985.

[82] John Harer. The fourth homology group of the moduli space of curves. Unpublished.

[83] John Harer. The second homology group of the mapping class group of an orientable surface. *Invent. Math.*, 72(2):221–239, 1983.

[84] John Harer. The cohomology of the moduli space of curves. In *Theory of moduli (Montecatini Terme, 1985)*, volume 1337 of *Lecture Notes in Mathematics*, pages 138–221. Springer, Berlin, 1988.

[85] John Harer. The third homology group of the moduli space of curves. *Duke Math. J.*, 63(1):25–55, 1991.

[86] John L. Harer. Stability of the homology of the mapping class groups of orientable surfaces. *Ann. of Math. (2)*, 121(2):215–249, 1985.

[87] John L. Harer. The virtual cohomological dimension of the mapping class group of an orientable surface. *Invent. Math.*, 84(1):157–176, 1986.

[88] W. J. Harvey. Boundary structure of the modular group. In *Riemann surfaces and related topics: (Stony Brook, NY, 1978)*, volume 97 of *Annals of Mathematics Studies*, pages 245–251, Princeton, NJ, 1981. Princeton University Press.

[89] A. Hatcher and W. Thurston. A presentation for the mapping class group of a closed orientable surface. *Topology*, 19(3):221–237, 1980.

[90] Allen Hatcher. On triangulations of surfaces. *Topology Appl.*, 40(2):189–194, 1991.

[91] Allen Hatcher. *Algebraic topology*. Cambridge University Press, Cambridge, 2002.

[92] Allen Hatcher, Pierre Lochak, and Leila Schneps. On the Teichmüller tower of mapping class groups. *J. Reine Angew. Math.*, 521:1–24, 2000.

[93] John Hempel. *3-manifolds*. AMS Chelsea Publishing, Providence, RI, 2004. Reprint of the 1976 original.

[94] Susumu Hirose. A complex of curves and a presentation for the mapping class group of a surface. *Osaka J. Math.*, 39(4):795–820, 2002.

[95] Morris W. Hirsch. *Differential topology*, volume 33 of *Graduate Texts in Mathematics*. Springer-Verlag, New York, 1994. Corrected reprint of the 1976 original.

[96] John Hubbard and Howard Masur. Quadratic differentials and foliations. *Acta Math.*, 142(3-4):221–274, 1979.

[97] John Hamal Hubbard. *Teichmüller theory and applications to geometry, topology, and dynamics. Vol. 1.* Matrix Editions, Ithaca, NY, 2006. Teichmüller theory, with contributions by Adrien Douady, William Dunbar, Roland Roeder, Sylvain Bonnot, David Brown, Allen Hatcher, Chris Hruska, and Sudeb Mitra, and with forewords by William Thurston and Clifford Earle.

[98] Heinz Huber. Zur analytischen Theorie hyperbolischen Raumformen und Bewegungsgruppen. *Math. Ann.*, 138:1–26, 1959.

[99] Heinz Huber. Zur analytischen Theorie hyperbolischer Raumformen und Bewegungsgruppen. II. *Math. Ann.*, 142:385–398, 1960/1961.

[100] Heinz Huber. Zur analytischen Theorie hyperbolischer Raumformen und Bewegungsgruppen. II. *Math. Ann.*, 143:463–464, 1961.

[101] Stephen P. Humphries. Generators for the mapping class group. In *Topology of low-dimensional manifolds (Chelwood Gate, 1977)*, volume 722 of *Lecture Notes in Mathematics*, pages 44–47. Springer, Berlin, 1979.

[102] A. Hurwitz. Ueber Riemann'sche Flächen mit gegebenen Verzweigungspunkten. *Math. Ann.*, 39(1):1–60, 1891.

[103] Atsushi Ishida. The structure of subgroup of mapping class groups generated by two Dehn twists. *Proc. Japan Acad. Ser. A Math. Sci.*, 72(10):240–241, 1996.

[104] N. V. Ivanov. Coefficients of expansion of pseudo-Anosov homeomorphisms. *Zap. Nauchn. Sem. Leningrad. Otdel. Mat. Inst. Steklov. (LOMI)*, 167(*Issled. Topol.* 6):111–116, 191, 1988.

[105] N. V. Ivanov. Residual finiteness of modular Teichmüller groups. *Sibirsk. Mat. Zh.*, 32(1):182–185, 222, 1991.

[106] Nikolai V. Ivanov. *Subgroups of Teichmüller modular groups*, volume 115 of *Translations of Mathematical Monographs*. American Mathematical Society, Providence, RI, 1992. Translated from the Russian by E. J. F. Primrose and revised by the author.

[107] Nikolai V. Ivanov. Mapping class groups. In *Handbook of geometric topology*, pages 523–633. North-Holland, Amsterdam, 2002.

[108] Nikolai V. Ivanov and Lizhen Ji. Infinite topology of curve complexes and non-Poincaré duality of Teichmueller modular groups. arXiv:0707.4322v1, 2007.

[109] Dennis Johnson. An abelian quotient of the mapping class group $\mathcal{I}_g$. *Math. Ann.*, 249(3):225–242, 1980.

[110] Dennis Johnson. Conjugacy relations in subgroups of the mapping class group and a group-theoretic description of the Rochlin invariant. *Math. Ann.*, 249(3):243–263, 1980.

[111] Dennis Johnson. The structure of the Torelli group. I. A finite set of generators for $\mathcal{I}$. *Ann. of Math. (2)*, 118(3):423–442, 1983.

[112] Dennis Johnson. A survey of the Torelli group. In *Low-dimensional topology (San Francisco, CA, 1981)*, volume 20 of *Contemporary Mathematics*, pages 165–179. American Mathematical Society, Providence, RI, 1983.

[113] Dennis Johnson. The structure of the Torelli group. II. A characterization of the group generated by twists on bounding curves. *Topology*, 24(2):113–126, 1985.

[114] Dennis Johnson. The structure of the Torelli group. III. The abelianization of $\mathcal{T}$. *Topology*, 24(2):127–144, 1985.

[115] Dennis L. Johnson. Homeomorphisms of a surface which act trivially on homology. *Proc. Amer. Math. Soc.*, 75(1):119–125, 1979.

[116] Michael Kapovich. *Hyperbolic manifolds and discrete groups*, volume 183 of *Progress in Mathematics*. Birkhäuser, Boston, MA, 2001.

[117] Martin Kassabov. Generating mapping class groups by involutions. math.GT/0311455, 2003.

[118] Christian Kassel and Vladimir Turaev. *Braid groups*, volume 247 of *Graduate Texts in Mathematics*. Springer, New York, 2008. With the graphical assistance of Olivier Dodane.

[119] Svetlana Katok. *Fuchsian groups*. Chicago Lectures in Mathematics. University of Chicago Press, Chicago, IL, 1992.

[120] Linda Keen. Collars on Riemann surfaces. In *Discontinuous groups and Riemann surfaces (College Park, MD., 1973)*, pages 263–268. Annals of Mathematics Studies, No. 79. Princeton Univ. Press, Princeton, NJ, 1974.

[121] Richard P. Kent, IV, Christopher J. Leininger, and Saul Schleimer. Trees and mapping class groups. *J. Reine Angew. Math.*, 637:1–21, 2009.

[122] Steven P. Kerckhoff. The Nielsen realization problem. *Ann. of Math. (2)*, 117(2):235–265, 1983.

[123] Rob Kirby. Problems in low-dimensional topology. Available at http://math.berkeley.edu/~kirby.

[124] Mustafa Korkmaz. Low-dimensional homology groups of mapping class groups: a survey. *Turkish J. Math.*, 26(1):101–114, 2002.

[125] Mustafa Korkmaz. Generating the surface mapping class group by two elements. *Trans. Amer. Math. Soc.*, 357(8):3299–3310, 2005 (electronic).

[126] Mustafa Korkmaz and András I. Stipsicz. The second homology groups of mapping class groups of oriented surfaces. *Math. Proc. Cambridge Philos. Soc.*, 134(3):479–489, 2003.

[127] Irwin Kra. On the Nielsen–Thurston–Bers type of some self-maps of Riemann surfaces. *Acta Math.*, 146(3–4):231–270, 1981.

[128] Ravi S. Kulkarni. Riemann surfaces admitting large automorphism groups. In *Extremal Riemann surfaces (San Francisco, CA, 1995)*, volume 201 of *Contemporary Mathematics*, pages 63–79. American Mathematical Society, Providence, RI, 1997.

[129] Michael Larsen. How often is $84(g - 1)$ achieved? *Israel J. Math.*, 126:1–16, 2001.

[130] M. A. Lavrentiev. Sur une classe des représentations continues. *Mat. Sb.*, 42:407–434, 1935.

[131] W. B. R. Lickorish. A finite set of generators for the homeotopy group of a 2-manifold. *Proc. Cambridge Philos. Soc.*, 60:769–778, 1964.

[132] Feng Luo. Torsion Elements in the Mapping Class Group of a Surface.

[133] Roger C. Lyndon and Paul E. Schupp. *Combinatorial group theory*. Classics in Mathematics. Springer-Verlag, Berlin, 2001. Reprint of the 1977 edition.

[134] A. M. Macbeath. On a theorem by J. Nielsen. *Quart. J. Math. Oxford Ser. (2)*, 13:235–236, 1962.

[135] Colin Maclachlan. Modulus space is simply-connected. *Proc. Amer. Math. Soc.*, 29:85–86, 1971.

[136] Wilhelm Magnus. Untersuchungen über einige unendliche diskontinuierliche Gruppen. *Math. Ann.*, 105(1):52–74, 1931.

[137] Wilhelm Magnus. Beziehungen zwischen Gruppen und Idealen in einem speziellen Ring. *Math. Ann.*, 111(1):259–280, 1935.

[138] Wilhelm Magnus, Abraham Karrass, and Donald Solitar. *Combinatorial group theory, 2nd ed.* Dover Publications Inc., Mineola, NY, 2004. Presentations of groups in terms of generators and relations.

[139] K. Mahler. On lattice points in n-dimensional star bodies. I. Existence theorems. *Proc. Roy. Soc. London. Ser. A.*, 187:151–187, 1946.

[140] Albert Marden and Kurt Strebel. A characterization of Teichmüller differentials. *J. Differential Geom.*, 37(1):1–29, 1993.

[141] Dan Margalit and Jon McCammond. Geometric presentations for the pure braid group. *J. Knot Theory Ramifications*, 18(1):1–20, 2009.

[142] J. Peter Matelski. A compactness theorem for Fuchsian groups of the second kind. *Duke Math. J.*, 43(4):829–840, 1976.

[143] Makoto Matsumoto. A simple presentation of mapping class groups in terms of Artin groups [translation of Sūgaku **52**, no. 1, 31–42, 2000; 1764273]. *Sugaku Expositions*, 15(2):223–236, 2002. Sugaku expositions.

[144] John D. McCarthy. Automorphisms of surface mapping class groups. A recent theorem of N. Ivanov. *Invent. Math.*, 84(1):49–71, 1986.

[145] James McCool. Some finitely presented subgroups of the automorphism group of a free group. *J. Algebra*, 35:205–213, 1975.

[146] Darryl McCullough and Andy Miller. The genus 2 Torelli group is not finitely generated. *Topology Appl.*, 22(1):43–49, 1986.

[147] Darryl McCullough and Kashyap Rajeevsarathy. Roots of Dehn twists. arXiv:0906.1601v1, 2009.

[148] Dusa McDuff and Dietmar Salamon. *Introduction to symplectic topology, 2nd ed.* Oxford Mathematical Monographs. Clarendon Press Oxford University Press, New York, 1998.

[149] H. P. McKean. Selberg's trace formula as applied to a compact Riemann surface. *Comm. Pure Appl. Math.*, 25:225–246, 1972.

[150] H. P. McKean. Correction to: "Selberg's trace formula as applied to a compact Riemann surface" (*Comm. Pure Appl. Math.* **25**: 225–246, 1972). *Comm. Pure Appl. Math.*, 27:134, 1974.

[151] Curtis T. McMullen. Polynomial invariants for fibered 3-manifolds and Teichmüller geodesics for foliations. *Ann. Sci. École Norm. Sup. (4)*, 33(4):519–560, 2000.

[152] Curtis T. McMullen. Personal communication. 2010.

[153] William H. Meeks, III and Julie Patrusky. Representing homology classes by embedded circles on a compact surface. *Illinois J. Math.*, 22(2):262–269, 1978.

[154] J. Mennicke. Zur Theorie der Siegelschen Modulgruppe. *Math. Ann.*, 159:115–129, 1965.

[155] Geoffrey Mess. The Torelli groups for genus 2 and 3 surfaces. *Topology*, 31(4):775–790, 1992.

[156] Werner Meyer. Die Signatur von Flächenbündeln. *Math. Ann.*, 201:239–264, 1973.

[157] Richard T. Miller. Geodesic laminations from Nielsen's viewpoint. *Adv. in Math.*, 45(2):189–212, 1982.

[158] J. Milnor. A note on curvature and fundamental group. *J. Differential Geometry*, 2:1–7, 1968.

[159] John Milnor. Construction of universal bundles. II. *Ann. of Math. (2)*, 63:430–436, 1956.

[160] Shigeyuki Morita. Characteristic classes of surface bundles. *Invent. Math.*, 90(3):551–577, 1987.

[161] Shigeyuki Morita. Casson's invariant for homology 3-spheres and characteristic classes of surface bundles. I. *Topology*, 28(3):305–323, 1989.

[162] Charles B. Morrey, Jr. On the solutions of quasi-linear elliptic partial differential equations. *Trans. Amer. Math. Soc.*, 43(1):126–166, 1938.

[163] Lee Mosher. Mapping class groups and $\mathrm{Out}(F_r)$. Available at http://aimath.org/pggt.

[164] David Mumford. Abelian quotients of the Teichmüller modular group. *J. Analyse Math.*, 18:227–244, 1967.

[165] David Mumford. Abelian quotients of the Teichmüller modular group. *J. Analyse Math.*, 18:227–244, 1967.

[166] David Mumford. A remark on Mahler's compactness theorem. *Proc. Amer. Math. Soc.*, 28:289–294, 1971.

[167] James Munkres. Obstructions to the smoothing of piecewise-differentiable homeomorphisms. *Ann. of Math. (2)*, 72:521–554, 1960.

[168] Jakob Nielsen. Untersuchungen zur Topologie der geschlossenen zweseitigen Flächen. *Acta Math.*, 50:189–358, 1927.

[169] Jakob Nielsen. Untersuchungen zur Topologie der geschlossenen zweiseitigen Flächen. II. *Acta Math.*, 53(1):1–76, 1929.

[170] Jakob Nielsen. Untersuchungen zur Topologie der geschlossenen zweiseitigen Flächen. III. *Acta Math.*, 58(1):87–167, 1932.

[171] Jakob Nielsen. Abbildungsklassen endlicher Ordnung. *Acta Math.*, 75:23–115, 1943.

[172] Jakob Nielsen. Surface transformation classes of algebraically finite type. *Danske Vid. Selsk. Math.-Phys. Medd.*, 21(2):89, 1944.

[173] Jean-Pierre Otal. Le théorème d'hyperbolisation pour les variétés fibrées de dimension 3. *Astérisque*, (235):x+159, 1996.

[174] Luis Paris. Personal communication. 2010.

[175] R. C. Penner. Bounds on least dilatations. *Proc. Amer. Math. Soc.*, 113(2):443–450, 1991.

[176] R. C. Penner and J. L. Harer. *Combinatorics of train tracks*, volume 125 of *Annals of Mathematics Studies*. Princeton University Press, Princeton, NJ, 1992.

[177] Robert C. Penner. A construction of pseudo-Anosov homeomorphisms. *Trans. Amer. Math. Soc.*, 310(1):179–197, 1988.

[178] Bernard Perron and Jean-Pierre Vannier. Groupe de monodromie géométrique des singularités simples. *C. R. Acad. Sci. Paris Sér. I Math.*, 315(10):1067–1070, 1992.

[179] Wolfgang Pitsch. Un calcul élémentaire de $H_2(\mathcal{M}_{g,1}, \mathbf{Z})$ pour $g \geq 4$. *C. R. Acad. Sci. Paris Sér. I Math.*, 329(8):667–670, 1999.

[180] Jerome Powell. Two theorems on the mapping class group of a surface. *Proc. Amer. Math. Soc.*, 68(3):347–350, 1978.

[181] Jerome Powell. Homeomorphisms of S^3 leaving a Heegaard surface invariant. *Trans. Amer. Math. Soc.*, 257(1):193–216, 1980.

[182] Józef H. Przytycki. History of the knot theory from Vandermonde to Jones. In *XXIVth National Congress of the Mexican Mathematical Society (Spanish) (Oaxtepec, 1991)*, volume 11 of *Aportaciones Matemáticas Comunicaciones*, pages 173–185. Sociedad Matemáticas Mexicana, México, 1992.

[183] Andrew Putman. Abelian covers of surfaces and the homology of the level l mapping class group. arXiv:0907.1718v2, 2009.

[184] Andrew Putman. Cutting and pasting in the Torelli group. *Geom. Topol.*, 11:829–865, 2007.

[185] Bernhard Riemann. Theorie der abelschen functionen. *J. Reine Angew. Math.*, Band 54, 1857.

[186] H. L. Royden. Automorphisms and isometries of Teichmüller space. In *Advances in the Theory of Riemann Surfaces (Stony Brook, NY, 1969)*, pages 369–383. Annals of Mathematics Studies, No. 66. Princeton University Press, Princeton, NJ, 1971.

[187] Walter Rudin. *Real and complex analysis*. McGraw-Hill, New York, 1966.

[188] Hermann Ludwig Schmid and Oswald Teichmüller. Ein neuer Beweis für die Funktionalgleichung der L-Reihen. *Abh. Math. Sem. Hansischen Univ.*, 15:85–96, 1943.

[189] Richard Schoen and Shing Tung Yau. On univalent harmonic maps between surfaces. *Invent. Math.*, 44(3):265–278, 1978.

[190] Peter Scott. The geometries of 3-manifolds. *Bull. London Math. Soc.*, 15(5):401–487, 1983.

[191] Peter Scott and Terry Wall. Topological methods in group theory. In *Homological group theory (Durham, 1977)*, volume 36 of *London Mathematical Society Lecture Note Series*, pages 137–203. Cambridge University Press, Cambridge, 1979.

[192] Herbert Seifert. Bemerkungen zur stetigen abbildung von flächen. *Abh. Math. Sem. Univ. Hamburg*, 12:23–37, 1937.

[193] Atle Selberg. On discontinuous groups in higher-dimensional symmetric spaces. In *Contributions to function theory (International Colloquium on Function Theory, Bombay, 1960)*, pages 147–164. Tata Institute of Fundamental Research, Bombay, 1960.

[194] J.-P. Serre. Rigidité de foncteur d'Jacobi d'échelon $n \geq 3$. *Sem. H. Cartan, 1960/1961*, appendix to Exp. 17, 1961.

[195] Jean-Pierre Serre. *Trees*. Springer Monographs in Mathematics. Springer-Verlag, Berlin, 2003. Translated from the French original by John Stillwell, corrected second printing of the 1980 English translation.

[196] Ya. G. Sinai. *Introduction to ergodic theory*. Mathematical Notes, No. 18. Princeton University Press, Princeton, NJ, 1976. Translated by V. Scheffer.

[197] Stephen Smale. Diffeomorphisms of the 2-sphere. *Proc. Amer. Math. Soc.*, 10:621–626, 1959.

[198] George Springer. *Introduction to Riemann surfaces*. Addison-Wesley, Reading, MA, 1957.

[199] John C. Stillwell. *Classical topology and combinatorial group theory*, volume 72 of *Graduate Texts in Mathematics*. Springer-Verlag, New York, 1980.

[200] Kurt Strebel. *Quadratic differentials*, volume 5 of *Ergebnisse der Mathematik und ihrer Grenzgebiete (3) [Results in Mathematics and Related Areas (3)]*. Springer-Verlag, Berlin, 1984.

[201] Toshikazu Sunada. Riemannian coverings and isospectral manifolds. *Ann. of Math. (2)*, 121(1):169–186, 1985.

[202] A. S. Švarc. A volume invariant of coverings. *Dokl. Akad. Nauk SSSR (N.S.)*, 105:32–34, 1955.

[203] Oswald Teichmüller. Extremale quasikonforme Abbildungen und quadratische Differentiale. *Abh. Preuss. Akad. Wiss. Math.-Nat. Kl.*, 1939(22):197, 1940.

[204] Carsten Thomassen. The Jordan–Schönflies theorem and the classification of surfaces. *Amer. Math. Monthly*, 99(2):116–130, 1992.

[205] W. P. Thurston. Hyperbolic structures on 3-manifolds, II: Surface groups and 3-manifolds which fiber over the circle. arXiv:math.GT/9801045, 1986.

[206] William P. Thurston. *The geometry and topology of 3-manifolds*. Princeton University Notes, Princeton University, Princeton, NJ, 1980.

[207] William P. Thurston. On the geometry and dynamics of diffeomorphisms of surfaces. *Bull. Amer. Math. Soc. (N.S.)*, 19(2):417–431, 1988.

[208] William P. Thurston. *Three-dimensional geometry and topology. Vol. 1*, volume 35 of *Princeton Mathematical Series*. Princeton University Press, Princeton, NJ, 1997. Edited by Silvio Levy.

[209] Bronislaw Wajnryb. A simple presentation for the mapping class group of an orientable surface. *Israel J. Math.*, 45(2-3):157–174, 1983.

[210] Bronislaw Wajnryb. An elementary approach to the mapping class group of a surface. *Geom. Topol.*, 3:405–466, 1999 (electronic).

[211] B. A. F. Wehrfritz. *Infinite linear groups. An account of the group-theoretic properties of infinite groups of matrices*. Springer-Verlag, New York, 1973. Ergebnisse der Matematik und ihrer Grenzgebiete, Band 76.

[212] André Weil. On discrete subgroups of Lie groups. *Ann. of Math. (2)*, 72:369–384, 1960.

[213] J. H. C. Whitehead. Manifolds with transverse fields in euclidean space. *Ann. of Math. (2)*, 73:154–212, 1961.

[214] A. Wiman. Uber die hyperelliptischen Curven und diejenigan vom Geschlechte $p = 3$, welche eindeutigen Transformationen in sich zulassen. *Bihang Kongl. Svenska Vetenskaps-Akademiens Handlingar*, 1895-1896.

[215] Scott Wolpert. The length spectra as moduli for compact Riemann surfaces. *Ann. of Math. (2)*, 109(2):323–351, 1979.

[216] Scott A. Wolpert. *Families of Riemann surfaces and Weil-Petersson geometry*, volume 113 of *CBMS Regional Conference Series in Mathematics*. Published for the Conference Board of the Mathematical Sciences, Washington, DC, 2010.

[217] Tatsuhiko Yagasaki. Homotopy types of homeomorphism groups of noncompact 2-manifolds. *Topology Appl.*, 108(2):123–136, 2000.

[218] Heiner Zieschang, Elmar Vogt, and Hans-Dieter Coldewey. *Surfaces and planar discontinuous groups*, volume 835 of *Lecture Notes in Mathematics*. Springer, Berlin, 1980. Translated from the German by John Stillwell.

Index